AF805408

Reviews in Fluorescence

Series Editors
Chris D. Geddes, University of Maryland Baltimore County, Baltimore, MD, USA
Joseph R. Lakowicz, University of Maryland, Baltimore, MD, USA

More information about this series at http://www.springer.com/series/6946

Chris D. Geddes
Editor

Reviews in Fluorescence 2015

Springer

Editor
Chris D. Geddes
Institute of Fluorescence
University of Maryland Baltimore County
Baltimore, MD, USA

ISSN 1573-8086
Reviews in Fluorescence
ISBN 978-3-319-24607-9 ISBN 978-3-319-24609-3 (eBook)
DOI 10.1007/978-3-319-24609-3

Springer Cham Heidelberg New York Dordrecht London
© Springer International Publishing Switzerland 2016
This work is subject to copyright. All rights are reserved by the Publisher, whether the whole or part of the material is concerned, specifically the rights of translation, reprinting, reuse of illustrations, recitation, broadcasting, reproduction on microfilms or in any other physical way, and transmission or information storage and retrieval, electronic adaptation, computer software, or by similar or dissimilar methodology now known or hereafter developed.
The use of general descriptive names, registered names, trademarks, service marks, etc. in this publication does not imply, even in the absence of a specific statement, that such names are exempt from the relevant protective laws and regulations and therefore free for general use.
The publisher, the authors and the editors are safe to assume that the advice and information in this book are believed to be true and accurate at the date of publication. Neither the publisher nor the authors or the editors give a warranty, express or implied, with respect to the material contained herein or for any errors or omissions that may have been made.

Printed on acid-free paper

Springer International Publishing AG Switzerland is part of Springer Science+Business Media (www.springer.com)

Preface

To date, seven volumes have been both published and well received by the scientific community, the very first volume 11 years ago in 2004. Since that time we have seen the continued growth of fluorescence techniques, as well as recognition for two fluorescence-based Nobel Prizes, namely, "GFP" (Nobel Prize in Chemistry in 2008) and, more recently, "Super Resolution Microscopy" (Nobel Prize in Chemistry in 2014).

In this 2015 volume we are pleased again with the broad and timely fluorescence content. We subsequently thank the authors for their very timely and exciting contributions again this year. We hope you all will find this volume as useful as past volumes.

In closing, I would like to thank both Tanja Koppejan and Meran Owen at Springer for their help is compiling this volume.

Baltimore, MD, USA
July 27, 2015

Chris D. Geddes Ph.D., F.R.S.C.

Contents

1 **Studying Protein Misfolding and Aggregation by Fluorescence Spectroscopy** 1
Mily Bhattacharya and Samrat Mukhopadhyay

2 **Time-Dependent Spectral Shifts in Tryptophan Fluorescence: Bridging Experiments with Molecular Dynamics Simulations** 29
Dmitri Toptygin

3 **Directional Fluorescence Based on Surface Plasmon-Coupling** 71
Yao-Qun Li, Shuo-Hui Cao, Wei-Peng Cai, Qian Liu, Xiao-Qing Liu, and Yu-Hua Weng

4 **Fluorescence Analysis of Thermoresponsive Polymers** 97
Cheryl Morris and Alan G. Ryder

5 **Principles of Fluorogenic Reagent Design for Forensics. Recent Progress Towards New Reagents to Develop Fingerprints in Blood and on Variable Surfaces** 127
Martha Sibrian-Vazquez and Robert M. Strongin

6 **Indicators for Ionic Copper in Biology** 147
Richard B. Thompson and Hui Hui Zeng

7 **Heterogeneous Lipid Distributions in Membranes as Revealed by Electronic Energy Transfer** 171
Radek Šachl and Lennart B.-Å. Johansson

8 **Fluoimaging Determination of Poisons, Pollutants, Narcotics and Drugs with No Added Reagents** 189
Natalia V. Strashnikova, Shlomo Mark, Yehoshua Kalisky, and Abraham H. Parola

9 Intramolecular Mechanisms for the Occurrence of Fluorescence from Upper Excited States of Aromatic Molecules and Linear Polyenes 213
Takao Itoh

10 Developments in the Photonic Theory of Fluorescence 235
Jamie M. Leeder, David S. Bradshaw, Mathew D. Williams, and David L. Andrews

11 Luminescent Lanthanide Sensors and Lanthanide Doped Upconversion Nanoparticles: Current Status and Future Expectations 269
Garima Sharma, Preeti Sehgal, and Anudeep Kumar Narula

12 Interplay Between DNA-Binding/Catalytic Functions and Oligomerization of Retroviral Integrases Studied by a Combination of Time-Resolved Fluorescence Anisotropy, Fluorescence Correlation Spectroscopy and Resonance Energy Transfer 301
Olivier Delelis and Eric Deprez

13 Dual Fluorescence Phenomenon in 'Push-Pull' Stilbenes 337
Dina Pines, Ehud Pines, Terry W.J. Steele, and Vladislav Papper

14 Molecular Anatomy of an Ion Channel Explored Utilizing Fluorescence Spectroscopy 353
Arunima Chaudhuri and Amitabha Chattopadhyay

Index 369

Chapter 1
Studying Protein Misfolding and Aggregation by Fluorescence Spectroscopy

Mily Bhattacharya and Samrat Mukhopadhyay

Abstract Protein misfolding leading to aggregation and amyloid fibril formation has been implicated in a variety of neurodegenerative disorders. Under suitably designed *in vitro* conditions, intermolecular contacts between polypeptide chains mediated by various non-covalent interactions result in the formation of oligomeric species that are eventually sequestered into β-sheet-rich amyloid fibrils. Owing to the inherent heterogeneity and complexity of protein aggregation processes, detection and structural characterization of the early, transiently-populated cytotoxic oligomeric intermediates during the amyloid fibrillation cascade still poses a formidable challenge. Fluorescence spectroscopy is an extremely sensitive multiparametric technique that provides simultaneous information about the conformational- and size changes for both early oligomeric species as well as for the large-sized aggregates. In this review, we emphasize recent and selected examples on the application of various fluorescence spectroscopic techniques in the study of protein aggregation. Additionally, we also summarize the recent results on protein aggregation studies using fluorescence spectroscopy from our laboratory.

Keywords Fluorescence spectroscopy • Protein misfolding • Protein aggregation • Amyloid fibrils • Oligomers • Single-molecule fluorescence

1.1 A Brief Introduction to Protein Misfolding and Aggregation

According to the energy landscape theory, an unfolded polypeptide chain folds into its native state by sampling through an ensemble of conformations as described by the protein folding funnel [1]. During the folding process, interplay of a variety of non-covalent interactions such as hydrogen bonding, hydrophobic interactions,

M. Bhattacharya • S. Mukhopadhyay (✉)
Indian Institute of Science Education and Research (IISER) Mohali, S.A.S. Nagar, Sector 81, Mohali 140306, Punjab, India
e-mail: mukhopadhyay@iisermohali.ac.in;
http://www.iisermohali.ac.in/html/faculty/mukhopadhyay.html

© Springer International Publishing Switzerland 2016
C.D. Geddes (ed.), *Reviews in Fluorescence 2015*,
DOI 10.1007/978-3-319-24609-3_1

electrostatic interactions play key roles in the formation of a correctly folded native structure. However, it has been conjectured that the protein energy landscape is significantly complex accompanied by a "dramatic increase" in its ruggedness due to the localization of the protein in a crowded cellular milieu [2]. Consequently, at higher protein concentrations in the cells, this confinement results in frequent collisions between the polypeptide chains initiating both intra- and intermolecular contacts. Under suitable conditions, when intermolecular interactions predominate over the intramolecular interactions responsible for correct folding, aberrant protein folding occurs that lead to the misfolding and aggregation. Protein aggregation that results in the formation of amyloid fibrils has been implicated in a variety of neurodegenerative disorders (Alzheimer's, Parkinson's, Huntington's and prion diseases), localized and systemic amyloidoses (type II diabetes and dialysis-related, respectively) [3]. The primary agent in protein aggregation and amyloid assembly is identified to be a partially destabilized conformer of a *natively folded* or a partially stabilized conformer of a *natively unfolded* protein that accumulates, undergoes conformational changes and ultimately results in the formation of ordered (amyloid) aggregates or self-associate non-specifically to form disordered (amorphous) aggregates depending on the solution conditions namely, pH, ionic strength, temperature etc. [4–6]. The fact that every protein can form aggregates without having a predisposition towards amyloid assembly indicates that the protein aggregation is a generic phenomenon and is likely to be independent of the native structure [7, 8]. However, it has been documented that the amino acid sequence and environmental conditions have a significant impact on the fibrillation kinetics, fibril morphology and architecture along with the fibril stability [9].

It is now recognized that tailoring the solution conditions conducive to protein aggregation *in vitro* results in the accumulation of a multitude of non-native conformers that serve as precursors to amorphous aggregates and amyloid fibrils. Detection and structural characterization of these non-native oligomeric species in a heterogeneous and complex mixture along with the elucidation of a molecular mechanism, especially during the early stages of protein aggregation still proves to be a formidable task. A number of experimental approaches have been used to detect and characterize the early intermediates since the oligomers have been suggested to be more cytotoxic than the fibrils [10, 11] but an in-depth comprehension of the initial structural details and protein aggregation mechanisms at the molecular level still remain elusive. In this review, we discuss various fluorescence spectroscopic techniques that are utilized to study protein aggregation. Additionally, our efforts directed towards investigation and elucidation of aggregation mechanisms in a few model proteins such as, lysozyme, serum albumins and casein using fluorescence spectroscopy will also be described.

1.2 Experimental Techniques to Study Protein Aggregation

Various biophysical techniques that are used to investigate protein aggregation have indicated that the fibril formation is a stochastic process preceded by the generation of transient species possessing heterogeneous conformations. These transient species, that are eventually sequestered into β-sheet-rich amyloids, have been characterized as oligomeric intermediates that share structural similarities with the amyloid fibrils but are shown to possess higher cytotoxicity than the fibril themselves [10, 11]. Therefore, there is a pressing need to detect and characterize these oligomers that undergo conformational changes and size growth enroute to amyloid fibrillation (Fig. 1.1). Here, we describe several *in vitro* spectroscopic techniques very briefly that are either used in isolation or more routinely, in combination with other techniques. Solid-state nuclear magnetic resonance (NMR) [12, 13] and H/D-exchange NMR experiments [14] are extensively used to investigate the structural details of the amyloid fibrils both at the residue-specific as well as at the core levels. Dynamic light scattering (DLS) allows the estimation of the size (hydrodynamic radius) of a protein (and/or protein complexes) [15]. Circular dichroism (CD) is another essential spectroscopic technique that is routinely used to analyze the secondary structural elements in protein monomers and aggregates. As amyloid fibrils are composed of a cross-β sheet-rich structure, CD is commonly utilized to monitor the conformational change from any native conformation (α, α + β etc.) to a predominantly β-sheet amyloid at 215–218 nm. Recently, an extension of CD namely, vibrational circular dichroism (VCD) has been shown to be extremely sensitive towards the detection of amyloid fibrils whereby the inherent chirality of the supramolecular amyloid fibrils is characterized [16, 17]. Another prevalent technique in structural amyloid biology is the X-ray diffraction method [18, 19] that reveals the cross-β structure of amyloids whereby the β-sheets are oriented parallel to the fiber axis but the constituent β-strands are oriented perpendicular to the axis. The X-ray diffraction pattern of amyloid fibrils exhibit two distinct intense reflections, namely, 4.7–4.8 Å meridional and 10 Å equatorial which denote the spacing between the adjacent β-strands and the "face-to-face separation" between the β-sheets, respectively. The X-ray diffraction technique serves as a direct and definitive proof of amyloid fibrils. Additionally, electrospray ionization mass spectrometry (ESI-MS) has been shown to successfully detect various stages of fibril formation whereby the temporarily-formed oligomeric species can be directly observed [20, 21]. Transmission electron microscopy (TEM) and Atomic force microscopy (AFM) are imaging techniques that are used to shed light into the morphologies of the protein aggregates and fibrils on the nanometer length-scale [22, 23]. TEM gives information about the sample surface in two-dimensions and involves staining of the sample to enhance the signal whereas AFM does not require any extraneous markers and provides a three-dimensional nanoscale morphology of aggregates and fibrils.

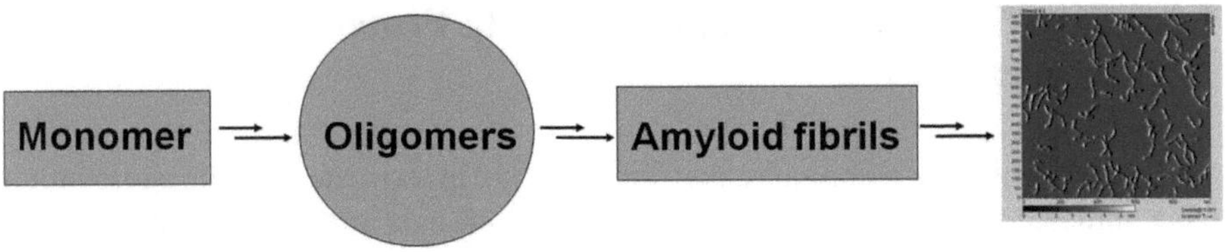

Fig. 1.1 A schematic of protein aggregation process leading amyloid fibrils

1.3 Fluorescence Spectroscopy and Protein Aggregation

Fluorescence spectroscopy has proved to be a powerful methodology to extract information about the structural and mechanistic characterization of protein aggregates and amyloid fibrils. The uniqueness of fluorescence spectroscopy lies in the fact that in addition to extreme sensitivity, the protein conformational changes occurring *concurrently* with size changes can be observed which is very important for monitoring the structural variations occurring during early misfolding and assembly events as well as during the later stages of aggregation (Table 1.1 and Fig. 1.2). Probing the conformational changes in the early oligomeric precursors is particularly useful and important since the fluorescence lifetime of the probes is in the nanosecond timescale which is much faster than the timescales of most conformational changes and interconversions. Information gathered through these observations aid in gaining insights into the complex mechanistic pathway of amyloid formation. Moreover, site-specific fluorophore labeling strategies allow one to selectively label a protein of interest (at ε–amino group of lysine or thiol group of cysteine; details are discussed later) covalently with a desired fluorescent probe to perform the aggregation studies either in the wild-type protein or in its mutants. The following sections describe the utilization of several oligomer- and amyloid-sensitive fluorescent probes followed by various fluorescence spectroscopic techniques that have exhibited their potential in the systematic investigation of protein aggregation.

1.4 Fluorescence Techniques to Study Protein Oligomers and Amyloids During Protein Aggregation

The advantage of fluorescence spectroscopy also lies in the fact that it is a multiparametric technique. This means that alterations in various parameters e.g. fluorescence intensity, fluorescence anisotropy, spectral shifts, lifetimes etc. of fluorophore(s) can be successfully utilized as distinct structural determinants of the monomeric, oligomeric and large-sized aggregated species of a protein under investigation. These possibilities enable one to monitor a number of different observables from the same reaction mixture populated by an ensemble of

Table 1.1 Readouts of fluorescence measurements

Fluorescence readouts	Observed phenomena
Fluorescence spectrum	Conformational changes
Fluorescence intensity, lifetime and quenching	Conformational changes
Fluorescence anisotropy	Size change (also local flexibility and/or microviscosity changes)
FRET	Conformational changes and distributions
Fluorescence autocorrelation	Size changes & distributions and conformational dynamics
ANS fluorescence	Hydrophobic pockets
Thioflavin-T fluorescence	Amyloid formation
Pyrene vibronic band (I_3/I_1)	Local dielectric constants at binding pockets
Pyrene excimer	Conformational changes and association

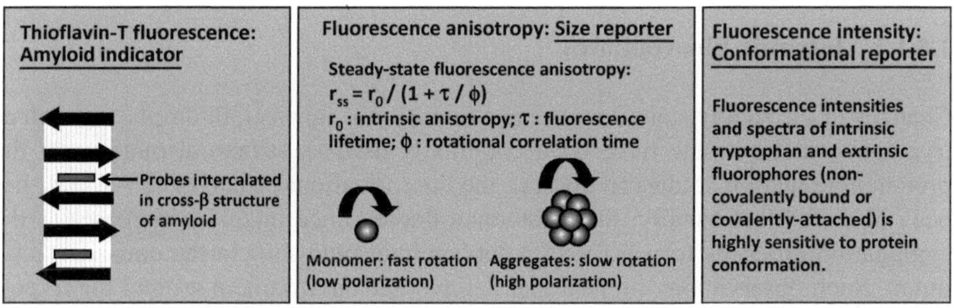

Fig. 1.2 A schematic showing different fluorescence reporters that are relevant in protein aggregation and amyloid fibril formation

structurally diverse molecular species. Such multiparametric data collection offers an unambiguous way to monitor the molecular events in protein aggregation that is commonly endowed with high variability. It is important to mention here that the measurements of fluorescence properties can be carried out either in the steady state or in the time-resolved format. The fluorescence intensity, emission maxima, spectral shifts and fluorescence anisotropy of intrinsic/extrinsic fluorophores can be observed and analyzed conveniently using steady-state fluorescence spectroscopy. In the time-resolved format, it is possible to monitor distributions and subpopulations of molecular species that are on the pathway to amyloid. Additionally, the time-resolved measurements are independent of concentrations of the fluorophores. The time-resolved fluorescence spectroscopy measurements are particularly useful in anisotropy, fluorescence resonance energy transfer (FRET) and quenching experiments. In the following paragraphs, we will discuss briefly about the various fluorescence observables and the usefulness of such parameters in extracting the molecular details during amyloid aggregation. As per the scope of the review, we have restricted to recent results available in the literature with an emphasis on detection of early oligomeric intermediates that are known to be more cytotoxic than the matured fibrils.

1.4.1 Fluorescence Intensity, Spectrum and Lifetime

The fluorescence intensity and lifetime are related to the fluorescence quantum yield and are sensitive to microenvironments [24]. This implies that any change in the protein conformation that affects the surroundings of the intrinsic and/or extrinsic fluorophores, that are sensitive to environmental changes, will be reflected as a change in the fluorescence intensity. Moreover, a shift in the fluorophore emission spectrum is observed depending on whether the fluorophore(s) are exposed or buried due to the protein conformational changes.

1.4.1.1 Intrinsic Fluorophores

Changes in the steady state fluorescence intensity of intrinsic fluorophores such as tryptophan and tyrosine have been commonly used as structural markers in the protein aggregation studies to observe the conformational changes [25–31]. It has been observed that usually, the tryptophan fluorescence intensity increases as the aggregation proceeds forward with a concomitant blue-shift in the emission maximum. Such observations suggest that the average environment around the tryptophan residue(s) progressively becomes non-polar as a function of aggregation. In few cases, a decrease in tryptophan fluorescence intensity has been reported during aggregation suggesting quenching of tryptophan(s) by proximal histidines, phenylalanines, disulfide bonds etc upon protein association [31]. Tyrosine is generally used as an aggregation marker in those cases where the protein is devoid of any tryptophan residue. Though the quantum yield of tyrosine is much lower than that of tryptophan, the absence of tryptophan residues limits any possibility of energy transfer from tyrosine to tryptophan and hence, the fluorescence emission of tyrosine is not quenched. Like tryptophan, the fluorescence intensity of tyrosine is dependent on the polarity of its surroundings but its emission maximum at ~305 nm remains unchanged unless there is a change in pH. An increase in pH results in the formation of tyrosinate ion and a red-shifted emission at ~340 nm. Quenching of the tyrosine fluorescence intensity occurs upon exposure to solvent or due to the presence of aspartic and/or glutamic acid carboxylate side chains in its vicinity [28]. In addition to fluorescence emission from intrinsic tryptophan and tyrosine, an intrinsic blue fluorescence emanating from protein aggregates and fibrils has been observed [32, 33]. As protein aggregates and β-sheet-rich amyloid fibrils comprise of extensive backbone hydrogen bonding networks (>C=O---H-N), it has been suggested that electron delocalization along the peptide backbone due to the presence of these intra- and intermolecular hydrogen bonds give rise to such fluorescence emission and the aromatic side chains do not play any role in the observed blue fluorescence. It was also shown that the fluorescence emission depends largely on the "retention" of water molecules under ambient conditions [33].

1.4.1.2 Extrinsic Fluorophores

A large number of extrinsic fluorophores (either non-covalently bound or covalently attached to proteins) have been used to investigate protein aggregation and amyloid formation (Figs. 1.2 and 1.3) [34]. Generally, thioflavin-T (ThT; amyloid-specific dye) fluorescence assay is routinely used to determine the formation of cross-β-rich amyloid fibrils whereby an increase in the fluorescence intensity of ThT at ~480 nm is observed [35, 36]. It has been reported in a few studies that ThT binding assay is unresponsive to prefibrillar species and the fluorescence enhancement is solely due to the binding of ThT to mature fibrils [37]. Also, ThT binding assay is carried out at neutral or slightly higher pH wherein a few protein amyloids formed at low pH have been reported to dissociate thus, limiting the efficiency of fibril detection [38]. Nile red has been reported as a better amyloid reporter compared to ThT, especially, if the aggregation studies are carried out under acidic conditions [39]. Unbound Nile red emission band exhibits a blue shift with a concomitant increase in the fluorescence intensity upon binding to amyloid fibrils. However, like ThT, Nile red is also more sensitive towards amyloid fibrils compared to the oligomeric species. Therefore, attempts have been made to synthesize new fluorophores to detect the oligomeric species formed at initial stages. DCVJ (4-(dicyanovinyl)-julolidine) [37] and a pentameric oligothiophene derivative (4′,3‴-bis(carboxymethyl) [2,2′;5′,2″;5″,2‴;5‴,2⁗] quinquethiophene-5,5⁗-dicarboxylic acid; p-FTAA) [40] have been shown to successfully detect and bind efficiently to oligomeric, prefibrillar intermediates even under low pH conditions (Fig. 1.3) [41]. DCVJ is a molecular rotor and an "intrinsically quenched" fluorophore wherein the fluorescence from the julolidine group is quenched by the freely rotating dicyano functionality in solution [37]. When it binds to prefibrillar oligomeric species, the torsional flexibility of the dicyano moiety is lost that results in an increase in DCVJ's quantum yield. Consequently, a blue-shift in the emission maximum (from ~510 to ~500 nm) of DCVJ is observed with a substantial increase in the fluorescence intensity. The free oligothiophene derivative, p-FTAA, forms π-stacked aggregates in solution which shows a blue-shift of ~100 nm (from 630 to 530 nm) upon binding to prefibrillar species implying a disruption of the p-FTAA π-aggregates [40]. Moreover, binding of p-FTAA to the aggregates causes hindrance in the conformational flexibility of the oligothiophene backbone, thus, affecting the fluorescence emission properties of the dye which can serve as an indirect readout for conformational changes of protein aggregates. Additionally, both the probes are found to be efficient for investigating aggregation kinetics without influencing the rate of amyloid aggregation even at higher concentration of the dye.

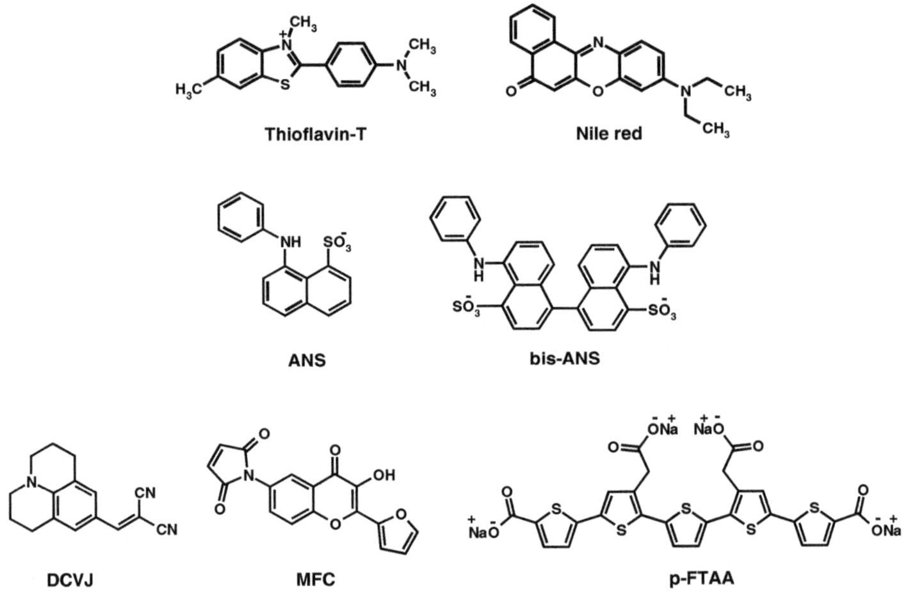

Fig. 1.3 The chemical structures of various extrinsic fluorescent probes for studying protein aggregation

In addition to the conjugated fluorescent dyes discussed above, non-covalently bound naphthalene-based dyes such as ANS (1,8-anilinonaphthalene-sulfonate) and bis-ANS (Fig. 1.3) have been utilized in probing prefibrillar and fibrillar species during protein aggregation [30, 31, 37, 42–44]. ANS is weakly fluorescent in aqueous environment but fluoresces strongly (with a concurrent blue-shift in its emission maximum from ~510 to ~475 nm) when located in a hydrophobic environment [45]. Therefore, as expected, an increase in ANS fluorescence intensity accompanied by a blue-shift in the emission is observed upon oligomerization and fibril formation. ANS has also been found to be a good reporter of the aggregation kinetics without perturbing the aggregation rate.

Another strategy of monitoring the emergence of oligomeric species and fibrils during protein aggregation studies involves covalent modification of specific sites of a protein (or its mutants) by a desired fluorophore. The site-specific fluorophore labeling approach offers a major advantage of precisely determining different segments or domains of the polypeptide chain that either participate in forming the hydrophobic interior of aggregates at different stages of fibril formation or remain isolated throughout the whole process. However, caution must be exercised during the external fluorophore labeling procedure to ensure that the secondary structure remains unaltered, especially in the mutant protein, as well as the attached fluorophore should not affect the kinetics of overall aggregation process. The former can be easily checked by collecting CD spectra of both unmodified and covalently-modified protein whereas the latter can be examined by observing any change in fluorescence properties in a concentration-dependent manner by altering the ratio of the fluorescently-labeled to the unlabeled protein. The labeling of ε-amino ($-NH_2$) side chain of lysine by the isothiocyanate ($-N{=}C{=}S$) and thiol ($-SH$) group of cysteine by the maleimide or iodoacetamide derivatives of fluorescein, rhodamine B, pyrene, acrylodan, Alexa-488, Cy3 etc. are quite common practices. Pyrene is a rigid, polycyclic aromatic hydrocarbon fluorophore whose fluorescence emission spectrum is dependent on the polarity of its microenvironment [46]. In a non-polar medium, the fluorescence emission spectrum of pyrene exhibits five vibronic bands, namely, I_1, I_2, I_3, I_4 and I_5 at 373 nm, 378 nm, 384 nm, 389 nm and 394 nm, respectively which reduces to three bands in a polar medium (I_1, I_3 and I_5) [47]. A qualitative estimate of the polarity of the pyrene microenvironment can be obtained from the ratio of the fluorescence intensities at I_3 and I_1. Higher I_3/I_1 ratio indicates that the environment surrounding the pyrene molecules is more hydrophobic. Additionally, when two pyrene molecules come closer within a distance of ~5 Å, they stack together and form excited state dimer or excimer exhibiting an emission in the range of 450–480 nm. The fluorescence properties of pyrene have been successfully used to detect the formation of oligomers and aggregates whereby a high I_3/I_1 ratio of pyrene accompanied by a blue-shift in the excimer emission and an enhancement in the excimer intensity are observed [48–50]. Additionally, these covalently bound dyes have proven to be effective in accurate determination of the specific roles of different protein domains (either sequential or concurrent involvement) in the amyloid fibril assembly kinetics [51–53]. Very recently, a new fluorescent probe called MFC, based on 2-(2-furyl)-3-hydroxychromone (maleimide derivative), has been reported to be extremely sensitive towards the detection of oligomers during initial stages of protein aggregation [54]. The probe works on the principles of ESIPT (excited-state intramolecular proton transfer) whereby upon excitation, there is a rapid proton transfer within the molecule (keto-enol tautomerization) at subnanoseconds timescale generating a tautomer. Distinct dual emission profiles are observed from both the "normal" (N*) and the tautomeric form (T*); the latter being red-shifted. Since the probe is environmentally-sensitive, the ratio of emission intensities I_{T*}/I_{N*} is a more convenient way to express the changes in polarity of the microenvironment. An increase in I_{T*}/I_{N*} ratio suggests a reduction in polarity and hydrogen bonding capability of the environment. These properties of the probe have been successfully utilized to monitor the formation of oligomers from α-synuclein wherein a 15-fold rise in the T* emission was observed compared to a 2-fold rise in the N* emission and these changes were detected prior to that observed by a rise in ThT intensity [54]. Using these different fluorescent probes, it is now possible to detect and distinguish between the prefibrillar oligomers and mature amyloid fibrils unambiguously leading to a better understanding of the structural diversity of different species involved in the sequence of events leading to amyloid formation.

1.4.2 Fluorescence Polarization (Anisotropy)

The fluorescence anisotropy measurements offer information about the overall size and the rigidity of the probe attached to biomolecules [24], a marker of the rate of reorientation (tumbling) of the fluorophore(s) (Fig. 1.2). This implies that larger the size, slower is the tumbling, and higher is the anisotropy. The steady-state fluorescence anisotropy is related to the overall size of the protein assuming that the overall size changes are predominant compared to the changes in internal (local) dynamics. The concentration dependence of fluorescence anisotropy indicates oligomer formation prior to fibril formation (Fig. 1.4) [31]. In the fluorescence anisotropy methodology, the fluorophore is excited by a vertically polarized light from a continuous light source following which the emission is collected using parallel and perpendicular geometry of the polarizers at a constant emission wavelength either in L- or in T-format. The steady-state anisotropy is estimated by the ratio of difference between vertically and horizontally polarized light to the total light intensity using the following equation:

$$r = \left(I_{\parallel} - GI_{\perp}\right)/\left(I_{\parallel} + 2GI_{\perp}\right) \tag{1.1}$$

where I_{II} and $I_{\perp}$ are parallel and perpendicular fluorescence intensities, respectively and the perpendicular components are corrected using corresponding G-factors. A continuous rise in fluorescence anisotropy of intrinsic and/or extrinsic (covalent and non-covalently bound) fluorophore(s) is expected as aggregation progresses suggesting the formation and growth of oligomeric and large-sized aggregates [25, 30, 31, 55]. As mentioned earlier, the steady-state anisotropy gives an average description of the system whereas time-resolved anisotropy can be particularly advantageous in differentiating between the local (residue-specific) and the global dynamics of the fluorophores (Fig. 1.5) [51]. In this technique, the time-dependent fluorescence anisotropy r(t) decay of the fluorescent species at a fixed emission wavelength is monitored upon excitation by vertically polarized laser pulse followed by fitting and analysis of the anisotropy decay in terms of the rotational correlation time φ. Quantitative analysis results in the determination of the rate of rotational motion of the fluorophore by using the following equation:

$$r(t) = r_0 \exp(-t/\varphi) \tag{1.2}$$

where r_0 is the intrinsic fluorescence anisotropy at time zero for a given fluorophore (usually close to 0.4). Incidentally, both local (fast rotation) and global (slow rotation) dynamics of a fluorophore attached to a protein molecule can contribute to the decay of intrinsic fluorescence anisotropy r_0 of the fluorophore which can be characterized by the following equation (Fig. 1.5):

$$r(t) = r_0\left[\beta_{fast} \exp(-t/\varphi_{fast}) + \beta_{slow} \exp(-t/\varphi_{slow})\right] \tag{1.3}$$

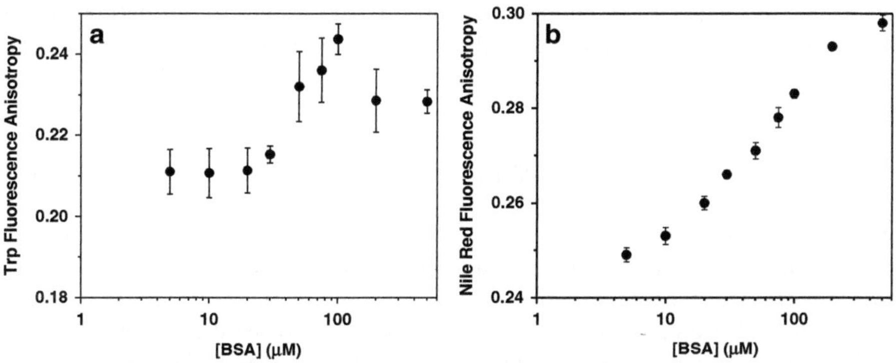

Fig. 1.4 The change in fluorescence anisotropy of (**a**) intrinsic tryptophans and (**b**) extrinsic Nile red as a function of BSA concentration at pH 3 and 50 mM NaCl at room temperature. All the observations suggest that BSA forms soluble oligomeric aggregates under our experimental condition at which fibrillation experiments were carried out ([BSA] = 100 μM, pH = 3) (Adapted from Ref. [31] with permission from the American Chemical Society)

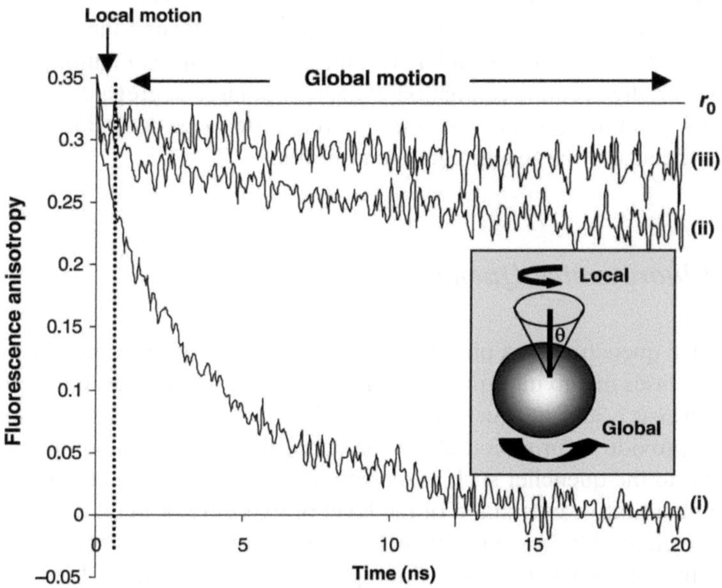

Fig. 1.5 Time-resolved fluorescence anisotropy decay of IAEDANS-labeled barstar in native form (i), low-pH oligomeric form (ii) and amyloid fibrils (iii). The intrinsic anisotropy (r_0) was measured by immobilizing the probe in glycerol. The *inset* shows the wobbling-in-cone model to illustrate the local and global rotational dynamics of fluorophores covalently attached to a protein (This figure was reproduced with permission from Elsevier (Ref. [51]))

where φ_{fast} is related to the fluorophore which undergoes fast local rotation. On the other hand, φ_{slow} is associated with the global motion of the aggregate which is directly proportional to the overall aggregate size (V) as given by the Stokes–Einstein–Debye equation:

$$\varphi_{slow} = \eta V/kT \tag{1.4}$$

where η is the viscosity and $V = 4/3\pi r_h^3$ where r_h is the hydrodynamic radius. β_{fast} denotes the amplitude of the local motion which is a measure of the rotational freedom (or the "degree of orientational constraint") of the fluorescent probe. Larger the volume of the molecular species, slower will be the anisotropy decay and therefore longer will be the rotational correlation time (φ). The utility of time-resolved fluorescence anisotropy measurements as a tool to detect protein aggregation was demonstrated [56] whereby an external fluorescein label was employed to study the aggregation of amyloid-beta peptide. The time-resolved fluorescence anisotropy measurements have been effectively used to investigate protein aggregation using both intrinsic and extrinsic fluorescent probes in addition to studying the participation of different residues of a protein at various stages of aggregation [26, 51, 57, 58]. In these studies the local dynamics indicated the conformational flexibility/rigidity whereas the global dynamics allowed the estimation of aggregate size. However, the accurate estimation of the size of large aggregates is not possible using nanosecond lifetime probes. One would require to use long lifetime probes to determine the size of large aggregates and fibrils.

1.4.3 Fluorescence Quenching

Fluorescence quenching is a phenomenon by which the fluorescence intensity of a sample decreases due to intermolecular interaction between a fluorophore (intrinsic and/or extrinsic) and a quencher [24]. The measurements on fluorescence quenching provide information about the accessibility of the fluorophore (buried or exposed) to the quencher such as oxygen, iodide, acrylamide etc. which in turn, provide a quantitative estimation of the level of exposure of the fluorophore on the polypeptide chain. Fluorescence quenching can be of two types namely, static and dynamic. In static quenching, the fluorophore and the quencher form a complex which does not fluoresce and the distance between the two is fixed. In dynamic or collisional quenching which is distance-dependent, the quencher diffuses and forms a complex with the fluorophore in the excited-state whereby the fluorophore does not emit any photon upon relaxation to the ground-state. Depending on the fluorescence lifetime of the fluorophore, diffusion can occur even at large distances. Also, by choosing an appropriate quencher, one can selectively quench a fluorophore of interest. Quenching analysis is generally performed by Stern-Volmer plots whereby the fluorescence intensities in the presence and absence of a quencher are collected independently and the ratio of the intensities is plotted as a function of varying quencher concentration from which Stern-Volmer quenching constants are estimated. A high quenching constant would indicate that the fluorophore is exposed to solvent whereas a low quenching constant would imply that the fluorophore is buried inside. There are few reports where fluorescence quenching, in addition to other fluorescence parameters, has been employed in protein aggregation studies [25, 29, 59]. As expected, low quenching efficiency was observed upon protein oligomer and aggregate formation wherein the fluorophore got buried inside the interior of the aggregates. Additionally, based on the quenching experiments, site-specific fluorescent protein mutants gave information about solvent accessibility of various segments of the protein during oligomerization and aggregation [25]. Additionally, internal (inbuilt) amino acid quenchers (Cys, Met, His etc) can also act as conformational reporters of fluorescently labeled proteins.

1.4.4 Fluorescence Resonance Energy Transfer (FRET)

FRET, also known as Förster resonance energy transfer, is a photophysical phenomenon providing a quantitative estimate of the biomolecular distance in the range of 1–10 nm [24]. The FRET technique is used as a reporter of alterations in biomolecular distance as a function of conformational changes and dynamics. It involves energy transfer from a donor (D) fluorophore in the excited state to an acceptor (A) fluorophore in the ground state Consequently, the acceptor emission intensity increases at the expense of donor emission (in steady-state mode) or the donor lifetime decreases (in the time-resolved mode). The FRET efficiency (E) is expressed as:

$$E = R_0^6/\left(R_0^6 + r^6\right) \tag{1.5}$$

where R_0 is the Förster distance (the distance at which energy transfer efficiency is 50 %) for a given donor-acceptor pair and r is the distance between donor and acceptor whereby the transfer efficiency is inversely proportional to the sixth power of interchromophore distance (r). The Förster distance (R_0) is simplistically expressed as:

$$R_0 = 0.211\left(\kappa^2 n^{-4} Q_D J(\lambda)\right)^{1/6} \tag{1.6}$$

R_0 shows dependence on the following factors: (i) extent of spectral overlap ($J(\lambda)$) of the donor emission spectrum with the acceptor absorption spectrum (ii) distance (r) between the donor and acceptor moieties (iii) the donor's quantum yield (Q_D) in the absence of acceptor and (iv) relative orientation (κ^2) of the donor and acceptor transition dipoles. n is the refractive index which is generally assumed to be 1.4 for biomolecules in aqueous solution and κ^2 is assumed to be 2/3 since the rotational motion of the donor and acceptor can "randomize the relative orientation" prior to energy transfer [24]. The FRET technique has been extended to protein aggregation studies, especially to oligomer detection, as reported by several groups. By suitably labeling the protein with donor and acceptor fluorophores, FRET measurements are performed to investigate and monitor conformational fluctuations in the monomeric protein as a consequence of changes in solution conditions conducive to aggregation and oligomeric intermediate formation along with conformational rearrangement preceding amyloid fibrillation in combination with other fluorescence techniques [25, 60, 61]. Earlier efforts in α-synuclein aggregation studies using FRET from tyrosine to tryptophan indicated the presence of early oligomeric species during the lag phase which eventually formed amyloid fibrils. Interestingly, few "partly oligomeric" intermediates were still present in the solution even after the completion of fibril formation [25]. Another study using FRET in confocal microscopy during the investigation of polyglutamine (polyQ) oligomer cytotoxicity revealed that polyQ oligomers are soluble and assemble via "length-dependent manner" inside cells and permeabilize the cell membranes [62]. Very recently, *in vivo* FRET imaging studies have been carried out in a living multicellular organism *C. Elegans* to probe α-synuclein aggregation that have shed light into the nature of the aggregated species which were found to be less ordered than the amyloid fibrils and the aggregation kinetics was also monitored [63].

1.4.5 Single-Molecule Fluorescence Studies

Single-molecule fluorescence methodologies are incredibly powerful tools to investigate the complex conformational behavior of aggregation prone amyloidogenic proteins [64, 65]. Using single molecule fluorescence, it is possible to detect, interrogate and analyze individual fluorescently labeled protein or protein complexes. One of us was involved in characterizing natively unfolded or intrinsically disordered yeast prion determinant of Sup35 protein [66]. Single-molecule FRET (SM-FRET) was used to monitor the conformational properties of dual-labeled protein. These experiments indicated that the protein adopts a collapsed conformer under native condition. Additionally, fluorescence correlation spectroscopy (FCS) measurements revealed the presence of nanosecond conformational fluctuations. These results established that the monomeric form of the yeast prion determinant of Sup35 protein adopts an ensemble of relatively unordered collapsed states with rapid conformational fluctuations [66]. Using single-molecule fluorescence, the mechanism of prion propagation has been suggested [67]. Recently, amyloidogenic oligomers have been elegantly detected and characterized by single-molecule fluorescence coincidence [68] and single-molecule photobleaching methods [69].

1.4.6 Fluorescence Correlation Spectroscopy (FCS)

FCS is an extremely sensitive technique which can be performed at nanomolar concentration and is sensitive down to single molecules. The basic principle of this methodology is as follows. As fluorescently-labeled molecules diffuse in-and-out through an illuminated confocal volume (≤1 femtoliter) defined by a focussed laser beam and pinhole, the fluorescence intensity fluctuates spontaneously as a result of the translational diffusion. Statistical analysis of these time-dependent fluctuations in fluorescence intensity using an autocorrelation function yields an autocorrelation fit/curve from which an average number of molecules in the observation volume and the average diffusion coefficient of the molecule are determined [70, 71]. The normalized autocorrelation function for the fluorescence intensity (F(t)) fluctuations is expressed as follows:

$$G(\tau) = \langle F(t)F(t+\tau)\rangle / \langle F(t)\rangle^2 \qquad (1.7)$$

which can be further denoted by:

$$G(\tau) = 1/V_{eff}\langle C\rangle \left[1/1 + (\tau/\tau_D) \times 1/\left\{\left(1 + \left(r_0^2/z_0^2\right)\ (\tau/\tau_D)\right\}^{1/2}\right] \qquad (1.8)$$

for diffusion-induced intensity fluctuations wherein V_{eff} is the effective volume (illuminated area), $\langle C\rangle$ is the average concentration of molecules, $V_{eff}\langle C\rangle = \langle N\rangle$ where N is the average number of molecules, τ_D is the translational diffusion time of the molecule and is given by:

$$\tau_D = r_0^2/4D \qquad (1.9)$$

where r_0 is the detection volume and D: diffusion coefficient and r_0^2/z_0^2 is the structure parameter related to the Gaussian observation volume. From the diffusion coefficient (D), the average size (or hydrodynamic radius; r_h) of the molecule can be estimated using the following equation: [72, 73]

$$D = kT/6\pi\eta r_h \qquad (1.10)$$

However, apart from diffusion, fast conformational dynamics and triplet-state dynamics also contribute to the fluctuations in the observed fluorescence intensity which typically occur at faster timescales compared to the diffusion and can be explained by a simple exponential decay function. Hence, in order to incorporate these factors, the autocorrelation function can be rewritten as:

$$G(\tau) = G_D(\tau)\, G_{fast}(\tau) = G_D(\tau)\{1 + (T/1 - T)\exp(-\tau/\tau_{fast})\} \qquad (1.11)$$

where T is the number of molecules occupying the dark triplet state [74, 75]. As a consequence of fast timescales of these processes, a shoulder is commonly observed in the measured curve. FCS has been widely used to probe protein aggregates and their size distribution either solely or in conjunction with FRET both in solution [61, 65, 76–81] as well as in the cells [82–84]. In a recent report on probing the effect of lipid on tau protein aggregation by FCS at various pH, it was shown that anionic lipid vesicles induce aggregation of the K18 fragment of tau protein when the latter reaches a critical aggregation concentration (CAC). The CAC was found to increase with an increase in pH of the solution. Both FCS and ThT-assay indicated the presence of tau-lipid co-aggregates and absence of the vesicles did not promote aggregation [78]. FCS has also been successfully applied to monitor inhibition of protein aggregation [85]. In another interesting study, FCS in combination with TEM has been utilized to investigate the size, molecular mass distribution and morphology of Aβ protein aggregates formed by two distinct routes. It was observed that at low pH, Aβ forms a dimer and possibly 12-mer species (spherical intermediates) that serve as precursors to amyloid fibrils whereas it forms trimers prior to the formation of toxic amylospheroids (ASPDs) which do not culminate in amyloid fibrils [79]. In another observation, it was pointed out from diffusion coefficient measurements that the Sup35 protein oligomers (diffused in the yeast cell cytoplasm) are actually fibrillar-shaped instead of being globular-shaped [84]. All the FCS measurements generally involve one-photon excitation of extrinsically-labeled fluorescent species, but recently, two-photon FCS using intrinsic tryptophan has been shown to be efficient in probing the formation of soluble oligomers pertinent to protein aggregation [86]. Additionally, the tryptophan lifetime as a function of aggregation could be monitored simultaneously with two-photon FCS technique. An extension of FCS is FCCS (fluorescence cross-correlation spectroscopy) whereby instead of a single-wavelength-labeled fluorescent species, the temporal fluctuations in fluorescence intensity of dual-wavelength-labeled fluorescent species are observed whereby the two colors are cross-correlated [72, 87] By this technique, one can observe the cross-correlation pattern arising out of "joint fluctuations" when the dual-colored molecules diffuse through the observation volume simultaneously [88]. The working principles of both FCS and FCCS have been used in image correlation spectroscopy. Raster image correlation spectroscopy (RICS) [73, 82] is one such technique which allows one to generate spatio-temporal autocorrelation curves from the images of the fluorescent species by raster-scanning the laser beam across the surface to acquire the image.

1.5 Case Studies from our Laboratory at IISER Mohali

In this section, we will briefly discuss our efforts directed towards the elucidation of aggregation mechanisms of three classes of model proteins, namely lysozyme (α+β) and bovine serum albumin (all α-helical) and κ-casein (intrinsically disordered protein) using primarily fluorescence spectroscopy in combination with other biophysical techniques [30, 31, 89, 90]. In all these studies, steady-state fluorescence spectroscopy was employed to monitor the simultaneous changes in both conformation and size. Information about the monomeric as well as oligomeric intermediates and early stages of the aggregation process were extracted by monitoring the aggregation kinetics using a variety of fluorescence readouts.

1.5.1 Lysozyme Aggregation

We have recently studied sodium dodecyl sulfate (SDS)-induced aggregation of lysozyme [30]. In our study, we investigated the aggregation kinetics in real time by monitoring the changes in tryptophan, ANS, fluorescein and dansyl fluorescence intensity and anisotropy with an emphasis towards detection and characterization of the early stages involved in the fibril assembly. The negatively-charged SDS, at a concentration (150 μM) much below its critical micellar concentration (8.6 mM), initiates aggregation of lysozyme (at a very low protein concentration ~5 μM) at pH 9.2 at room temperature (Fig. 1.6). During aggregation kinetics experiments, an increase in both fluorescence intensity and anisotropy of all the fluorescent probes were observed as a function of time suggesting conformational- and size changes leading to aggregation. The rate of change in fluorescence intensity of the fluorophores could be fitted to a biexponential function indicating that the conformational change during lysozyme aggregation is a biphasic process whereas the rate of change in fluorescence anisotropy could be satisfactorily explained by a single exponential function indicating an overall size growth. Comparison of the apparent rate constants obtained during intensity and anisotropy kinetics suggested that conformational changes precede the formation of large-sized aggregates (Fig. 1.7). It was proposed that interaction between the positively-charged lysozyme and negatively charged SDS results in conformational expansion with increased hydrophobicity and such conformational expansion facilitates the formation of soluble oligomers.

1.5.2 Serum Albumin Aggregation

In another study on serum albumin aggregation, we first identified and characterized a 'molten-globule-like' intermediate of bovine serum albumin (BSA) at pH 3 that served as an amyloidogenic precursor. Using various fluorescence techniques involving tryptophan fluorescence intensity and anisotropy, ANS fluorescence intensity and anisotropy and pyrene I_3/I_1 measurements, we established that BSA adopts an expanded conformational state at pH 3 [89]. Figure 1.8 shows the plot of tryptophan fluorescence anisotropy for both bovine and human serum albumins as a function of pH indicating multiple conformational transitions. At pH 3 and elevated temperature, BSA fibrillation was observed [31] in the presence of salt whereby ThT fluorescence assay indicated a dramatic rise in the ThT fluorescence intensity as aggregation proceeded. The fibrillation occurred instantaneously without any

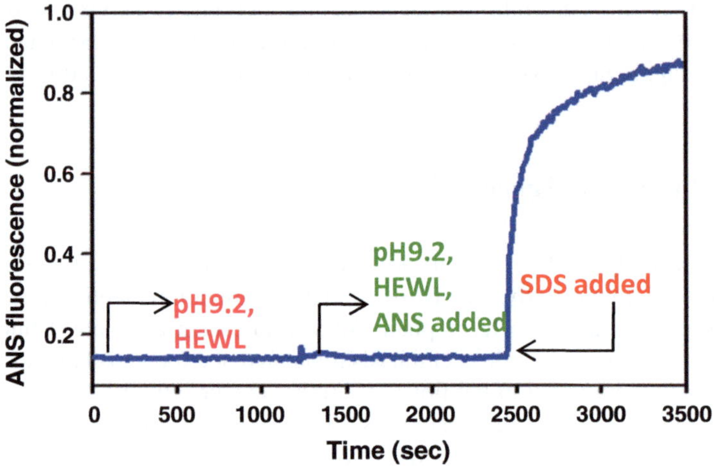

Fig. 1.6 The time course of ANS fluorescence intensity during lysozyme aggregation upon addition of SDS. A triggering experiment shows that SDS is essential for lysozyme aggregation at pH 9.2 (Adapted from Ref. [30])

lag-phase. A comparative analysis of BSA fibrillation kinetics (monitored by ThT) at various pH indicated that the rate of fibrillation was the fastest at pH 3. After establishing the formation of amyloid fibrils at low pH, attempts were made to extract detailed information about the alterations in conformation and size of the aggregates by observing the rate of changes in fluorescence intensity and anisotropy of both intrinsic (tryptophan) and extrinsic (ANS, IAEDANS) fluorophores as a function of time. Prior to heating, the presence of oligomers was detected by all the fluorescent probes whereby higher fluorescence anisotropy was observed compared to that of the monomeric protein indicating an increase in size (Fig. 1.4). After heating, as the aggregation progressed, the tryptophan fluorescence intensity decreased as a function of time but an increase in the ANS and AEDANS fluorescence intensity with a simultaneous blue-shifted emission were observed. It was suggested that as the hydrophobically associated oligomers accumulated and assembled into amyloid fibrils, the tryptophans were quenched due to proximal amino acid residues such as histidines, phenylalanines and disulfides whereas ANS and AEDANS served as fluorescent markers for enhanced hydrophobicity. The fluorescence anisotropy of all the probes showed a monotonic increase as a function of aggregation suggesting the formation of large-sized aggregates. Both the rate of change in fluorescence intensity and anisotropy of the fluorescent reporters could be fitted to a mono-exponential function to extract the apparent rate constants. Comparison of the rate constants obtained from various probes yielded a unified observation that the conformational conversion in the preformed molten oligomers occurs much faster than the overall size growth. Interestingly, the average rate constant obtained from ThT-fluorescence kinetics was similar to that observed for fluorescence anisotropy, thus, re-confirming the fact that ThT is more sensitive to

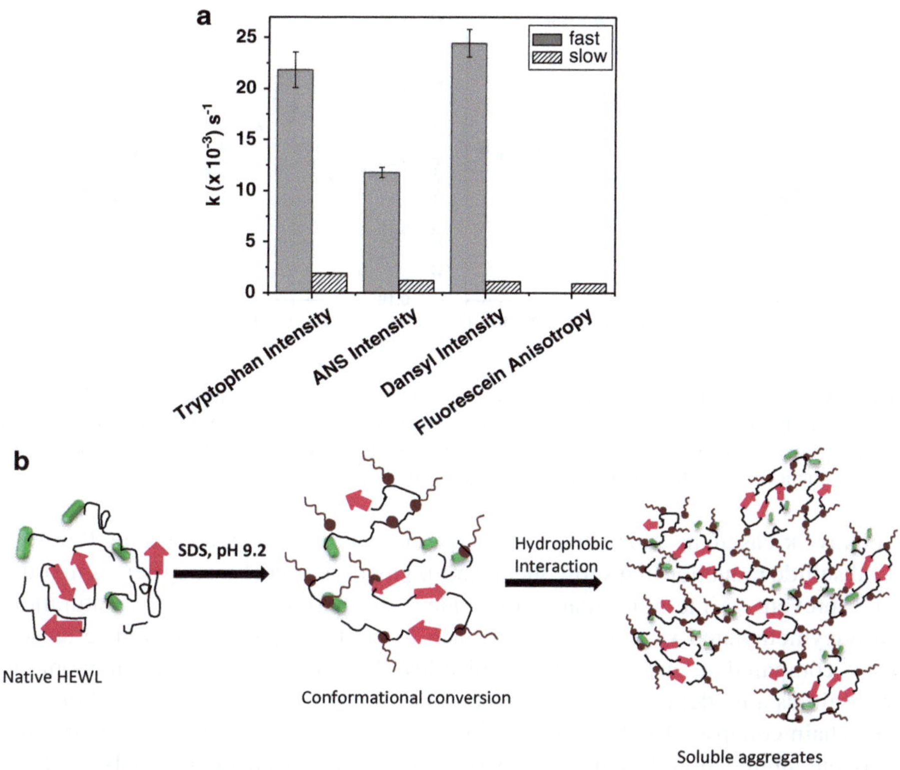

Fig. 1.7 (**a**) The observed rate constants obtained during SDS-induced lysozyme aggregation using various fluorescence readouts indicate that the conformational changes in lysozyme precedes overall size growth of the aggregates. (**b**) A plausible model of SDS-induced lysozyme aggregation (Adapted from Ref. [30])

mature β-sheet-rich amyloid fibrils. It was proposed that the 'molten-globule-like' conformer of BSA forms soluble oligomers readily at higher protein concentration assisted by the salt. At an increased temperature, conformational rearrangements occur in these molten oligomers that culminate in fibrils formation (Fig. 1.9).

1.5.3 Conformational Property of κ-Casein – A Model Intrinsically Disordered Protein

Caseins are flexible milk proteins which belong to the class of intrinsically disordered proteins (IDPs) that lack specific secondary structure. They form protein micelles under native conditions. We have used bovine κ-casein as a model IDP to study conformational and aggregation behaviors. Using pyrene fluorescence vibronic band ratio (I_3/I_1), we show concentration dependent micellar aggregation

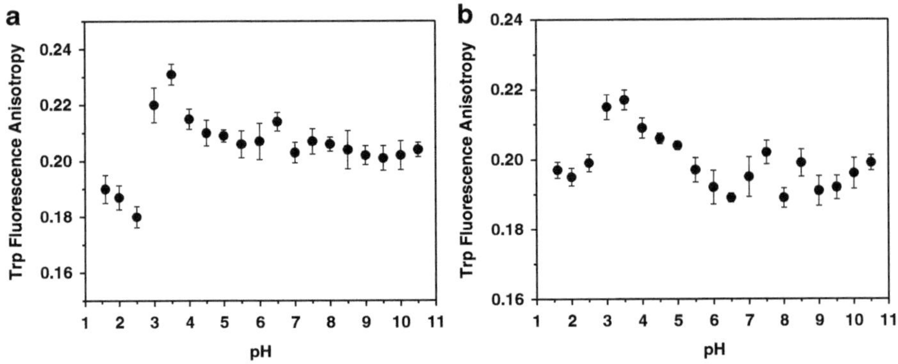

Fig. 1.8 pH-induced conformational isomerism using intrinsic protein fluorescence: Changes in tryptophan fluorescence anisotropy of conformational isomers of (**a**) BSA and (**b**) HSA at room temperature. A molten-globule-like expanded state of serum albumins at pH 3–3.5 is characterized by an enhancement in the tryptophan anisotropy (Adapted from Ref. [89])

of native κ-casein in water (Fig. 1.10a). We monomerized it by denaturing followed by disulfide reduction and subsequent carboxymethylation of free cysteines in order to prevent intermolecular disulfide formation. Upon renaturation, carboxymethylated κ-casein showed much lower tryptophan fluorescence anisotropy compared to native κ-casein micelles (Fig. 1.10b). We used monomeric κ-casein as a model protein to study conformational behavior of an IDP. Polypeptide chain collapse of IDPs is believed to play a key role in protein misfolding and amyloid aggregation. Using a variety of fluorescence spectroscopic tools, we have first established that monomeric κ-casein adopts a collapsed 'pre-molten globule' like conformers under native condition. We then took the advantage of two free cysteines that are separated by 77-amino acid residues and then covalently labeled the cysteines using thiol-reactive pyrene maleimide. This dual-labeled protein demonstrated a strong excimer formation upon renaturation from urea- and acid-denatured states both under equilibrium and kinetic conditions providing a compelling evidence of polypeptide chain collapse under physiological conditions (Fig. 1.10c) [90]. We believe that our pyrene excimer fluorescence-based methodology will be applied to other IDPs. Currently, investigations of aggregation behavior of IDPs are in progress in our laboratory.

1.6 Conclusions and Future Directions

In this review, we have provided an overview of protein misfolding and aggregation studies using fluorescence spectroscopy at various levels of sophistication from steady-state fluorimeter measurements to time-resolved and single molecule fluorescence measurements. It is expected that this review will serve as a benchmark to the practitioners in the field of protein misfolding and aggregation. Monitoring

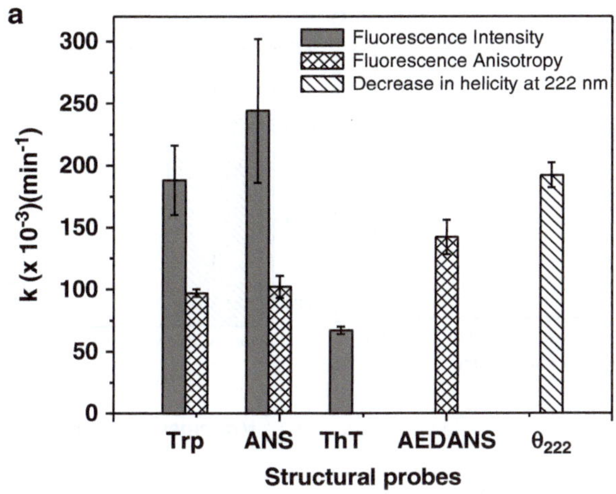

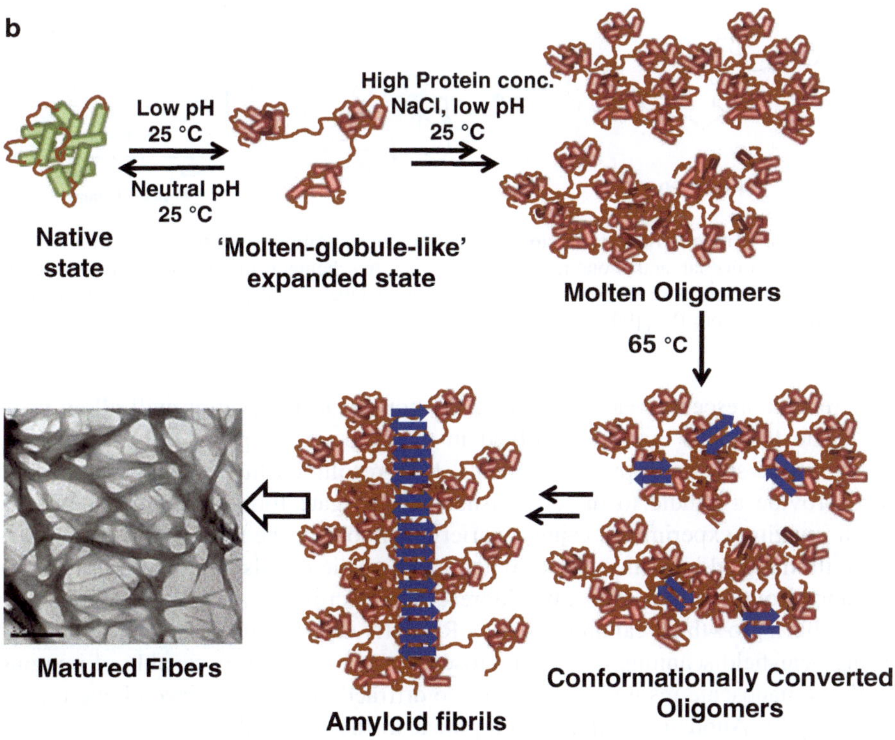

Fig. 1.9 (**a**) The rate constants of BSA fibrillation kinetics of multiple structural probes used to monitor the changes in conformation and size to delineate the steps involved in the fibrillation event. (**b**) A proposed model for fibril formation from serum albumins based on aggregation kinetics (Reproduced from Ref. [31] with permission from the American Chemical Society)

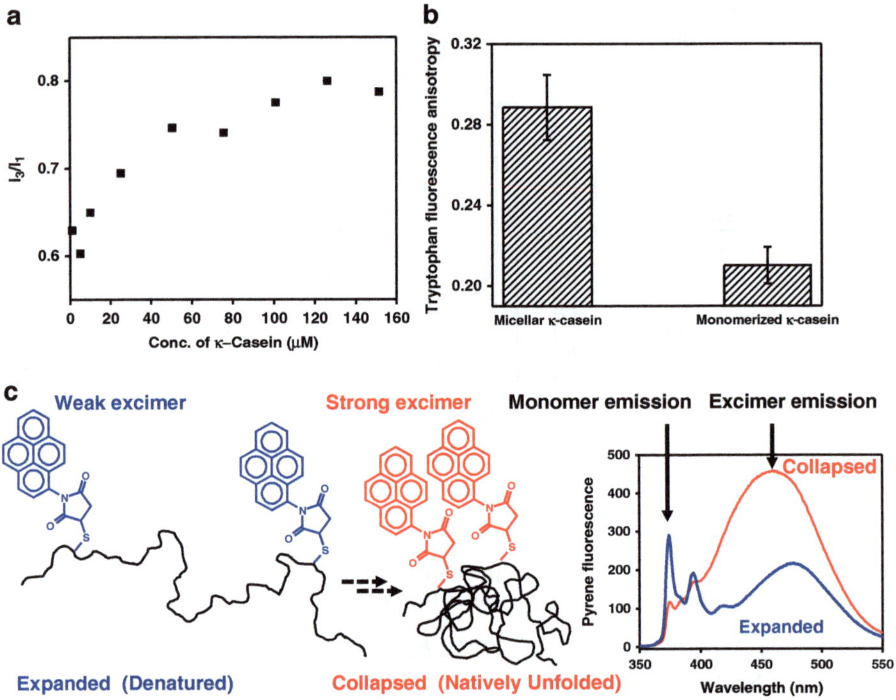

Fig. 1.10 (**a**) Pyrene I_3/I_1 as a function of κ-casien concentration. (**b**) Tryptophan fluorescence anisotropy of micellar and monomeric κ-casien. (**c**) A schematic of excimer formation upon polypeptide chain collapse of monomeric κ-casien dually labeled with pyrene maleimide (Adapted with permission from Ref. [90])

different fluorescence readouts during the course of aggregation will allow us to delineate the steps that are involved in the aggregation process. In addition to fluorescence spectroscopic experiments, high-resolution fluorescence microscopy will provide a handle to directly visualize aggregate formation and maturation. Such imaging experiments using a variety of fluorescence microscopy techniques have illuminated intricate structural and mechanistic details of amyloid fibrils. For instance, total internal reflection fluorescence microscopy has been employed to watch fibril growth in real time [67, 91]. Recently, amyloid fibrils have been imaged using near-field scanning optical microscopy technique that is capable of imaging with the nanoscale resolution (beyond the diffraction limit of conventional microscope) [92]. Another exciting development in the field will be to directly watch aggregation and fibril formation in living cells. A recent work describes the methodology to probe fibrillation using FRET imaging in living cells [93]. We anticipate that in the near future, developments in fluorescent probe chemistry, protein engineering and (unnatural) fluorescent amino acid incorporation, cell biology and fluorescence microscopy techniques will further push the boundaries to drive the amyloid biology field in new and exciting directions.

Acknowledgments We thank IISER Mohali for financial support and the past and present members of the Mukhopadhyay laboratory for their studies on protein aggregation using fluorescence spectroscopy. M.B. thanks the Department of Science and Technology (DST) Women Scientists' Scheme and S.M. thanks the Council of Scientific and Industrial Research (CSIR), India for research grants.

References

1. Dill KA, Chan HS (1997) From Levinthal to pathways to funnels. Nat Struct Biol 4:10–19
2. Jahn TR, Radford SE (2008) Folding vs aggregation: polypeptide conformations on competing pathways. Arch Biochem Biophys 469:100–117
3. Dobson CM (2006) Protein aggregation and its consequences for human disease. Protein Pept Lett 13:219–227
4. Uversky VN, Fink AL (2004) Conformational constraints for amyloid fibrillation: the importance of being unfolded. Biochim Biophys Acta 1698:131–153
5. Morel B, Casares S, Conejero-Lara FA (2006) Single mutation induces amyloid aggregation in the α-spectrin SH3 domain: analysis of the early stages of fibril formation. J Mol Biol 356:453–468
6. Kumar S, Udgaonkar JB (2010) Mechanisms of amyloid fibril formation by proteins. Curr Sci 98:639–656
7. Guijarro JI, Sunde M, Jones JA, Campbell ID, Dobson CM (1998) Amyloid fibril formation by an SH3 domain. Proc Natl Acad Sci U S A 95:4224–4228
8. Chiti F, Webster P, Taddei N, Clark A, Stefani M, Ramponi G, Dobson CM (1999) Designing conditions for in vitro formation of amyloid protofilaments and fibrils. Proc Natl Acad Sci U S A 96:3590–3594
9. Chiti F, Stefani M, Taddei N, Ramponi G, Dobson CM (2003) Rationalization of the effects of mutations on peptide and protein aggregation rates. Nature 424:805–808
10. Bucciantini M, Giannoni E, Chiti F, Baroni F, Formigli L, Zurdo J, Taddei N, Ramponi G, Dobson CM, Stefani M (2002) Inherent toxicity of aggregates implies a common mechanism for protein misfolding diseases. Nature 416:507–511
11. Glabe CG (2008) Structural classification of toxic amyloid oligomers. J Biol Chem 283:29639–29643
12. Tycko R (2006) Molecular structure of amyloid fibrils: insights from solid state NMR. Q Rev Biophys 39:1–55
13. Sivanandam VN, Jayaraman M, Hoop CL, Kodali R, Wetzel R, van der Wel PCA (2011) The aggregation-enhancing Huntingtin N-terminus is helical in amyloid fibrils. J Am Chem Soc 133:4558–4566
14. Carulla N, Zhou M, Arimon M, Gairí M, Giralte E, Robinson CV, Dobson CM (2009) Experimental characterization of disordered and ordered aggregates populated during the process of amyloid fibril formation. Proc Natl Acad Sci U S A 106:7828–7833
15. Fabian H, Gast K, Laue M, Misselwitz R, Uchanska-Ziegler B, Ziegler A, Naumann D (2008) Early stages of misfolding and association of β_2_microglobulin: insights from infrared spectroscopy and dynamic light scattering. Biochemistry 47:6895–6906
16. Ma S, Cao X, Mak M, Sadik A, Walkner C, Freedman TB, Lednev IK, Dukor RK, Nafie LA (2007) Vibrational circular dichroism shows unusual sensitivity to protein fibril formation and development in solution. J Am Chem Soc 129:12364–12365
17. Measey TJ, Schweitzer-Stenner R (2011) Vibrational circular dichroism as a probe of fibrillogenesis: the origin of the anomalous intensity enhancement of amyloid-like fibrils. J Am Chem Soc 133:1066–1076

18. Sunde M, Blake CCF (1998) From the globular to the fibrous state: protein structure and structural conversion in amyloid formation. Q Rev Biophys 31:1–39
19. Makin OS, Atkins E, Sikorski P, Johansson J, Serpell LC (2005) Molecular basis for amyloid fibril formation and stability. Proc Natl Acad Sci U S A 102:315–320
20. Larson JL, Ko E, Miranker AD (2000) Direct measurement of islet amyloid polypeptide fibrillogenesis by mass spectrometry. Protein Sci 9:427–431
21. Smith AM, Jahn TR, Ashcroft AE, Radford SE (2006) Direct observation of oligomeric species formed in the early stages of amyloid fibril formation using electrospray ionisation mass spectrometry. J Mol Biol 364:9–19
22. Goldsbury C, Baxa U, Simon MN, Steven AC, Engel A, Wall JS, Aebi U, Müller SA (2011) Amyloid structure and assembly: insights from scanning transmission electron microscopy. J Struct Biol 173:1–13
23. Gosal WS, Myers SL, Radford SE, Thomson NH (2006) Amyloid under the atomic force microscope. Protein Pept Lett 13:261–270
24. Lakowicz JR (2006) Principles of fluorescence spectroscopy, 3rd edn. Springer, New York
25. Dusa A, Kaylor J, Edridge S, Bodner N, Hong D-P, Fink AL (2006) Characterization of oligomers during α-synuclein aggregation using intrinsic tryptophan fluorescence. Biochemistry 45:2752–2760
26. Padrick SB, Miranker AD (2002) Islet amyloid: phase partitioning and secondary nucleation are central to the mechanism of fibrillogenesis. Biochemistry 41:4694–4703
27. Rolinski OJ, Amaro M, Birch DJS (2010) Early detection of amyloid aggregation using intrinsic fluorescence. Biosens Bioelectron 25:2249–2252
28. Maji SK, Amsden JJ, Rothschild KJ, Condron MM, Teplow DB (2005) Conformational dynamics of Amyloid β-protein assembly probed using intrinsic fluorescence. Biochemistry 44:13365–13376
29. van Rooijen BD, van Leijenhorst-Groener KA, Claessens MMAE, Subramaniam V (2009) Tryptophan fluorescence reveals structural features of α-synuclein oligomers. J Mol Biol 394:826–833
30. Jain N, Bhattacharya M, Mukhopadhyay S (2011) Kinetics of surfactant-induced aggregation of lysozyme studied by fluorescence spectroscopy. J Fluoresc 21:615–625
31. Bhattacharya M, Jain N, Mukhopadhyay S (2011) Insights into the mechanism of aggregation and fibril formation from bovine serum albumin. J Phys Chem B 115:4195–4205
32. Shukla A, Mukherjee S, Sharma S, Agrawal V, Radhakishan KV, Guptasarma P (2004) A novel UV laser-induced visible blue radiation from protein crystals and aggregates: scattering artifacts or fluorescence transitions of peptide electrons delocalized through hydrogen bonding? Arch Biochem Biophys 428:144–153
33. del Mercato LL, Pompa PP, Maruccio G, Torre AD, Sabella S, Tamburro AM, Cingolani R, Rinaldi R (2007) Charge transport and intrinsic fluorescence in amyloid-like fibrils. Proc Natl Acad Sci U S A 104:18019–18024
34. Hawe A, Sutter M, Jiskoot W (2008) Extrinsic fluorescent dyes as tools for protein characterization. Pharm Res 25:1487–1499
35. Levine H III (1999) Quantification of β-sheet amyloid fibril structures with thioflavin T. Methods Enzymol 309:274–284
36. Biancalana M, Koide S (2010) Molecular mechanism of thioflavin-T binding to amyloid fibrils. Biochim Biophys Acta 1804:1405–1412
37. Lindgren M, Sörgjerd K, Hammarström P (2005) Detection and characterization of aggregates, prefibrillar amyloidogenic oligomers, and protofibrils using fluorescence spectroscopy. Biophys J 88:4200–4212
38. Platt GW, Radford SE (2009) Glimpses of the molecular mechanisms of β_2-microglobulin fibril formation in vitro: aggregation on a complex energy landscape. FEBS Lett 583:2623–2629

39. Mishra R, Sörgjerd K, Nyström S, Nordigården A, Yu Y-C, Hammarström P (2007) Lysozyme amyloidogenesis is accelerated by specific nicking and fragmentation but decelerated by intact protein binding and conversion. J Mol Biol 366:1029–1044
40. Hammarström P, Simon R, Nyström S, Konradsson P, Aslund A, Nilsson KPR (2010) A fluorescent pentameric thiophene derivative detects in vitro-formed prefibrillar protein aggregates. Biochemistry 49:6838–6845
41. Lindgren M, Hammarström P (2010) Amyloid oligomers: spectroscopic characterization of amyloidogenic protein states. FEBS J 277:1380–1388
42. Bjorndahl TC, Zhou G-P, Liu X, Perez-Pineiro R, Semenchenko V, Saleem F, Acharya S, Bujold A, Sobsey CA, Wishart DS (2011) Detailed biophysical characterization of the acid-induced PrP^{c} to PrP^{β} conversion process. Biochemistry 50:1162–1173
43. Giurleo JT, He X, Talaga DS (2008) β-lactoglobulin assembles into amyloid through sequential aggregated intermediates. J Mol Biol 381:1332–1348
44. Bolognesi B, Kumita JR, Barros TP, Esbjorner EK, Luheshi LM, Crowther DC, Wilson MR, Dobson CM, Favrin G, Yerbury JJ (2010) ANS binding reveals common features of cytotoxic amyloid species. ACS Chem Biol 5:735–740
45. Daniel E, Weber G (1966) Cooperative effects in binding by bovine serum albumin I: the binding of 1-anilino-8-naphthalenesulfonate fluorimetric titrations. Biochemistry 5:1893–1900
46. Glushko V, Thaler MSR, Karp CD (1981) Pyrene fluorescence fine structure as a polarity probe of hydrophobic regions: behavior in model solvents. Arch Biochem Biophys 210:33–42
47. Dong DC, Winnik MA (1982) The P*y* scale of solvent polarities. Solvent effects on the vibronic fine structure of pyrene fluorescence and empirical correlations with the E_T and Y values. Photochem Photobiol 35:17–21
48. Christensen PA, Pedersen JS, Christiansen G, Otzen DE (2008) Spectroscopic evidence for the existence of an obligate pre-fibrillar oligomer during glucagon fibrillation. FEBS Lett 582:1341–1345
49. Krishnan R, Lindquist SL (2005) Structural insights into a yeast prion illuminate nucleation and strain diversity. Nature 435:765–772
50. Thirunavukkuarasu S, Jares-Erijman EA, Jovin TM (2008) Multiparametric fluorescence detection of early stages in the amyloid protein aggregation of pyrene-labeled α-synuclein. J Mol Biol 378:1064–1073
51. Mukhopadhyay S, Nayak PK, Udgaonkar JB, Krishnamoorthy G (2006) Characterization of the formation of amyloid protofibrils from barstar by mapping residue-specific fluorescence dynamics. J Mol Biol 358:935–942
52. Sun Y, Breydo L, Makarava N, Yang Q, Bocharova OV, Baskakov IV (2007) Site-specific conformational studies of prion protein (PrP) amyloid fibrils revealed two cooperative folding domains within amyloid structure. J Biol Chem 282:9090–9097
53. Yap TL, Pfefferkorn CM, Lee JC (2011) Residue-specific fluorescent probes of α-synuclein: detection of early events at the N- and C-termini during fibril assembly. Biochemistry 50:1963–1965
54. Yushchenko DA, Fauerbach JA, Thirunavukkuarasu S, Jares-Erijman EA, Jovin TM (2011) Fluorescent ratiometric MFC probe sensitive to early stages of α-synuclein aggregation. J Am Chem Soc 132:7860–7861
55. Koo BW, Miranker AD (2005) Contribution of the intrinsic disulfide to the assembly mechanism of islet amyloid. Protein Sci 14:231–239
56. Allsop D, Swanson L, Moore S, Davies Y, York A, El-Agnaf OMA, Soutar I (2001) Fluorescence anisotropy: a method for early detection of Alzheimer β-peptide (A-β) aggregation. Biochem Biophys Res Commun 285:58–63
57. Luk KC, Hyde EG, Trojanowski JQ, Lee VM-Y (2007) Sensitive fluorescence polarization technique for rapid screening of α-synuclein oligomerization/fibrillization inhibitors. Biochemistry 46:12522–12529

58. Jha A, Udgaonkar JB, Krishnamoorthy G (2009) Characterization of the heterogeneity and specificity of interpolypeptide interactions in amyloid protofibrils by measurement of site-specific fluorescence anisotropy decay kinetics. J Mol Biol 393:735–752
59. Kayser V, Turton DA, Aggeli A, Beevers A, Reid GD, Beddard GS (2004) Energy migration in novel pH-triggered self-assembled β-sheet ribbons. J Am Chem Soc 126:336–343
60. Teoh CL, Bekard IB, Asimakis P, Griffin MDW, Ryan TM, Dunstan DE, Howlett GJ (2011) Shear flow induced changes in apolipoprotein C-II conformation and amyloid fibril formation. Biochemistry 50:4046–4057
61. Nath S, Meuvis J, Hendrix J, Carl SA, Engelborghs Y (2010) Early aggregation steps in α-synuclein as measured by FCS and FRET: evidence for a contagious conformational change. Biophys J 98:1302–1311
62. Takahashi T, Kikuchi S, Katada S, Nagai Y, Nishizawa M, Onodera O (2008) Soluble polyglutamine oligomers formed prior to inclusion body formation are cytotoxic. Hum Mol Genet 17:345–356
63. Schierle GSK, Bertoncini CW, Chan FTS, van der Goot AT, Schwedler S, Skepper J, Schlachter S, van Ham T, Esposito A, Kumita JR, Nollen EAA, Dobson CM, Kaminski CF (2011) A FRET sensor for non-invasive imaging of amyloid formation in vivo. Chem Phys Chem 12:673–680
64. Mukhopadhyay S, Deniz AA (2007) Biomolecular structure and dynamics using fluorescence from single diffusing molecules. J Fluoresc 17:775–783
65. Hillger F, Nettels D, Dorsch S, Schuler B (2007) Detection and analysis of protein aggregation with confocal single molecule fluorescence spectroscopy. J Fluoresc 17:759–765
66. Mukhopadhyay S, Krishnan R, Lemke EA, Lindquist S, Deniz AA (2007) A natively unfolded yeast prion monomer adopts an ensemble of collapsed and rapidly fluctuating structures. Proc Natl Acad Sci U S A 104:2649–2654
67. Collins SR, Douglass A, Vale RD, Weissman JS (2004) Mechanism of prion propagation: amyloid growth occurs by monomer addition. PLoS Biol 2(e321):1582–1590
68. Orte A, Birkett NR, Clarke RW, Devlin GL, Dobson CM, Klenerman D (2008) Direct characterization of amyloidogenic oligomers by single-molecule fluorescence. Proc Natl Acad Sci U S A 105:14424–14429
69. Ding H, Wong PT, Lee EL, Gafni A, Steel DG (2009) Determination of the oligomer size of amyloidogenic protein β-Amyloid(1-40) by single-molecule spectroscopy. Biophys J 97:912–921
70. Magde D, Webb WW, Elson E (1972) Thermodynamic fluctuations in a reacting system-measurement by fluorescence correlation spectroscopy. Phys Rev Lett 29:705–708
71. Schwille P, Haustein E (2002) Fluorescence correlation spectroscopy: an introduction to its concepts and applications. Biophysics textbook online
72. Sahoo H, Schwille P (2011) FRET and FCS-friends or foes? Chem Phys Chem 12:532–541
73. Kolin DL, Wiseman PW (2007) Advances in image correlation spectroscopy: measuring number densities, aggregation states, and dynamics of fluorescently labeled macromolecules in cells. Cell Biochem Biophys 49:141–164
74. Widengren J, Mets U, Rigler R (1995) Fluorescence correlation spectroscopy of triplet states in solution: a theoretical and experimental study. J Phys Chem 99:13368–13379
75. Mütze J, Ohrt T, Schwille P (2011) Fluorescence correlation spectroscopy in vivo. Laser Photonics Rev 5:52–67
76. Tjernberg LO, Pramanik A, Bjijrling S, Thyberg P, Thyberg J, Nordstedt C, Berndt KD, Terenius L, Rigler R (1999) Amyloid β-peptide polymerization studied using fluorescence correlation spectroscopy. Chem Biol 6:53–62
77. Kitamura A, Kubota H, Pack C-G, Matsumoto G, Hirayama S, Takahashi Y, Kimura H, Kinjo M, Morimoto RI, Nagata K (2006) Cytosolic chaperonin prevents polyglutamine toxicity with altering the aggregation state. Nat Cell Biol 8:1163–1169
78. Elbaum-Garfinkle S, Ramlall T, Rhoades E (2010) The role of the lipid bilayer in tau aggregation. Biophys J 98:2722–2730

79. Matsumura S, Shinoda K, Yamada M, Yokojima S, Inoue M, Ohnishi T, Shimada T, Kikuchi K, Masui D, Hashimoto S, Sato M, Ito A, Akioka M, Takagi S, Nakamura Y, Nemoto K, Hasegawa Y, Takamoto H, Inoue H, Nakamura S, Nabeshima Y, Teplow DB, Kinjo M, Hoshi M (2011) Two distinct amyloid β Protein (Aβ) assembly pathways leading to oligomers and fibrils identified by combined fluorescence correlation spectroscopy, morphology, and toxicity analyses. J Biol Chem 286:11555–11562
80. Sutter M, Oliveira S, Sanders NN, Lucas B, van Hoek A, Hink MA, Visser AJWG, De Smedt SC, Hennink WE, Jiskoot W (2007) Sensitive spectroscopic detection of large and denatured protein aggregates in solution by use of the fluorescent dye Nile red. J Fluoresc 17:181–192
81. Sengupta P, Garai K, Balaji J, Periasamy N, Maiti S (2003) Measuring size distribution in highly heterogeneous systems with fluorescence correlation spectroscopy. Biophys J 84:1977–1984
82. Vetri V, Ossato G, Militello V, Digman MA, Leone M, Gratton E (2011) Fluctuation methods to study protein aggregation in live cells: concanavalin A oligomers formation. Biophys J 100:774–783
83. Takahashi Y, Okamoto Y, Popiel HA, Fujikake N, Toda T, Kinjo M, Nagai Y (2007) Detection of polyglutamine protein oligomers in cells by fluorescence correlation spectroscopy. J Biol Chem 282:24039–24048
84. Kawai-Noma S, Pack C-G, Kojidani T, Asakawa H, Hiraoka Y, Kinjo M, Haraguchi T, Taguchi H, Hirata A (2010) In vivo evidence for the fibrillar structures of Sup35 prions in yeast cells. J Cell Biol 190:223–231
85. Ghosh R, Sharma S, Chattopadhyay K (2009) Effect of arginine on protein aggregation studied by fluorescence correlation spectroscopy and other biophysical methods. Biochemistry 48:1135–1143
86. Sahoo B, Balaji J, Nag S, Kaushalya SK, Maiti S (2008) Protein aggregation probed by two-photon fluorescence correlation spectroscopy of native tryptophan. J Chem Phys 129:0751031–0751035
87. Hwang LC, Wohland T (2007) Recent advances in fluorescence cross-correlation spectroscopy. Cell Biochem Biophys 49:1–13
88. Bieschke J, Schwille P (1998) Aggregation of prion protein investigated by dual-color fluorescence cross-correlation spectroscopy. In: Slavik J (ed) Fluorescence microscopy and fluorescent probes, vol 2. Plenum Press, New York
89. Bhattacharya M, Jain N, Bhasne K, Kumari V, Mukhopadhyay S (2011) pH-induced conformational isomerization of bovine serum albumin studied by extrinsic and intrinsic protein fluorescence. J Fluoresc 21:1083–1090
90. Jain N, Bhattacharya M, Mukhopadhyay S (2011) Pyrene excimer formation during polypeptide chain collapse of an amyloidogenic intrinsically disordered protein. Biophys J 101:1720–1729
91. Ban T, Yamaguchi K, Goto Y (2006) Direct observation of amyloid fibril growth, propagation, and adaptation. Acc Chem Res 39:663–670
92. Kitts CC, Bout DAV (2009) Near-field scanning optical microscopy measurements of fluorescent molecular probes binding to insulin amyloid fibrils. J Phys Chem B 113:12090–12095
93. Roberti MJ, Bertoncini CW, Klement R, Jares-Erijman EA, Jovin TM (2007) Fluorescence imaging of amyloid formation in living cells by a functional, tetracysteine-tagged α-synuclein. Nat Methods 4:345–351

Chapter 2
Time-Dependent Spectral Shifts in Tryptophan Fluorescence: Bridging Experiments with Molecular Dynamics Simulations

Dmitri Toptygin

Abstract Molecular dynamics (MD) simulations are widely used to model protein motions. Although the time resolution of MD simulations is virtually unlimited, simulated MD is seldom compared with experimental data on the picosecond time scale because few experimental techniques can probe molecular vibrations in the frequency range between 300 MHz and 300 GHz. Time-dependent spectral shift (TDSS, also known as dynamic Stokes shift) in fluorescence emission from solvatochromic dyes has long been used to study relaxation dynamics of polar solvents on the picosecond time scale. This chapter reviews the use of TDSS in connection with protein dynamics. Different methods of calculating TDSS from the non-equilibrium and equilibrium MD are compared and the limits of their applicability are defined. Methods for separating the contributions of water and protein to TDSS are considered. Two relaxation modes of bulk water are described and their effects on the TDSS in proteins are examined. The rates of water relaxation near interfaces and inside protein pockets are evaluated. A method for identifying conformational changes responsible for the TDSS on different time scales is introduced.

Keywords Non-equilibrium protein dynamics • Time-dependent spectral shift • Solvent relaxation • Longitudinal dielectric relaxation • Transverse dielectric relaxation

2.1 Introduction

Empirical forcefields, such as AMBER [1], CHARMM [2], and GROMOS [3], derive the values of the bonded parameters from the experimentally measured vibrational frequencies in the infrared and Raman spectra, which cover the

D. Toptygin (✉)
Department of Biology, Johns Hopkins University, 3400 N. Charles St., Baltimore, MD 21218, USA
e-mail: toptygin@jhu.edu

© Springer International Publishing Switzerland 2016

C.D. Geddes (ed.), *Reviews in Fluorescence 2015*,
DOI 10.1007/978-3-319-24609-3_2

frequency range from 300 GHz to 400 THz. This ensures that simulated molecular dynamics (MD) is always in good agreement with experimental measurements at frequencies between 300 GHz and 400 THz. MD simulations have also been extensively compared to Nuclear Magnetic Resonance (NMR) [4–6], in which the bandwidth of the recorded dynamic signal usually does not exceed 300 MHz. The frequency range from 300 MHz to 300 GHz did not receive as much attention as the frequencies outside of this range. Using the uncertainty principle $\Delta f \Delta t \sim 1/(2\pi)$ where Δf is the bandwidth on the frequency scale and Δt is the pulse width on the time scale, the 300 MHz to 300 GHz frequency range translates to the pulse width range from ~0.5 ps to ~0.5 ns on the time scale. The time scale between ~0.5 ps and ~0.5 ns is here referred to as the "picosecond time scale".

MD simulations are widely used to study protein folding and ligand docking, to estimate thermodynamic stabilities and binding affinities of proteins and other biomolecules [7–10]. Much less often MD simulations are used to study nonequilibrium protein dynamics, such as the dynamics of conformational changes in response to an instantaneous change in one parameter. How well the dynamics obtained from MD simulations agrees with experimental observations on the picosecond time scale is still an open question. The picosecond time scale and the corresponding 300 MHz to 300 GHz frequency range are of critical importance for two reasons. First, all elementary conformational changes in proteins (such as flipping of just one dihedral angle to a different energy minimum) take place on the picosecond time scale. Biologically significant conformational changes, such as folding and unfolding, binding of enzymes to coenzymes and substrates, etc., consist of a multitude of elementary conformational steps and thus take much longer time, but the rates of the big conformational changes are determined by the rates of the elementary conformational steps, which occur on the picosecond time scale.

The second reason why the picosecond time scale is so important has its origin in quantum physics. All existing MD packages are based on the classical (Newtonian) description of nuclear motion, where the rotation of small molecules (like H_2O) about their principal axes of inertia and small groups (like $-CH_3$, $-NH_3^+$, $-CO_2^-$, $-C_6H_5$) about the single bonds connecting these groups to the rest of the biomolecule is not quantized. In quantum mechanics the projection of the angular momentum on the rotation axis can assume only discrete values, which are multiples of $\hbar$ (where $\hbar$ is the Planck's constant divided by 2π), the rotational energies are multiples of $\hbar^2/(2I)$ (where I is the moment of inertia for the rotation about the selected axis), and the rotational transition frequencies are multiples of $\hbar/(4\pi I)$ [11]. Using atomic masses, bond lengths and angles it is not difficult to calculate the fundamental rotational transition frequencies $\hbar/(4\pi I)$ for small groups and molecules: 5.6 GHz for the phenyl group ($-C_6H_5$), 11.5 GHz for the carboxyl group ($-CO_2^-$), 157 GHz for the methyl group ($-CH_3$), 186 GHz for the amino group ($-NH_3^+$), and either 286 GHz or 822 GHz for TIP3P water molecule (depending on the choice of the rotation axis). The biggest difference between classical and quantum mechanics will be observed at frequencies near $\hbar/(4\pi I)$, most of which, as we just saw, fall in the 300 MHz to 300 GHz range.

Classical description of nuclear motion also fails to explain the operation of the ammonia maser [12]. Amplification of microwave radiation by ammonia molecules is due to a transition between two vibrational states [12, 13], which are characterized by different wavefunctions. The most interesting property of these wavefunctions is that the probability density for the normal vibrational coordinate has two equal peaks, i. e., the nitrogen nucleus is on both sides of the plane containing the three hydrogen nuclei [14, 15]. The excited-state wavefunction changes its sign when the nitrogen nucleus crosses the hydrogen plane, whereas the ground-state wavefunction has the same sign on both sides of the plane [14, 15]. Classical description of the nuclear motion cannot handle the case where some (or all) nuclei are delocalized between two (or more) locations at the same time, therefore some of the terms in the interaction energy between molecules with delocalized nuclei are missing in every forcefield that is based on the classical description of nuclear motion. The missing energy terms may play an important role in protein folding. The difference between the classical and quantum description of the nuclear motion is most apparent at frequencies close to those at which masers operate: 24 GHz for the ammonia maser, 22 GHz and 96 GHz for the water maser, etc.. Again, all these frequencies fall between 300 MHz and 300 GHz, where simulated MD has not been adequately tested against experimental data.

Consider the experimental techniques that can be used to study protein dynamics in the 300 MHz to 300 GHz frequency range. Microwave spectroscopy can be used to study protein molecules in vacuum but not in aqueous solution, since liquid water is not transparent in the frequency range between 300 MHz and 300 GHz. The signals measured in dielectric relaxation studies of aqueous protein solutions are overwhelmed by the relaxation of bulk water and contain only a small contribution from the protein molecules, which is largely attributable to the rotation of the protein as a whole. Thus, it is difficult to extract information about vibrational and conformational dynamics in specific protein regions from the dielectric relaxation data. To study the dynamics of elementary conformational changes in proteins and other biomolecules in their native aqueous environment it is necessary to use electromagnetic radiation within the window of transparency of water, which extends over the wavelengths from 200 nm to 900 nm and covers all the visible range, a significant part of the UV range, and a small part of the near-IR range. To convert the spectrum of the electromagnetic signal from the optical range to the low-frequency range ($f < 300$ GHz), and then back from the low-frequency range to the optical range, there must be a nonlinear electro-optical element inside the biomolecule. Any solvatochromic fluorescent dye can play the role of the nonlinear element. Excitation of the solvatochromic dye by a short laser pulse results in a redistribution of its π-orbitals electron density, which, in turn, results in an abrupt change of the magnitude and/or direction of its permanent electric dipole moment. This instantly changes the electrostatic forces acting between the solvatochromic dye and the partial charges on the atoms in its environment. If the system was in the state of equilibrium before the laser pulse, then what we observe after the laser pulse is known as the nonequilibrium dynamics. As the system relaxes to a new equilibrium, the charged atoms in the vicinity of the solvatochromic dye move in a systematic (non-random) manner, generating a time-variant electric field that acts back on the solvatochromic dye and modulates the energy gap between its ground and excited state via the linear Stark effect. The time variation of the electronic energy gap results in the Time-Dependent Spectral Shift (TDSS) in fluorescence emission.

TDSS has been experimentally observed in the fluorescence of solvatochromic dyes in polar solvents [16–24] and in the fluorescence of solvatochromic fluorophores incorporated in biological macromolecules, such as proteins [25–52] and DNA [53, 54]. While there is a broad consensus that the TDSS in the emission from dyes in polar solvents reflects the relaxation dynamics of the solvent [17], there is still no agreement on the origins of TDSS in biological macromolecules. Some authors attributed experimentally-observed TDSS in proteins to the relaxation of the protein matrix [25–28, 30, 33, 38, 41, 47, 48]. Others argue that the relaxation of the solvent (water) is entirely responsible for TDSS in proteins [32, 35, 36, 39, 42, 46, 49–51]. There are also several reports in which the authors found experimental evidence for contribution from both water and protein matrix to the TDSS [29, 34, 40, 43–45, 52].

One approach that may help to separate the contributions of water and protein matrix to TDSS in proteins is based on the different time scales of these contributions. In bulk water TDSS occurs on the sub-picosecond time scale [18]. TDSS in proteins covers a wide range of time scales, from femtoseconds to at least tens of nanoseconds [25–52]. Abbyad et al. [43] measured TDSS at seven different sites within the same protein. Both sub-picosecond and slower (picosecond and nanosecond) relaxations were found to be contributing to TDSS at every site, however, the sub-picosecond relaxations were found to be dominant at those sites where the fluorophore was in contact with water, and the slower relaxations were found to be dominant at the sites buried deep inside the protein. Based on these findings Abbyad et al. [43] attributed the sub-picosecond TDSS to the relaxation of water, and the slower TDSS to the relaxation of the protein matrix. In accordance with this point of view, in those experiments where the time resolution was much slower than 1 ps, the experimentally observed TDSS would be entirely due to protein relaxation, which is in agreement with the conclusions in most studies where the time resolution was insufficient to detect the relaxation of bulk water [25–27, 33, 38, 41, 48].

A different point of view on the origin of slow TDSS in proteins was expressed by a group of authors [32, 35, 36, 39, 42, 45, 46, 49–52]. According to the latter point of view, the motion of water molecules in the vicinity of the protein is highly constrained, and this "biological water" layer is the sole origin of the slow TDSS in proteins [32, 35, 36, 39]. In later papers it is suggested that the fluctuations of water and protein atoms are coupled and TDSS is attributed the motion of water molecules in the protein hydration layer [42, 45, 46, 49–52]. Thus, there are two diametrically opposite points of view on the origin of slow TDSS in proteins, and each one has serious implications. If the slow TDSS represents the relaxation of the protein matrix, then it can be used to study the dynamics of elementary conformational changes on the picosecond time scale. However, if the slow TDSS represents the relaxation of "biological water", then it calls for developing a physically viable theory that would explain the properties of this form of H_2O. Unfortunately, experimental evidence alone cannot distinguish the contribution to TDSS due to the relaxation of the protein matrix from that due to the relaxation of the solvent.

MD simulations offer more than one way to separate the contributions of water and protein atoms to TDSS. Numerous MD simulations of TDSS in proteins have been reported [35, 44, 45, 55–62]. In some of these reports the contributions of water and protein atoms to TDSS were separated [45, 56, 57, 59–62], and here the points of view regarding the contributions of water and protein to slow TDSS diverged again. According to Nilsson and Halle [57] water cannot contribute to slow TDSS. According to Golosov and Karplus [60], depending on the location of the solvatochromic fluorophore, TDSS may be due to water, due to protein, or due to both water and protein. In addition, several groups of authors found independently that in those cases where both protein and solvent contribute to the spectral shift, the contributions from protein and water have a negative correlation, i. e., when protein shifts the emission spectrum to the red, water shifts it to the blue and vice versa [56–59, 62]. Halle and Nilsson [63] proposed a simple explanation for the negative correlation based on a dielectric continuum model.

It is likely that the differences in the methods of converting MD trajectories to spectral shifts contributed in part to the differences in the conclusions regarding the relative contributions of protein and water. Non-equilibrium hybrid quantum mechanics - molecular dynamics (QM-MD) simulations were used to calculate TDSS in all papers that came out of the group of Callis [55, 56, 64]. A much faster approach, based on Coulomb's equation, was employed by Hassanali et al. [59] to estimate TDSS from non-equilibrium MD trajectories. A large group of authors used the linear-response method to estimate TDSS [44, 57, 58, 60]. This method makes it possible to estimate TDSS from correlations of random fluctuations in an equilibrium MD trajectory. Some reports describe the application of both the non-equilibrium MD and the linear-response method [45, 61, 62].

This chapter reviews the explicit and implicit assumptions on which different methods of calculating TDSS from the non-equilibrium and equilibrium MD are based and defines the limits of applicability for each method. Also it reviews methods of separating the contributions of water and protein to the TDSS. Two different dielectric relaxation modes of bulk water are considered and their effects on the TDSS of fluorophores in bulk water and those incorporated in protein molecules are examined. The rates of dielectric relaxation of bulk water near interfaces and of isolated water molecules in protein pockets are evaluated. This chapter also describes how one can identify the mode of protein conformational changes responsible for the TDSS on every time scale.

2.2 Relationship Between Spectral Shifts and Electronic Energy Levels

The energies E_0 and E_1 of the singlet electronic states S_0 and S_1 vary with the electric field (Stark effect). Motions of charged and polar chemical groups in the environment of the fluorophore result in a time variation of the energies E_0 and E_1, which will be denoted below as $E_0(t)$ and $E_1(t)$ to emphasize their time dependence. The energy gap $E_1(t)$-$E_0(t)$ defines the frequency ν_{00} of the vibrationless electronic transition (the 0-0 transition):

$$h\nu_{00}(t) = E_1(t) - E_0(t) \tag{2.1}$$

In UV-visible absorption and emission spectra of solvatochromic fluorophores in polar environments, at T ~ 300 K, the vibrational structure cannot be resolved, therefore the value of ν_{00} cannot be determined from experimental data, which makes it impossible to obtain $E_1(t)$-$E_0(t)$ using Eq. (2.1). In most experimental studies the time variation of $E_1(t)$-$E_0(t)$ is estimated from either the peak or the mean frequency in the absorption/emission spectra using one of the following relationships:

$$h\nu_{ab,c}(t) = E_1(t) - E_0(t) + h\nu_{ex,c}^{vib} \tag{2.2}$$

$$h\nu_{ab,\,pk}(t) = E_1(t) - E_0(t) + h\nu_{ex,\,pk}^{vib} \tag{2.3}$$

$$h\nu_{em,c}(t) = E_1(t) - E_0(t) - h\nu_{gr,c}^{vib} \tag{2.4}$$

$$h\nu_{em,\,pk}(t) = E_1(t) - E_0(t) - h\nu_{gr,\,pk}^{vib} \tag{2.5}$$

In these equations h is the Planck constant, ν^{vib} denote vibrational frequencies, ν without a superscript denote frequencies in UV-visible absorption and emission spectra, the subscripts *ab* and *em* refer to absorption and emission, respectively, the subscripts *gr* and *ex* refer to the vibrational frequencies in the ground-state and in the excited-state electronic configuration, respectively, subscript c indicates the "center of gravity frequency", i. e., the Franck-Condon factor weighted mean frequency, and the subscript *pk* refers to the peak position in the Franck-Condon factor envelope. The shapes of the Franck-Condon factor envelopes are not identical to the shapes of the absorption and emission spectra, since for transitions with equal Franck-Condon factors the probabilities of emission vary proportionally to ν^3, and the probabilities of absorption vary proportionally to ν^1. To obtain Franck-Condon factor envelopes one has to divide the extinction coefficient spectrum $\varepsilon(\nu)$ by ν and the emitted photon density spectrum $F(\nu)$ by ν^3. Thus, the mean frequencies for the absorption and emission spectra should be calculated as follows:

$$\nu_{ab,c}(t) = \frac{\int_{\nu_1}^{\nu_2} \varepsilon(\nu,t)d\nu}{\int_{\nu_1}^{\nu_2} \varepsilon(\nu,t)\nu^{-1}d\nu} \tag{2.6}$$

$$\nu_{em,c}(t) = \frac{\int_{\nu_3}^{\nu_4} F(\nu,t)\nu^{-2}d\nu}{\int_{\nu_3}^{\nu_4} F(\nu,t)\nu^{-3}d\nu} \tag{2.7}$$

The frequency intervals (ν_1,ν_2) and (ν_3,ν_4) must completely include the absorption band $S_1 \leftarrow S_0$ and the emission band $S_0 \leftarrow S_1$, respectively. The importance of dividing $F(\nu)$ by ν^3 prior to the calculation of the center of gravity has been emphasized in previous work [33, 65], where an expression for emission center of gravity similar to that in Eq. (2.7) was derived.

Equations (2.2, 2.3, 2.4, and 2.5) are similar in one respect: the right-hand side in each of them equals $E_1(t)$-$E_0(t)$ plus a constant term. This means that no matter whether the absorption or emission spectrum is measured and whether the central frequency or the peak frequency is calculated from the instantaneous spectra, essentially the same equation can be used; the only difference is a constant term. A somewhat more complicated situation exists in the case of the time variation of the Stokes shift. Stokes shift is defined as the difference in frequency between the absorption and the emission peak. By subtracting Eq. (2.5) from Eq. (2.3) one can obtain

$$\begin{aligned} h\left[\nu_{ab,\,pk}(t') - \nu_{em,\,pk}(t'')\right] = {} & E_1(t') - E_1(t'') - E_0(t') \\ & + E_0(t'') + h\left[\nu^{vib}_{gr,\,pk} + \nu^{vib}_{ex,\,pk}\right] \end{aligned} \tag{2.8}$$

Here t' and t'' are the instances when the absorption spectrum and the emission spectrum are measured, respectively, and the difference in square brackets on the left-hand side of Eq. (2.8) represents the Stokes shift. If we assume, for instance, that $t' = t''$, then the Stokes shift equals $\nu^{vib}_{gr,\,pk} + \nu^{vib}_{ex,\,pk}$, i. e. the Stokes shift is not time-dependent. The latter appears to be in contradiction with the concept of the time-dependent Stokes shift. However, the condition $t' = t''$ cannot be achieved in reality, since the absorption spectrum can be measured only when the fluorophore is in the ground state, and the emission spectrum can be measured only when it is in the excited state. In time-resolved fluorescence emission experiments the instance of excitation is usually taken for $t = 0$, therefore in Eq. (2.8) $t' = 0$ and $t'' > 0$. In pump-dump-probe experiments [30] one can take the instance of dumping for $t = 0$, therefore in Eq. (2.8) $t' > 0$ and $t'' = 0$. In general, however, the Stokes shift is a function of two variables rather than one. This makes the terms Time-Dependent Stokes Shift (TDSS) and Dynamic Stokes Shift (DSS) somewhat more confusing than the term Time-Dependent Spectral Shift (TDSS), which represents a function of just one variable. From the second law of thermodynamics it follows that in the case of a homogeneous fluorophore population the TDSS in fluorescence emission is always the Time-Dependent Red Shift (TDRS), and in pump-dump-probe experiments (where the time-resolved absorption spectra after stimulated emission are measured) the TDSS is always a Time-Dependent Blue Shift (TDBS).

2.3 Effect of the Electric Field on the Electronic Energy Levels

Quantum-mechanical perturbation theory has long been used to describe the Stark effect in atoms and molecules [11]. Here we use the perturbation theory to derive a simple equation relating the energy gap $E_1(t)$-$E_0(t)$ that enters in each of Eqs. (2.1, 2.2, 2.3, 2.4, and 2.5) to the local electric field acting on the fluorophore. Within the framework of the perturbation theory, the Hamiltonian operator $\hat{H}$ of the fluorophore (in the fixed-nuclei approximation) must be represented as a sum of an unperturbed Hamiltonian $\hat{H}\,(0)$and a perturbation $\hat{V}$,

$$\hat{H} = \hat{H}\,(0) + \hat{V} \tag{2.9}$$

The theory of the Stark effect is usually applied to an atom or a molecule in vacuum. In this case the Hamiltonian operator in the absence of the external electric field is taken for $\hat{H}^{(0)}$, and $\hat{V}$ describes the additional potential energy of interaction between the electrons and the external electric field. It is shown below (see Sect. 2.4, "Nonlinear Stark Effect") that this approach may yield inaccurate results in the case of a solvatochromic fluorophore in a polar environment, especially if the fluorophore possesses two close electronic energy levels. Here $\hat{H}^{(0)}$is defined as the Hamiltonian operator for the fluorophore in a uniform electric field [66] $\mathbf{E}^{(0)}$, and $\hat{V}$ is defined as the additional potential energy due to the difference between the actual electric field (which does not have to be uniform) and $\mathbf{E}^{(0)}$. If $\mathbf{E}^{(0)}$ does not differ too much from the mean [67] electric field in the environment of the fluorophore, then the errors resulting from the use of the method described here will be insignificant, see Sect. 2.4, "Nonlinear Stark Effect". The perturbed energies E_n of the two lowest singlet electronic levels ($n = 0,1$) can be expressed in the form suggested by Landau [11]:

$$E_n = E_n^{(0)} + E_n^{(1)} + E_n^{(2)} + \ldots \tag{2.10}$$

$$E_n^{(1)} = V_{nn} \tag{2.11}$$

$$E_n^{(2)} = \sum_{m \neq n} \frac{|V_{mn}|^2}{E_n^{(0)} - E_m^{(0)}} \tag{2.12}$$

Here $E_n^{(0)}$ denote the unperturbed energy levels, i. e. the eigenvalues of the unperturbed Hamiltonian $\hat{H}^{(0)}$, $E_n^{(1)}$ denote the first-order corrections, $E_n^{(2)}$ denote the second-order corrections, and V_{nn} and V_{mn} represent the diagonal and off-diagonal matrix elements of the perturbation Hamiltonian $\hat{V}$. In a uniform electric field **E** the matrix elements of the perturbation Hamiltonian can be conveniently expressed in terms of the matrix elements of the electric dipole operator:

$$V_{mn} = -\boldsymbol{\mu}_{mn}^{(0)} \cdot (\mathbf{E} - \mathbf{E}^{(0)}) \tag{2.13}$$

Here $\cdot$ denotes a scalar product of two vectors. Vectors $\boldsymbol{\mu}_{mn}^{(0)}$ are the matrix elements of the electric dipole operator, defined on the basis set of eigenfunctions $\Psi_n^{(0)}$ of the Hamiltonian $\hat{H}^{(0)}$, which represents the fluorophore in the uniform electric field $\mathbf{E}^{(0)}$.

The electric field acting on a fluorophore in a polar environment is generated mostly by charged and polar groups in close proximity to the fluorophore; this field is not expected to be uniform. For our purposes it is more convenient to describe the non-uniform electric field by its scalar potential ϕ rather than by the vector **E**. The uniform electric field $\mathbf{E}^{(0)}$ corresponds to the following potential,

$$\phi^{(0)}(\mathbf{r}) = -\ \mathbf{r} \cdot \mathbf{E}^{(0)} \tag{2.14}$$

Thus, the deviation of the electric field from the uniform field is described by the potential

$$\delta\phi(\mathbf{r}) = \phi(\mathbf{r}) - \phi^{(0)}(\mathbf{r}) = \phi(\mathbf{r}) + \mathbf{r} \cdot \mathbf{E}^{(0)} \tag{2.15}$$

The perturbation Hamiltonian is defined as follows,

$$\hat{V} = \sum_k q_k \delta\phi(\mathbf{r}_k) \tag{2.16}$$

Here the summation is carried out over all charged elementary particles, including electrons and nuclei, q_k is the charge of the particle k and $\mathbf{r}_k$ is the radius-vector describing its coordinates. Matrix elements of this perturbation Hamiltonian are calculated as usual [11], which involves integration over the coordinates of all electrons, but not over the coordinates of the nuclei, because here we are using the fixed-nuclei approximation. For the diagonal matrix elements the result of this integration can be reduced to the following expression:

$$V_{nn} = \sum_j Q_{jn}^{(0)} \delta\phi(\mathbf{r}_j) \tag{2.17}$$

Here the summation is carried out over all atoms of the fluorophore, $\mathbf{r}_j$ is the radius-vector of the center of atom j, and $Q_{jn}^{(0)}$ is Mulliken atomic charge [68] on atom j when the fluorophore is in the state S_n. The partial charges $Q_{jn}^{(0)}$ must be calculated using eigenfunctions $\Psi_n^{(0)}$ of the Hamiltonian $\hat{H}(0)$, which represents the fluorophore in the uniform electric field $\mathbf{E}^{(0)}$. In transition from Eq. (2.16) to Eq. (2.17) we have assumed that the electron density can be attributed to individual atoms as suggested by Mulliken [68] and that the electron density attributed to each atom is symmetrically distributed about the atom's center (spherical symmetry). The latter assumption is accurate only for the inner-shell electrons; the distribution of valence electrons is not spherically symmetrical. However, the errors associated with deviations from the spherical symmetry are smaller than the errors associated with using the electric dipole approximation in a non-uniform electric field, therefore Eq. (2.17) is likely to give more accurate results than Eq. (2.13).

From Eqs. (2.10, 2.11, and 2.17) follows the first-order estimate for the energy E_n:

$$E_n = E_n^{(0)} + \sum_j Q_{jn}^{(0)} \mathbf{r}_j \cdot \mathbf{E}^{(0)} + \sum_j Q_{jn}^{(0)} \phi(\mathbf{r}_j) \tag{2.18}$$

or

$$E_n = G_n^{(0)} + \sum_j Q_{jn}^{(0)} \phi(\mathbf{r}_j) \tag{2.19}$$

where the constant energies G_n are defined as follows

$$G_n^{(0)} = E_n^{(0)} + \sum_j Q_{jn}^{(0)} \mathbf{r}_j \cdot \mathbf{E}^{(0)} \tag{2.20}$$

Now E_n from Eq. (2.19) can be substituted for E_1 and E_0 in each of the Eqs. (2.2, 2.3, 2.4, and 2.5). The following equation was obtained from Eq. (2.5), however, similar results can be obtained from Eqs. (2.2), (2.3), or (2.4):

$$h\nu_{em,\,pk}(t) = G_1^{(0)} - G_0^{(0)} - h\nu_{gr,\,pk}^{vib} + \sum_j \left(Q_{j1}^{(0)} - Q_{j0}^{(0)}\right)\phi(\mathbf{r}_j, t) \tag{2.21}$$

Note, that $Q_{j1}{}^{(0)} - Q_{j0}{}^{(0)}$ represents the difference between the partial atomic charge on atom j in the excited and in the ground state. This difference will be denoted $\Delta Q_j{}^{(0)}$:

$$\Delta Q_j^{(0)} = Q_{j1}^{(0)} - Q_{j0}^{(0)} \tag{2.22}$$

Using this new notation we can rewrite the sum on the right-hand side of Eq. (2.21) in the form

$$\Delta E(t) = \sum_j \Delta Q_j^{(0)} \phi(\mathbf{r}_j, t) \tag{2.23}$$

Replacing the first three constant terms on the right hand side of Eq. (2.21) with a new constant $h\nu_0$ and the last term with $\Delta E(t)$ from Eq. (2.23) yields

$$h\nu(t) = h\nu_0 + \Delta E(t) \tag{2.24}$$

Here we have dropped the subscripts *em* and *pk* on the left-hand side because Eq. (2.24) equally applies to absorption and emission spectra and it also equally applies to peak frequencies and center-of-gravity frequencies; only the definition of the constant term $h\nu_0$ is different in these cases.

2.4 Nonlinear Stark Effect

In transition from Eqs. (2.10, 2.11, 2.17) to Eq. (2.18) we have omitted the second-order and higher correction terms that appear on the right-hand side of Eq. (2.10). A sum of the terms that have been omitted represents the error arising from the use of Eq. (2.23). This error will be considered now.

Equation (2.12) gives the second-order correction term, which is usually the most significant of the terms that have been omitted. The off-diagonal matrix elements V_{mn} in Eq. (2.12) cannot be expressed in terms of Mulliken partial atomic charges as we did it with the diagonal matrix elements in Eq. (2.17). This means that in estimating the errors we will have to limit ourselves to the case of a uniform electric field, where Eq. (2.13) can be used. Substituting V_{mn} from Eq. (2.13) into Eq. (2.12) yields

$$E_n^{(2)} = \sum_{m \neq n} \frac{|\boldsymbol{\mu}_{mn}^{(0)} \cdot (\mathbf{E} - \mathbf{E}^{(0)})|^2}{E_n^{(0)} - E_m^{(0)}} \tag{2.25}$$

Equation (2.25) explicitly shows that the second-order correction term $E_n{}^{(2)}$ is quadratic in $\mathbf{E} - \mathbf{E}^{(0)}$. Furthermore, it can be shown that every other term $E_n{}^{(k)}$ also varies with the field as the k-th power of $\mathbf{E} - \mathbf{E}^{(0)}$. Thus, the series in Eq. (2.10) represents an expansion of the energy E_n in powers of $\mathbf{E}-\mathbf{E}^{(0)}$. This series has a finite convergence domain. The convergence domain is limited by a convergence boundary. When the point $\mathbf{E}^{(0)}$, in the neighborhood of which the series expansion is made, is at the convergence boundary, the convergence is lost completely. This will be used to identify the convergence boundaries for the expansion in Eq. (2.10). If $E_m = E_n$ for some $m \neq n$, then one of the denominators in the sum in Eq. (2.25) equals zero. This is also true for all correction terms $E_n^{(k)}$ with $k \geq 2$. The set of all points where the condition $E_m = E_n$ is met for at least one $m \neq n$, represents the convergence boundary. Since we are only interested in the series expansions for E_0 and E_1, the convergence domain is limited by the boundaries $E_m = E_0$ and $E_m = E_1$. For most fluorophores, the ground state is separated from the lowest electronically-excited state by a large energy gap, therefore the condition $E_m = E_0$ is not likely to be reached in a realistic experimental setting. On the contrary, the condition $E_m = E_1$ is commonly achieved when two excited electronic states "cross" each other. For example, 3-methylindole, which plays the role of the fluorophore in tryptophan (Trp), has two excited electronic configurations commonly referred to as ^1L_a and ^1L_b [69]. In vacuum ^1L_b is the lowest excited state and ^1L_a is higher in energy [64]. In a polar solvent the lowest excited state is ^1L_a and ^1L_b has a greater energy [64]. This means that in some electric field of intermediate strength between that in a polar environment and in vacuum the energies E_1 and E_2 must be equal, and this defines the convergence boundary. If we choose $\mathbf{E}^{(0)}$ on one side of that boundary, then the series expansion of E_1 will diverge on the other side of that boundary. This means that if we choose $\mathbf{E}^{(0)} = \mathbf{0}$, which corresponds to vacuum, then the series in Eq. (2.10) is going to diverge for the fluorophore in a polar environment. However, if we choose $\mathbf{E}^{(0)}$ equal to the Onsager reaction field in a polar solvent, then the series in Eq. (2.10) are going to diverge in vacuum.

In those cases where the first-order estimate from Eq. (2.23) is used for $\Delta E(t)$, the question of convergence becomes irrelevant, however, the question of accuracy still comes down to the correct choice of $\mathbf{E}^{(0)}$. Since the second- and higher-order correction terms $E_n^{(k)}$ vary with the field as the k-th power of $\mathbf{E}-\mathbf{E}^{(0)}$, minimizing the absolute value of the difference $\mathbf{E}-\mathbf{E}^{(0)}$ is the obvious way of reducing the errors. To achieve this goal, the value of $\mathbf{E}^{(0)}$ should be chosen close to the mean [67] value of $\mathbf{E}$ experienced by the fluorophore in the specific environment where the TDSS is studied. This choice of $\mathbf{E}^{(0)}$ will ensure the most accurate values of the differential partial charges $\Delta Q_j^{(0)}$ and it will ultimately result in accurate first-order estimates of TDSS.

2.5 Hybrid QM-MD Versus Classical MD

Quantum mechanical (QM) calculations provide the most accurate way of calculating $\Delta E(t)$ for a fluorophore in an electric field. Hybrid QM-MD simulations of TDSS in proteins and peptides have been reported by the group of Callis [55, 56, 70]. Hybrid QM-MD simulations usually cover only a short period of time, from

2 ps [55] to 30 ps [56] for proteins, and up to 60 ps for small peptides[70]. In those cases where several nanoseconds of TDSS must be calculated, the use of QM-MD simulations is not desirable because of their slow speed. This has encouraged many authors [44, 45, 57, 59–62] to use a faster method of calculating $\Delta E(t)$ from MD trajectories, which is based on Eq. (2.26) or its equivalent,

$$\Delta E(t) = \frac{1}{4\pi\varepsilon_0} \sum_i \sum_j \frac{Q_i \; \Delta Q_j^{(0)}}{\left|\mathbf{r}_j(t) - \mathbf{r}_i(t)\right|} \tag{2.26}$$

Here ε_0 is the dielectric permittivity of vacuum, index i counts all atoms except those that belong to the fluorophore, Q_i is the partial electric charge on atom i, $\mathbf{r}_i(t)$ is the radius-vector of this atom at time t, index j counts only the fluorophore atoms, $\Delta Q_j^{(0)}$ is defined in Eq. (2.22), $\mathbf{r}_j(t)$ is the radius-vector of atom j at time t, and $|\mathbf{a} - \mathbf{b}|$ denotes the magnitude of the difference between the vectors **a** and **b**.

Although Eq. (2.26) may be obvious to those who use it, it is important to understand how it is obtained and what are the errors associated with the use of this equation. Here Eq. (2.26) is obtained by substituting into Eq. (2.23) the following expression for the electric potential $\phi(\mathbf{r},t)$:

$$\phi(\mathbf{r}, t) = \frac{1}{4\pi\varepsilon_0} \sum_i \frac{Q_i}{\left|\mathbf{r} - \mathbf{r}_i(t)\right|} \tag{2.27}$$

Equation (2.27) represents the electrostatic potential generated by the partial electric charges of all atoms except those that belong to the fluorophore. The same electrostatic potential is used in hybrid QM-MD simulations, therefore any differences in accuracy between the hybrid QM-MD simulations and the faster method based on Eq. (2.26) must arise from Eq. (2.23) and not from Eq. (2.27). Equation (2.23) was obtained from the first-order quantum-mechanical perturbation theory [11]. The use of the perturbation theory is inevitable when we deal with a quantum system containing more than two elementary particles: analytical solution of the Schrodinger equation exists only for the hydrogen atom, while for any atom with more than one electron as well as for any molecule the use of the perturbation theory is unavoidable. The QM portion of the hybrid QM-MD simulations is entirely based on the perturbation theory. Thus, the errors associated with the faster method of calculating $\Delta E(t)$ do not result from the use of the perturbation theory, but they result from the use of the first-order perturbation theory. The errors are equal to the sum of the higher-order correction terms that were originally included in Eq. (2.10), but were then dropped. These terms are discussed in Sect. 2.4, "Nonlinear Stark Effect", where it is concluded that in order to minimize the omitted higher-order correction terms one has to use the differential partial charges $\Delta Q_j^{(0)}$ corresponding to the fluorophore in the electric field $\mathbf{E}^{(0)}$ that closely resembles the mean local field acting on the fluorophore. The mean local field can be obtained from a hybrid QM-MD simulation, which does not have to be very long. Thus, the best way to evaluate $Q_{j1}^{(0)}$, $Q_{j0}^{(0)}$, and $\Delta Q_j^{(0)}$ is to run a short hybrid

QM-MD simulation and to average the instantaneous partial charges on the fluorophore atoms over a time period of several picoseconds. Callis and co-workers [56, 64] ran QM-MD simulations (using CHARMM forcefield) for 3-methylindole in ^1L_a excited state in a drop of 1100 explicit TIP3 waters and averaged the charges over the last 4 ps of the trajectory. Then all $Q_{j1}^{(0)}$, $Q_{j0}^{(0)}$, and $\Delta Q_j^{(0)}$ were scaled by a factor of 0.80 [56], which was necessary to obtain the best match between the simulated and experimental ΔE values. It is still unclear why the calculated ΔE values were not accurate without the scale factor. From Eq. (2.26) it follows that the scaling of all $\Delta Q_j^{(0)}$ by some factor has the same effect on ΔE as the scaling of all Q_i by the same factor, therefore one possible explanation is that in CHARMM the partial charges Q_i on protein and solvent atoms are exaggerated on average by a factor of 1/(0.80). The other possible explanation is that CHARMM is not a polarizable forcefield; using a polrizable forcefield would probably decrease the electric field on average by a factor of 0.80, which would make the calculated ΔE values closer to the experimental ones. The charges $Q_{j1}^{(0)}$, $Q_{j0}^{(0)}$, and $\Delta Q_j^{(0)}$ that were originally obtained by Callis and co-workers [56, 64] and then used by others to calculate TDSS from MD trajectories [62] are given in Table 2.1. This appears to be the best set of charges to be substituted in Eq. (2.26) in combination with the partial charges Q_i from the CHARMM forcefiled. The use of $\Delta Q_j^{(0)}$ from Table 2.1 and Eq. (2.26) makes possible to accurately calculate TDSS from a non-equilibrium MD trajectory without the use of hybrid QM-MD simulations.

2.6 Direct-Response Versus Linear-Response Method

Non-equilibrium MD is commonly triggered by switching the partial charges of the fluorophore atoms from $Q_{j0}^{(0)}$ to $Q_{j1}^{(0)}$ at $t = 0$. TDSS is then calculated using Eq. (2.26) directly from a non-equilibrium MD trajectory; this approach will be called the direct-response method to distinguish it from the linear-response method. TDSS calculated using the direct-response method from just one MD trajectory is overwhelmed by random noise, therefore averaging over about 100 trajectories is usually necessary. An alternative approach, known as the linear-response method, makes it possible to obtain the same information from the autocorrelation of random fluctuations in just one very long equilibrium trajectory. The method is described below.

Consider some interaction with the energy $\Delta E(t)$ that can be instantaneously turned on. For example, Coulombic interaction between the additional charges $\Delta Q_j^{(0)}$ on fluorophore atoms and the common partial charges of other atoms instantaneously turns on when the fluorophore jumps from the ground state to the excited state. Equation (2.26) gives the energy of this interaction. According to the fluctuation-dissipation theorem, in the linear-response approximation, the ensemble average response $\langle \Delta E(t) \rangle$ averaged over an infinite number of non-equilibrium

Table 2.1 Trp side chain atom charges in units of positive electron charge *e*. Reprinted with permission from Ref. [62], copyright 2010 American Chemical Society

Trp atom	$Q_{j0}^{(0)}$	$Q_{j1}^{(0)}$	$\Delta Q_j^{(0)}$
CB	−0.003	+0.004	+0.007
HB1	+0.013	+0.017	+0.004
HB2	+0.016	+0.015	−0.001
CG	−0.046	+0.184	+0.230
CD1	+0.011	+0.122	+0.111
HD1	+0.030	+0.047	+0.017
NE1	−0.130	+0.017	+0.147
HE1	+0.125	+0.132	+0.007
CE2	+0.052	+0.082	+0.030
CD2	−0.009	−0.087	−0.078
CE3	−0.027	−0.214	−0.187
HE3	+0.027	+0.028	+0.001
CZ3	−0.033	−0.077	−0.044
HZ3	+0.005	−0.003	−0.008
CZ2	−0.039	−0.243	−0.204
HZ2	+0.017	+0.021	+0.004
CH2	−0.026	−0.063	−0.037
HH2	+0.017	+0.018	+0.001

trajectories, in which the interaction is turned on at $t=0$, is connected to the autocorrelation function $C(t)$ defined below by a simple relation:

$$\langle \Delta E(t) \rangle - \langle \Delta E(\infty) \rangle = \beta\ C(t) \tag{2.28}$$

Here β is the inverse temperature, $\beta = 1/(k_B T_A)$, where k_B is Boltzmann constant and T_A is absolute temperature. The autocorrelation function $C(t)$ can be defined either as ensemble average, i. e. the average over infinite number of trajectories, or as a time average over one infinitely long trajectory. We will use the second approach, because in practice the autocorrelation function is always obtained from one trajectory [44, 45, 57, 58, 60–62]. Time-averaged $C(t)$ is defined as the limit at $T \to \infty$ of the autocorrelation function obtained from a finite-length trajectory,

$$C(t) = \lim_{T \to \infty} C_T(t) \tag{2.29}$$

Here $C_T(t)$ is calculated from a trajectory of length T; it is defined as follows:

$$C_T(t) = \frac{1}{T-t} \int_0^{T-t} \left[\Delta E(t') - \overline{\Delta E_T}\right]\left[\Delta E(t'+t) - \overline{\Delta E_T}\right] dt' \tag{2.30}$$

$$\overline{\Delta E_T} = \frac{1}{T}\int_0^T \Delta E(t')dt' \tag{2.31}$$

$\Delta E(t')$ must be calculated from an equilibrium trajectory, i. e. the interaction should not be turned on or turned off during the MD simulation that generates $\Delta E(t')$ to be used in Eqs. (2.30 and 2.31).

The results obtained using the linear-response method contain both random errors and systematic errors. The random errors originate from substituting $C_T(t)$ instead of $C(t)$ into Eq. (2.28). For a short trajectory the difference between $C_T(t)$ and $C(t)$ can be substantial, which can easily lead to false interpretations. The most common artifacts in $C_T(t)$ resulting from averaging over a finite-length trajectory are illustrated in Fig. 2.1.

The random noise in panel 1A of Fig. 2.1 is a sum of the random fluctuations in 16 noise-driven damped harmonic oscillators with different frequencies and different damping ratios (Langevin dynamics) [62]. This random noise visually resembles $\Delta E(t)$ from a real MD simulation, yet, computing this random noise is millions of times faster than a real MD simulation for a real protein, which makes it easy to study the evolution of $C_T(t)$ up to $T = 10{,}000$ ns. Panel 1A shows only the first 10 ns of $\Delta E(t)$, however, the "trajectory" was actually generated all the way up to $t = 10{,}000$ ns. In panel 1B of Fig. 2.1 the black line depicts $C_{10\text{ns}}(t)$, calculated from the part of the trajectory in panel 1A. The red line depicts $C_{10000\text{ns}}(t)$, which is practically identical to $C(t)$, since it was established that the increase in T from 5000 to 10000 ns produces no visible changes in $C_T(t)$ [62]. The difference between $C_{10\text{ns}}(t)$ and $C_{10000\text{ns}}(t)$ is the low-frequency noise, which is a common artifact associated with the use of the linear-response method. Theoretically, it should be possible to decrease the amplitude of the low-frequency noise to a desired level, however, in practice this would require a very long MD trajectory, since the amplitude of the noise in $C_T(t)$ decreases as $T^{-1/2}$. The low-frequency noise, also known as pink noise, is potentially more dangerous than the white noise in direct-response TDSS, because the low-frequency noise is not subjectively perceived as noise.

The noise pattern in panel 1A of Fig. 2.1 resembles near-equilibrium atomic fluctuations in a protein that does not change its conformation at all. If the entire protein or just one side chain in close proximity to the fluorophore jumps between two discrete isomers, then the noise pattern may look like the ones shown in panels 2A or 3A. $\Delta E(t)$ in panels 2A and 3A were generated using a random switch that jumps between two discrete levels [62]. The switch in panel 2A spends equal times (on average) in both states; the average time between jumps for this switch equals 5 ns. The switch in panel 3A spends 90 % of time in the lower energy state and 10 % in the upper energy state, with the residence times of 0.33 ns for the upper energy state and 3 ns for the lower energy state. The random noise from the 16 damped harmonic oscillators was also added to the signals depicted in panels 2A and 3A. $\Delta E(t)$ curves were generated for a total of 10000 ns in all cases.

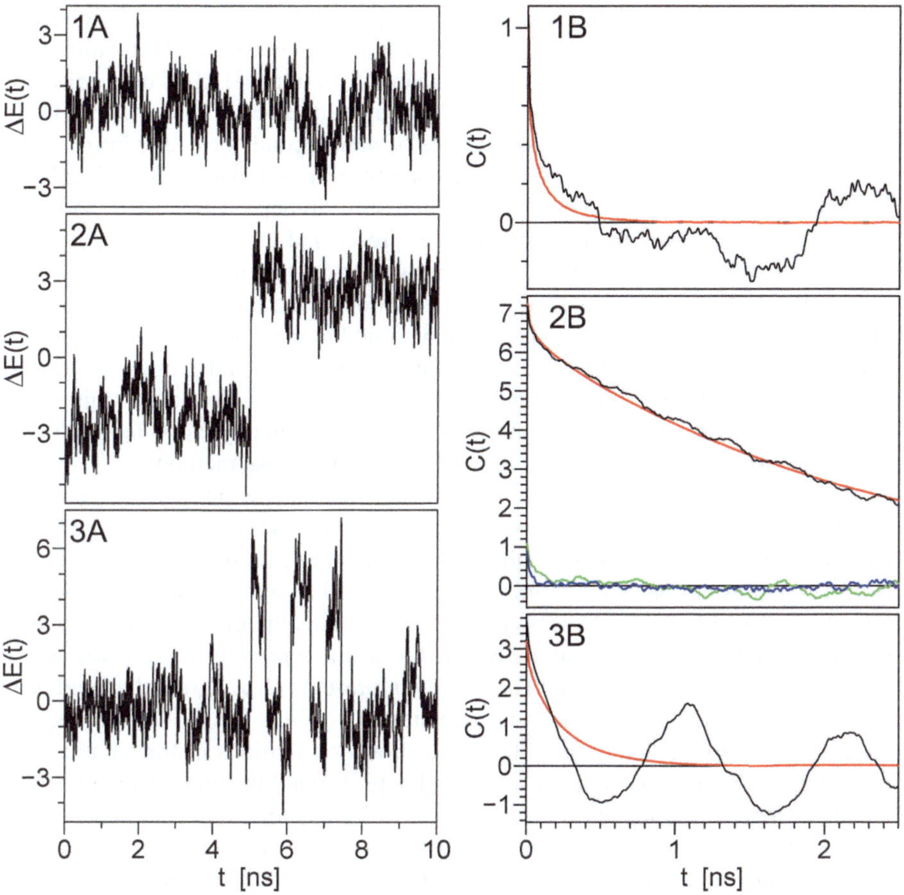

Fig. 2.1 Random noise patterns (panels **1A, 2A, 3A**) and corresponding autocorrelation functions $C_T(t)$ (panels **1B, 2B, 3B**). The *black* line in each #B panels represents $C_{10\mathrm{ns}}(t)$ calculated from the 10 ns trajectory in the corresponding #A panel using Eqs. (2.30 and 2.31). The *red* line represents $C_{10000\mathrm{ns}}(t)$ calculated from the 10000 ns trajectory generated by the same random process as the 10 ns trajectory in the corresponding #A panel. The *green* and the *blue lines* in panel **2B** depicts $C_{5\mathrm{ns}}(t)$ calculated from for the first 5 ns and from the last 5 ns of the trajectory in panel **2A**, respectively. Reprinted with permission from Ref. [62], copyright 2010 American Chemical Society

Panel 2A of Fig. 2.1 shows a selected 10 ns window during which the protein spent the first 5 ns as isomer 1 (low ΔE) and the following 5 ns as isomer 2 (high ΔE). The autocorrelation $C_{10\mathrm{ns}}(t)$ calculated from the selected window (black line in panel 2B) and the autocorrelation $C_{10000\mathrm{ns}}(t)$ calculated from the full trajectory (red line in panel 2B) are close to each other, and this is for a good reason: the average times the protein spends as isomer 1 or isomer 2 equal 5 ns, and during the selected 10 ns window it spent exactly 5 ns as isomer 1 and 5 ns as isomer 2. Random switching between isomers makes a large contribution to $C(t)$ and it is solely responsible for the difference between the red line in panel 2B and the red line in panel 1B. This switching increases the full amplitude of the autocorrelation function from 1.0 in panel 1B to 7.2 in panel 2B. It also increases the correlation time from less than 0.5 ns in panel 1B to more than 2.5 ns in panel 2B. Thus, switching between two (or more) isomers could be the main source of the slow TDSS observed experimentally on the time scales of nanoseconds or even tens of nanoseconds [25–27, 33, 38, 41, 48]. The green and blue lines in panel 2B depict $C_{5\mathrm{ns}}(t)$ calculated from for the first 5 ns (during which only isomer 1 was present) and from the last 5 ns (during which only isomer 2 was present) of the trajectory in panel 2A. Since there were no transitions between the isomers during the time windows used to calculate either $C_{5\mathrm{ns}}(t)$, the slow TDSS is not observed in this case. An example of splitting MD trajectory in two parts can be found in the work of Li et al. [45, 61], who separately analyzed the trajectories for the isomers 1 and 2 using the linear-response method. No nanosecond-scale relaxation was found in this work [45, 61]. This example clearly shows how insufficient trajectory length T can radically alter the results obtained using the linear-response method.

The example shown in panels 3A and 3B illustrates a possible but unlikely situation. During the 10 ns window selected for panel 3A the protein jumped from the low ΔE state (isomer 1) to the high ΔE state (isomer 2) and back three times in a row. $\Delta E(t)$ does not behave like this during every 10 ns window, thus, the piece of trajectory shown in panel 3A is atypical. This explains the big difference between the black and the red line in panel 3B. The black line represents $C_{10\mathrm{ns}}(t)$ calculated from the selected 10 ns window. The red line represents $C_{10000\mathrm{ns}}(t)$ calculated from the full trajectory, and it closely approximates $C(t)$. While the red line approaches the zero level aperiodically, the black line shows oscillations characteristic of an underdamped oscillator. Golosov and Karplus [60] found the underdamped behavior in just one of the eleven linear-response MD trajectories they simulated. This can be the case of a real underdamped oscillator or the case where the low-frequency noise resembles the behavior of an underdamped oscillator, similar to that shown in panels 3A and 3B. To find out which is the case, one would have to generate a much longer linear-response MD trajectory.

The results obtained using the linear-response method may also contain systematic errors. The systematic errors have their origin in deviations from linearity. Equation (2.28) is valid only in the linear-response approximation, therefore in the case where $\beta|\Delta E|>1$, i. e., $|\Delta E|>k_B T_A$, the use of Eq. (2.28) can result in significant systematic errors. The systematic errors will be further investigated here using a very simple model system, which consists of just one TIP3P water molecule in a uniform electric field that is turned on at $t=0$. The orientation of the water molecule is described here using Euler angles (ϕ,θ,ψ) [71]. The Z' axis of the molecular frame is chosen parallel to the permanent electric dipole moment $\boldsymbol{\mu}$ of the water molecule, and the Z axis of the laboratory frame is chosen parallel to the electric field $\mathbf{E}$. With this choice of axes, the interaction energy depends on only one Euler angle:

$$\Delta E(t) = -\boldsymbol{\mu}\cdot\mathbf{E} = -|\boldsymbol{\mu}||\mathbf{E}|\cos\theta \qquad (2.32)$$

Using the fact that in the absence of the electric field the orientation of the water molecule is random, and therefore $\langle\cos^2\theta\rangle = 1/3$, we can calculate the value of $C(t)$ for $t=0$:

$$C(0) = \langle \Delta E^2(t) \rangle = |\boldsymbol{\mu}|^2 |\mathbf{E}|^2 \langle \cos^2\theta \rangle = \frac{1}{3} |\boldsymbol{\mu}|^2 |\mathbf{E}|^2 \tag{2.33}$$

According to the linear-response method,

$$\langle \Delta E(0) \rangle - \langle \Delta E(\infty) \rangle = \beta C(0) = \frac{1}{3} \beta |\boldsymbol{\mu}|^2 |\mathbf{E}|^2 \tag{2.34}$$

Taking into account that $\langle \Delta E(0) \rangle = 0$ and $\langle \Delta E(\infty) \rangle = |\boldsymbol{\mu}||\mathbf{E}|\langle \cos\theta \rangle$, one can obtain the following expression for the equilibrium value of $\langle\cos\theta\rangle$ in the presence of the electric field:

$$\langle \cos\theta \rangle = \frac{1}{3} \beta |\boldsymbol{\mu}||\mathbf{E}| \tag{2.35}$$

The estimate in Eq. (2.35) is based on the linear-response approximation, therefore it is accurate in weak electric fields only. A general expression for the equilibrium value of the first-rank order parameter $\langle\cos\theta\rangle$ in the electric field of any strength can be obtained using the equilibrium orientational distribution of the molecule, which is essentially a Boltzmann distribution in Euler angles:

$$d^3 P(\phi, \theta, \psi) = A \exp(\beta |\boldsymbol{\mu}||\mathbf{E}| \cos\theta)\, d\phi \, \sin\theta d\theta \, d\psi \tag{2.36}$$

The constant A in Eq. (2.36) must be chosen so that the integral over all orientations equals unity. Multiplying the distribution in Eq. (2.36) by $\cos\theta$ and integrating over all orientations yields

$$\langle \cos\theta \rangle = \coth(\beta |\boldsymbol{\mu}||\mathbf{E}|) - (\beta |\boldsymbol{\mu}||\mathbf{E}|)^{-1} \tag{2.37}$$

In Fig. 2.2 the estimate obtained using the linear-response method, Eq. (2.35), is depicted by the red line, and the value from Eq. (2.37), which is accurate at any level of the electric field, is depicted by the blue line. The curves in Fig. 2.2 were calculated using $\beta = 1/(k_B \times 300\ \text{K})$, which corresponds to the temperature of 300 K, and $|\boldsymbol{\mu}| = 7.829 \cdot 10^{-30}$ C·m, which is the magnitude of the permanent electric dipole moment of a TIP3P water molecule. The relative deviation of the linear-response estimate from the actual value reaches 10 % at the electric field strength of 7 MV/cm. The additional excited-state charges $\Delta Q_j^{(0)}$ on the Trp side chain (see Table 2.1) generate an electric field of 7 MV/cm magnitude at distances up to 8.5 Å from the center of the fluorophore. Thus, if the distance between the center of the indole moiety and the center of at least one water molecule is less than 8.5 Å, then the amplitude of the TDSS obtained using the linear-response method is likely to be

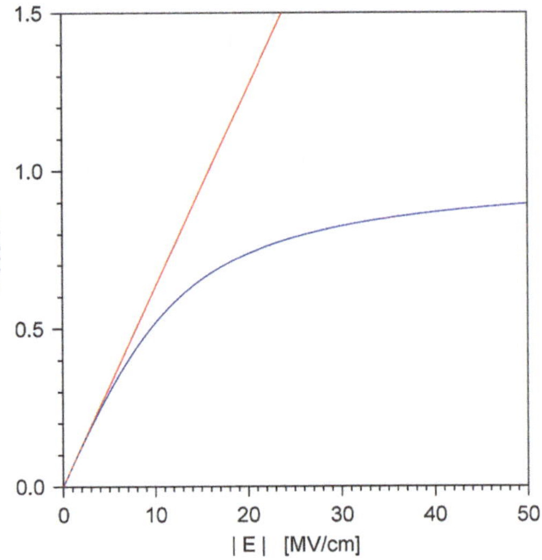

Fig. 2.2 The first-rank order parameter ⟨cosθ⟩ for a TIP3P water molecule in a uniform electric field **E**. *Red line*: the estimate obtained using the linear-response method, Eq. (2.35). *Blue line*: the exact value from Eq. (2.37). The curves were calculated for 300 K temperature and $7.829 \cdot 10^{-30}$C·m dipole moment. Reprinted with permission from Ref. [62], copyright 2010 American Chemical Society

overestimated. A water molecule in van der Waals contact with the Trp side chain may experience electric fields up to 50 MV/cm; in electric fields of this magnitude the estimate obtained using the linear-response method is 3.5-fold greater than one obtained using the direct-response method. An overestimated amplitude is always accompanied by a distorted shape of the TDSS, therefore re-normalization does not solve the problem, it just makes it more difficult to be acknowledged. Maroncelli and Fleming [72] questioned the validity of the linear-response method when they discovered that in linear-response MD simulations the solvation response to a charge jump in ST2 water does not vary linearly with the magnitude of the charge jump. Comparing the theoretical estimates in Eqs. (2.35) and (2.37) makes it possible to determine in advance whether the results of a future linear-response MD simulation will be valid or not.

2.7 Separation of Contributions from Different Motions

A trivial method for separating the protein and solvent contributions to TDSS is based on splitting the summation over the index i in Eq. (2.26) into two partial sums. The first partial sum covers the range of i that corresponds to protein atoms; this partial sum represents the protein contribution to TDSS. The second partial sum involves only those i values that correspond to solvent atoms and represents the solvent contribution to TDSS. This method of separating the protein and solvent contributions to TDSS was used several times [45, 56, 57, 59, 60]. The main drawback of this method is that it attributes the TDSS resulting from the motion of the fluorophore itself in part to the protein and in part to the solvent. In rigid proteins where the fluorophore is tightly packed in the protein core and cannot rotate relative to the protein this drawback is unimportant and the use of the trivial separation method is justifiable. On the other hand, if the protein is not rigid and/or the fluorophore is on the surface, then the use of the trivial separation method may result in significant errors. A method allowing the separation of contributions from the motion of the fluorophore itself, from the motion of other protein atoms, and from the motion of solvent atoms was recently proposed [62]. A description of this method is given below.

According to Eq. (2.23) the spectral shift depends on the radius-vectors $\mathbf{r}_j$ of the fluorophore atoms and also on the electric potential $\phi(\mathbf{r},t)$ generated by the partial charges of all atoms except those of the fluorophore. Thus, the motion of the fluorophore as well as the motion of the surrounding atoms contribute to the spectral shift. To be able to tell which atoms moved and how far, we will have to choose a reference frame. It makes little sense to consider atomic motions in the laboratory frame, since in this frame the rotation and translation of the protein as a whole results in much greater atomic displacements than those relative atomic motions that actually produce TDSS. It makes even less sense to use the reference frame attached to a small atomic group (e. g., one side chain) of the protein molecule, or to a water molecule. The only reasonable choice is to attach the reference frame to the protein molecule. Since the molecule is flexible, the question how this can be practically accomplished is not trivial. An algorithm (based on successive iterations) for defining the reference frame attached to the protein is described in Ref. [62]. Coordinate transformations between the laboratory reference frame and that attached to the protein molecule are described by the following equations:

$$\mathbf{r}^{lab} = \mathbf{R}\mathbf{r}^{mol} + \mathbf{u} \tag{2.38}$$

$$\mathbf{r}^{mol} = \mathbf{R}^T\left(\mathbf{r}^{lab} - \mathbf{u}\right) \tag{2.39}$$

Here $\mathbf{r}^{lab}$ denotes the column vector of three Cartesian coordinates x^{lab}, y^{lab}, z^{lab} in the laboratory reference frame; these are the coordinates used in MD simulations. Likewise $\mathbf{r}^{mol}$ denotes the column vector of three Cartesian coordinates x^{mol}, y^{mol}, z^{mol} in the reference frame attached to the protein molecule. Throughout this chapter equations containing radius-vectors $\mathbf{r}$ without superscripts are valid regardless of the choice of the reference frame; for example Eq. (2.26) is valid both in the laboratory and in the molecular reference frame. If radius-vectors from more than one reference frame appear in the same equation, then each radius-vector is provided with a superscript to identify its reference frame, as in Eqs. (2.38 and 2.39). $\mathbf{R}$ denotes a 3×3 rotation matrix of the special orthogonal group SO(3); its transpose $\mathbf{R}^T$ is equal to its inverse. $\mathbf{u}$ is the translation vector, which describes the position of the origin of the molecular reference frame in the laboratory reference frame. The rotation and translation of the protein molecule results in the time variation of the matrix $\mathbf{R}$ and vector $\mathbf{u}$.

For every protein atom n (including fluorophore atoms) the vector of mean equilibrium coordinates $\langle \mathbf{r}_n{}^{mol} \rangle$ is defined as the ensemble averages of $\mathbf{r}_n{}^{mol}$ at long times after excitation. In practice one can average $\mathbf{r}_n^{mol}$ over all MD trajectories and also over a time period that starts after the bulk of the TDSS is over and continues to the end of the trajectory. Note, that averaging is carried out on the coordinates of protein atoms only and in the reference frame attached to the protein; it makes no sense to average any Cartesian coordinates in the laboratory reference frame or to average the coordinates of solvent atoms in the protein reference frame.

Using the mean coordinate vectors $\langle \mathbf{r}_j^{mol} \rangle$ for the fluorophore atoms it is possible to split the energy $\Delta E(t)$ in Eq. (2.23) into the part $\Delta E_f(t)$ that varies due to the motion of the fluorophore itself and the part $\Delta E_e(t)$ that varies due to the changes in the electric field:

$$\Delta E(t) = \Delta E_f(t) + \Delta E_e(t) \tag{2.40}$$

$$\Delta E_f(t) = \sum_j \Delta Q_j^{(0)} \left[\phi\left(\mathbf{r}_j^{mol}, t\right) - \phi\left(\left\langle \mathbf{r}_j^{mol} \right\rangle, t\right) \right] \tag{2.41}$$

$$\Delta E_e(t) = \sum_j \Delta Q_j^{(0)} \phi\left(\left\langle \mathbf{r}_j^{mol} \right\rangle, t\right) \tag{2.42}$$

Here $\Delta Q_j^{(0)}$ is the charge difference defined in Eq. (2.22) and ϕ is the electric potential defined in Eq. (2.27). If the fluorophore does not move in the reference frame attached to the protein, then all the differences in square brackets on the right-hand side of Eq. (2.41) equal zero, and therefore $\Delta E_f(t)$ also equals zero. On the other hand, in the hypothetical situation where all protein atoms except those of the fluorophore itself and all water atoms are fixed in the reference frame attached to the protein, the potentials $\phi(\langle \mathbf{r}_j^{mol} \rangle, t)$ are time-invariant, and therefore $\Delta E_e(t)$ is also time-invariant. In other words, without the motion of the fluorophore there is no time variation in $\Delta E_f(t)$, and without the motions of other atoms there is no time variation in $\Delta E_e(t)$. This justifies treating $\Delta E_f(t)$ as the contribution of the fluorophore atoms and $\Delta E_e(t)$ as the contribution of non-fluorophore protein atoms and solvent atoms.

To split $\Delta E_e(t)$ further into the contributions of protein and solvent atoms, the electric potentials $\phi(\langle \mathbf{r}_j^{mol} \rangle, t)$ that appear in Eq. (2.42) must be expressed explicitly in terms of the coordinates of these atoms,

$$\phi(\langle \mathbf{r}_j^{mol} \rangle, t) = \frac{1}{4\pi\varepsilon_0} \sum_i \frac{Q_i}{|\mathbf{r}_i^{mol}(t) - \langle \mathbf{r}_j^{mol} \rangle|} \tag{2.43}$$

Here index i counts all atoms except those that belong to the fluorophore, Q_i is the partial charge on atom i, and $\mathbf{r}_i^{mol}$ the coordinate vector for this atom. The time variation of $\phi(\langle \mathbf{r}_j^{mol} \rangle, t)$ results only from the time variation of $\mathbf{r}_i^{mol}$, since the partial atomic charges Q_i and the mean coordinates $\langle \mathbf{r}_j^{mol} \rangle$ are time-invariant. The contribution of one non-fluorophore atom i to $\Delta E_e(t)$ and also to $\Delta E(t)$ can be obtained by taking just one term from the sum in Eq. (2.43) and substituting it in Eq. (2.42),

$$\Delta E_i(t) = \frac{1}{4\pi\varepsilon_0}\sum_j \frac{\Delta Q_j^{(0)} Q_i}{\left|\mathbf{r}_i^{mol}(t) - \left\langle \mathbf{r}_j^{mol} \right\rangle\right|} \tag{2.44}$$

The contribution of one fluorophore atom j to $\Delta E_f(t)$ and also to $\Delta E(t)$ can be obtained by taking just one term from the sum in Eq. (2.41). The solvent contribution can be determined by summing up $\Delta E_i(t)$ from Eq. (2.44) over all water atoms. The contribution of the protein can be determined by summing up $\Delta E_i(t)$ from Eq. (2.44) over all non-fluorophore protein atoms; in the case where the fluorophore (e. g., tryptophan) is considered to be a part of the protein it is also necessary to add $\Delta E_f(t)$ from Eq. (2.41) to the sum. The protein contribution, water contribution, and contributions of individual atoms must be then averaged over all nonequilibrium trajectories (ensemble averaging).

2.8 Dielectric Relaxation of Bulk Solvent

Dielectric relaxation of the solvent usually has a significant contribution to the TDSS of fluorophores in proteins. Dielectric relaxation of polar solvents occurs on multiple time scales [73]. The Debye model of dielectric relaxation in polar liquids contains only one characteristic time constant τ_D, which is commonly called the Debye relaxation time [17, 73–76]. An exponential term of the form $\exp(-t/\tau_D)$ is directly observed in the time variation of the electric displacement **D** after a jump in the electric field **E** in the form of Heaviside step function $\theta(t)$ [76], as shown in the top panel of Fig. 2.3. However, if the step-function jump is applied to the electric displacement **D**, then the relaxation of the electric field **E** does not contain the term $\exp(-t/\tau_D)$. Instead, it contains a faster exponential term $\exp(-t/\tau_L)$ [76], as shown in the bottom panel of Fig. 2.3. The relationship between τ_L and τ_D is well known [17, 73–76],

$$\tau_L = \frac{\varepsilon_\infty}{\varepsilon_S}\tau_D \tag{2.45}$$

In Eq. (2.45) ε_S is the static (low-frequency) dielectric permittivity of the solvent, ε_∞ is the high-frequency dielectric permittivity of the solvent, τ_L is the longitudinal relaxation time, and τ_D is the transverse relaxation time and also the Debye relaxation time. The terms "transverse" and "longitudinal" are relevant in the case of a continuous homogeneous polar liquid without borders or foreign objects. Using Helmholtz decomposition the dielectric polarization density **P** can be separated into two independent parts: (i) the longitudinal part, for which the divergence of **P** is non-zero ($\nabla\cdot\mathbf{P}\neq 0$) and the curl of **P** equals the zero vector ($\nabla\times\mathbf{P}=\mathbf{0}$), and (ii) the transverse part, for which the divergence of **P** equals zero ($\nabla\cdot\mathbf{P}=0$) and the curl of **P** is not equal to the zero vector ($\nabla\times\mathbf{P}\neq 0$) [73]. If the dielectric response is described by the Debye model, then both the longitudinal and

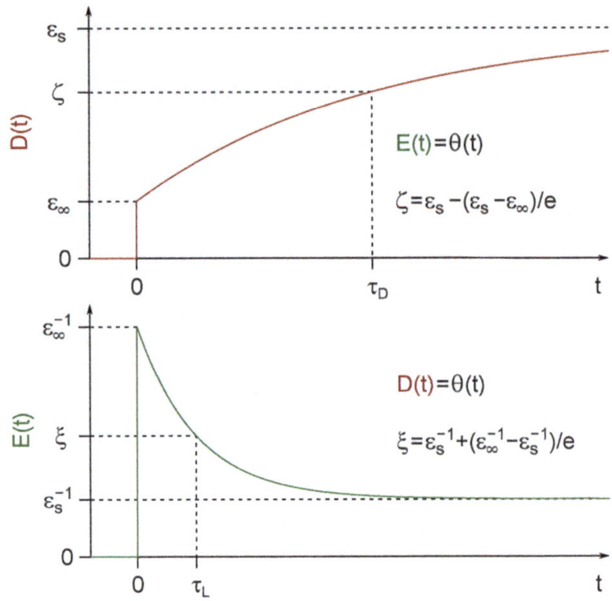

Fig. 2.3 Graphical definitions of the transverse relaxation time τ_D (*top panel*) and longitudinal relaxation time τ_L (*bottom panel*). $\mathbf{E}(t)$ is the electric field, $\mathbf{D}(t)$ is the electric displacement, and $\theta(t)$ is Heaviside step function. Adapted with permission from Ref. [62], copyright 2010 American Chemical Society

the transverse component relax exponentially, but with different relaxation times τ_L and τ_D [73, 76].

The dielectric polarization near a spherical ion immersed in a continuous homogeneous polar solvent consists of the longitudinal part only, therefore a step-function jump in the charge of the ion would induce an exponential relaxation process with the characteristic time τ_L. Such a charge jump contradicts the principle of electric charge conservation, therefore it cannot be achieved in practice. A step-function jump in the electric dipole moment is practically achievable; it occurs when a solvatochromic fluorophore jumps from the ground state to the excited state. Maroncelli and Fleming [74] have shown that for a point-dipole centered inside a spherical cavity a step-function jump in the dipole moment induces a relaxation process containing a single exponential term $\exp(-t/\tau_F)$, where [74, 75]

$$\tau_F = \frac{2\varepsilon_\infty + \varepsilon_C}{2\varepsilon_S + \varepsilon_C}\tau_D \tag{2.46}$$

In Eq. (2.46) ε_C is the dielectric permittivity of the spherical cavity, which is intended to represent fluorophore polarizability. From Eqs. (2.45 and 2.46) it follows that $\varepsilon_C = 0$ results in $\tau_F = \tau_L$, while $\varepsilon_C = \infty$ results in $\tau_F = \tau_D$. According to realistic estimates [17, 74, 75], for spherical fluorophores in polar solvents τ_F is only slightly greater than τ_L, while τ_D is much greater then both τ_L and τ_F.

The above example with a dipole jump in a spherical cavity shows that the observed dielectric relaxation cannot be always separated in two parts with the characteristic relaxation times τ_L and τ_D. Complete separation of the transverse and longitudinal polarization components can be accomplished only in the absence of borders and foreign objects (such as the spherical cavity in the above example). In the case of a parallel-plate capacitor the concepts "transverse" and "longitudinal" become completely irrelevant. The electric field **E** between the plates of such a capacitor is uniform, and so is the dielectric polarization density **P**, therefore the divergence of **P** equals zero ($\nabla \cdot \mathbf{P} = 0$) and the curl of **P** equals the zero vector ($\nabla \times \mathbf{P} = \mathbf{0}$), which fits neither the definition of longitudinal polarization nor the definition of transverse polarization [73]. Yet, with a parallel-plate capacitor one can observe both relaxation times τ_L and τ_D. If a step-function jump is applied to the voltage across the capacitor, then the charge on each plate of the capacitor will change with time as $\mathbf{D}(t)$ in the top panel of Fig. 2.3. However, if a step-function jump is applied to the charge on the plates of the same capacitor, then the voltage across the capacitor will change with time as $\mathbf{E}(t)$ in the bottom panel of Fig. 2.3.

In all previous examples the relaxation curve contained only one exponential term, which could be $\exp(-t/\tau_L)$, or $\exp(-t/\tau_F)$, or $\exp(-t/\tau_D)$. Now the question is whether two or more exponential terms can be observed simultaneously in a relaxation curve if the solvent is described by the Debye relaxation model. One physical structure in which both $\exp(-t/\tau_L)$ and $\exp(-t/\tau_D)$ can be observed simultaneously is depicted in the top panel of Fig. 2.4. A layer of a non-polar solid dielectric (such as polyethylene or diamond), shown by the yellow color in the top panel of Fig. 2.4, is immersed in a polar liquid (such as water), shown by the blue color. A narrow cylindrical channel, whose diameter is much smaller than the thickness of the layer, is drilled through the solid dielectric and is also filled with the polar liquid. The whole structure is placed between the plates of a parallel-plate capacitor (shown by black color, with the + and − labels) and a step-function jump is applied to the charge on the plates of this capacitor. This creates a step-function jump in $\mathbf{D}_1$ (the electric displacement in the polar liquid). Since $\mathbf{D}_1$ and $\mathbf{D}_2$ are both normal to the interface between the two dielectrics, $\mathbf{D}_1 = \mathbf{D}_2$, and therefore there is also a step-function jump in $\mathbf{D}_2$ (the electric displacement in the non-polar dielectric). For the non-polar dielectric the static dielectric constant and the high-frequency dielectric constant is the same quantity, which will be denoted ε_{NP}. The relationship between the electric field and the electric displacement in the non-polar dielectric is instantaneous, i. e. $\mathbf{D}_2 = \varepsilon_{NP}\mathbf{E}_2$, which results in a step-function jump in $\mathbf{E}_2$ (the electric field in the non-polar dielectric). Since $\mathbf{E}_2$ and $\mathbf{E}_3$ are both parallel to the walls of the cylindrical channel, $\mathbf{E}_2 = \mathbf{E}_3$, and therefore there is also a step-function jump in $\mathbf{E}_3$ (the electric field in the polar liquid inside the channel). A step-function jump in the electric field in the polar liquid results in a relaxation process that contains the exponential term $\exp(-t/\tau_D)$. The relaxation of the polar liquid outside the channel will result in the term $\exp(-t/\tau_L)$. Therefore, both τ_L and τ_D can be observed at the same time in one dielectric relaxation curve.

Using a conformal transformation the planar structure shown in the top panel of Fig. 2.4. can be transformed to a near-spherical structure shown in the left bottom panel of Fig. 2.4. Since conformal transformations preserve the solutions of Maxwell-Lorentz equations, the dielectric relaxation curves observed with the structures in the bottom left panel and in the top panel of Fig. 2.4 are identical; exponential terms with both τ_L and τ_D will be observed in the dielectric relaxation

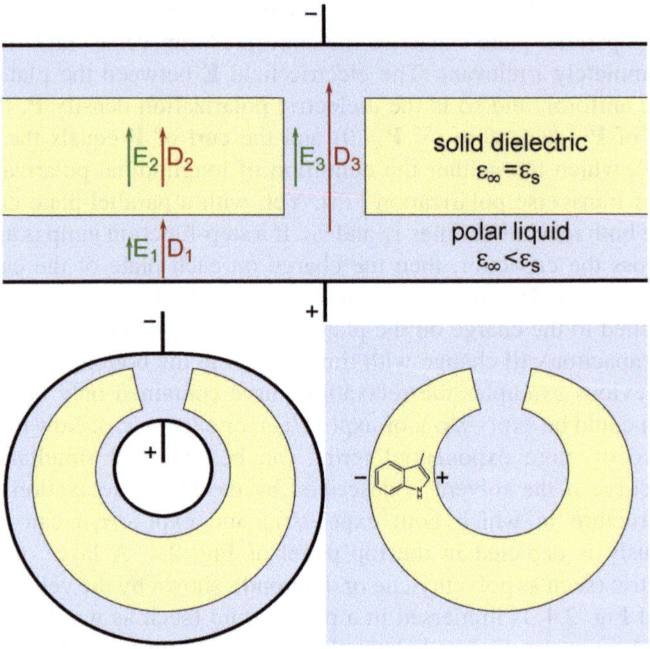

Fig. 2.4 Examples of simple structures in which both $\exp(-t/\tau_D)$ and $\exp(-t/\tau_L)$ are present in the dielectric relaxation curve (*top panel* and *left bottom panel*) or in the TDSS (*right bottom panel*)

curve using either of these structures. Furthermore, the size of the structure shown in the bottom left panel of Fig. 2.4 is not important, it can be reduced to that of a protein. In the next step we remove the capacitor plates (shown by black color, with the + and − labels) and replace them by a solvatochromic fluorophore, such as the sidechain of tryptophan. The result is shown in the bottom right panel of Fig. 2.4. In the TDSS of the fluorophore shown in the bottom right panel one will observe two exponential terms, one with τ_L, and one with τ_D.

Generally speaking, it should be also possible to observe two exponential terms, one with a τ close to τ_L, and one with a τ close to τ_D, in an experiment where a solvatochromic fluorophore is embedded in a non-polar dielectric of irregular shape and immersed in a polar liquid. The hydrophobic core of a protein can play the role of the non-polar dielectric, however, an irregular shape made of polyethylene or diamond can play this role equally well. An exponential term with a τ close to τ_D is expected only if the non-polar structure contains at least one narrow channel filled with the polar solvent and if the cross-section of the channel is significantly smaller than the surface area of each of the two cavities that are connected by the channel (one of these cavities can be the outside solvent). If the structure contains no internal solvent pockets and no channels, then no slow relaxation terms with characteristic times close to τ_D will be observed, but there can still be more than one exponential term with τ between τ_L and τ_D. When the shape of the dielectric structure approaches a perfect sphere, the number of exponential terms in the TDSS reduces to one with $\tau = \tau_F$ given by Eq. (2.46).

The value of τ_D for liquid water at 25 °C equals 8.27 ps [77]. This value, as well as the experimental values [77] of $\varepsilon_S = 78.36$ and $\varepsilon_\infty = 5.2$ can be substituted in Eq. (2.45), which yields a 550 fs estimate for the longitudinal relaxation time τ_L. Experimental TDSS of coumarin 343 in bulk water was reported to contain two close exponentials with $\tau = 126$ fs and $\tau = 880$ fs [18], with the amplitude-weighted mean τ_F of 606 fs. This, as expected, is not far from $\tau_L = 550$ fs. According to Eq. (2.46) the value of τ_F increases with the polarizability of the fluorophore, therefore τ_F values as high as 1 ps are expected in the case of highly polarizable fluorophores, such as 3-methylindole, in water. Solvent relaxation times longer than about 1 ps are not expected for solvatochromic dyes immersed directly in water. Using the example of Fig. 2.4. it was demonstrated here that if the solvatochromic dye is embedded in an irregular-shaped nonpolar dielectric (or in a protein) that is immersed in water, then a relaxation time close to $\tau_D = 8.27$ ps can contribute to TDSS in some cases. This relaxation time represents a different relaxation mode of the solvent and has nothing to do with the biological aspects of the irregular-shaped nonpolar dielectric. The addition of free ions like K^+ or Cl^- to water can further increase the value of τ_D. The addition of organic hydrogen-bonding co-solvents, such as ethanol, ethylene glycol, glycerol, etc. results in non-exponential solvent relaxation with some very slow components that can be much slower than τ_D of pure H_2O.

2.9 Relaxation of Water Molecules Near Protein Surface and Inside Protein

In the previous section we considered the relaxation of the bulk solvent, where it was possible to treat the solvent as a continuous dielectric. Now we will consider the solvent at the level of individual molecules. A rigid solvent molecule can participate in two kinds of motion: the translational motion and the rotational motion. First of all, we will show that the contribution from the translational motion of a solvent molecule to dielectric relaxation and to TDSS is always insignificant in incompressible liquids. The contribution of one solvent molecule to the energy of a capacitor in a dielectric relaxation experiment or to $\Delta E(t)$ in a TDSS measurement equals $-\boldsymbol{\mu}_S \cdot \mathbf{E}$, where $\boldsymbol{\mu}_S$ is the electric dipole moment of the solvent molecule and $\mathbf{E}$ is the electric field. In a dielectric relaxation experiment $\mathbf{E}$ is generated by the charges on the capacitor plates, and in a TDSS experiment it is generated by the charges $\Delta Q_j^{(0)}$ defined in Eq. (2.22). In a uniform electric field (the case of dielectric relaxation) the energy $-\boldsymbol{\mu}_S \cdot \mathbf{E}$ is independent of the location of the solvent molecule, therefore the translational motion has zero contribution to the dielectric relaxation curve. The electric field generated by $\Delta Q_j^{(0)}$ is not uniform, and it can be decomposed into the dipole, quadrupole, octupole, and higher multipole terms.

The significance of these terms decreases in the order in which they are listed here. For a rough estimate we will keep only the most significant dipole term and disregard the smaller terms. The dipole moment of the fluorophore is

$$\boldsymbol{\mu}_F = \sum_j \Delta Q_j^{(0)} \mathbf{r}_j \tag{2.47}$$

and it generates the electric field

$$\mathbf{E} = \frac{1}{4\pi\varepsilon_0}\left(\frac{3(\boldsymbol{\mu}_F \cdot \mathbf{r})\mathbf{r}}{|\mathbf{r}|^5} - \frac{\boldsymbol{\mu}_F}{|\mathbf{r}|^3}\right) \tag{2.48}$$

where **r** is now measured from the center of the fluorophore. For a rough estimate we will assume that the shape of the fluorophore is spherical and that the dipole $\boldsymbol{\mu}_F$ is pointing in the direction from the "south pole" to the "north pole". A solvent molecule located near either pole of the fluorophore sphere has the largest possible contribution to TDSS, which equals $-\xi\cos\theta$, where $\xi = |\boldsymbol{\mu}_S||\boldsymbol{\mu}_F|/(2\pi\varepsilon_0 R^3)$, θ is the angle between $\boldsymbol{\mu}_S$ and $\boldsymbol{\mu}_F$, and R is the sum of the van der Waals radii of the fluorophore and the solvent. If the solvent molecule moves from the pole to the equator, then its contribution to TDSS changes from $-\xi\cos\theta$ to $+(\xi/2)\cos\theta$, i. e. 1.5-fold, whereas if it just rotates in place from $\theta = 0°$ to $\theta = 180°$, then its contribution to TDSS changes from $-\xi$ to $+\xi$, i. e. twofold. To move from the pole to the equator the solvent molecule would have to travel the distance of $\pi R/2$, which would require several exchanges of places between the solvent molecule under consideration and other solvent molecules. The orientation of two molecules cannot be preserved when they switch places in a liquid, therefore the likelihood of the event that a solvent molecule would travel all the way from the pole to the equator without changing its orientation is negligible, and so is the contribution of the translational motion to the TDSS.

The exchange between a water molecule hydrogen-bonded to protein surface and a free water molecule represents a special case of the translational motion. This motion does not contribute to TDSS either, because (as a result of the hydrogen bonding) after such exchange the second water molecule will be in the same orientation in which the first one was before the exchange, while the orientation of the free molecule (the second one before the exchange, the first one after the exchange) changes much faster than the rate of exchange between the two molecules. While the contribution to the TDSS from each molecule changes significantly during their exchange, the net contribution to TDSS from both molecules equals that of the rotational relaxation of the free molecule. This basically means that the residence times of the molecules bound to protein surfaces are not involved in the exponential terms observed in the TDSS.

Once we have established that only the rotational motion of solvent molecules contributes to TDSS, it is necessary to explain how this one type of motion can be responsible for both the longitudinal relaxation mode of the bulk solvent with the relaxation time τ_L and the transverse relaxation mode with the relaxation time τ_D. During the relaxation process the electric field generated by the relaxing solvent molecules superimposes on the external electric field (which triggers the relaxation in the first place), and this represents a feedback mechanism. In the case of the longitudinal relaxation the feedback is negative, and this is what makes the relaxation dynamics faster. In the case of the transverse relaxation the feedback is zero (the net electric field **E** does not change during the transverse relaxation, which was emphasized in Fig. 2.3 and in the corresponding discussion). Thus, τ_D can be considered the intrinsic relaxation time of a solvent molecule in the absence of the feedback. Using Eq. (2.35) and the expression for the dipole-dipole interaction energy it is not difficult to derive the following expression for the feedback coefficient between two solvent molecules in a weak uniform electric field:

$$f_{mn} = \frac{|\boldsymbol{\mu}_S|^2}{4\pi\varepsilon_0 k_B T_A R_{mn}^3} \left(\cos^2\theta_{mn} - \frac{1}{3} \right) \tag{2.49}$$

Here $|\boldsymbol{\mu}_S|$ is the magnitude of the electric dipole moment of each solvent molecule, ε_0 is dielectric permittivity of vacuum, k_B is Boltzmann constant, T_A is absolute temperature, R_{mn} is the distance between the centers of the solvent molecules m and n, θ_{mn} is the angle between the electric field **E** and the line connecting the centers of the molecules m and n. The additional electric field acting on molecule n from the molecule m equals f_{mn}**E**. The feedback coefficient is positive for $0° < \theta_{mn} < 54.7°$ and $125.3° < \theta_{mn} < 180°$; it is negative for $54.7° < \theta_{mn} < 125.3°$. Averaging of the feedback from Eq. (2.49) over all solvent molecules in the bulk solvent results in a large negative feedback, which reduces the relaxation time from τ_D to τ_L. If the solvent molecule is at the planar interface between the bulk solvent and a nonpolar solid dielectric, then exactly one half of the positive feedback and exactly one half of the negative feedback disappears regardless of the angle between **E** and the plane of the interface, which results in a relaxation time close to $2\tau_L$. This shows that near a planar interface with a non-hydrogen-binding substance the longitudinal relaxation time roughly doubles, and this has nothing to do with the biological aspects of the nonpolar dielectric.

Now consider a linear chain of water molecules connecting a reaction center in the protein core with external water [78, 79]. For the electric field component parallel to the linear chain the feedback coefficients between all water molecules are positive, which may significantly slow down the relaxation dynamics. A solvent relaxation component significantly slower than τ_D may be observed in this case.

The relaxation time for an isolated solvent molecule in the protein core is expected to be comparable to τ_D if the molecule has about as much rotational freedom as one in the bulk solvent. If the solvent molecule is directly adjacent to a charged chemical group, such as $-NH_3^+$ or $-CO_2^-$, then its dipole moment is always aligned with the strong electric field from the charged group; the contribution from this water molecule to TDSS is expected to be small in amplitude and very fast (a few femtoseconds), like that of the stoichiometric water in ionic crystals, e. g. $CuSO_4{\cdot}5H_2O$.

2.10 Separation of Relaxation Modes by Their Time Scales

In Sect. 2.5 it was shown how the total TDSS energy $\Delta E(t)$ can be calculated from nonequilibrium MD trajectories. In Sect. 2.7 it was shown how the fluorophore contribution $\Delta E_f(t)$, the protein contribution $\Delta E_p(t)$, and the solvent contribution $\Delta E_s(t)$ can be separated. The total TDSS and the contributions must be then ensemble-averaged over a large number of nonequilibrium MD trajectories. The next step is the separation of the relaxation modes by their time scales. This can be accomplished by globally fitting the total TDSS as well as the contributions to it from the fluorophore, solvent and protein by the following model function,

$$\Delta E_c(t) = \alpha_{c,0} + \sum_{n=1}^{N_{\exp}} \alpha_{c,n} \exp[-\max(t,0)/\tau_n] \tag{2.50}$$

In Eq. (2.50) c represents the name of the component (f for fluorophore contribution, s for solvent, p for protein, and blank for the total effect), n is the number of the exponential term, $\alpha_{c,n}$ is the corresponding amplitude, and τ_n is the corresponding relaxation time. The values of all fitting parameters $\alpha_{c,n}$ and τ_n must be determined simultaneously in the course of a global weighted nonlinear least square minimization. Due to the well known negative correlation [56–59, 62, 63] between the protein and the solvent contribution to TDSS it is easy to miss some of the relaxation components when only the total TDSS is fitted. Separately fitting the curves $\Delta E(t)$, $\Delta E_f(t)$, $\Delta E_p(t)$, and $\Delta E_s(t)$ results in four different sets of relaxation times τ_n that cannot be reconciled after the fitting.

Global fitting of the component TDSS curves calculated from 100 nonequilibrium MD trajectories for the tryptophan fluorophore in GB1 protein has been described [62]. The ensemble-averaged component TDSS is presented in Figs. 2.5 and 2.6. A minimum of five exponential terms in Eq. (2.50) was required for an adequate fit to the data. The values of the five relaxation times τ_n resulting in the best global fit to the data are: $\tau_1 = 36.1 \pm 1.6$ fs, $\tau_2 = 384 \pm 26$ fs, $\tau_3 = 5.63 \pm 0.33$ ps, $\tau_4 = 131 \pm 5$ ps, and $\tau_5 = 2.58 \pm 1.06$ ns. The amplitudes associated with the five exponential terms and their decomposition into the contributions from the solvent, protein, and fluorophore are shown in Fig. 2.7 in the form of a histogram.

The histogram in Fig. 2.7 shows how much energy protein and solvent contribute to each relaxation mode, but it says nothing about the mechanical character of the motions associated with these modes. To reveal the mechanical nature of the relaxation modes the following operations have been carried out [62]. All protein internal coordinates (bond lengths, bond angles, dihedral angles) and also all

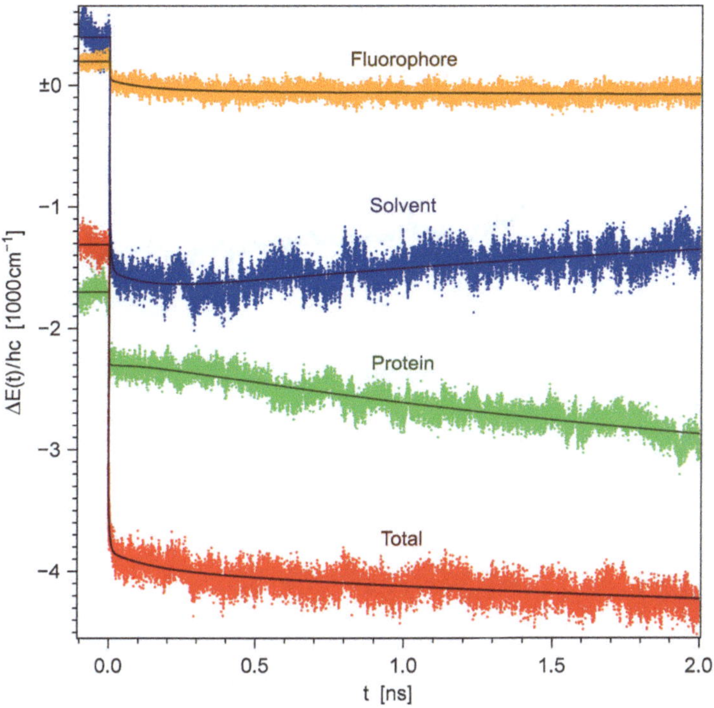

Fig. 2.5 TDSS obtained using the direct-response method, ensemble-averaged over 100 trajectories. Different colors are used to depict the total TDSS (*red*), the solvent contribution to TDSS (*blue*), the contribution from all protein atoms, including fluorophore (*green*), and the contribution from the fluorophore only (*orange*). *Dots* represent the ensemble mean values of $\Delta E(t)/hc$, *solid lines* represent the best global fits by the multiexponential model function from Eq. (2.50) with $N_{exp} = 5$. Reprinted with permission from Ref. [62], copyright 2010 American Chemical Society

pairwise distances between the α-carbons of the protein amino-acids were calculated from each of the 100 trajectories and then ensemble-averaged. The resulting functions of time were fit by the model function from Eq. (2.50) with the fixed values of the five relaxation times τ_n (the values had been obtained in the global fitting of the component TDSS). The amplitudes $\alpha_{c,n}$ (where the subscript c now represents the serial number of the internal coordinate or distance between α-carbons) and the standard deviations of these amplitudes were screened to detect those amplitudes that were statistically significant (i. e., the absolute value of the amplitude exceeds the standard deviation at least threefold). This gave an idea which internal coordinates or distances between α-carbons change in correlation with each of the five relaxation modes. The results are listed below.

On the time scale of $\tau_1 = 36.1$ fs there were significant changes in the bond angles and dihedral angles of several groups in close proximity to the fluorophore, for example, the backbone NH group of Trp-43 (the very residue whose sidechain plays the role of the fluorophore) and the backbone CO group of Gly-41, which is

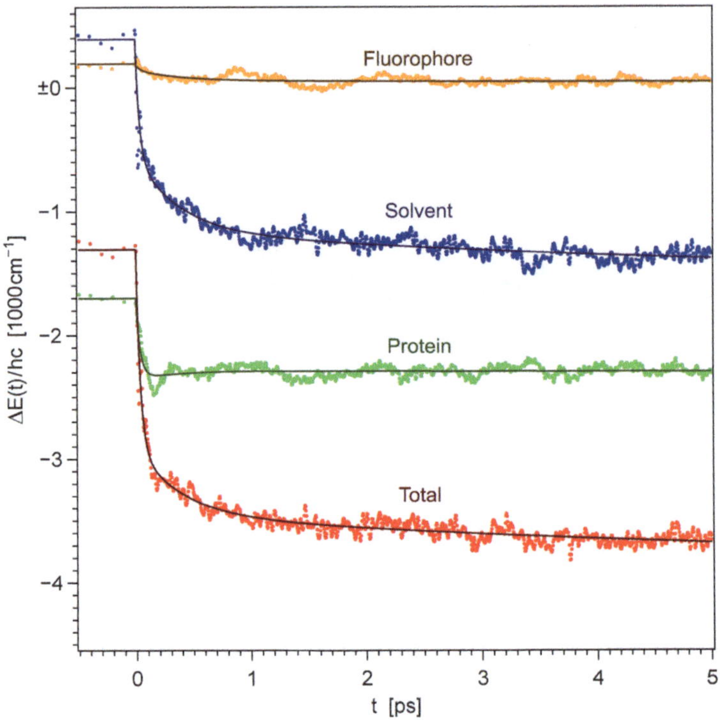

Fig. 2.6 Direct-response TDSS data from Fig. 2.5, shown on a 400-fold expanded time scale to reveal the early stages of the relaxation process. The color scheme and the roles of the *dots* and *solid lines* are the same as in Fig. 2.5. Reprinted with permission from Ref. [62], copyright 2010 American Chemical Society

only about 3 Å away from the fluorophore. The systematic changes in the bond angles and dihedral angles of these groups in close proximity to the fluorophore are quite small, but they still contribute about 656 cm^{-1} or 1.87 kcal/mol to the TDSS on this ultrafast time scale. Slightly more than that (944 cm^{-1} or 2.70 kcal/mol) is contributed by the librational relaxation of water molecules. The librational relaxation mode involves very small adjustments in the orientations of the solvent molecules, which do not result in breaking of the hydrogen bonds. This type of relaxation was observed both in the experiment [18] and in MD simulations [64]; the time scale of this solvent relaxation mode is usually faster than 50 fs.

On the time scale of $\tau_2 = 384$ fs and $\tau_3 = 5.63$ ps none of the protein internal coordinates or distances between α-carbons undergo statistically significant changes. Furthermore, as it can be clearly seen in Fig. 2.7, neither the motion of protein atoms nor the motion of the fluorophore significantly contributes to the second and the third relaxation component. These relaxation components represent two modes of solvent relaxation. We already know that for real water $\tau_L = 550$ fs and $\tau_D = 8.3$ ps. The value of τ_2 from MD simulations is 30 % shorter than the experimental value of τ_L and the value of τ_3 from MD simulations is 30 % shorter

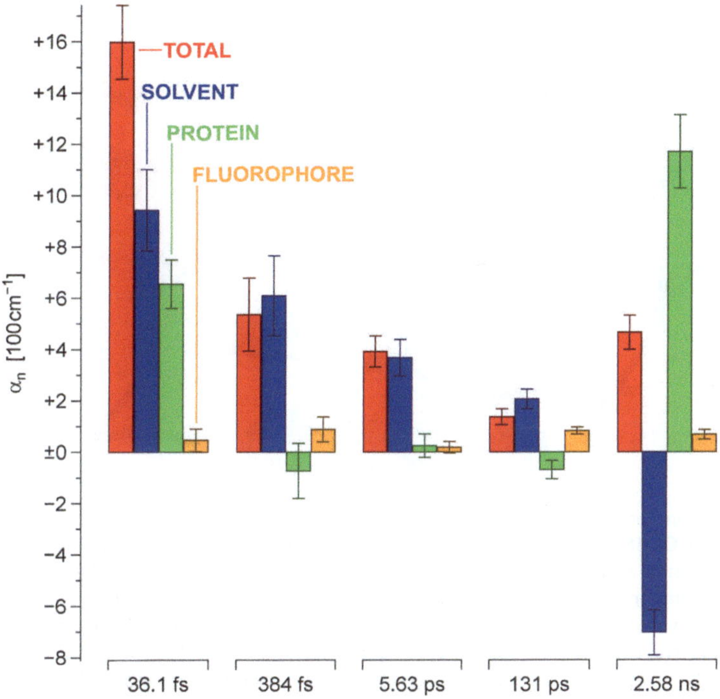

Fig. 2.7 A histogram of the amplitudes corresponding to the five exponential terms recovered in the global fitting of the direct-response TDSS data by the model function in Eq. (2.50). *Red color* denotes the total amplitude α_n. *Blue color* denotes the solvent contribution $\alpha_{s,n}$. *Green color* denotes the protein contribution $\alpha_{p,n}$. *Orange color* denotes the fluorophore contribution $\alpha_{f,n}$. *Error bars* represent 95 % confidence intervals. Corresponding values of τ_n are shown at the bottom. Reprinted with permission from Ref. [62], copyright 2010 American Chemical Society

than the experimental value of τ_D. This is a well-known artifact of TIP3P water model, which was used in the MD simulations described here [62]. In a hybrid QM-MD simulations of 3-methylindole in TIP3P water [64] the longitudinal relaxation time was found to be 400 fs, which is also about 30 % less than the experimental value of τ_D. All this points to the fact that the second and the third relaxation component represent the longitudinal and the transverse relaxation modes of bulk water, with the amplitudes of 536 cm^{-1} or 1.53 kcal/mol and 393 cm^{-1} or 1.12 kcal/mol, respectively.

On the time scale of $\tau_4 = 131$ ps statistically significant but modest changes (not exceeding 3°) in the ensemble mean values of some backbone dihedral angles ϕ and ψ are observed on the time scale of τ_4, and most of these changes are observed in the turn regions, residues 9–12, 20–23, and 36–42, see Fig. 2.8. This indicates that the secondary-structure elements (one α-helix and four β-strands) do not change their conformation, but an adjustment of the tertiary structure takes place on the time scale of τ_4. The nature of this adjustment becomes clear when we look at the changes in the distance between the α-carbon of each residue and the α-carbon of

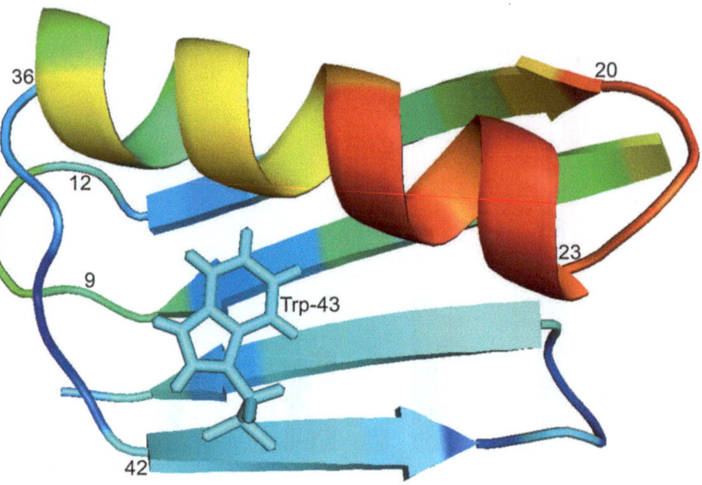

Fig. 2.8 Representative relaxed excited-state structure of GB1, depicted using cartoon representation in PyMOL [80]. The color represents the change in the ensemble mean distance between the α-carbon of each residue and the α-carbon of Trp-43 that takes place on the time scale of the relaxation component τ_4. A residue that does not move closer or further away from Trp-43 is depicted by *cyan color*. The residues that move away from Trp-43 are depicted by *green*, *yellow*, *orange*, and *red colors*, with the *red color* corresponding to the largest positive change in the distance (about +0.23 Å) on the time scale of τ_4. The residues that move closer to Trp-43 are depicted by the shades of *blue color*, with the *dark blue color* corresponding to the largest negative change in the distance (about −0.06 Å) on the time scale of τ_4. Reprinted with permission from Ref. [62], copyright 2010 American Chemical Society

Trp-43. The time scale of τ_4 is the only time scale on which the changes in the distances between α-carbons are statistically significant. On the time scales of τ_1, τ_2, τ_3, and τ_5 there is no statistically-significant changes in distances between α-carbons. As it can be clearly seen from Fig. 2.8, on the time scale of τ_4 the α-helix pulls away from the β-strand containing Trp-43, and this allows a slightly greater water access to the fluorophore. Note, that all the energy of the red shift comes from water, which contributes 206 cm^{-1} or 0.59 kcal/mol, whereas the contribution of the protein equals −69 cm^{-1} or −0.20 kcal/mol (the negative amplitude corresponds to the blue shift). There is a negative correlation between the contributions from the solvent and the protein on the time scale of τ_4. Although the interaction between the solvent and the fluorophore provides all the energy for the motion on this time scale, the relaxation process on the time scale of τ_4 reflects the protein dynamics, specifically the dynamics of small adjustments in the tertiary structure. This shows that the conclusions based solely on the energy contributions to TDSS from the protein atoms and solvent atoms can be misleading.

On the time scale of $\tau_5 = 2.58$ ns statistically significant changes are observed in the dihedral angles χ_1 and χ_2 of Glu-42 sidechain. The nature of these changes becomes obvious from Fig. 2.9, where each color point represents one Glu-42 sidechain conformation saved during the MD simulation described elsewhere

[62]. A total 2.1 million points, representing 100 MD trajectories are shown in Fig. 2.9. The color of each point is related to the time from the instance when the charges on the fluorophore (the Trp-43 sidechain) were changed from $Q_{j0}^{(0)}$ to $Q_{j1}^{(0)}$ (this time is taken for t=0). At t<0 the color is deep blue. During the time between 0 and 1 ns the color gradually shifts from blue to green. During the time between 1 and 2 ns the color gradually shifts from green to red. This makes it possible to see that the populations of the Glu-42 sidechain rotamers marked D and E in Fig. 2.9 are high in the ground state, but they decrease after the excitation of Trp-43. On the other hand, the populations of the Glu-42 sidechain rotamers marked C and H in Fig. 2.9 are low in the ground state and they increase after the excitation of Trp-43. In configurations C and H the negatively-charged $-CO_2^-$ group at the end of Glu-42 sidechain is very close to the positively-charged end of the excited-state fluorophore, which is not the case in configurations D and E. The conformational change of the Glu-42 sidechain results in a 1170 cm^{-1} or 3.33 kcal/mol contribution to the TDSS, which is opposed by the −703 cm^{-1} or −2.00 kcal/mol contribution from the solvent.

The negative contribution of water to the relaxation component τ_5 can be explained in terms of the dielectric continuum model [63]. If $\Delta E = 1170$ cm^{-1} is the energy contribution to TDSS from the motion of Glu-42 sidechain in vacuum, and if we neglect the physical dimensions of the Glu-42 sidechain as well as the fluorophore and the rest of the protein and consider the motion of point-charges immersed in continuous solvent of the dielectric constant ε_S, then in the solvent the TDSS amplitude will decrease from ΔE to $\Delta E/\varepsilon_S$, which means that the solvent contribution to the TDSS equals $\Delta E/\varepsilon_S - \Delta E$, and this is always a negative number. If we take $\varepsilon_S = 78.36$, then the solvent contribution to the TDSS estimated using the formula $\Delta E/\varepsilon_S - \Delta E$ equals −1155 cm^{-1}; the value actually obtained from the MD simulations is only −703 cm^{-1}. The difference in the magnitude shows that it was a crude approximation to neglect the physical dimensions of the Glu-42 sidechain as well as the fluorophore and the rest of the protein; in spite of this crude approximation the negative sign of the solvent contribution was explained correctly, at least on the qualitative level.

The net amplitude of the TDSS associated with the conformational change of the Glu-42 sidechain obtained from MD simulations equals 466 cm^{-1} [62]. The amplitude of the slow TDSS measured experimentally equals 706 cm^{-1} [41], however, the experimental amplitude should be compared to the sum of the amplitudes corresponding to τ_4 and τ_5 [41], which equals 603 cm^{-1}. The difference between 603 and 706 cm^{-1} is within experimental errors. The main problem is that the value of $\tau_5 = 2.58 \pm 1.06$ ns determined from MD simulations [62] is about 30-fold greater than the experimentally measured τ value of 77 ps [41]. Several explanations of this disagreement are possible. First, it is possible that in CHARMM the potential barriers separating different sidechain conformations are higher than in the real life. Second, it is possible that the sidechain can tunnel through the potential barrier like the nitrogen nucleus in NH_3 molecule can tunnel through the plane of the three hydrogen nuclei (see the description of the ammonia maser in the

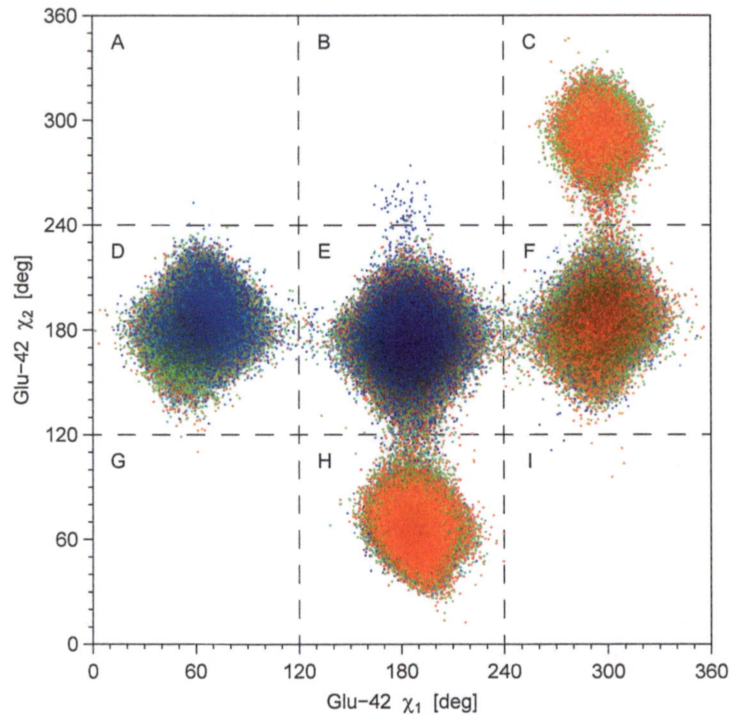

Fig. 2.9 The set of transient Glu-42 side chain conformations, depicted by $2.1 \cdot 10^6$ *color dots* on a plot of χ_2 versus χ_1. The internal coordinates (χ_1,χ_2) were saved at 0.1 ps intervals along 100 trajectories. The color of each dot is blue at early times (ground state and the early part of the excited-state trajectory). During the first 1 ns in the excited state the dot color gradually shifts from blue to green, and during the second 1 ns the color shifts from green to red. Overlaps between dots are colored using pigment mixing rules rather than intensity addition rules. Variable brightness enhancement prevents areas where too many dots of different colors overlap from going completely black. Broken lines divide the χ_1-χ_2 space into nine rotamers, labeled with letters A through I. Each rotamer represents a topologically connected area of high dot density on this plot. To prevent splitting of the connected areas at the artificial boundaries where $\chi_1 = \pm 180°$ or $\chi_2 = \pm 180°$, the angles χ_1 and χ_2 have been redefined so that their domain is $0° \leq \chi < 360°$ rather than $-180° < \chi \leq 180°$. Reprinted with permission from Ref. [62], copyright 2010 American Chemical Society

introduction to this chapter). Finally, it is possible that in the 77 ps relaxation time observed in the experiment describes the relaxation of the counter-ions (such as K^+) that are likely to be found near the $-CO_2^-$ group, but were not included in the MD simulation [62]. The hypothesis regarding counter-ions can be verified experimentally by running the experiments in solvents containing different ions (Li^+ or Na^+ instead of K^+), of different ionic strengths, and of different pH. If the slow relaxation rate or the TDSS amplitude associated with it varies with any of these experimental parameters, then this will show that the TDSS can detect the motion of counterions near charged groups on protein surfaces.

2.11 Conclusions

The TDSS in fluorescence emission of tryptophan residues in proteins can be utilized to obtain valuable information regarding protein dynamics in the frequency range between 300 MHz and 300 GHz, where non-optical experimental methods cannot provide sufficient information. This chapter describes how the results obtained in the experimental studies can be interpreted using computer MD simulations. It has been shown how the experimentally-observed TDSS curves can be obtained theoretically from MD simulations, how the simulated TDSS curves can be separated into multiple relaxation modes, and how the physical nature of the motions associated with each relaxation mode can be identified. Two relaxation modes for the bulk solvent have been described, one of which can be observed only if the protein contains internal water channels or pockets. The methods described in this chapter can be used to study the dynamics of soft vibration modes in proteins as well as the dynamics of internal water molecules in the protein core and the motions of counterions near the protein surface.

Acknowledgments This research was supported by the National Science Foundation awards MCB-0719248 and MCB-1051996.

References

1. Cornell WD, Cieplak P, Bayly CI, Gould IR, Merz KM Jr, Ferguson DM, Spellmeyer DC, Fox T, Caldwell JW, Kollman PAJ (1995) A second generation force field for the simulation of proteins, nucleic acids, and organic molecules. Am Chem Soc 117:5179–5197
2. MacKerell AD Jr, Bashford D, Bellott M, Dunbrack RL Jr, Evanseck J, Field MJ, Fischer S, Gao J, Guo H, Ha S, Joseph D, Kuchnir L, Kuczera K, Lau FTK, Mattos C, Michnick S, Ngo T, Nguyen DT, Prodhom B, Reiher IWE, Roux B, Schlenkrich M, Smith J, Stote R, Straub J, Watanabe M, Wiorkiewicz-Kuczera J, Yin D, Karplus M (1998) All-hydrogen empirical potential for molecular modeling and dynamics studies of proteins using the CHARMM22 force field. J Phys Chem B 102:3586–3616
3. Scott WRP, Huenenberger PH, Tironi IG, Mark AE, Billeter SR, Fennen J, Torda AE, Huber T, Krueger P, van Gunsteren WF (1999) The GROMOS biomolecular simulation program package. J Phys Chem A 103:3596–3607
4. Palmer AG III (2001) NMR probes of molecular dynamics: overview and comparison with other techniques. Annu Rev Biophys Biomol Struct 30:129–155
5. Banyai DR, Murakhtina T, Sebastiani D (2010) NMR chemical shifts as a tool to analyze first principles molecular dynamics simulations in condensed phases: the case of liquid water. Magn Reson Chem 48:S56–S60
6. Parkesh R, Fountain M, Disney MD (2011) NMR spectroscopy and molecular dynamics simulation of r(CCGCUGCGG)$_2$ reveal a dynamic UU internal loop found in myotonic dystrophy type 1. Biochemistry 50:599–601
7. Fersht AR, Daggett V (2002) Protein folding and unfolding at atomic resolution. Cell 108:1–20

8. Scheraga HA, Khalili M, Liwo A (2007) Protein-folding dynamics: overview of molecular simulation techniques. Annu Rev Phys Chem 58:57–83
9. Bolhuis PG (2009) Two-state protein folding kinetics through all-atom molecular dynamics based sampling. Front Biosci 14:2801–2828
10. Deng Y, Roux BJ (2009) Computations of standard binding free energies with molecular dynamics simulations. Phys Chem B 113:2234–2246
11. Landau LD, Lifschitz EM (1977) Quantum mechanics (non-relativistic theory), 3rd edn. Pergamon Press, Oxford
12. Gordon JP, Zeiger HJ, Townes CH (1955) The maser – New type of microwave amplifier, frequency standard, and spectrometer. Phys Rev 99:1264–1274
13. Gordon JP, Zeiger HJ, Townes CH (1954) Molecular microwave oscillator and new hyperfine structure in the microwave spectrum of NH_3. Phys Rev 95:282–284
14. Dennison DM, Uhlenbeck GE (1932) The two-minima problem and the ammonia molecule. Phys Rev 41:313–321
15. Rosen N, Morse PM (1932) On the vibrations of polyatomic molecules. Phys Rev 42:210–217
16. Ware WR, Chow P, Lee SK (1968) Time-resolved nanosecond emission spectroscopy: spectral shifts due to solvent-solute relaxation. Chem Phys Lett 2:356–358
17. Maroncelli M (1993) The dynamics of solvation in polar liquids. J Mol Liq 57:1–37
18. Jimenez R, Fleming GR, Kumar PV, Maroncelli M (1994) Femtosecond solvation dynamics of water. Nature 369:471–473
19. Horng ML, Gardecki JA, Papazyan A, Maroncelli M (1995) Subpicosecond measurements of polar solvation dynamics: Coumarin 153 revisited. J Phys Chem 99:17311–17337
20. Stratt RM, Maroncelli M (1996) Nonreactive dynamics in solution: the emerging molecular view of solvation dynamics and vibrational relaxation. J Phys Chem 100:12981–12996
21. Gardecki JA, Maroncelli M (1999) Comparison of the single-wavelength and spectral-reconstruction methods for determining the solvation-response function. J Phys Chem A 103:1187–1197
22. Kovalenko SA, Schanz R, Senyushkina TA, Ernsting NP (2002) Femtosecond spectroscopy of *P*-dimethylaminocyanostilbene in solution No evidence for dual fluorescence. Phys Chem Chem Phys 4:703–707
23. Arzhantsev S, Jin H, Baker GA, Maroncelli M (2007) Measurements of the complete solvation response in ionic liquids. J Phys Chem B 111:4978–4989
24. Sajadi M, Obernhuber T, Kovalenko SA, Mosquera M, Dick B, Ernsting NP (2009) Dynamic polar solvation is reported by fluorescing 4-aminophthalimide faithfully despite H-bonding. J Phys Chem A 113:44–55
25. Brand L, Gohlke JR (1971) Nanosecond time-resolved fluorescence spectra of a protein-dye complex. J Biol Chem 246:2317–2324
26. Gafni A, DeToma RP, Manrow RE, Brand L (1977) Nanosecond decay studies of a fluorescence probe bound to apomyoglobin. Biophys J 17:155–168
27. Pierce DW, Boxer SG (1992) Dielectric relaxation in a protein matrix. J Phys Chem 96:5560–5566
28. Riter RR, Edington MD, Beck WF (1996) Protein-matrix solvation dynamics in the α subunit of *C*-phycocyanin. J Phys Chem 100:14198–14205
29. Jordanides XJ, Lang MJ, Song X, Fleming GR (1999) Solvation dynamics in protein environments studied by photon echo spectroscopy. J Phys Chem B 103:7995–8005
30. Changenet-Barret P, Choma CT, Gooding EF, DeGrado WF, Hochstrasser RM (2000) Ultrafast dielectric response of proteins from dynamics stokes shifting of coumarin in calmodulin. J Phys Chem B 104:9322–9329
31. Vincent M, Gilles AM, de la Sierra IML, Briozzo P, Barzu O, Gallay J (2000) Nanosecond fluorescence dynamic stokes shift of tryptophan in a protein matrix. J Phys Chem B 104:11286–11295

32. Pal SK, Mandal D, Sukul D, Sen S, Bhattacharyya K (2001) Solvation dynamics of DCM in human serum albumin. J Phys Chem B 105:1438–1441
33. Toptygin D, Savtchenko RS, Meadow ND, Brand L (2001) Homogeneous spectrally- and time-resolved fluorescence emission from single-tryptophan mutants of IIAGlc protein. J Phys Chem B 105:2043–2055
34. Cohen BE, McAnaney TB, Park ES, Jan YN, Boxer SG, Jan LY (2002) Probing protein electrostatics with a synthetic fluorescent amino acid. Science 296:1700–1703
35. Pal SK, Peon J, Bagchi B, Zewail AH (2002) Biological water: femtosecond dynamics of macromolecular hydration. J Phys Chem B 106:12376–12395
36. Peon J, Pal SK, Zewail AH (2002) Hydration at the surface of the protein monellin: dynamics with femtosecond resolution. Proc Natl Acad Sci U S A 99:10964–10969
37. Mataga N, Chosrowjan H, Taniguchi S, Hamada N, Tokunaga F, Imamoto Y, Kataoka M (2003) Ultrafast photoreactions in protein nanospaces as revealed by fs fluorescence dynamics measurements on photoactive yellow protein and related systems. Phys Chem Chem Phys 5:2454–2460
38. Lampa-Pastirk S, Beck WF (2004) Polar solvation dynamics in Zn(II)-substituted cytochrome *C*: diffusive sampling of the energy landscape in the hydrophobic core and solvent-contact layer. J Phys Chem B 108:16288–16294
39. Qiu W, Zhang L, Kao Y-T, Lu W, Li T, Kim J, Sollenberger GM, Wang L, Zhong D (2005) Ultrafast hydration dynamics in melittin folding and aggregation: helix formation and tetramer self-assembly. J Phys Chem B 109:16901–16910
40. Guha S, Sahu K, Roy D, Mondal SK, Roy S, Bhattacharyya K (2005) Slow solvation dynamics at the active site of an enzyme: implications for catalysis. Biochemistry 44:8940–8947
41. Toptygin D, Gronenborn AM, Brand L (2006) Nanosecond relaxation dynamics of protein GB1 identified by the time-dependent Red shift in the fluorescence of tryptophan and 5-fluorotryptophan. J Phys Chem B 110:26292–26302
42. Qiu WH, Kao YT, Zhang LY, Yang Y, Wang LJ, Stites WE, Zhong DP, Zewail AH (2006) Protein surface hydration mapped by site-specific mutations. Proc Natl Acad Sci U S A 103:13979–13984
43. Abbyad P, Shi XH, Childs W, McAnaney TB, Cohen BE, Boxer SG (2007) Measurement of solvation responses at multiple sites in a globular protein. J Phys Chem B 111:8269–8276
44. Halder M, Mukherjee P, Bose S, Hargrove MS, Song XY, Petrich JW (2007) Solvation dynamics in protein environments: comparison of fluorescence upconversion measurements of Coumarin 153 in monomeric hemeproteins with molecular dynamics simulations. J Chem Phys 127:055101
45. Li T, Hassanali AA, Kao Y-T, Zhong D, Singer SJ (2007) Hydration dynamics and time scales of coupled water-protein fluctuations. J Am Chem Soc 129:3376–3382
46. Zhang L, Wang L, Kao Y-T, Qiu W, Yang Y, Okobiah O, Zhong D (2007) Mapping hydration dynamics around a protein surface. Proc Natl Acad Sci U S A 104:18461–18466
47. Abbyad P, Childs W, Shi X, Boxer SG (2007) Dynamic stokes shift in green fluorescent protein variants. Proc Natl Acad Sci U S A 104:20189–20194
48. Jesenska A, Sykora J, Olzynska A, Brezovsky J, Zdrahal Z, Damborsky J, Hof M (2009) Nanosecond time-dependent stokes shift at the tunnel mouth of haloalkane dehalogenases. J Am Chem Soc 131:494–501
49. Zhang LY, Yang Y, Kao YT, Wang LJ, Zhong DP (2009) Protein hydration dynamics and molecular mechanism of coupled water-protein fluctuations. J Am Chem Soc 131:10677–10691
50. Othon CM, Kwon OH, Lin MM, Zewail AH (2009) Solvation in protein (un)folding of melittin tetramer-monomer transition. Proc Natl Acad Sci U S A 106:12593–12598
51. Kwon O-H, Yoo TH, Othon CM, Van Deventer JA, Tirrell DA, Zewail AH (2010) Hydration dynamics at fluorinated protein surfaces. Proc Natl Acad Sci U S A 107:17101–17106
52. Zhong D, Pal SK, Zewail AH (2011) Biological water: a critique. Chem Phys Lett 503:1–11

53. Brauns EB, Madaras ML, Coleman RS, Murphy CJ, Berg MA (1999) Measurement of local DNA reorganization on the picosecond and nanosecond time scales. J Am Chem Soc 121:11644–11649
54. Andreatta D, Lustres JLP, Kovalenko SA, Ernsting NP, Murphy CJ, Coleman RS, Berg MA (2005) Power-law solvation dynamics in DNA over six decades in time. J Am Chem Soc 127:7270–7271
55. Callis PR, Burgess BK (1997) Tryptophan fluorescence shifts in proteins from hybrid simulations: an electrostatic approach. J Phys Chem B 101:9429–9432
56. Vivian JT, Callis PR (2001) Mechanisms of tryptophan fluorescence shifts in proteins. Biophys J 80:2093–2109
57. Nilsson L, Halle B (2005) Molecular origin of time-dependent fluorescence shifts in proteins. Proc Natl Acad Sci U S A 102:13867–13872
58. Bandyopadhyay S, Chakraborty S, Balasubramanian S, Bagchi B (2005) Sensitivity of polar solvation dynamics to the secondary structures of aqueous proteins and the role of surface exposure of the probe. J Am Chem Soc 127:4071–4075
59. Hassanali AA, Li TP, Zhong DP, Singer SJ (2006) A molecular dynamics study of Lys-Trp-Lys: structure and dynamics in solution following photoexcitation. J Phys Chem B 110:10497–10508
60. Golosov AA, Karplus M (2007) Probing polar solvation dynamics in proteins: a molecular dynamics simulation analysis. J Phys Chem B 111:1482–1490
61. Li T, Hassanali AA, Singer SJ (2008) Origin of slow relaxation following photoexcitation of W7 in myoglobin and the dynamics of its hydration layer. J Phys Chem B 112:16121–16134
62. Toptygin D, Woolf TB, Brand L (2010) Picosecond protein dynamics: the origin of the time-dependent spectral shift in the fluorescence of the single Trp in the protein GB1. J Chem Phys B 114:11323–11337
63. Halle B, Nilsson L (2009) Does the dynamic stokes shift report on slow protein hydration dynamics? J Phys Chem B 113:8210–8213
64. Muino PL, Callis PR (1994) Hybrid simulations of solvation effects on electronic spectra: indoles in water. J Chem Phys 100:4093–4109
65. Toptygin D, Brand L (2000) Spectrally- and time-resolved fluorescence emission of indole during solvent relaxation: a quantitative model. Chem Phys Lett 322:496–502
66. We use bold typeface **E** to denote the electric field vector and light typeface italic E to denote energy. To avoid confusion, the magnitude of vector **E** is denoted as $|\mathbf{E}|$ rather than E
67. In this context the term "mean" implies averaging of the three Cartesian components of the electric field vector in the reference frame that rotates together with the fluorophore
68. Mulliken RS (1955) Electronic population analysis on LCAO-MO molecular wave functions. J Chem Phys 23:1833–1840
69. Callis PR (1991) Molecular orbital theory of the ^1L_b and ^1L_a states of indole. J Chem Phys 95:4230–4240
70. Pan C-P, Callis PR, Barkley MD (2006) Dependence of tryptophan emission wavelength on conformation in cyclic hexapeptides. J Phys Chem B 110:7009–7016
71. Landau LD, Lifschitz EM (1976) Mechanics, 3rd edn. Pergamon Press, Oxford
72. Maroncelli M, Fleming GR (1988) Computer simulation of the dynamics of aqueous solvation. J Chem Phys 89:5044–5069
73. Hubbard J, Onsager L (1977) Dielectric dispersion and dielectric friction in electrolyte solutions. J Chem Phys 67:4850–4857
74. Maroncelli M, Fleming GR (1987) Picosecond solvation dynamics of Coumarin 153: the importance of molecular aspects of solvation. J Chem Phys 86:6221–6239
75. Castner EW, Fleming GR, Bagchi B, Maroncelli M (1988) The dynamics of polar solvation: inhomogeneous dielectric continuum models. J Chem Phys 89:3519–3534
76. Kivelson D, Friedman H (1989) Longitudinal dielectric relaxation. J Phys Chem 93:7026–7031
77. Kaatze U (1993) Dielectric relaxation of H_2O/D_2O mixtures. Chem Phys Lett 203:1–4

78. Martinez SE, Huang D, Ponomarev M, Cramer WA, Smith JL (1996) The heme redox center of chloroplast cytochrome *F* is linked to a buried five-water chain. Protein Sci 5:1081–1092
79. Pomès R, Roux B (1998) Free energy profiles for H^+ conduction along hydrogen-bonded chains of water molecules. Biophys J 75:33–40
80. DeLano W (2009) Pymol home page. www.pymol.org

Chapter 3
Directional Fluorescence Based on Surface Plasmon-Coupling

Yao-Qun Li, Shuo-Hui Cao, Wei-Peng Cai, Qian Liu, Xiao-Qing Liu, and Yu-Hua Weng

Abstract Fluorescence technology, including the optical sensing and microscopic imaging, has been playing important roles in biology research and medical diagnosis. However it still remains a great challenge to meet the increasing needs of sensitivity and applicability. Surface plasmon-coupled emission (SPCE) is a novel technique that can significantly improve the ability of fluorescence technology. In SPCE, the excited fluorophores will couple with surface plasmons on a continuous thin metal film, which in turn radiate into the higher refractive index media with a narrow angular distribution. Attributed to the direction emission, the sensitivity can be highly improved with the high collection efficiency. This review will summarize the unique features of SPCE that are important in analytical researches, in particular, with a focus on the recent advancements in the strategies for improving the SPCE performance. The optical imaging based on SPCE and some examples of the analytical applications of SPCE are also highlighted. Recent achievements in SPCE suggest that it could provide new technical platforms with widespread potential applications in various areas, such as nucleic acid, protein and other biochemical sensing.

Keywords Surface plasmon-coupled emission • Directional fluorescence • Polarization • Biosensing • Imaging

3.1 Introduction

Due to its high sensitivity and the nondestructive feature, fluorescence technology has obtained great success in *in vivo* detection. Fluorescence technology has become one of the most important tools in the fields of bioanalysis, genetic engineering, clinical diagnosis and drug screening. Yet, there is still a dramatic increase in the need of increased sensitivity with the rapid development in these

Y.-Q. Li (✉) • S.-H. Cao • W.-P. Cai • Q. Liu • X.-Q. Liu • Y.-H. Weng
Department of Chemistry and the MOE Key Laboratory of Spectrochemical Analysis & Instrumentation, College of Chemistry and Chemical Engineering, Xiamen University, Xiamen 361005, China
e-mail: yaoqunli@xmu.edu.cn

© Springer International Publishing Switzerland 2016
C.D. Geddes (ed.), *Reviews in Fluorescence 2015*,
DOI 10.1007/978-3-319-24609-3_3

applications. Recently, a novel fluorescence technique referred as surface plasmon-coupled emission (SPCE) or surface plasmon-coupled directional emission (SPCDE), first described by Lakowicz et al. [1–3], has attracted more and more attentions. It has been shown that the sensitivity could be increased greatly in SPCE and its unique properties were useful in sensing applications [4].

The typical configuration for SPCE is illustrated in Fig. 3.1. The fluorophores which can couple with surface plasmons and result in the directional emission are positioned above a continuous metallic nanofilm, which is about 20–50 nm thick, and the distance between the metal and the fluorophores is about 5-200 nm [5]. Remarkably, the emission can be seen through a prism attached to the film and it occurs only at a unique angle θ_F measured from the normal of the interface. When a hemisphere prism is used, the emission is observed as a cone around **z**-axis. This phenomenon is quite different from the normal fluorescence, where the spatial distribution is isotropic. The emission in the cone has the same emission spectrum as the fluorophores, so it can be considered that it originates from the fluorophores. However, the emission is highly *p*-polarized no matter what kind of the polarization of the excitation light, indicating that the emission should reflect the property of the surface plasmons. In fact, the complex emission involved both in fluorophores and surface plasmons can be regarded as the reverse process of surface plasmon resonance (SPR). The most important prerequisite is the matching of wavevectors belonging to fluorophores and surface plasmons, which can be satisfied with a high index prism attached to the metal surface.

Based on the prism coupling measurement, there are two modes which can be used in SPCE [4]. As shown in Fig. 3.2, the fluorophores can be excited with light incident from either the prism side or the sample side. Through the prism, the incident light at SPR angle can induce the evanescent field and the oscillation of surface plasmons. The fluorophores within near-field will be excited by the evanescent field. In this mode, the device is similar to SPR and called as Kretschmann (KR) configuration. On the other hand, the incident light can directly excite the fluorophores from the sample side and the dipoles in near-field will couple to surface plasmons, inducing directional emission through the prism. This mode is called as reverse Kretschmann (RK) configuration. In the case of KR configuration, the excited field is enhanced greatly in a resonance with evanescent field [6], so the SPCE intensity will be much higher than that in RK configuration [7]. However, thanks to the direct excitation in RK configuration, there will be more available devices [4]. For examples, a light-emitting diode (LED) has been used in low-cost designs [8], and Xenon light takes on the advantages of convenient choices of wavelengths in the experiments [9]. Collective efficiency can be further enhanced with the setup of a conical mirror around the prism, collecting all of the signals of the SPCE ring reflected by a conical mirror into a point [10]. A miniaturized device can be achieved through modifying the design of a spectrofluorometer [11]. Besides the prism coupling, the grating can also induce the directional emission [12]. It should be pointed out that the optical excitation is not essential for directional emission. Chemiluminescence and electrochemiluminescence species are also useful applications for SPCE [13, 14]. Besides, the directional emission has been observed as forms of phosphorescence and two photon excited fluorescence as well [15, 16].

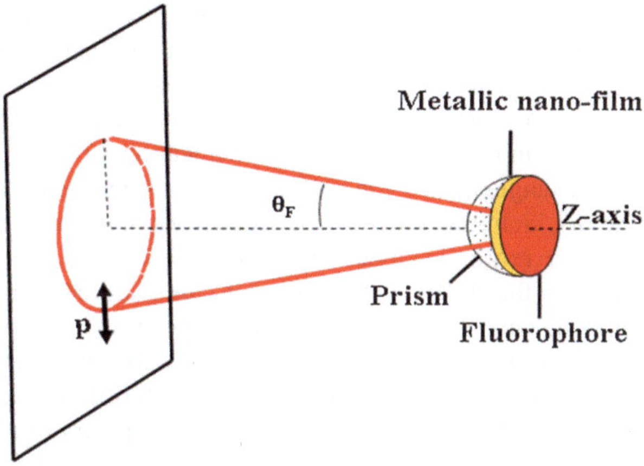

Fig. 3.1 Scheme of surface plasmon coupled directional emission

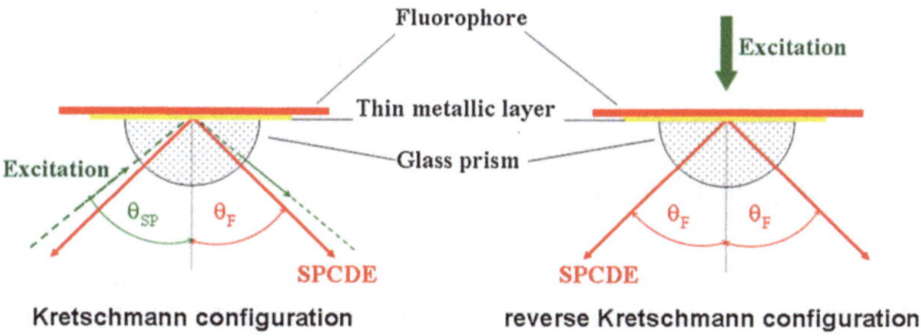

Fig. 3.2 The excitation and emission of SPCE in Kretschmann (KR) configuration and reverse Kretschmann (RK) configuration

3.2 Properties of SPCE

The phenomena of SPCE are unique and have attracted more and more interests for recent years [17]. Here, we briefly summarize and analyze the main properties of SPCE which have been explored. Those potentially important for developing novel analytical methods will be especially concerned.

3.2.1 Directional Emission

The directional emission is an important and interesting characteristic of SPCE. It means the signals that should have emitted all around the space are almost all concentrated to a direction. Thus, high collection can be conveniently achieved just with a detector in the defined direction. It brings great significance in increasing sensitivity, where 50-fold increase will be expected with the high collective efficiency compared to the isotropic spontaneous fluorescence. Besides, if KR configuration is used, 10 – 40-fold enhanced excitation field should be additionally considered, resulting in an overall increase of up to 1000-fold [4]. Such a significant sensitivity enhancement made it possible to develop ultrasensitive fluorescence assays to monitor extremely low concentrations of analytes which are often required in biochemistry and biomedical research.

The angle θ_F that the emission radiates directionally is relevant to many optical factors, including the dielectric properties of metal, the wavelength of fluorophores, and the refractive index of environment [18]. Therefore, the directional angle displays the diversity for different cases and contains important analytical information, which can be an important index to reflect the state of system and the progress of reaction. Besides, it provides the opportunity of designing new apparatuses. For gold film, the fluorophores with different emission maxima will display SPCE at different angles. As a result, the spectra of a mixture of fluorophores will be distinct, reflecting only one fluorophore at a defined angle [5, 19]. It can be seen as a format of wavelength resolution. Wavelengths are distinguished directly through angular separation without filters, which may be a promising design for miniaturized instruments by reducing optical components.

3.2.2 Background Suppression

The generation of SPCE is from the interaction of fluorophores and nearby metal film. Therefore, the signal comes from the substance near the interface, while that more distant from the surface will not be produced. It suggests that SPCE present an approach with the ability of spatial resolution to selectively detect analytes located in near field, and the interference from the bulk solution can be eliminated [5]. As shown in Fig. 3.3, Rhodamine B was located in the near field of gold film through spin cast. Chlorophyll B in the bulk solution was added as the background. In RK configuration, the incident light excited both Rhodamine B (RhB) and Chlorophyll B (ChlB) directly, so the both signals can be observed in the detection of free space emission (FSE) from the sample side. However, it can be seen that only the signal of RhB was observed in the detection of SPCE, even with the high background that ChlB in the chosen concentrations was dominant in the FSE signal. Selective observation of the fluorophores near the metal surface can be useful for various types of surface-bound assay including immunoassays and DNA arrays.

3.2.3 P-Polarized Emission

The polarization of SPCE is the most convincing fact that proves the coupling with surface plasmons. For isotropic orientated fluorophore layer within about 100 nm, the emission is almost *p*-polarization no matter what the polarization of the incident excitation is, while free-space emission displays similar intensities for both *p*- and

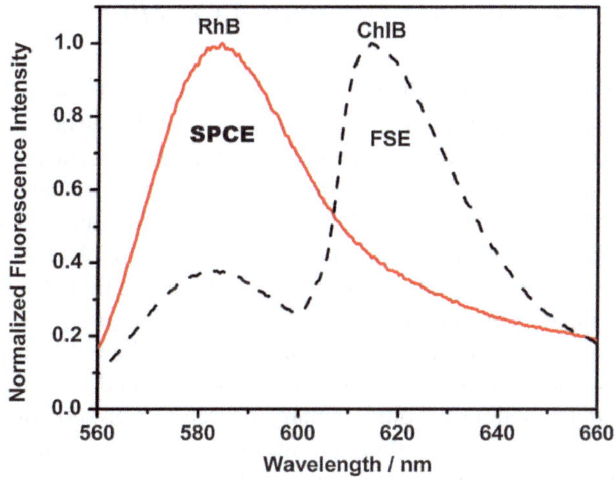

Fig. 3.3 Normalized SPCDE and free space emission (FSE) spectra of 1 mM Rhodamine B (RhB) in 30 nm poly (vinyl alcohol) film spin cast to gold film immerged with 1 mM Chlorophyll B (ChlB) in ethanol as background. RK configuration is used (Unpublished work)

s-polarization [1, 5]. The characteristic of *p*-polarized emission provides an approach to identify the coupling of surface plasmons and exclude the background scattering by a polarizer. The anisotropy emission has been theoretically and experimentally proven to be relative to the orientation of dipoles, i.e. the excited fluorophores regarded as points [20, 21]. The radiation of vertical dipoles standing up on the surface is pure *p*-polarized which can totally contribute to the excitation of surface plasmons. For the horizontal dipoles along surface, the radiation contains the components of both *s*-polarized and *p*-polarized [22], but only *p*-polarized part is the resource of SPCE. As a result, the asymmetry of the SPCE cone will be observed, with higher signal in the direction parallel to the dipoles and weak signal in the direction perpendicular to the dipoles, when the dipoles are orientated horizontally [21]. In most cases, the orientation of dipoles is isotropic and the SPCE cone can be seen as symmetrical.

When referring to the situation with fluorophore layer thicker than 100 nm, the phenomena seem to be more complex. More than one emission ring can be observed in SPCE at one wavelength. That means the emission direction is no longer only one in the case. The polarization and the intensity will show subverted properties, where high directional signal with *s*-polarized can be observed and the polarization is alternating for rings with different dielectric thickness [23]. The origin of these effects can be explained that there are additional plasmon modes to exist on the metal surface [24]. Considering the thickness of dielectric film comparable to the wavelength, it is possible to have the admittance to reverse the phase shift of *s*-polarized to fit the mode of waveguide [25, 26] or the waveguide can be coupled to plasmon resonance [27]. The multi-polarization is valuable as an obviously analytical parameter to dielectric thickness which usually undergoes great alternation in biomedicine through biomolecular binding.

3.2.4 Distance-Determined Coupling

The coupling of SPCE is closely related to the distance between the fluorophore and the metal surface. When it is very close to the surface, especially within 10 nm, the energy of fluorophores is transferred to metal surface as radiationless transition, so the quenching effect is dominant. While at distance longer than 500 nm, the interference between the far-field emission and its reflection dominates [28–30]. As a result, the appropriate distance for SPCE is beyond the range of quenching but close enough to the surface, usually 20 – 200 nm, where the excited dipoles can couple to the surface plasmons radiating into the prism [4]. It should be noted that the coupling efficiency of fluorophores distributed at the effective coupling distance is gradient, displaying a rapidly increased trend with the increase of distance within 50 nm, and undergoing dramatic decrease over 200 nm [31, 32]. It indicates that the position where the fluorophores are located plays an important role in determining the intensity of the directional radiation. The maximum of the coupling efficiency can be expected at the optimized position. These properties bring both great opportunity and challenge in practical analysis, especially in quantitative analysis, since it is really difficult to affirm the position of fluorophores near the surface, but once it can be completed, accurate spatial resolution can be expected, which will be emphasized again in the following parts. The emission dynamics is also related to the distance. At short distance around 10 nm, SPCE decays faster than free space emission; while at distance larger than 20 nm, the lifetime of both is nearly the same, which is supposed to be attributed to the coupling sensitive to the orientation of dipole within a few nanometers because of the different components of the dipoles orientating between SPCE and FSE [33].

3.3 The Strategies for Improving SPCE

Since SPCE has a potential to become a promising platform in the field of analytical chemistry and biochemistry, increased efforts have been made to enhance the capabilities for the applications. In this part, the strategies for improving SPCE are discussed.

3.3.1 Enlarging the Optical Window

The generality of SPCE for kinds of fluorophores with different excitation and emission wavelengths is significant in facilitating the development of the technique platforms. Gold and silver are usually used as the films in SPR and SPCE, because their free electrons can form a coherent oscillation at the electromagnetic radiation, while the resonance of other metals with large imaginary part of dielectric constants will be depressed [34]. But the appropriate spectral range that gold and silver can be applied to is concentrated in the red spectral region and visible spectral region, respectively [18, 35], which limits the generality of the technique in some assays, such as the protein with fluorescence at ultraviolet. On the other hand, some metals including transition metals have also been proven exciting surface enhancement, though not very strong [36].

The investigations of various metals have been operated in search of a chance to enlarge the optical window where SPCE can be applied. Geddes et al. have summarized a general procedure for the selection of a metal thin film to be employed in applications of SPCE [37]. Reflectivity curves calculated by performing Fresnel equations are the useful tools to indicate the potential surface plasmon resonance. Firstly, the situations of various thicknesses of metal films should be simulated to check the optimum one. For the metals with a large imaginary part of their dielectric function, the optimum thickness is around 20 nm, which is less than the usual thickness around 50 nm for gold and silver films. Then the wavelength range matching to chosen metal can be evaluated by the minimum reflectivity of *p*-polarized light at the SPR angle, where the value usually should be smaller than 0.2. After the simulation, the possibility of the utilization with chosen metal can be validated in practice. It has been reported that copper [38], aluminum [39], zinc [40, 41], iron [19, 42, 43], nickel [44], platinum [45], palladium [46], chromium [37] and rhodium [37] can be used in SPCE. It is worth noting that the whole spectral regions from ultraviolet to infrared can now have the corresponding selections of metal films to be employed, as shown in Table 3.1 [37]. The enlarged optical window makes the approaches more available concerning the use of optional fluorophores, and alternative metals can therefore be expected.

An additional advantage with iron and palladium is the fixed angle observation, which is convenient to the potential combination of present spectroscopic measurements [37, 46]. It gives the opportunity to observe directional signal of multi-fluorophores by one spectral scan, which is attractive in labeled systems in biological applications [19]. Actually, we found that iron thin film performed well with the effective enhancement nearly three-fold for the mixture of fluorophores in the visible wavelength. It also worked in the whole blood sample with the multi-fluorophores as the labels, displaying excellent optical background suppression. As shown in Fig. 3.4, the clear and distinguished spectra of labeled fluorophores were observed from the SPCDE channel, while the signal obtained in FSE channel was submerged in the strong background from the whole blood sample. Besides, it was found that the penetration depth, distance that evanescent wave reaches into the dielectric, can be manipulated by using different metal films [47]. So it becomes more flexible in the establishment of assays. If the analytes are close to surface and the background is strong, the metal with short penetration depth should be used to selectively detect the analytes. While referring to the analytes with big sizes, especially in the bioassays, metal films with long penetration depth should be chosen to effectively excite the analytes.

Table 3.1 List of metals used in SPCE with the appropriate applications [37]

No.	Metal	Metal thickness, nm	Wavelength range, nm	Observation angle
1	Chromium	20	276–517	Variable: 50–80
2	Aluminum	20	280–600	Variable: 42–46
3	Palladium	15/20	300–800	Variable: 45–70
4	Zinc	30	305–545	Variable: 45–68
5	Nickel	15/20	344–1240	Variable: 50–70
6	Rhodium	15	344–1240	Variable: 45–75
7	Silver	40	413–620	Variable: 45–60
8	Gold	50	496–866	Variable: 45–60
9	Iron	15	496–866	Fixed: 65
10	Copper	40	521–659	Variable: 55–70
11	Platinum	15	800–1500	Variable: 50–70

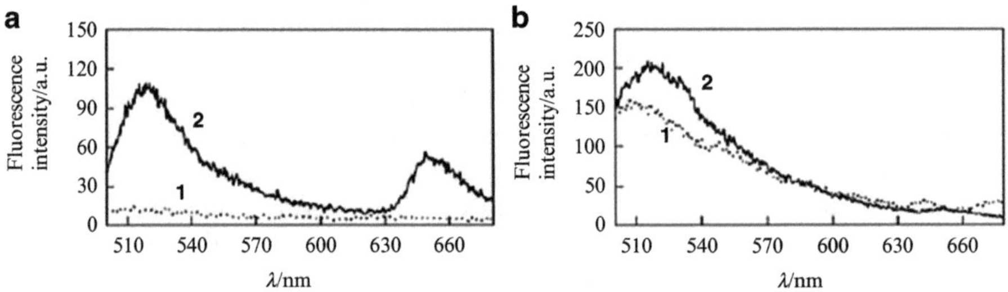

Fig. 3.4 SPCDE (**a**) and FSE (**b**) spectra of blood before (1) and after (2) added with fluorescein and TPPS mixture [19]

3.3.2 *Improving the Applicability and Sensitivity*

To further drive SPCE close to the actual applications, some innovations in this technique have been implemented to improve the applicability and sensitivity. One technique introduced here is called as microwave-accelerated surface plasmon-coupled directional emission (MA-SPCE) [48]. Both the speed and the sensitivity of assays are paramount in clinical analysis. But the molecular reactions in bio-assays usually take a long time, from a few minutes to hours, which has become the current bottleneck in the fast detection. Microwaves can be employed to accelerate the kinetics of molecular reactions in bioassays [49], offering the opportunity to develop fast and sensitive assays combining both elements of microwave and SPCE. The feasible hyphenated method has been achieved with the metal surface limited to a small area with the diameter significantly less than the wavelength of the microwaves, where the surface sparking and arcing due to the surface charge buildup under microwaves can be avoided. The running time of bioassays can be significantly reduced [48, 50]. In addition, SPCE offers the detection of directional and polarized emission with high sensitivity. Thus, both speed and sensitivity are improved in this technique. It has also found that some special geometrical structures with tips can be used to further enhance the microwave effect due to the intense field distributions near the tips [51].

Considering the increasing demand for improving sensitivity and detectability in assays, efforts on amplifying the signal have been made. Metal enhanced fluorescence (MEF) is playing an important role in the biological applications of fluorescence. The electromagnetic field is enhanced for the localized surface plasmon resonance (LSPR) effect of metal nanoparticles in MEF [52]. Though it is different from the mechanism of SPCE with the propagating of surface plasmon along smooth metal surface, it still inspires researchers to attempt the combination of both MEF and SPCE, to check whether the two populations of near-field induced plasmons can occur in one system. Fortunately, the assumption was supported by experiments [53]. Silver colloids with different radiuses were put into substrates coated with gold films or not, and then the fluorophores were immobilized to cover the substrates. Emissions from the prism side and the sample side were collected. The presence of silver colloids not only enhanced the free space emission, but also increased the intensity of SPCE. It was clear that the intensity was much higher in the case of 80 nm colloids from the back of gold substrate, compared to that of 20 nm colloids where the radius was within the quenching limit and the control sample without colloids. In further researches, it showed that colloidal nanoparticles can be used to enhance the coupling efficiency of excited state fluorophores to surface plasmons on smooth metal films. The optimal colloid size is 40 nm for silver colloids-silver films system, where excited fluorophores will create the LSPR on nanoparticles which subsequently couples with the surface plasmons on the metal films thus leading to significant enhanced electromagnetic field and increases in SPCE signal [54]. It was also found that silver islands deposited onto cooper films resulted in enhancement in SPCE [55], which indicates a possible way to enhance the SPCE signal from various metal films by combining the effect of MEF. Therefore, the novel systems of MEF-SPCE provide dramatic signal enhancement and improved photostability of both free space emission and plasmon-coupled emission, which can be potentially employed to increase the detectablity of assays based on SPCE.

3.3.3 Obtaining Controllable Coupling with Spatial Resolution

As we have mentioned above, the coupling effect close to the metal surface is dramatically sensitive to the spatial position along the normal to the metal surface, which is quite different from the traditional fluorescence technique demanding homogeneous distribution in the entire space. It has become a huge challenge that we cannot but confront, finding approaches to ensure the fluorophores under test can be uniformly distributed at the expected spatial position. Otherwise the efficiency that fluorophores coupled to plasmons will be unpredictable and not uniform, thus resulting in false response in assays. On the other hand, the distinct distance-dependent profile indicates the possibility to monitor the molecules around surface with spatial resolution through evaluating the coupling effect relative to spatial position.

Langmuir-Blodgett (LB) films were introduced to investigate the distance-dependence of SPCE [32]. They can be used to obtain the uniform coupling effect for molecules at defined spatial distribution. The distance between fluorophores and the metal surface was defined by depositing stearic acid with different numbers of layers. The long chain of amphiphilic cyanine dye DiI was deposited as the top monolayer. All dyes were put in a uniform position which could be controlled by designing the layers of LB films as spacers. Another advantage in this system is that all of the dyes are structured with the orientation of transition dipoles parallel to the metal surface. Although the coupling effect for the horizontal orientation is not high, it is still a great progress to offer the important information about the interactions between fluorophores and metal surface, eliminating the affection of dipole orientations. It showed clearly that about 20 nm from surface was the position with the maximum intensity of SPCE in the situation. LB spacing is a promising way to confirm the uniform distribution and the uniform coupling effect of fluorophores along the normal to metal surface, thus the reliable signals in assays. Furthermore, controlling the coupling effect between fluorophores and metal surface through depositing LB films as spacers can be achieved.

The electric field at the interface is also highly localized and distance-dependent. The effective distance is almost at the level of nanometers overlapping the range that SPCE occurs. It is possible to utilize the electric field to control the distribution or conformation of charged molecules near surface to satisfy the uniform and optimum coupling effect in assays. Based on these considerations, we proposed the concept of electric field assisted SPCDE (E-SPCDE) [56]. In this example, we chose DNA, a natively negative charged biomolecule, as a model to illustrate the spatial coupling control in E-SPCDE. The signals of fluorophores labeled to DNA duplexes attached to gold surface were detected in E-SPCDE. With the assistance of an external electric field, the interfacial molecular conformation was controlled with DNA standing at a negative potential and lying at a positive potential, so the coupling efficiency between dyes and metal surface was affected by the potential. It was observed that the signal was higher at more negative potential within the defined range, suggesting that the coupling efficiency was increased with the more perpendicular orientation of DNA layers. As shown in Fig. 3.5, the enhancing effect was improved more than three-fold when extending the oligomers from 36 bases to 48 bases. The signal was increased more for the longer DNA chain at the same potential, further confirming that the signal change was correlated to the conformation of DNA. Coupling efficiency increases with the labeled dyes departing from surface as DNA duplexes orientated perpendicularly at negative potential. Thus, larger conformation changes with longer chain produce higher signal responses in the distance-sensitive detection. On the other hand, we also proved the low fluorescence at positive potential with labeled DNA adsorbed to the metal surface. So the coupling efficiency can be controlled spatially by an active

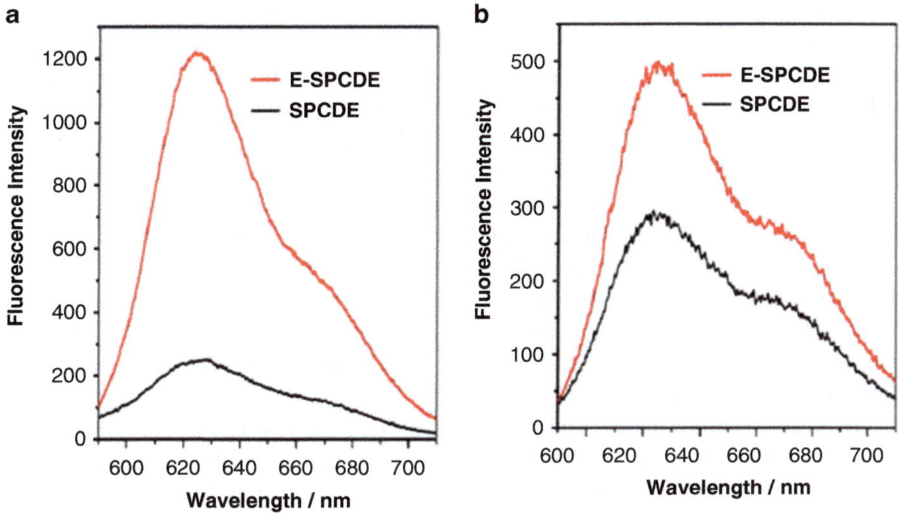

Fig. 3.5 E-SPCDE and SPCDE spectra of 48-base hairpin (**a**) and 36-base hairpin (**b**) hybridized with matched DNA target [56]

way, providing a good opportunity to make the manipulation of the nanometer-scale processes more accessible and detectable.

3.4 SPCE Optical Imaging

The attractive characteristics of SPCE with high collection efficiency and low background allow it proceeding excellent performance in optical imaging. In fact, there have been increased interests in combing the imaging technique with SPCE. In this part, the optical imaging based on SPCE will be introduced. Both the technique and the relevant researches will be illustrated.

3.4.1 SPCE Imaging System Based on Microscope Technique

Considering the ripened optical components in microscopic equipment, the SPCE imaging can be accomplished in a proper way. As well known, in total internal reflection fluorescence (TIRF) system, a TIRF objective with high numerical aperture (NA) is used like a prism, which can also assure the high refractive index and the light incident over the critical angle. Therefore the TIRF microscope can be rebuilt to SPCE microscope. The light beam was incident to excite the fluorophores at the SPR angle through the objective, and the emission was collected also through the same objective at the SPCE angle. As seen in Fig. 3.6a, this optical design of the inverted microscopic detection can be regarded as KR configuration [57, 58]. To avoid the precise incident angle adjustment in KR-SPCE microscope, RK-SPCE microscope was proposed as the simple implement [59]. In RK configuration, the metal-coated glass substrate was placed on the objective of inverted microscopy with the sample side up to air. The light directly illuminated the sample side without the need of angle adjustment or polarization, reducing the loss of incident intensity as well. Then the directional emission from the opposite side was collected by the objective (Fig. 3.6b). The excitation from the sample side is easy to be realized in practice and the metal film makes the background rejection obviously. If standing-wave was used as the excitation, with two beams propagating through the objective lens interfered to excite the sample on the metal-coated cover slip, the lateral resolution was improved [60].

In other designs, leakage radiation microscope (LRM) was used to get the direct image of SPCE [61]. In LRM, a high-NA objective was used to focus the exciting beams, so that the excitation area is significantly smaller than the lateral spread of the SP, and the leakage radiation can be observed (Fig. 3.6c). A clear SPCE ring can be observed with the Fourier image plane. Furthermore, the propagation of the surface plasmon polaritons (SPPs) can be imaged, which cannot be realized in the prism-based setup. For the fluorophores dispersed in a planar thin film with random orientations, SPCE was observed as a circular ring of uniform intensity. Using SPCE-LRM, it was found that SPCE ring displayed different intensity patterns with different dielectric shapes, and the intensity distribution was affected by the illumination location on the shaped films, which offers a new way to control the collection efficiency at imaging and sensing [62].

The platforms of supercritical angle fluorescence (SAF) can also be modified for SPCE [63]. The transparent paraboloid polymer elements coated with thin metal layer was used with the cooperation of the annular aperture to ensure the defined excitation and the selective observation of SPCE as the KR configuration in inverted microscopic system (Fig. 3.6d). It can be performed as arrays on biochip and work better in bimetal (silver/gold) layer [64]. The geometry of the paraboloid biochip element matches the SPR excitation and the SPCE emission, hence establishing the feasibility of combining the advantages of SPCE with enhanced SAF light collection to produce the paraboloid SPCE array biochip. The key features of SPCE, namely enhanced plasmon excitation, high directionality of emission and surface selectivity, have the potential to be incorporated in a disposable array biochip for applications in high-throughput analysis of biomolecular interactions.

3.4.2 Point Spread Function of SPCE Microscope

The image formation processes of SPCE microscopy have been investigated, and the point spread function (PSF)engineering approaches were performed to evaluate the resolution in imaging. Both theoretical analysis and experiment results indicate

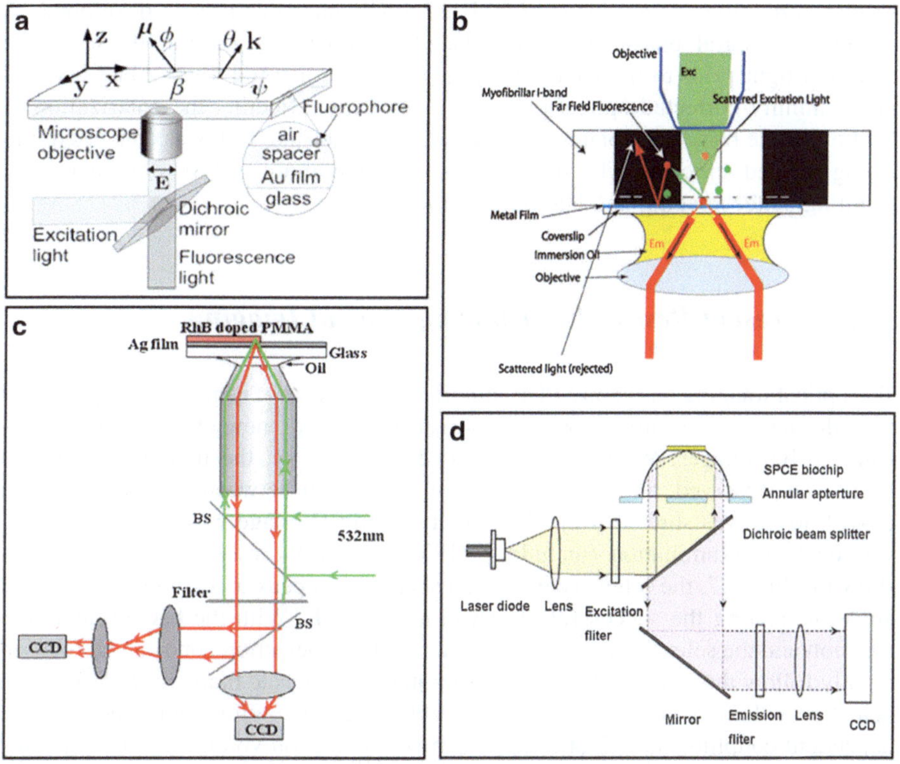

Fig. 3.6 Sketches of the experimental setups of (**a**) KR-SPCE imaging [57], (**b**) RK-SPCE imaging [59], (**c**) SPCE imaging by leakage radiation microscopy [61], and (**d**) SPCE imaging based on paraboloid polymer elements in supercritical angle fluorescence [63]

that the PSF of SPCE microscopy has a rotationally symmetric annular-like structure and a valley in the central region [65]. It differs from the PSFs of conventional wide-field microscopy and confocal microscopy, which typically assume an airy-disc structure with the resolution defined using the Rayleigh criterion. So it becomes hard to clearly distinguish how the resolution should be defined and what is its relation to wavelength and numerical aperture in SPCE microscopy. It suggests that caution should be carried when interpreting images captured by this method.

Fortunately, some means have been proposed as the response. A nonlinear deconvolution algorithm can be applied to convert the doughnut-shape PSF into a more-conventional Airy disk shape PSF. Based on the numerical model, the deconvolved image is clear, and the standing-wave SPCE can even achieve the lateral resolution approaching 100 nm [60]. But there are still some limitations for the numerical method, such as the inadvertent enhancement of noise artifacts, and the requirement for prior measurement of the system PSF, leading to practical difficulties especially in biological experiments. Therefore, a novel method with optical correction of the distorted PSF of SPCE microscopy was developed. A single-order spiral phase plate was placed to superpose a spiral phase to the emission light prior to the CCD imaging [66]. With a spiral phase plate inserted in the infinity-corrected optical path of the detection affecting the polarization of the field at the back focal plane of objective, the donut-like PSF was converted into a single lobed structure. The full-width at half-maximum of PSF was decreased, resulting better imaging and lateral resolution.

3.4.3 Present Researches Utilizing Optical Imaging

The optical imaging based on SPCE integrates the effects of near-field coupling and the information from microcosmic observation, creating new channels to investigate the bio-images. With the aid of SPCE microscopy, the imaging of muscle fibrils was observed, and the dynamics of interaction between actin and myosin cross-bridges was studied [67]. The quality of SPCE image was superior to conventional epi-illumination (including TIRF and epi-fluorescence excitation). As shown in Fig. 3.7, the intensity increased sharply to a peak as varying the incident angle, indicating the successful observation of SPCE. With the background suppression and the selective excitation in near field, the detection zone can be focused to only follow the motion of 12 actin promoters in a muscle filament. In SPCE, the detection volume is a product of evanescent wave penetration depth and distance dependent coupling, so 1–2 attoliter minimized detection volume can be expected supposing 50 nm effective coupling depth and 200 nm lateral diffraction limit.

The shallow detection volume of SPCE is just convenient to the study of molecular dynamics based on fluorescence correlation spectroscopy (FCS). Moreover, the background is very low and the changes of signals upon axial movement of molecules are more observable with the gradient of evanescent wave intensity and the quenching effect of the metal surface. As a result, it runs well to study the molecular diffusion with microscopic SPCE-FCS [68]. Based on the microscopic RK-SPCE-FCS, the kinetics of a single cross-bridge in familial hypertrophic cardio-myopathy heart muscle was measured through monitoring the fluorescently labeled actin with directional and *p*-polarized radiation [59]. The statistically significant differences between wild and mutated heart muscle were observed. The rate of change of orientation is significantly faster in contracting cardiac myofibrils of transgenic mice than the wild type, which is consistent with the fact that mutated heart muscle myosin translates actin faster than the wild type in the motility assays *in vitro*.

Two-photon excitation was also induced to the imaging process. The colorful multiple SPCE circles were measured with highly collective efficiency through optimizing the silver thickness, surface smoothness, the protective dielectric spacer [69]. The separation of incident and emission angle is large in two-photon excited SPCE, which made the SPCE collection easier and more effective than that with one-photon excitation. Besides, two-photon excitation imaging takes the

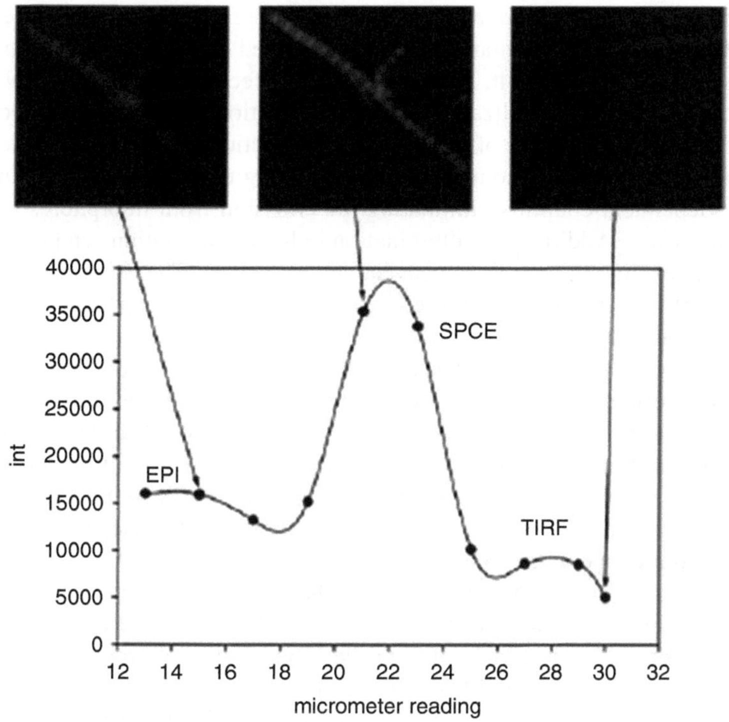

Fig. 3.7 SPCE signal from skeletal myofibrils as a function of the illumination angle. The *top panel* shows photographs taken from the microscope for various illumination angles [67]

advantages of suppressing background signal, reducing photodamage and extending detection depth. Therefore, it is believed that the two-photon excited SPCE will be an effective imaging technique towards biological systems.

3.5 The Applications of SPCE in Analysis

The great demands for reliable and sensitive detection of various objects, especially biological entities of complexity, have fueled up the development of analytical research. SPCE can be utilized in broad applications, including nucleic acid sensing, protein sensing and other chemical sensing.

3.5.1 DNA Sensing

The DNA hybridization was measured foremost between fluorophore-labeled DNA targets and surface-bound capture oligomers in SPCE [70]. If the fluorophore-labeled DNA oligomers were complementary, they localized near the metal surface after hybridization, and a dramatic increase of fluorescence could be observed. In contrast, the non-complementary fluorophore-labeled DNA oligomers did not cause changes in SPCE. As a result, SPCE has shown effective detection of DNA based on the change in probe localization upon hybridization inducing efficient coupling of metal. Another advantage of SPCE in DNA detection is the effective rejection of background coming from the molecules not binding to the surface. It is due to the distance-dependent coupling eliminating the emission from fluorphores more distant from surface. Additionally, illumination in KR configuration can preferentially localize the excitation to regions near the metal surface. The evanescent field was further amplified and resulted in the magnitude of signal much larger than that in RK configuration [7].

Microwave-accelerated surface plasmon-coupled directional emission (MA-SPCE) offers an alternative approach to current DNA based detection, satisfying the requirements of speed and sensitivity [50]. The hybridization of fluorophore-labeled target oligo with anchor probe on the gold films under microwave heating resulted in obvious fluorescence in MA-SPCE within 1 min, comparing to the fact that the reaction usually needs about 4 h to reach the comparative signal at room temperature incubation in SPCE. The high speed in DNA sensing is contributed to the design of small gold disk that will not be affected with dielectric breakdown using low power microwaves, contrarily creating a temperature gradient along the disk, which in turn results in the larger influx of complementary DNA towards an anchor probe-modified surface. The detection was directly finished in whole blood samples without extra separation steps. The DNA was not perturbed under low power microwaves, providing for identical fluorescence intensities after melting and re-hybridization, thus ideal reversibility.

To improve the sensitivity and selectivity in DNA hybridization assays, we have developed a DNA sensor with label-hairpin DNA as capture in electric field assisted surface plasmon coupled directional emission (E-SPCDE) system for the highly sensitive detection of label-free DNA target [56]. 'Hairpin' DNA probes labeled with fluorescent dyes were attached to a gold surface, and then exposed to a solution of single-stranded DNA chains. If the strand has a sequence that matches the probe, the hairpin will be unfolded and form a twinned duplex that will position the fluorescent dye a dozen nanometers away from the surface. Therefore, the dyes should strongly interact with oscillating surface plasmons, and fire off an amplified fluorescent signal. But, in practice, DNA duplexes anchored to the surface may form a slanted or random orientation, leading to the labeled fluorophores close to the surface, which limits the efficiency of enhancement and results in potential signal deviation with the unpredictable orientation. In addition, the duplex formation involving a single base mismatch may also enhance the fluorescence signal in SPCDE, resulting in false positive detection. In our work, we found that E-SPCDE could be used to resolve the above problems effectively. With the assistance of appropriate negative potential, the hybrids with matched DNA stood straight up in a uniform way, ensuring all the labeled fluorophores locate in the enhancement zone to the largest extent. On the other hand, most fluorophores were still close to the surface in the mismatched situation, because the mismatched duplexes were less stable and the reaction was hindered at the repulsive potential.

We found that −400 mv (vs. AgCl/Ag) was the appropriate potential for E-SPCDE. At this potential, the matched duplexes were pulled straight; extending labeled fluorophores enhanced coupling effect sufficiently, while the mismatched hybridization was prevented. By actively modulating the conditions locally at interface with the synergistic effect of intelligently amplifying the right signal and suppressing the wrong signal, E-SPCDE has shown a successful application in sensitive DNA sensing with high discriminatory capacity up to 20-fold for a single base mutation in a large concentration range (Fig. 3.8). This technique is easy to be applied to a variety of distance-sensitive surface assays. It has the potential to create a new generation of miniaturized high-performance sensing platforms, especially in chip-based microarray assays.

3.5.2 Aptamers-Based Protein Sensing

Aptamers are single-stranded nucleic acids possessing unique binding characteristics to their targets. Numerous high-affinity and highly specific aptamers have been selected against a wide variety of target molecules including proteins. The configurations of aptamers are specific for the process of binding targets [71]. Considering the conformational changes of aptamers before and after binding proteins, which would cause distance changes of labeled fluorophores, the protein assays based on the distance-dependent coupling effect in SPCDE can be expected. We designed a molecular beacon aptamer with fluorophore labeled on the end. In the absence of the target thrombin, the immobilized aptamer formed the stem, and thereby the fluorophore was brought in close proximity to the metal, resulting in the quenching of fluorescence and no signal was detected by SPCDE. Upon thrombin binding, the distance between the fluorophore and the gold surface increased due to the conformational change of aptamer (Fig. 3.9a) [72]. A detail that should be concerned is that sometimes the change of DNA conformation may not be large enough to extend from quenching zone to enhancement zone. In this case, the conformational change after binding target was just about 2 nm, while the quenching effect dominates around 5 nm near the surface. Therefore, we subtly designed a monomolecular layer of avidin to link the aptamer, warranting the labeled fluorophore placed on the critical zone between quenching and coupling. Consequently, a sensitive response was obtained after binding to target with the conformational change up to the enhancement zone totally. The fluorescence was sharply distributed at a direction, and the SPCDE signal was about six-fold greater than the spontaneous free space emission with undesired scattering of light from the light source suppressed (Fig. 3.9b, c). It proved that the combination of quenching and coupling for sensing purpose was effective and successful. The design is promising in developing efficient biosensors based on conformational-switching signaling aptamers.

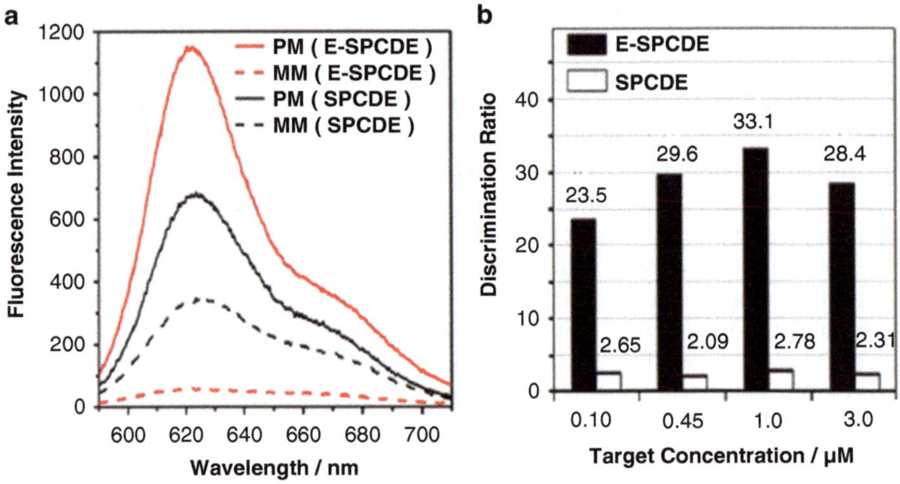

Fig. 3.8 (**a**) SPCDE and E-SPCDE fluorescence spectra of DNA hairpin responding to 3.0 μM perfect matches (PM) and single base mismatches. (**b**) Discrimination ratios with target concentrations ranging from 0.10 to 3.0 μM PM measured in SPCDE and E-SPCDE [56]

3.5.3 Immunological Detection

Immunoassay based on fluorescence is one important approach to highly sensitive detection of biomarkers through the specific recognition of antigens with antibodies. SPCE provides increased sensitivity and background rejection, enhancing the capability of detection in immunoassays. In SPCE, the successful affinity assays have been reported in the format of antigen-reporter antibody (fluorophore labeled) and the sandwich assay of capture antibody-antigen-reporter antibody (fluorophore labeled) [73–75]. The protein interactions near the interface can be sensitively monitored with the coupling effect. Moreover, the protein layer posed on the interface effectively acted as a spacer layer as the average protein dimensions can be taken as a few nanometers, which natively eliminated the quenching effect [76]. After reaction and the critical rinsing, the directional and polarized emission was observed with the measured intensities indicating the different concentrations of antigens, while the binding kinetic was also able to be detected in the defined direction, because only the fluorophores labeled to antibodies bound to the antigens on the surface can be coupled with surface plasmons. The background of the unbound reagents and the autofluorecence from sample were suppressed due to being distant from surface, and the optically dense media in sample matrixes such as serum or whole blood cannot attenuate the SPCE signal in assays [77], providing the possibility of performing immunoassays without separation steps to complex sample. If microwave is applied, a fast and sensitive bioassay can be expected. Using a low power microwave heating, a 30-fold increase in assay kinetics compared to the identical bioassay were achieved [48].

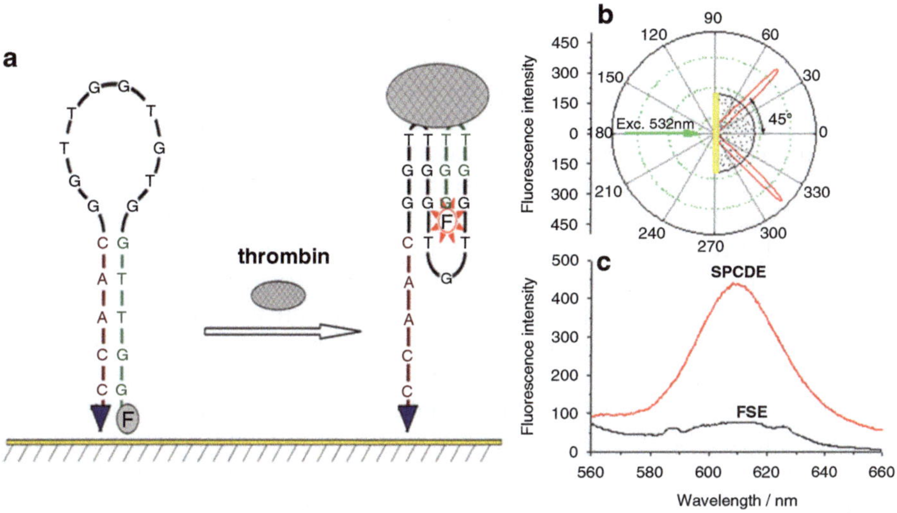

Fig. 3.9 (**a**) The working principle of SPCDE based on a conformational switching signaling aptamer. When the target thrombin is added, the fluorophore is displaced into the enhancement zone. (**b**) The angular distribution of signaling aptamer after the addition of thrombin. (**c**) The spectra of SPCDE and FSE after the addition of thrombin [72]

According to the directional emission in silver films, the angle at which the radiation propagates through the prism depends on the surface plasmon angle for the relevant wavelength. So, it will allow the measurement of multiple analytes using multiple emission wavelengths that will be detected at individual angle. Model experiments have been taken to demonstrate the multi-wavelength immunoassays in SPCE [78]. The antibodies were labeled with two fluorophores radiating at 595 and 665 nm. These antibodies were directed against antigen proteins bound to the silver surface. The emission from both labeled antibodies was strongly directional at different angles on the prism. The emission from 595 nm peaked at 71° and that from 665 nm at 68°. The emission from each labeled antibody occurred at a different angle on the glass prism, allowing independent measurement of surface binding of each antibody.

To develop the parallel assay platform for rapid and high-throughput analysis, array and even microarray based biochip is a promising technology. Actually, researchers have performed the relevant investigation in SPCE [63]. Arrays of transparent paraboloid polymer elements were coated with a thin gold layer to facilitate SPCE in the modified system of supercritical angle fluorescence (SAF). As a model assay, the sandwich assay for human IgG was carried out on a gold-coated 3×3 array biochip. The response was observed in the concentrations ranging from 2 to 200 μg/ml with a limit of detection of 20 ng/ml. The limit of detection reduced to 10 pg/ml using optimum bimetallic layer with the composition of 36 nm silver and 10 nm gold [64].

3.5.4 Other Fluorescence Analysis

Based on SPCE, other chemical fluorescence analysis can also be operated very well. Owing to the high collection efficiency in SPCE, single molecule detection has been achieved [52]. The simplest way to detect fluorescent molecules is to lay the objects near the metal surface, such as by spin-coating or dropping [79, 80], and the SPCE signal can be obtained. The thin spacers including silicon dioxide, LB films or polyelectrolyte had better be engineered between fluorophores and metal surface to protect the metal surface and minimize the metal quenching effects [32, 40, 63]. The special properties of SPCE present more information in detection.

The strong wavelength dependence of surface plasmon coupling at specific angles of observation offers a great potential to investigate complex systems. The mixture of monomer, dimer, and higher order aggregates of Rhodamine-6G were resolved utilizing SPCE [81]. Fluorescence emission from closely located protonated, deprotonated, and excimer species of 8-hydroxy-1,3,6-pyrene trisulfonic acid coupling into surface plasmons were easily separated and observed with 11–14 fold intensity enhancement [82].

Fluorescence resonance energy transfer efficiency were improved and relative transfer rate can be increased by two-fold in the presence of continuous silver films [83]. It was observed that the rise time of acceptor fluorescence intensity upon donor excitation was 10 times shorter in the presence of SPCE and acceptor emission in SPCE was totally linearly polarized [84].

The orientation of the transition dipole momentum of fluorophores can be detected in SPCE, and the degree of horizontal or vertical orientation can be determined through the comparison with simulations [85]. Because the wavevector of surface plasmon parallel to the molecular orientation takes a larger value than that perpendicular to the molecular orientation, the spectral peak for the parallel configuration is located at a longer wavelength than that for the perpendicular configuration at the same observation angle [86].

Besides the fluorophores directly detected in SPCE, they can also be designed to behave as the sensing layer to reflect the stimulants outside. In a report on oxygen sensing [87], the ruthenium fluorophore was electrostatically attached to the SPCE surface as an ultra-thin sensing layer for sensing. The response of the probes is due to collisional quenching by oxygen. Both fluorescence intensity and lifetime were detected in this work for oxygen sensing. The responses for the two measurements were sensitive in SPCE and Lifetime-based sensing was more reproducible than the intensity-based measurement, while the lifetime of the probes under free space emission could not be resolved due to the weak signal from the thin sensing layer. It proved that SPCE-based fluorescence and lifetime sensing measurements could lead to the development of chemical sensing devices with low cost but high sensitivity.

3.6 Conclusions

The combination of fluorescence with surface plasmon resonance has resulted in a paradigm shift for both techniques. There is a steady growth in terms of instruments, devices, and sensors based on SPCE. The processes described for fluorescence emission are readily applicable to other sources of molecular luminescence, such as phosphorescence, chemiluminescence and electrochemiluminescence. The optical geometry of SPCE also attracts attentions from large scopes. For example, stimulated emission has been investigated in the similar device for potential laser design [88], the directional surface enhanced Raman scattering has been observed [89], and enhanced scanning tunneling microscopic light emission has been obtained through the prism at a defined angle [90]. So SPCE creates the platform for the intensive investigation of nanoscale optics via plasmon coupling.

However, it is worth noting the complexity of fluorophore-metal interactions, which involves both near- and far-field effects. The coupling between fluorophores and nanostructures with complex geometry is lacking of enough understanding [52, 91]. As a result, there is still a long way to go, demanding arduous exploring both in fundamentals and applications. The introduction of novel nano-materials, nano-structures, and nano-optical devices into SPCE would further improve the analytical merits of SPCE and expedite the design of low-cost and high-performance sensors. SPCE based imaging will be an emerging tendency. Integrating the unique properties of the directional emission into microarray imaging would reduce the crosstalk interference and improve the imaging quality [92]. It is no doubt that the development of SPCE platforms will redound to the field of chip-based biomolecular detection and significantly further the progress in analytical researches.

Acknowledgements We would like to acknowledge the fund supports from the NSFC (21375111, 21127005, 20975084), the 973 Program of China (2013CB933703, 2007CB935600) and the Ministry of Education of China (20110121110011, PCSIRTIRT13036). We thank Dr. Yu-Luan Chen for critical reading of this manuscript.

References

1. Lakowicz JR, Malicka J, Gryczynski I, Gryczynski Z (2003) Directional surface plasmon-coupled emission: a new method for high sensitivity detection. Biochem Biophys Res Commun 307:435–439
2. Lakowicz JR (2005) Radiative decay engineering 5: metal-enhanced fluorescence and plasmon emission. Anal Biochem 337:171–194
3. Geddes CD, Gryczynski I, Malicka J, Gryczynski Z, Lakowicz JR (2004) Directional surface plasmon coupled emission. J Fluoresc 14:119–123
4. Lakowicz JR (2004) Radiative decay engineering 3. Surface plasmon-coupled directional emission. Anal Biochem 324:153–169

5. Gryczynski I, Malicka J, Gryczynski Z, Lakowicz JR (2004) Radiative decay engineering 4. Experimental studies of surface plasmon-coupled directional emission. Anal Biochem 324:170–182
6. Neumann T, Johansson ML, Kambhampati D, Knoll W (2002) Surface-plasmon fluorescence spectroscopy. Adv Funct Mater 12:575–586
7. Joanna Malicka IG, Gryczynski Z, Lakowicz JR (2004) Use of surface plasmon-coupled emission to measure DNA hybridization. J Biomol Screen 9:208–215
8. Smith DS, Kostov Y, Rao G, Gryczynski I, Malicka J, Gryczynski Z, Lakowicz JR (2005) First observation of surface plasmon-coupled emission due to LED excitation. J Fluoresc 15:895–900
9. Cao SH, Xie TT, Cai WP, Li YQ (2009) Observation of surface plasmon-coupled directional emission at a fixed angle using thin iron films. The 13th Beijing conference and exhibition on instrumental analysis (BCEIA 2009), C52
10. Smith DS, Kostov Y, Rao G (2008) Signal enhancement of surface plasmon-coupled directional emission by a conical mirror. Appl Opt 47:5229–5234
11. Li YQ, Xie TT, Cai WP (2009) Surface plasmon coupled fluorescence detection device. Patent. CN ZL200910111882.8
12. Hung YJ, Smolyaninov II, Davis CC, Wu HC (2006) Fluorescence enhancement by surface gratings. Opt Express 14:10825–10830
13. Chowdhury MH, Malyn SN, Aslan K, Lakowicz JR, Geddes CD (2007) First observation of surface plasmon-coupled chemiluminescence (SPCC). Chem Phys Lett 435:114–118
14. Zhang J, Gryczynski Z, Lakowicz JR (2004) First observation of surface plasmon-coupled electrochemiluminescence. Chem Phys Lett 393:483–487
15. Previte MJR, Aslan K, Zhang YX, Geddes CD (2006) Surface plasmon coupled phosphorescence (SPCP). Chem Phys Lett 432:610–615
16. Gryczynski I, Malicka J, Lakowicz JR, Goldys EM, Calander N, Gryczynski Z (2005) Directional two-photon induced surface plasmon-coupled emission. Thin Solid Films 491:173–176
17. Mondal PP, Gilbert RJ, So PTC (2008) Plasmon enhanced fluorescence microscopy below quantum noise limit with reduced photobleaching effect. Appl Phys Lett 93:093901
18. Gryczynski I, Malicka J, Gryczynski Z, Lakowicz JR (2004) Surface plasmon-coupled emission with gold films. J Phys Chem B 108:12568–12574
19. Cao SH, Xie TT, Cai WP, Li YQ (2010) Observation of surface plasmon-coupled directional fluorescence using thin iron films. Chem J Chin Univ 31:61–63
20. Calander N (2004) Theory and simulation of surface plasmon-coupled directional emission from fluorophores at planar structures. Anal Chem 76:2168–2173
21. Hiep HM, Fujii M, Hayashi S (2007) Effects of molecular orientation on surface-plasmon-coupled emission patterns. Appl Phys Lett 91:183110
22. Penninck L, Mladenowski S, Neyts K (2010) The effects of planar metallic interfaces on the radiation of nearby electrical dipoles. J Opt 12:075001
23. Gryczynski I, Malicka J, Nowaczyk K, Gryczynski Z, Lakowicz JR (2004) Effects of sample thickness on the optical properties of surface plasmon-coupled emission. J Phys Chem B 108:12073–12083
24. Gryczynski I, Malicka J, Nowaczyk K, Gryczynski Z, Lakowicz JR (2006) Waveguide-modulated surface plasmon-coupled emission of Nile blue in poly(vinyl alcohol) thin films. Thin Solid Films 510:15–20
25. Salamon Z, Macleod HA, Tollin G (1997) Coupled plasmon-waveguide resonators: a new spectroscopic tool for probing proteolipid film structure and properties. Biophys J 73:2791–2797
26. Salamon Z, Tollin G (2001) Optical anisotropy in lipid bilayer membranes: coupled plasmon-waveguide resonance measurements of molecular orientation, polarizability, and shape. Biophys J 80:1557–1567

27. Calander N (2005) Surface plasmon-coupled emission and Fabry-Perot resonance in the sample layer: a theoretical approach. J Phys Chem B 109:13957–13963
28. Barnes WL (1998) Fluorescence near interfaces: the role of photonic mode density. J Mod Opt 45:661–699
29. Hellen EH, Axelrod D (1987) Fluorescence emission at dielectric and metal-film interfaces. J Opt Soc Am B 4:337–350
30. Weber WH, Eagen CF (1979) Energy transfer from an excited dye molecule to the surface plasmons of an adjacent metal. Opt Lett 4:236–238
31. Ford GW, Weber WH (1984) Electromagnetic interactions of molecules with metal surfaces. Phys Rep 113:195–287
32. Ray K, Szmacinski H, Enderlein J, Lakowicz JR (2007) Distance dependence of surface plasmon-coupled emission observed using Langmuir-Blodgett films. Appl Phys Lett 90:251116
33. Wang YK, Yang TY, Pourmand M, Miller JJ, Tuominen MT, Achermann M (2010) Time-resolved surface plasmon polariton coupled exciton and biexciton emission. Opt Express 18:15560–15568
34. Willets KA, Van Duyne RP (2007) Localized surface plasmon resonance spectroscopy and sensing. Annu Rev Phys Chem 58:267–297
35. Sun YG, Xia YN (2003) Gold and silver nanoparticles: a class of chromophores with colors tunable in the range from 400 to 750 nm. Analyst 128:686–691
36. Tian ZQ, Ren B, Li JF, Yang ZL (2007) Expanding generality of surface-enhanced Raman spectroscopy with borrowing SERS activity strategy. Chem Commun 43:3514–3534
37. Aslan K, Geddes CD (2009) Directional surface plasmon coupled luminescence for analytical sensing applications: which metal, what wavelength, what observation angle? Anal Chem 81:6913–6922
38. Previte MJR, Zhang YX, Aslan K, Geddes CD (2007) Surface plasmon coupled fluorescence from copper substrates. Appl Phys Lett 91:151902
39. Gryczynski I, Malicka J, Gryczynski Z, Nowaczyk K, Lakowicz JR (2004) Ultraviolet surface plasmon-coupled emission using thin aluminum films. Anal Chem 76:4076–4081
40. Asian K, Previte MJR, Zhang YX, Geddes CD (2008) Surface plasmon coupled fluorescence in the ultraviolet and visible spectral regions using zinc thin films. Anal Chem 80:7304–7312
41. Aslan K, Geddes CD (2009) Surface plasmon coupled chemiluminescence from zinc substrates: directional chemiluminescence. Appl Phys Lett 94:073104
42. Aslan K, Weisenberg M, Hortle E, Geddes CD (2009) Surface plasmon coupled chemiluminescence from iron thin films: directional and approaching fixed angle observation. J Appl Phys 106:014131
43. Aslan K, Zhang YX, Geddes CD (2009) Directional, broad, and fixed angle surface plasmon coupled fluorescence from iron thin films. J Phys Chem C 113:20535–20538
44. Aslan K, Zhang YX, Geddes CD (2009) Surface plasmon coupled fluorescence in the visible to near-infrared spectral regions using thin nickel films: application to whole blood assays. Anal Chem 81:3801–3808
45. Ray K, Chowdhury MH, Lakowicz JR (2008) Observation of surface plasmon-coupled emission using thin platinum films. Chem Phys Lett 465:92–95
46. Aslan K, Weisenberg M, Hortle E, Geddes CD (2009) Fixed-angle observation of surface plasmon coupled chemiluminescence from palladium thin films. Appl Phys Lett 95:123117
47. Weisenberg M, Aslan K, Hortle E, Geddes CD (2009) Directional surface plasmon coupled chemiluminescence from nickel thin films: fixed angle observation. Chem Phys Lett 473:120–125
48. Aslan K, Malyn SN, Geddes CD (2007) Microwave-accelerated surface plasmon-coupled directional luminescence: application to fast and sensitive assays in buffer, human serum and whole blood. J Immunol Methods 323:55–64
49. Aslan K, Geddes CD (2005) Microwave-accelerated metal-enhanced fluorescence: platform technology for ultrafast and ultrabright assays. Anal Chem 77:8057–8067

50. Aslan K, Previte MJR, Zhang Y, Geddes CD (2008) Microwave-accelerated surface plasmon-coupled directional luminescence 2: a platform technology for ultra fast and sensitive target DNA detection in whole blood. J Immunol Methods 331:103–113
51. Previte MJR, Geddes CD (2007) Microwave-triggered surface plasmon coupled chemiluminescence. J Am Chem Soc 129:9850–9851
52. Lakowicz JR, Ray K, Chowdhury M, Szmacinski H, Fu Y, Zhang J, Nowaczyk K (2008) Plasmon-controlled fluorescence: a new paradigm in fluorescence spectroscopy. Analyst 133:1308–1346
53. Previte MJR, Aslan K, Zhang YX, Geddes CD (2007) Metal-enhanced surface plasmon-coupled phosphorescence. J Phys Chem C 111:6051–6059
54. Chowdhury MH, Ray K, Geddes CD, Lakowicz JR (2008) Use of silver nanoparticles to enhance surface plasmon-coupled emission (SPCE). Chem Phys Lett 452:162–167
55. Aslan K, McDonald K, Previte MJR, Zhang YX, Geddes CD (2008) Silver island nanodeposits to enhance surface plasmon coupled fluorescence from copper thin films. Chem Phys Lett 464:216–219
56. Cao SH, Xie TT, Cai WP, Liu Q, Li YQ (2011) Electric field assisted surface plasmon-coupled directional emission: an active strategy on enhancing sensitivity for DNA sensing and efficient discrimination of single base mutation. J Am Chem Soc 133:1787–1789
57. Stefani FD, Vasilev K, Bocchio N, Stoyanova N, Kreiter M (2005) Surface-plasmon-mediated single-molecule fluorescence through a thin metallic film. Phys Rev Lett 94
58. Gryczynski Z, Borejdo J, Calander N, Matveeva EG, Gryczynski I (2006) Minimization of detection volume by surface-plasmon-coupled emission. Anal Biochem 356:125–131
59. Mettikolla P, Calander N, Luchowski R, Gryczynski I, Gryczynski Z, Borejdo J (2010) Kinetics of a single cross-bridge in familial hypertrophic cardiomyopathy heart muscle measured by reverse Kretschmann fluorescence. J Biomed Opt 15:017011
60. Chung E, Kim YH, Tang WT, Sheppard CJR, So PTC (2009) Wide-field extended-resolution fluorescence microscopy with standing surface-plasmon-resonance waves. Opt Lett 34:2366–2368
61. Zhang DG, Yuan XC, Bouhelier A (2010) Direct image of surface-plasmon-coupled emission by leakage radiation microscopy. Appl Opt 49:875–879
62. Zhang DG, Moh KJ, Yuan XC (2010) Surface plasmon-coupled emission from shaped PMMA films doped with fluorescence molecules. Opt Express 18:12185–12190
63. Yuk JS, Trnavsky M, McDonagh C, MacCraith BD (2010) Surface plasmon-coupled emission (SPCE)-based immunoassay using a novel paraboloid array biochip. Biosens Bioelectron 25:1344–1349
64. Yuk JS, MacCraith BD, McDonagh C (2011) Signal enhancement of surface plasmon-coupled emission (SPCE) with the evanescent field of surface plasmons on a bimetallic paraboloid biochip. Biosens Bioelectron 26:3213–3218
65. Tang WT, Chung E, Kim YH, So PTC, Sheppard CJR (2007) Investigation of the point spread function of surface plasmon-coupled emission microscopy. Opt Express 15:4634–4646
66. Tang WT, Chung E, Kim YH, So PTC, Sheppard CJR (2010) Surface-plasmon-coupled emission microscopy with a spiral phase plate. Opt Lett 35:517–519
67. Borejdo J, Gryczynski Z, Calander N, Muthu P, Gryczynski I (2006) Application of surface plasmon coupled emission to study of muscle. Biophys J 91:2626–2635
68. Borejdo J, Calander N, Gryczynski Z, Gryczynski I (2006) Fluorescence correlation spectroscopy in surface plasmon coupled emission microscope. Opt Express 14:7878–7888
69. Kuo-Chih Chiu C-YL, Dong CY, Chen S-J (2011) Optimizing silver film for surface plasmon-coupled emission induced two-photon excited fluorescence imaging. Opt Express 19:5386–5396
70. Malicka J, Gryczynski I, Gryczynski Z, Lakowicz JR (2003) DNA hybridization using surface plasmon-coupled emission. Anal Chem 75:6629–6633
71. Cho EJ, Lee JW, Ellington AD (2009) Applications of aptamers as sensors. Annu Rev Anal Chem 2:241–264

72. Xie TT, Liu Q, Cai WP, Chen Z, Li YQ (2009) Surface plasmon-coupled directional emission based on a conformational-switching signaling aptamer. Chem Commun 45:3190–3192
73. Lakowicz JR, Malicka J, Matveeva E, Gryczynski I, Gryczynski Z (2005) Plasmonic technology: novel approach to ultrasensitive immunoassays. Clin Chem 51:1914–1922
74. Matveeva E, Gryczynski Z, Gryczynski I, Lakowicz JR (2004) Immunoassays based on directional surface plasmon-coupled emission. J Immunol Methods 286:133–140
75. Matveeva E, Gryczynski Z, Gryczynski I, Malicka J, Lakowicz JR (2004) Myoglobin immunoassay utilizing directional surface plasmon-coupled emission. Anal Chem 76:6287–6292
76. Yuk JS, McDonagh C, MacCraith BD (2010) Demonstration of a surface plasmon-coupled emission (SPCE)-based immunoassay in the absence of a spacer layer. Anal Bioanal Chem 398:1947–1954
77. Matveeva EG, Gryczynski I, Barnett A, Calander N, Gryczynski Z (2007) Red blood cells do not attenuate the SPCE fluorescence in surface assays. Anal Bioanal Chem 388:1127–1135
78. Matveeva E, Malicka J, Gryczynski I, Gryczynski Z, Lakowicz JR (2004) Multi-wavelength immunoassays using surface plasmon-coupled emission. Biochem Biophys Res Commun 313:721–726
79. Gryczynski I, Malicka J, Jiang W, Fischer H, Chan WCW, Gryczynski Z, Grudzinski W, Lakowicz JR (2005) Surface-plasmon-coupled emission of quantum dots. J Phys Chem B 109:1088–1093
80. Kostov Y, Smith DS, Tolosa L, Rao G, Gryczynski I, Gryczynski Z, Malicka J, Lakowicz JR (2005) Directional surface plasmon-coupled emission from a 3 nm green fluorescent protein monolayer. Biotechnol Prog 21:1731–1735
81. Sathish S, Kostov Y, Rao G (2009) High-resolution surface plasmon coupled resonant filter for monitoring of fluorescence emission from molecular multiplexes. Appl Phys Lett 94:223113
82. Sathish RS, Kostov Y, Rao G (2009) Spectral resolution of molecular ensembles under ambient conditions using surface plasmon coupled fluorescence emission. Appl Opt 48:5348–5353
83. Szmacinski H, Ray K, Lakowicz JR (2009) Effect of plasmonic nanostructures and nanofilms on fluorescence resonance energy transfer. J Biophoto 2:243–252
84. Jankowski D, Bojarski P, Kwiek P, Rangelowa-Jankowska S (2010) Donor-acceptor nonradiative energy transfer mediated by surface plasmons on ultrathin metallic films. Chem Phys 373:238–242
85. Frischeisen J, Yokoyama D, Adachi C, Brutting W (2010) Determination of molecular dipole orientation in doped fluorescent organic thin films by photoluminescence measurements. Appl Phys Lett 96:073302
86. Ghazali FAM, Fujii M, Hayashi S (2009) Anisotropic propagation of surface plasmon polaritons caused by oriented molecular overlayer. Appl Phys Lett 95:033303
87. Smith DS, Kostov Y, Rao G (2007) SPCE-based sensors: ultrafast oxygen sensing using surface plasmon-coupled emission from ruthenium probes. Sens Actuators B 127:432–440
88. Noginov MA, Zhu G, Mayy M, Ritzo BA, Noginova N, Podolskiy VA (2008) Stimulated emission of surface plasmon polaritons. Phys Rev Lett 101:226806
89. Meyer SA, Le Ru EC, Etchegoin PG (2011) Combining Surface Plasmon Resonance (SPR) Spectroscopy with Surface-Enhanced Raman Scattering (SERS). Anal Chem 83:2337–2344
90. Ahamed JU, Sanbongi T, Katano S, Uehara Y (2010) Prism-coupled scanning tunneling microscope light emission spectroscopy of Au film covered with self-assembled alkanethiol monolayer. Jpn J Appl Phys 49:08LB09
91. Fort E, Gresillon S (2008) Surface enhanced fluorescence. J Phys D Appl Phys 41:013001
92. Cai WP, Liu Q, Cao SH, Weng YH, Liu XQ, Li YQ (2012) Prism-based surface plasmon coupled emission imaging. ChemPhysChem 13: 3848-3851

Chapter 4
Fluorescence Analysis of Thermoresponsive Polymers

Cheryl Morris and Alan G. Ryder

Abstract The use of microscale thin polymer films is widespread in biomedical science and engineering, with applications in areas such as tissue engineering, drug delivery, microfluidic devices, bio-adhesion mediators, and bio-actuators. Much attention is devoted to the use of functional polymers that display stimuli-responsive behavior with the intention of providing "smart" coatings. One potential example is the use of thin thermoresponsive polymer films as drug eluting coatings on medical devices, where not only does the polymer function as a drug reservoir but it also acts as a biocompatibility modulator to improve device performance.

Often these thin polymer coatings have to be applied to complex geometries, which can cause problems for *in-situ* analysis. Another important consideration is the fact that these films have large surface area to mass ratios and thus water uptake can be significant. This is serious because coating stability, device efficacy, and long-term storage are influenced by the physiochemical properties of the polymer which are modulated by water content. Thus, there is a need for a rapid, non-contact, non-destructive, analytical method capable of analyzing thermoresponsive polymers in solution, and *in-situ* of the solid-state on medical devices. Fluorescence spectroscopy based methods can deal with both sample types and provide additional benefits in terms of high sensitivity and low probe concentrations, which provide for minimal sample disruption. This article gives a brief overview of the application of various fluorescence methods for the physicochemical characterization of thermoresponsive polymers such as poly (N-isopropylacrylamide), PNIPAm.

Keywords Fluorescence • Spectroscopy • Polymer • Thermoresponsive • Biomedical

---

C. Morris • A.G. Ryder (✉)
Nanoscale Biophotonics Laboratory, School of Chemistry, National University of Ireland Galway, Galway, Ireland
e-mail: alan.ryder@nuigalway.ie

© Springer International Publishing Switzerland 2016

C.D. Geddes (ed.), *Reviews in Fluorescence 2015*,
DOI 10.1007/978-3-319-24609-3_4

4.1 Introduction

The use of microscale thin polymer films is widespread in biomedical science and engineering, with applications in tissue engineering, drug delivery systems, microfluidic devices, bio-adhesion mediators and bio-actuators [1–14]. The choice of polymer for such applications is very important, and one area of significant interest has been the development functional polymers that display stimuli-responsive behavior with the intention of providing "smart" applications (Fig. 4.1) in the biomedical field [8, 15, 16]. One such potential use of thin thermoresponsive polymer films is as coatings on drug eluting coronary stents, where not only does the polymer function as a drug reservoir (providing anti-restenosis therapy) but it also acts as a biocompatibility modulator to improve device performance [5, 10, 17, 18].

In many applications, these polymer coatings are very thin (from μm to nm) and are formed on complex geometries, which may cause problems for *in-situ* analysis. Another important consideration is the fact that these films have large surface area to mass ratios and water uptake is an important factor to consider. This is serious because issues such as device manufacturing, coating stability, device efficacy, and long-term storage are influenced by the physiochemical properties of the polymer [19]. Thus, there is a need for the non-contact, non-destructive, analysis of these types of thermoresponsive polymers in solution and *in-situ* for fabricated films/devices. Optical spectroscopy, and in particular fluorescence based methods, which offer the combination of high sensitivity and low probe concentrations, provide the best solution for these analytical challenges.

This article gives a brief overview of the application of fluorescence methods to the characterization of thermoresponsive polymers and in particular poly (N-isopropylacrylamide), PNIPAm. It is not meant to provide a comprehensive or detailed review of the use of fluorescence but rather an insight into how various fluorescence methods can be employed for studying the physical and chemical properties of thermoresponsive polymers in a variety of states and forms.

4.2 Stimuli-Responsive Polymers

The design, synthesis, and development of functional polymers that respond to external stimuli is an area of significant interest [8, 15, 16, 20]. These synthetic polymer systems are often designed to mimic natural biopolymers, and a variety of functional forms have been developed to meet various specific biomedical and scientific applications [21]. These polymers are variously described as being "environment sensitive polymers" [22], "stimuli-responsive polymers" [23], "intelligent polymers" [24, 25], or "smart polymers" [15, 26]. "Smart" polymers can be defined as materials that undergo strong conformational changes in response to small changes in the surrounding environment [15, 21].

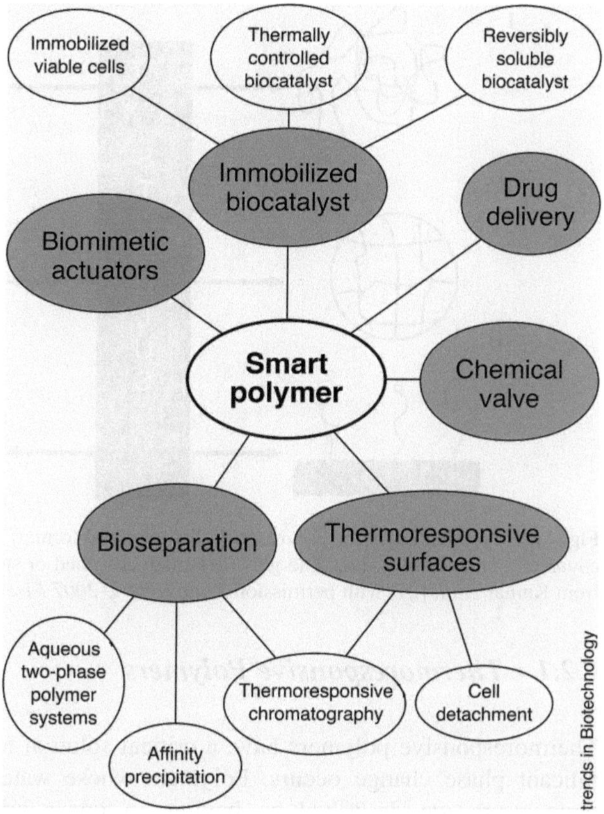

Fig. 4.1 Potential uses of stimuli-responsive polymers in biotechnology and medicine (Adapted with permission from Ref. [15]. *Copyright © 1999 Elsevier Science Ltd., all rights reserved*)

In many of the smart polymers used for biomedical applications, the sharp, large, reversible non-linear conformational changes can be attributed to the balance between hydrophilic and hydrophobic groups within the polymer system [8, 15, 16, 23, 27–29]. The consequent responses may be observed as changes in shape, solubility, and/or surface characteristics [21]. "Smart" polymers can be classified into three types depending on their physical forms (Fig. 4.2): *(i)* linear free chains in solution (following the application of an external stimulus the polymer undergoes a reversible collapse), *(ii)* covalently cross-linked gels and reversible or physical gels (for which swelling or shrinking behavior is environmentally triggered), and *(iii)* chain adsorbed or surface-grafted form (following a change of the external parameter, the polymer reversibly swells or collapses on a surface) [21].

The response of these "smart" polymers can be induced by a variety of environmental triggers, such as ionic strength, light, magnetic field, pH, electric field and temperature [1, 8, 15, 16, 20, 27, 29–37]. From a biomedical standpoint, the favored "smart" polymers are generally those sensitive to pH and/or temperature changes [30]. The most widely studied class of stimuli responsive polymers are thermoresponsive polymer systems; as the name suggests, these systems undergo conformational changes in response to temperature [16, 30].

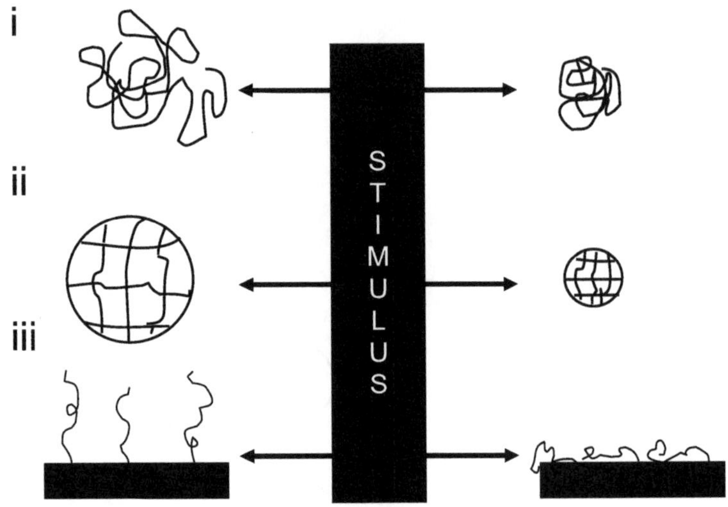

Fig. 4.2 Classification of the polymers by their physical form: (*i*) linear free chains in solution (*ii*) covalently cross-linked reversible gels (*iii*) chain adsorbed or surface grafted form (Reproduced from Kumar *et al.* [21] with permission. *Copyright © 2007 Elsevier Ltd, all rights reserved*)

4.2.1 Thermoresponsive Polymers

Thermoresponsive polymers have a critical solution temperature, at which a significant phase change occurs. Polymers whose water solubility increases with temperature are described as having an upper critical solution temperature (UCST) or a higher critical solution temperature (HCST) and solutions of these polymers appear biphasic below this critical temperature. Conversely, polymers for which the solubility decreases with increasing temperature often have a lower critical solution temperature (LCST) where the solutions transition to a biphasic state. This occurs because the polymer becomes less solvated as the temperature increases [16, 22, 30, 38]. Below the LCST, the polymer is soluble in aqueous solutions due to the domination of hydrophilic interactions (*i.e.* hydrogen bonding between the polymer and water) over hydrophobic (intramolecular) interactions, and thus it typically assumes a relaxed coil-like conformation. Raising the temperature above the LCST results in an increased dominance of the hydrophobic interactions, causing the collapse/contraction of the polymer. This leads to the adoption of a more globule-like conformation which minimizes the polymer-water contact and can eventually lead to precipitation from solution [30, 39, 40].

In the case of polymer solutions with a UCST, the entropy of mixing is usually large and positive but is dominated by enthalpic contributions at low temperatures. When the temperature increases, the entropic contribution increases, and eventually surpasses the enthalpic contribution at the UCST, resulting in a negative Gibb's free energy. Therefore in these polymer systems, higher temperatures enhance solubility [38]. For polymer solutions possessing a LCST, H-bonding between polymer polar groups and water molecules are the driving force for solvation at low temperatures, resulting in a large, dominant, negative enthalpy of mixing. In this state, the polymer is ordered, leading to an unfavorable negative entropy contribution, but overall the system is stable in this mixed form below its LCST due to the large enthalpic contribution. It has been suggested that the phase separation at the LCST of a polymer can be attributed to entropic effects [41]. At higher temperatures the contribution from entropy (displacement of water from the polymer matrix) surpasses the exothermic enthalpy contribution from hydrogen bonding between polar groups in the polymer and water molecules [8, 27, 30, 41, 42]. It is this balance between entropy and enthalpy that causes the polymer to assume a more hydrophobic state above the LCST.

A list of popular thermoresponsive polymer systems with their corresponding LCSTs is given in Table 4.1 PNIPAm and PVCL exhibit LCST behavior in a physiologically relevant temperature range rendering them viable for many biomedical applications [21, 40, 41]. Often these polymers are used as core elements to synthesize co-polymers with specific thermoresponsive and other desired properties. For example, PEO and PPO have been used to fabricate block copolymers, which possess an inverse thermoresponsive behavior. These block copolymers are available commercially under the names *Pluronics®* and *Tetronics®*, many of which have been approved by the FDA and EPA for use in food, pharmaceuticals and agriculture [8, 21, 22]. Alternatively copolymerization with hydrophilic or hydrophobic co-monomers can be used to increase or decrease the LCST respectively [36]. It is also possible to modulate or fine tune the LCST by the addition of additives for example, the addition of sodium dodecylsulfate increases the LCST, whereas addition of sodium chloride has the opposite effect [43, 44].

4.2.2 Poly(N-IsoPropylAcrylamide), PNIPAm

Of the family of poly(N-substituted acrylamides), PNIPAm (Fig. 4.3) is probably the most widely known and studied. PNIPAm is a chemical isomer of poly-leucine, in that it has the polar peptide group in its side chain rather than in the hydrocarbon backbone [45]. In aqueous solution, PNIPAm has a LCST of 32 °C; at this temperature it undergoes a sharp and reversible coil-to-globule phase transition from a hydrophilic to a more hydrophobic state, forcing water from the matrix [3, 7, 14, 16, 27, 46–48]. This phenomenon occurs due to the domination of entropic over enthalpic effects as the temperature increases above the LCST [30, 42]. Below the LCST, PNIPAm chains exist in an extended coil conformation, and solvation is driven by the enthalpic gain from intermolecular hydrogen bonding between the PNIPAm chains and water molecules [16, 49]. Solvation is further encouraged by a type of hydrophobic hydration, where the water molecules surround the a-polar isopropyl entities in a cage-like structure [50]. As the temperature is increased towards the LCST, intramolecular hydrogen bonding between carboxyl and amide groups on the PNIPAm chains result in the interruption of hydrogen bonding of these groups with water molecules,

Table 4.1 LCSTs of some common thermoresponsive polymers [40]

Abbreviation	Name	LCST (°C)
PPO	poly(propylene oxide)	10–20
PNPAm	poly(N-n-propylacrylamide)	25
PNIPAm	poly(N-isopropylacrylamide)	32
PEPA	poly(ethoxypropylacrylamide)	~32
PVME	poly(vinyl methyl ether)	33.8
PIPOZ	poly(2-isopropyl-2-oxazoline)	~36
PVCL	poly(N-vinylcaprolactam)	38
HPC	Hydroxypropylcellulose	42
PBMEAm	poly(N,N-bis(2-methoxyethyl) acrylamide)	49
MC	Methylcellulose	50
PDMA	poly((2-dimethylamino)ethyl methacrylate)	50
PEOZ	poly(2-ethyl-2-oxazoline)	~62
PMPAm	poly(N-(3-methoxypropyl)acrylamide)	>60
EHEC	Ethyl(hydroxyethyl)cellulose	65
PEMA	poly(N,N-ethylmethylacrylamide)	70
PEA	poly(N-ethylacrylamide)	82

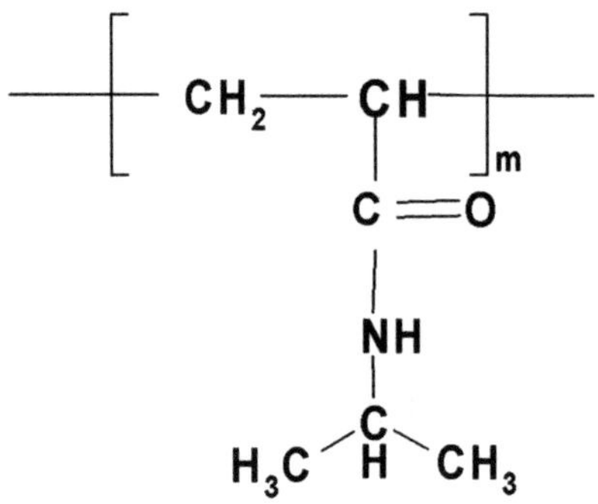

Fig. 4.3 Chemical structure of PNIPAm [22]

ultimately resulting in the chain adopting a collapsed conformation, driving out the water, and causing the polymer to precipitate out of solution [49]. These properties of PNIPAm and its copolymers make them applicable to a diverse range of pharmaceutical and biomedical applications [5, 13, 51, 52].

The thermoresponsive behavior of PNIPAm may be altered by the introduction of hydrophobic or hydrophilic groups into the polymer structure. By copolymerization with more hydrophobic monomers, like N-tBAAm (N-tert-butylacrylamide), the LCST of the polymer is shifted to lower temperatures while copolymerization with more hydrophilic monomers, like AAm (acrylamide), increases the LCST [26]. Rochev and co-workers have synthesized a series of copolymers based on N-isopropylacrylamide (NIPAm) and N-tBAAm. Increasing the amount of the hydrophobic monomer (from a mole ratio of 0 to 50 %) increases hydrophobicity and therefore lowers the LCST (from 33 to ~10 °C) [53]. It was also observed that cell adhesion, cell growth properties and drug elution from cast NIPAm/N-tBAAm copolymer films were all dependent on the copolymer composition [54, 55]. In contrast, by increasing the amount of a hydrophilic monomer AAm in the NIPAm/AAm copolymers (from a molar ratio of 100:0 to 75:25), the LCST is increased (from 33 to 47 °C) [56].

4.3 Polymer Characterisation by Fluorescence

The use of fluorescence spectroscopy for the evaluation of thermoresponsive polymers can provide useful insights into both the gross physical-dynamic processes and the more subtle physicochemical changes that occur in these polymer systems. Fluorescence spectroscopy offers high sensitivity for low probe loading and fast response times, thus minimizing perturbation of the polymer system. In thermoresponsive polymers like PNIPAm, the most common application of fluorescence measurements are for the study of micelle formation, aggregation dynamics, and the major phase changes that occur at the LCST. However, fluorophore choice needs to be very carefully considered because multiple factors (Fig. 4.4) are at play and the environment is considerably more complex than that encountered in solvents. This potential multi-factor environmental sensitivity may mean that the observed changes in absorption / fluorescence intensity, shifts in absorption / fluorescence spectra, anisotropy, or fluorescence lifetimes originate from a combination of effects [57–59]. For example, when studying the phase change behavior of thermoresponsive polymers one needs to ensure that the selected fluorophore does not have a significant intrinsic temperature dependence, which may be convoluted with the responses due to the temperature, induced changes in polymer conformation.

Another very significant environmental factor to consider when dealing with bulk or thin films of thermoresponsive polymers is the issue of water uptake. Many thermoresponsive polymers like PNIPAm are appreciably hydrophilic because of the presence of polar amide groups and will absorb water from the atmosphere [61]. Thus in many cases where one wishes to study the behavior of thermoresponsive polymer thin films one will need to control, or account for the presence of water in the thin films [62].

The potential range of fluorophores available for the analysis of thermoresponsive polymers is vast and a detailed analysis is outside the scope of this article. However, once a fluorophore has been selected there are only two general modes of employment: covalent attachment of the probe to the polymer [63], or deployment as a freely diffusing probe which is then introduced to polymer solutions or doped into thin films of the polymer. There are several drawbacks to the use of freely diffusing fluorophores ranging from the fact that they are free to diffuse out of the polymer structures, that they may aggregate, and that there is no control over where the probe interacts with the polymer. This obviously limits the potential applications; however, the approach is intrinsically very simple and lessens the risk of

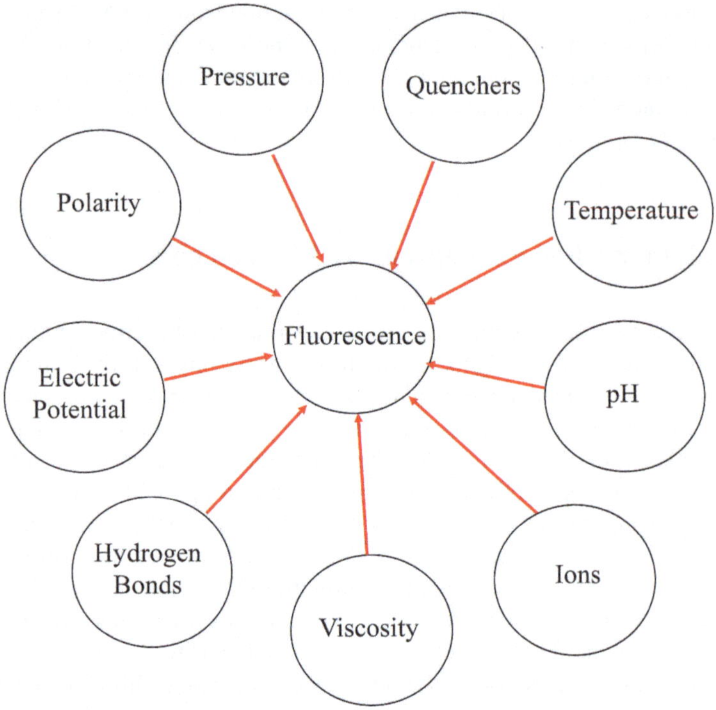

Fig. 4.4 Some of the environmental factors which may affect impact on emission properties of fluorophores in polymers [60]

modifying polymer structure. One can minimize/eliminate many of these issues by covalently attaching the probe to the polymer; however, this may not always be feasible or straightforward. One critical aspect of the covalent labeling approach is to decide if it is possible to introduce the fluorophore during or after the polymerization process. For characterization of thermoresponsive polymers there are a wide range of covalently labeled fluorophores described in the literature. Some examples include pyrene and napthalene [63–66], *N,N*-(dimethy1amino)naphthalenesulfonamide (dansyl) [67], and the cyanine Cy5/Cy5.5 pair for FRET studies [68].

4.4 Thermoresponsive Polymer Characterization

The fluorescence analysis of thermoresponsive polymer systems can be categorized, for the sake of textual organization, as either physiochemical or physical characterization. We note that there is a considerable degree of overlap and that in practice these factors should never be considered in isolation. The chemistry influences the physical factors and *vice-versa*. Here, we begin with physiochemical characterization and the crucial factor of water absorption, which then leads to the measurement of the chemical and polarity properties of thermoresponsive polymers. For the physical characterization, we start with Critical Micelle Concentration (CMC) and aggregation measurements and then progress to the study of phase transitions around the LCST, before finally showing how fluorescence can be utilized to study the assembly/behavior of thermoresponsive particles and surfaces.

When undertaking a physicochemical characterization of thermoresponsive polymers we have to consider first the chemical structure of the polymer and second the environment created by the polymer when it is in solution or deployed as a particle, or a thin film. In the first case Hydrogen-bonding, van der Waals interactions, and conformational changes are all significant, particularly in solution. In the second case where the polymer is fabricated into a higher density form, all these factors are again very important, but one must also take into account solvent/water infiltration which can mediate the chemical behavior of the polymer very significantly.

4.4.1 Water Sorption

Water uptake in a thin polymer film can lead to significant changes in the physicochemical properties of the polymer [69, 70]. In critical applications like medical devices, this may lead to such problems as reduced adhesion and mechanical properties, pronounced physical and chemical aging, and swelling and expansion, compromising the intended function of the polymer and also modulating biocompatibility. The situation will be exacerbated when using thin films because the surface to mass ratio is much larger which facilitates water uptake. One of the most important considerations with thermoresponsive polymers is that they can be appreciably hygroscopic above and below the LCST. For PNIPAm, thermal gravimetric measurements on bulk polymer, indicated that no water was adsorbed at 40 °C (or more correctly, it was not possible to measure the low amount of adsorbed water) [71]. However, when PNIPAm is fabricated as a thin film then one can observe significant water absorption both below and above the LCST. For instance, a 10 μm thick PNIPAm film in an environment with a relative humidity of 90 %RH, was measured to have absorbed 26.5 % by weight of water below the LCST (at 25 °C). Above the LCST (at 37 °C) the amount of adsorbed/absorbed water was still a very significant 6.1 % by weight [61]. While the water absorption can be described as a purely physical effect, it has consequences for the chemical properties of the polymer because it affects the integral hydrogen bonding within the polymer. Thus for the relatively hydrophilic polymers like PNIPAm in thin film form, there is a clear requirement to handle the polymers under conditions of controlled humidity.

Fluorescent 3-hydroxyflavone (3-HF) derivatives have been used to monitor the infiltration of water in PNIPAm thin films [61]. 3-HF emission is governed by an excited-state intramolecular proton transfer (ESIPT) process and these probes demonstrate very strong solvatochromism and electrochromism [72–74]. These

3-HF fluorophores thus exhibit dual band fluorescence emission which is sensitive to environmental factors, and as a sensor/probe they have a clear advantage over single band fluorophores [75–80]. One of these bands originates from the normal excited state (N*), and the other is due to the ESIPT reaction product tautomer (T*). This emission, in terms of the wavelengths of maximum emission and relative intensities of the two emission bands, is sensitive to various environmental factors [72–74, 81–83]. The most important emission parameter is the ratio of the emission intensities from the N* and T* excited states, I_{N*}/I_{T*}, which is associated with the relative populations of the N* and T* states and is a very sensitive indicator of solvent polarity [82]. The behavior of these emission bands as well as the relationship of their intensities depends strongly on probe structure. Changing the chemical structure at the fluorophore core can be used to adjust the sensitivity to a specific range of solvent polarity, to modulate hydrogen bonding, or to electric fields. 4'-Diethylamino-3-hydroxyflavone (FE), 5, 6-benzo-4'-diethylamino-3-hydroxyflavone (BFE), and 4'-diethylamino-3- hydroxy-7-methoxyflavone (MFE) are examples of 3-HF fluorophores, which are sensitive to the properties of the solvent environment. An increase in solvent polarity and hydrogen bonding ability of the solvent leads to an increase in the population of the N* form relative to T* form, which is due to a greater dielectric stabilization of the N* form [74]. It has been possible to correlate the ratio of the emission intensities of these two forms with changes in polarity and hydrogen bonding in various classes of solvents [74, 84]. BFE, unlike FE and MFE, is nearly insensitive to the H-bond donor ability of protic solvents because of the interference from the additional benzene ring [84]. MFE shows more solvent-dependent dual emission in more polar solvents when compared to FE [77].

For these three 3-HF probes (FE, MFE, and BFE) it was observed that the relative emission from the N* and T* bands (as measured by the increase in log (I_{N*}/I_{T*})) varied very significantly with water ingress (Fig. 4.5). The N* band was also significantly red shifted whereas the T* band maximum decreased only slightly with increasing humidity. Analysis of the fluorescence data indicated that water adsorption in the PNIPAm films followed a two-step process: First, at low relative humidity the incoming water molecules disrupted the specific polymer-fluorophore H-bond interaction giving rise to small changes in $\log(I_{N*}/I_{T*})$ ratios and the overall fluorescence intensity. Then at higher relative humidity (>50 %), as the amount of adsorbed water increases, these intensity parameters change dramatically indicating a larger change in the local polarity of the probe environment [61].

Emission measurements carried out above the LCST showed similar $\log(I_{N*}/I_{T*})$ changes, except that the magnitude was lower. This indicates that the PNIPAm films were still absorbing some water in this more hydrophobic state, interrupting the polymer-probe interaction (Fig. 4.6).

In the more hydrophobic, condensed state above the LCST one would have expected that the rate of water adsorption should be decreased. However, fluorescence analysis of PNIPAm thin films show clearly that in every case, the $\log(I_{N*}/I_{T*})$ value increases at a similar rate (Fig. 4.6). This method thus provides a

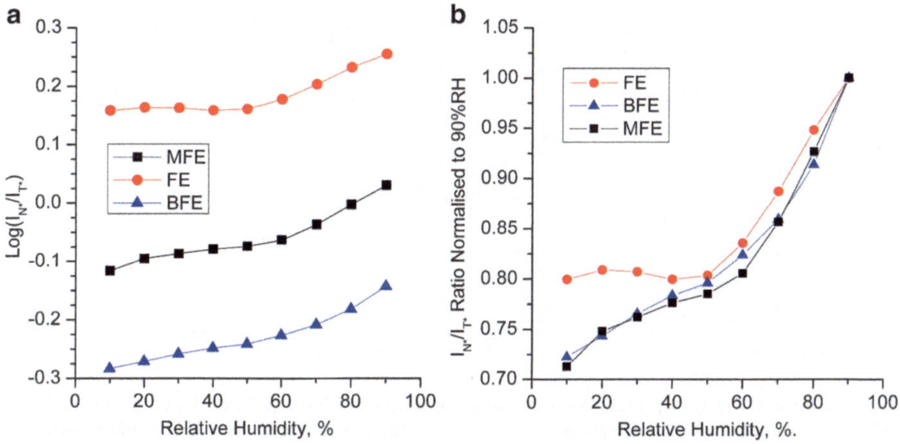

Fig. 4.5 (**a**) Plot of the log(I_{N*}/I_{T*}) ratio vs. increasing relative humidity at 25 °C for PNIPAm films doped with BFE (▲), MFE (■), and FE (●). (**b**) Plot of the I_{N*}/I_{T*} ratio normalized to the value recorded at 90 % RH. Plots show the two-phase water adsorption process (Taken from Ref. [61] and reproduced with permission. *Copyright © 2010 American Chemical Society, all rights reserved*)

relatively facile, non-contact methodology for assessing the ingress of water into thermoresponsive polymers when fabricated as thin films on solid surfaces.

4.4.2 Physicochemical Parameters by Solvatochromism

Measuring polarity and hydrogen bonding changes in thermoresponsive polymers is vital for understanding how these polymers will behave in the real world. For example, we know that many of the common thermoresponsive polymers are appreciably hygroscopic and will absorb appreciable amounts of water below and above the LCST thus influencing polarity within thin films [61]. The most widely based methods for assessing polarity is based on solvatochromism, and there is an extensive literature on solvent systems. Solvatochromism describes the change in shape, location and sometimes intensity of absorption or emission spectra, relative to the polarity of the medium. A hypsochromic (or blue) shift corresponds to negative solvatochromism, while a bathochromic (or red) shift corresponds to positive solvatochromism with increasing solvent polarity [85–87]. Solvatochromic methods are generally robust and reasonably easy to implement but do require the use of relatively high probe concentrations compared to fluorophore indicators [88]. The most common solvatochromic solvent polarity scales are the: $E_T(30)$ scale of Dimroth and Reichardt [85], α and β scales of Kamlet and Taft [89, 90], π* scale of Kamlet, Abboud, and Taft [91], and the pyrene (Py) scale of Dong and Winnik [92]. These solvatochromic methods are typically used to characterize solvent systems but they have also been extended to study polymers.

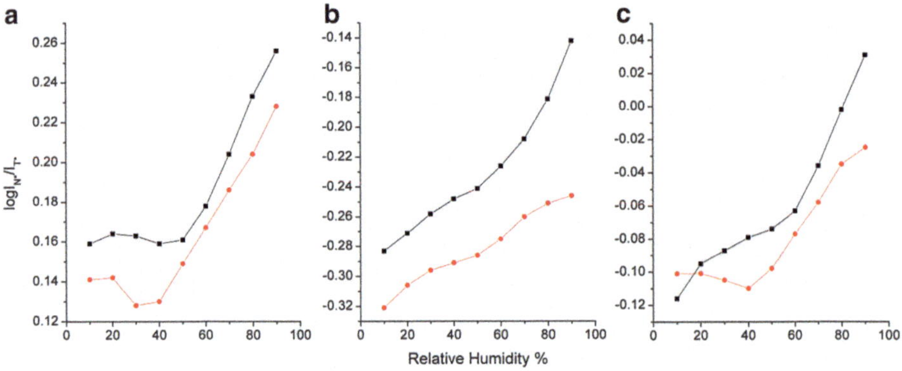

Fig. 4.6 Plot of log(I_{N*}/I_{T*}) ratio vs. increasing humidity of PNIPAm films doped with (**a**) FE, (**b**) BFE and (**c**) MFE at 25 °C (■) and 37 °C (•) (Taken from Ref. [61] and reproduced with permission. *Copyright © 2010 American Chemical Society, all rights reserved*)

The $E_T(30)$ scale uses the 2, 6-diphenyl- 4- (2,4,6- triphenyl- 1-pyridino)-phenolate betaine dye (also known as Reichardt's dye) [85, 93]. The $E_T(30)$ polarity parameter is based on the transition energy for the longest wavelength absorption band measured in the relevant environment and expressed in Kcal·mol^{-1} [85]. A normalized scale (E_T^N) was defined from 0 to 1 using tetramethylsilane and water as the extreme nonpolar and polar solvents respectively [85, 93]. The α and β scales provide a measurement of solvent hydrogen-bond donor (HBD) or a hydrogen-bond acceptor (HBA) ability. These scales are based on a solvatochromic comparison method (SCM) which involves the comparison of solvent induced shifts of the longest wavelength absorption band of two similar compounds. For the determination of α values one compound is capable of acting as a HBA towards HBD solvents (*e.g.* Reichardt's dye) whereas the other cannot (*e.g.* 4-nitroanisole) [89, 90, 93]. To determine β values, one compound will be capable of acting as a HBD towards solvents (*e.g.* 4-nitroaniline) whereas the other cannot (*e.g.* 4-nitro-N,N-dimethylaniline) [93]. The solvent dipolarity/polarizability π* scale provides a quantitative measure of the non-specific part of van der Waals interactions between solvents and solutes [91, 94, 95]. The original scale was based on the spectral properties of carefully selected aromatic molecules which contain both electron-acceptor and electron-donor groups [94, 95]. Dimethyl sulfoxide (DMSO) and cyclohexane (c-C_6H_{12}) were used as reference solvents by taking π* (c-C_6H_{12}) = 0 and π* (DMSO) = 1 [94, 95]. More recently, the scale has been updated/revised and is now based on the averaging of data from several solvatochromic indicators [96]. Laurence *et al.* re-determined π* values for 229 solvents using only two solvatochromic indicators, 4-nitroanisole and N, N-dimethylamino-4-nitroaniline [85, 96].

For the physicochemical characterization of polymers by optical spectroscopy, one generally uses either vibrational or electronic spectroscopies. One of the key advantages of electronic spectroscopy (either absorption or fluorescence) is the potential sensitivity, enabling the observation of subtle effects in condensed media. One of the simplest approaches is to pursue a solvatochromic approach and measure the UV-visible spectra of the appropriate indicators doped into the polymers [97–99]. For example, Matsuguchi *et al.* employed solvatochromic methods to characterize the water sorption behavior in polymer films [99]. They found that indicator band positions (and thus the solvatochromic parameters) were dependent on the polymer type, molecular weight, and the amount of absorbed water. On increasing relative water vapor pressures, all the solvatochromic parameters (α, β, π^*) were seen to increase except for EC, PEO, PVP where the β values were lower at the higher water vapor pressures. The authors also examined the relationship between the $E_T(30)$ scale and the π^*, α and β parameters and a linear correlation was observed indicating that the microenvironments experienced by the probe molecules in both wet and dry films was similar to that observed in liquid solvents.

Szczupak *et al.* evaluated the micro-polarity and H-bond donor/acceptor ability for a series of thermoresponsive N-isopropylacrylamide/N-tert-butylacrylamide (NIPAm/NtBA) copolymer films using the $E_T(30)$, α, β and π^* empirical solvatochromic parameters [19, 60]. For the dry NIPAm/NtBA copolymer films it was found that they are strong H-bond acceptors (β), moderate H-bond donors (α) and are strongly dipolar/polarizable (π^*). It was observed that $E_T(30)$, α and π^* values all decreased linearly on increasing the hydrophobic NtBA fraction in the copolymer films whereas the β parameter remained relatively unchanged. A good correlation was also found between experimental $E_T(30)$ and independently determined α, β and π^* values which confirms that the behavior of the solvatochromic indicators in the NIPAm/NtBA films was similar to that in solvents [19].

4.4.3 Physicochemical Parameters by Fluorescence

While the application of solvatochromic methods for the characterization of physicochemical properties is feasible, and can generate accurate data, there is still a need for a more flexible measurement methodology. Specifically, for many applications, the concentrations required of the solvatochromic probes for absorption spectroscopy in these condensed media is relatively high (typical weight ratios of 36–20 to 1) [99], and it would be preferable to utilize much lower concentrations such as the levels employed in fluorescence spectroscopy (typical weight ratios of 1900–2230 to 1) [100]. Another need is the requirement for standoff measurements in sealed environments and in such cases, fluorescence may be the only feasible option.

Pyrene is a common fluorophore used for polarity assessment because the intensities of various vibronic bands are very sensitive to solvent polarity [87, 101]. Karpovich and Blanchard explained that the relative changes in vibronic band intensities of the pyrene fluorescence spectrum are a result of vibronic coupling between the first (weakly allowed) and the second (strongly allowed) excited singlet states [102]. Their observations also suggest that solvent-dipole-solute induced dipole interactions play a major role in the pyrene scale. The extent to which an induced dipole moment is generated by vibrational distortions of pyrene is governed by the polarity of the solvent. The polarity of the pyrene environment may thus be estimated by measuring the ratio of the fluorescence intensities of the third and first vibronic bands (I_1/I_3) [87, 101]. Dong and Winnik developed the Py scale of solvent polarities, based on this ratio and the scale is relatively insensitive to the hydrogen bonding ability of protic solvents [92]. Py scale values range from 0.58 for n-hexane to 1.95 for dimethyl sulfoxide. Variations in the pyrene lifetime can also be used to monitor changes in thermoresponsive polymers. For aqueous solutions of PNIPAm one observes approximately a 60 ns increase in average lifetime from 115 ns to ~175 ns as the temperature increases from 30 to 40 °C [103]. Szczupak *et al.* assessed the polarity of poly(NIPAm-co-NtBA) copolymer films by means of fluorescence based methods, using pyrene and 3-HF derivatives as polarity sensitive fluorescent probes [100]. They reported a decrease in the I_1/I_3 ratio of pyrene with increasing NtBA fraction indicating a reduction in polarity with increasing NtBA content (Fig. 4.7), which is in agreement with previous results obtained in a solvatochromic study involving the same polymers [19, 100].

The pyrene I_1/I_3 ratio displayed poor correlation with measured solvatochromic polarity parameters [19], which was attributed to the fact that the polarity component sensed by pyrene is only related to dipole-induced dipole interactions. This ignores the very significant H-bonding effects, which are also present in these polymers. Furthermore, it was noted that the physiochemical changes that occur at the LCST of these copolymers did not greatly affect pyrene fluorescence. This was revealed by the small and linear decrease in the I_1/I_3 ratio (1.345 to 1.285) observed for these thin films as the temperature changed from 20 to 40 °C. This contrasts strongly with the situation reported for aqueous solutions of PNIPAm where the ratio changes decreases from 1.79 to 1.38 [104]. We also note that this magnitude of change is almost identical to that observed for solutions of pyrene in ethanol and 1-propanol [60]. The reason why the change is small originates from the lack of a distinct aqueous phase, so the pyrene probe remains dispersed throughout the solid polymer where there is much less change in probe environment and thus in emission properties. It should be noted that these measurements were made without rigorous humidity control [100], and so these films probably contained a significant amount of adsorbed water. The hydrogen-bonding effects inherent in the PNIPAm thin films also produces problems for the use of 3-HF derivatives for polarity assessment of thermoresponsive hydrophilic polymers in thin films, using simple emission parameters. This arises from the heterogeneity of the ground-state H-bonding and an ESIPT process that is not in equilibrium, both effects can be elucidated from the observed excitation wavelength dependence and a difference in the fluorescence lifetimes of the N* and T* bands [74, 100, 105].

One example where fluorescence was used to assess the polarity change at temperatures near the LCST exploited the solvatochromic shifts of the zwitterionic form of rhodamine X [106]. The authors employed a rhodamine X labeled oligonucleotide composed of 25-mers of thymine (dT_{25}-ROX). In dioxane/water

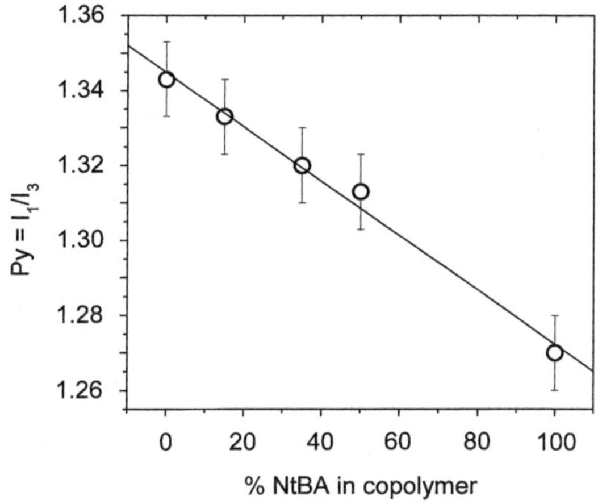

Fig. 4.7 Plot of pyrene emission ratio I_1/I_3 ratio *vs.* NtBA content for a series of copolymer films measured at 20 °C (Taken from Ref. [100] and reproduced with permission. *Copyright © 2010 Journal of Fluorescence Spectroscopy*)

mixtures this probe shows a linear relationship between $E_T(30)$ and shifts in the fluorescence emission maximum (in wavenumbers) for the $E_T(30)$ range between 50 and 65 Kcal.mol^{-1}. They used this probe to look at the polarity changes of the PNIPAm shell of PMMA/PNIPAm core-shell latex particles at temperatures near the LCST. For this system, the calculated $E_T(30)$ value for the PNIPAm shell decreases in a sigmoidal manner as the temperature increases from 15 to 45 °C. At the lower temperatures, the PNIPAm shell polarity is nearly identical to that of water, while above the transition it is equivalent to the polarity of a dioxane/water mixture (30 % (v/v)). In contrast to the smooth transition observed in the wavelength shifts, the lifetime of the dT_{25}-ROX probe showed a sharp drop at the LCST which is ascribed to the refractive index change, which accompanies the dehydration process during the phase change.

Derivatives of quinoxaline (N-(2,3-dimorpholinoquinoxalin-6-yl)acrylamide, *QxA* and N-(1-(2,3-dimorpholinoquinoxalin-6-ylamino)prop-2-yl)methacrylamide, *QxAlaMA*) have also been incorporated into PNIPAm [107]. These fluorophores when incorporated into the polymer showed intense solvatochromism in their fluorescence without perturbing the LCST. The wavelength at the maximum fluorescence intensity of the *QxAlaMA*-labeled PNIPAm dramatically blue-shifted (by ~20 nm) and the fluorescence intensity of the *QxA*-labeled PNIPAm significantly increased, by a factor of ~10, as the temperature increased from 30 to 34 °C.

4.4.4 Critical Micelle Concentration (CMC) and Aggregation Measurements

The CMC is a critical parameter that affects the macroscopic behavior of these thermoresponsive polymers. Measuring the CMC is thus an important facet of polymer science and can usually be achieved by conventional means using techniques like Dynamic Light Scattering (DLS). However, in some cases the CMC can be low and as such is not very amenable to conventional measurement methods. For example, the CMC of some amphiphilic block copolymers is of the order of ~10^{-6} M and cannot be easily determined by scattering methods. This problem can be overcome by using single molecule detection (SMD) methods like fluorescence correlation spectroscopy (FCS) where one can easily observe the very low concentration regime [108–110]. One straightforward method collects FCS data from samples which have a fixed, very low concentration (typically 50 nM or less) of a poorly water soluble fluorophore such as R6G in aqueous solutions of the thermoresponsive polymer with varying concentrations (micro-molar range). The polymer concentrations need to be selected such that they span a range above and below the suspected CMC value. At polymer concentrations below the CMC, the fluorophore should remain dissolved in water, and only the diffusion of free dye should be observed (free R6G has a hydrodynamic radius of 0.8 nm) [111]. As the polymer concentration increase above the CMC value, an increasingly significant fraction of the fluorophores will become associated with the micelles as they are formed. This association process should result in the observation of an additional slow diffusion process being incorporated in the FCS correlation curve.

For example, Adelselberger and co-workers used FCS to study the behavior of amphiphilic, symmetric tri-block thermoresponsive copolymers having short, deuterated polystyrene (PS) end blocks and a large PNIPAm middle block in aqueous solutions at very low concentration close to the CMC [110]. Using FCS they found that at polymer concentrations above 0.9 μM, a second, slower decay appeared in the correlation curves, indicating the onset of micelle formation. Fitting of the correlation curves yielded hydrodynamic radii of 20.8 ± 0.7 nm for PS_{11}-b-$PNIPAm_{280}$- b-PS_{11} and 25.8 ± 0.9 nm for PS_{11}-b-$PNIPAm_{370}$-- b-PS_{11}. These values were validated from DLS measurements made at dilute (0.1 mg/mL) concentrations. The authors do however, point out that with these free probes, micelles may still be present at very low concentrations (<0.9 μM), but that there may be too few to solubilize ample Rh6G molecules to generate sufficient signal that can be extracted from the correlation curves. A variation of this methodology is to use polymer covalently labeled with a fluorophore and measure the change in hydrodynamic radii via FCS [108, 109]. This can be a very sensitive method for determining the hydrodynamic radii of not only the micelles but also the unimers, and intermediate, unstable aggregates.

While these SMD methods are elegant, care needs to be taken with interpretation of results because of complications induced by the use of very low fluorophore and polymer concentrations. These include changes in polymer and probe concentrations due to surface binding to the container walls, which can be different for the free polymer and the various polymer assemblies. Consideration also needs to be given to variations in fluorophore photophysics (quantum yield, lifetime *etc.*) generated by the different physical environments of the solution and the micelle [112].

An extension of the FCS methodology was used by Wang and co-workers to study lateral diffusion on PNIPAm brushes attached to solid surfaces [113]. They used a poly(2-vinylpyridine) probe molecule that was covalently labeled with Alexa 488. Their studies enable the extraction of friction force data and also a better understanding of how the PNIPAm brushes interact with polyelectrolyte probe molecules. They showed that below the LCST, the decrease of viscosity of the solvent water brought about a decrease in the friction forces via coupling of the lateral probe diffusion with the motion of the brush chain. The LCST transition induced stiffening of the PNIPAm chain and the hardening of the PNIPAm brushes above the LCST generated a large increase in friction forces.

Another study compared the diffusion of probe molecules interacting with octadecyltriethoxysilane (OTE) monolayers and surface-tethered PNIPAm polymer chains of varying thickness and surface coverage [114]. An interesting observation from this study was that the R6G fluorophore interacted more strongly with PNIPAm compared to the OTE monolayer. This indicates that the coupling of interfacial interaction and polymer chain dynamics causes a greater slowing of probe diffusion. The sensitivity of the FCS methodology is very beneficial here because it allows the differentiation of many different effects.

Diffusion of large particles formed from thermoresponsive polymers has also been studied using more conventional fluorescence microscopy techniques like Fluorescence Recovery after Photobleaching (FRAP). In the FRAP technique the sample is observed under a microscope (usually a confocal laser scanning microscope), a region of interest (RoI) is rapidly bleached using relatively high power excitation, then the region is imaged over time to determine how long it takes for new fluorescent molecules/particles to diffuse back into the RoI. Relatively simple image analysis can be used then to recover useful information such as diffusion rates, which in turn can be used to ascertain changes in particle size. For thermoresponsive polymers one can use the FRAP method to observe and measure physical effects on the micro to macro size scale. One example used FRAP to quantify the variation in permeability of the walls of thermoresponsive hollow capsules with temperature [115]. The two di-block copolymers of PNIPAm studied were prepared as hollow capsules, and using standard confocal laser scanning microscopy one could monitor temperature induced changes in particle size. The permeability of the thermoresponsive shells could be assessed by comparing the degree of infiltration of two differently sized fluorophores, 6-carboxyfluorescein and fluorescein-labeled dextran.

ANS (1-anilino-8-naphthalene sulfonate) is one of the most widely used polarity probes because it is highly fluorescent in low polarity solvents but is weakly fluorescent in aqueous solution [87, 116]. This important feature enables one to visualize the hydrophobic regions of a given system with minimal influence from ANS molecules remaining in the aqueous environment, and thus this fluorophore has found widespread application in biological and material sciences [117–122]. Kujawa *et al.* utilized ANS to study the concentration (0.02 to 10 g/L range) and temperature dependent solution properties of telechelic PNIPAm (C_{18}-PNIPAM-C_{18}) [123]. As polymer concentration was increased a steep increase in

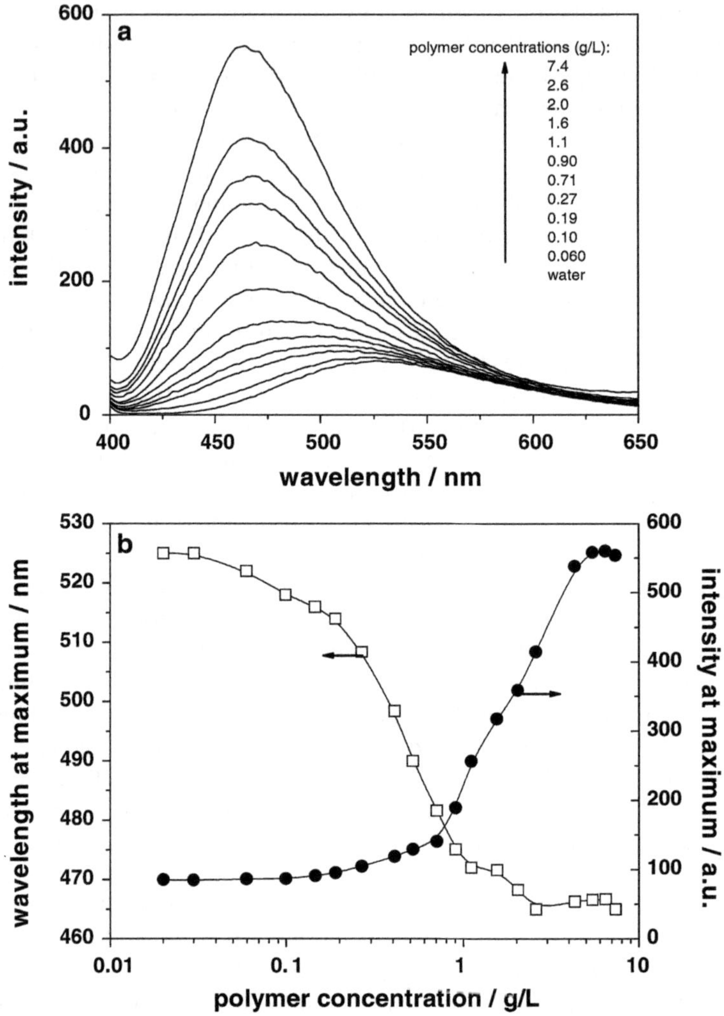

Fig. 4.8 (**a**) Emission spectra of ANS in water and in aqueous solution of telechelic PNIPAm at 20 °C. (**b**) Fluorescence intensity and position of emission band maxima as a function of polymer concentration at 20 °C (Reproduced with permission from Ref. [123]. *Copyright © 2005, Springer Berlin/Heidelberg*)

emission intensity of ANS was observed, accompanied by a blue shift of emission band maxima. These observations with increasing polymer concentration indicate the increased hydrophobicity of the probe (ANS) environment (Fig. 4.8).

These fluorescence results support the results from DLS that suggest for the telechelic PNIPAm in solution, the degree of aggregation within the rosettes increases as the concentration increases, and that the steric crowding results in expulsion/release of polymer bound water in order to accommodate the increased steric pressure.

4.4.5 Phase Transitions at the LCST

The phase transitions that occur at the LCST are the defining feature of thermoresponsive polymers. While the phase transition is often easily observed (with high concentration solutions or when the polymer has been fabricated into a film), there are cases where the use of sensitive fluorescence methods are warranted. This is particularly the case where the polymer concentration is low, or where a non-contact method is required. Both the simple freely diffusing probe and the more complex covalently bound probe approaches have been successfully employed. Winnik utilized pyrene labeled PNIPAm to study the temperature induced phase transitions in aqueous solutions [104, 124]. One study showed that the complex pyrene photophysics when covalently labeled with PNIPAm was dependent on the degree of labeling and involved emission from monomers and excimers [104]. At ambient temperature in water, the presence of ground-state pyrene dimers and higher aggregates was observed for the pyrene labeled PNIPAm. These aggregates can form between pyrene attached to the same chain or between pyrene probes located on different chains. Heating solutions of the labeled PNIPAm above the LCST results in disruption of the pyrene aggregates and the quantum yield of the monomeric species increases relative to that of the excimers. This dissociation is complete in the case of the sparsely labeled polymer(PNIPAm/Py/200) but only partial in solutions of the more highly labeled polymer (PNIPAm/Py/20). When trace amounts of the labeled PNIPAm were added to solutions of unlabelled PNIPAm, changes in pyrene fluorescence could be used to ascertain the interaction between chains. They found that below the LCST, there was no indication of interactions between labeled and unlabeled polymers, whereas above the LCST, the labeled polymers were incorporated into the PNIPAm-rich phase. These fluorescence studies also showed that in the low concentration limit (<1 ppm) for highly labeled polymers there was evidence for the formation of single-polymer chain micelles. More recently, pyrene was employed to study long-range polymer chain dynamics for PNIPAm using a variety of models including the Fluorescence Blob Model (FBM) [125, 126]. Chee *et al.* have also investigated in detail the interactions between PNIPAm and pyrene using time resolved fluorescence spectroscopy and concluded that above the LCST PNIPAm is capable of solubilizing hydrophobic guests such as pyrene but that below the LCST much of this capability is lost [127].

A more facile method for assessing the effects of the phase changes in aqueous solutions of telechelic PNIPAm (C_{18}-PNIPAM-C_{18}) involved the use of ANS and the measurement of steady-state emission spectra. When the emission was measured over the 10 to 50 °C temperature range one observed at approximately 29 °C a sharp increase in fluorescence intensity coupled with a blue shift in the emission band maximum (Fig. 4.9). This was clear evidence for a very significant change in the micropolarity sensed by ANS, as with increasing temperature ANS passes from a hydrophilic environment of "*highly hydrated rosettes*", formed at lower

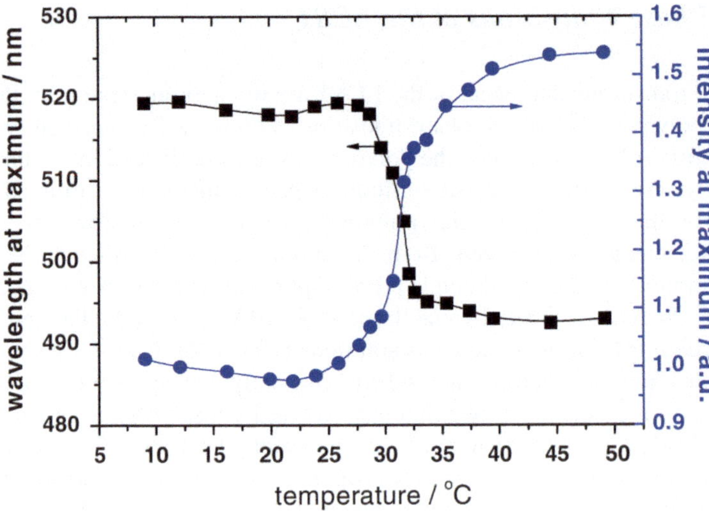

Fig. 4.9 Temperature dependence of the ANS fluorescence intensity and wavelength of emission band maximum in polymer solution (polymer concentration is 0.1 gL^{-1}) (Reproduced with permission from Ref. [123]. *Copyright © 2005, Springer Berlin/Heidelberg*)

temperatures to the "*hydrophobic medium of collapsed and associated polymeric micelles*" [123].

Another method for analyzing thermoresponsive polymer transitions near the LCST is to utilize polarization anisotropy in conjunction with fluorescence quenching studies. The use of acenaphthylene (ACE) covalently labeled PNIPAm for studying the effect of urea on polymer structure/dynamics in solution has been demonstrated several times [128, 129]. ACE was used because there is no rotation independent of the segment to which the ACE probe is attached. Steady-state anisotropy studies showed that below the LCST, anisotropy was small irrespective of the presence of urea (a quencher), which was consistent to the open structure of the loose coil polymer conformation [129]. Above the LCST, large *anisotropy* values were recorded in the absence of the quencher indicating that the polymer had adopted a compact conformation. However, when urea was added, the anisotropy decreased very significantly indicating that the polymer conformation had opened up.

A combination of anisotropy and FRET has been used to probe the behavior of mesoglobular phases formed when PNIPAm was heated in solution [66]. Anisotropy measurements conducted on napthyl labeled PNIPAm gave an indication of the rotational freedom of the pendant fluorophore and this was related to the microviscosity of the phase in which it was dissolved. The results indicated that solutions heated within the 31–36 °C temperature range consisted of fluid-like particles which were able to merge and grow in size. At higher temperatures (36–45 °C) the PNIPAm mesoglobules behaved like more rigid spheres that were unable to merge by collision into larger, slower rotating globules. This study also employed FRET measurements using pyrene (Py) and naphthalene (Np) labeled PNIPAm.

Monitoring the changes in Py and Np fluorescence shows the expected spike in the I_{Py}/I_{Np} fluorescence intensity ratio near the LCST. The authors ascribed this change as arising from two factors: *(i)* chain motion increased upon heating, enhancing the probability of a close encounter between the two different fluorophores, and *(ii)* the local fluorophore concentration increased as a result of solution demixing. However, at temperatures above ~33.5 °C, the I_{Py}/I_{Np} fluorescence intensity ratio decreased once more which was ascribed to the increased micro-viscosity within the mesoglobule. This restricted the motion of the polymer chains, which caused a reduction in the frequency of the Py-Np encounters.

The study of Quantum dot (QD)-PNIPAm hybrid particles by FCS has also been recently reported [130]. This straightforward study described how FCS can be used to monitor changes in hydrodynamic radii near the LCST for particles having different length PNIPAm chains attached. It also indicated that there were substantial decreases in the QD lifetime as the temperature increased through the LCST caused by PNIPAm chain collapse. This alteration in average lifetimes was caused by large changes in the fast lifetime component, which is generally assigned to trapping processes caused by surface defects or impurities.

4.4.6 Swelling and Assembly

Dansyl is a versatile probe for characterizing PNIPAm as it can provide information via steady-state and time-resolved measurements about the polarity and viscosity of the local environment [67, 131, 132]. Some of the advantages of the dansyl probe are that it is relatively insensitive to oxygen quenching, its absorption maximum is relatively independent of the medium and variation in the wavelength of maximum emission is directly related to changes in local polarity and/or viscosity.

The mechanism of shrinking in PNIPAm derived materials can be elucidated by measuring changes in the wavelength of the fluorescence emission maximum of a covalently attached dansyl fluorophore [132]. The rate of shrinkage of different PNIPAm gels in aqueous solution (normal, with grafted side chains, and semi-interpenetrating polymer network) was investigated by measuring the change in the position of the peak emission wavelength. These fluorescence measurements showed clearly that the grafted chains underwent the coil-to-globule transition at lower temperature (~25 °C) than the main polymer chains (~33 °C). Dansyl (covalently labeled) has also been used as a probe to look at the changes due to cross-linking PNIPAm [133]. In this case, cross-linking with glutaraldehyde caused an increase in hydrophobicity which could be evaluated by measuring the blue shift in the fluorescence emission spectra.

Dansyl labeled PNIPAm (PNIPAm-Da label content, 0.06 molar %) was also used to investigate the phenomenon of cononsolvency exhibited by the PNIPAm/water/methanol ternary system. The study involved both the PNIPAm-Da polymer and the cross-linked PNIPAm-DA gels [67]. The swelling of these polymers and gels decreased abruptly in aqueous solutions containing 7–25 mol % methanol and increased gradually in systems with a higher methanol concentration. Shifts in the dansyl wavelength of maximum emission, changes in the fluorescence lifetimes, and changes in the rotational diffusion coefficients could all be correlated with the macroscopic changes in swelling volume.

In some cases, the swelling behavior of PNIPAm can be affected by the presence of other materials in solution. When freely diffusing pyrene was used as a probe of PNIPAm behavior, the introduction of urea changes in the pyrene emission only above the LCST [129]. This arose from urea induced swelling of the PNIPAm compact coil conformation, caused by disruption of the intramolecular hydrophobic interaction. This caused the pyrene probe to experience a much more hydrophobic environment which was measured by the ratio of the emission intensities I_3/I_1. This study also showed that the coil structure was reasonably robust and could be observed at urea concentrations of up to 3 M.

The swelling behavior of thermoresponsive polymers can also be investigated using Förster Resonance Energy Transfer (FRET), where the degree of energy transfer provides information as to separation between donor and acceptor fluorophores [59]. For example, Jones and co-workers used FRET studies to analyze core-shell PNIPAm microgels (both the core and shell components were lightly cross-linked with *N,N′*-methylene(bisacrylamide)), where the core was doubly labeled with cyanine Cy5 (donor) and Cy5.5 (acceptor) [68]. In these structures, the PNIPAm shell can restrict the core from swelling to its native volume, and thus the extent of core expansion will be a function of shell thickness. To covalently attach the fluorophores, the core contained a small percentage of amine groups for post-polymerization modification with the cyanine fluorophores which were functionalized with *N*-hydroxysuccinimidyl ester. For the naked core, the degree of FRET was low when it was swollen to its maximum volume below the LCST (31 °C). The presence of a PNIPAm shell produces a significant degree of FRET under the same solution conditions, indicating that the polymer chains are more constrained relative to the fully swollen state. By monitoring the degree of energy transfer in the absence and presence of the shell, over a range of temperature values, the researchers observed the decreased swelling ability of the core in the presence of the added shell, and were thus able to estimate the shell thickness. This FRET method has obvious advantages when compared to conventional Photon Correlation Spectroscopy (PCS) measurements, which can only yield an apparent particle size, and not discriminate between changes in shell thickness and core compression associated with thicker shells.

The combination of thermoresponsive polymers and the FRET methodology has also been exploited for sensing applications. In one such example, PNIPAm microgels were modified to incorporate potassium ion recognizing 4-acrylamidobenzo18-crown-6 residues (B18C6Am) and then a FRET pair of fluorophores (4-(2-acryloyloxyethylamino) -7-nitro-2,1,3-benzoxadiazole (NBDAE), and rhodamine-B-based FRET acceptors (RhBEA)) [134]. The key operational feature is the fact that the polymer LCST is directly affected by the K^+ ion concentration, increasing by ~9 °C as the K^+ concentration increased from 0 to 300 μM. This reasonably rapid process (~4 second response time) can be monitored by measuring the FRET efficiency calculated from the fluorescence intensity ratio measured at 588 and 529 nm.

Many biomedical uses for PNIPAm involve the preparation of complex macro-, or meso-scale structures and the assembly process can be studied using fluorescence. For example, the 4-acrylamidofluorescein-modified poly(N-isopropylacrylamide-co-acrylic acid) (PNIPAm-co-AAc*) was used with simple fluorescence microscopy to observe layer-by-layer (LbL) deposition of microgel thin films [135]. The method is reasonably effective at showing coverage and layer formation, but because of its non-confocal nature the resolution normal to the surface is very poor. Another area in which fluorescence techniques can be useful for studying thermoresponsive polymers is in the LbL assembly of polyelectrolyte multilayers on soft and porous PNIPAm microgels. One facet of the LbL process is that the polyelectrolytes can interdigitate both with each other and with the microgel during multilayer formation. The problem is further compounded by the fact that the particles are often sub-micron in size and thus not amenable to conventional microscopy evaluation. Using FCS, however, one can easily distinguish between free, labeled polyelectrolytes and those that are bound to the microgel. Wong and co-workers used dual color FCS to confirm that two different polyelectrolytes were binding onto the same microgel particles (~400 nm in size) of PNIPAm [136].

One of the drawbacks with conventional FCS measurements is that one cannot generally get absolute diffusion coefficients from the data and thus one has to correlate with standards of known values. In dual-focus FCS (2*f*-FCS), one uses two focal volumes, which are a precisely known distance apart and generate an overlapping detection volume. This enables accurate and precise quantitative measurement of absolute diffusion coefficient values [137]. 2*f*-FCS measurements have been undertaken at different temperatures to determine the hydrodynamic radii of bare nanogels (p(NIPAM-co-AA-co-rhodamine)) and nanogels coated with various numbers of layers of polyelectrolytes [138]. These temperature dependent studies showed that the polyelectrolyte multilayer shell was still bound to the nanogel during the phase transition at the LCST.

4.5 Conclusions

Fluorescence spectroscopy offers a range of convenient methodologies for the analysis of thermoresponsive polymers. The most widespread application is for the monitoring of the phase and polarity changes at the LCST. The high sensitivity and low probe concentrations required ensures that the fluorescence analysis has a minimal impact on polymer structure or physical behavior. Furthermore, many of these analytical techniques can be performed using standard off-the-shelf, inexpensive fluorescence spectrometers. Of increasing importance is the use of single molecule detection methods to probe the dynamic processes that occur at very low polymer concentrations. This can provide unique information and insights into intra- and inter-chain interactions, allowing one to first examine the growth of polymer aggregates and other higher order structures in solution.

Acknowledgements The preparation of this work was supported by Science Foundation Ireland under a Research Frontiers Grant 07/RFP/MASF423 to A.G.R.

References

1. Philippova OE, Hourdet D, Audebert R, Khokhlov AR (1997) pH-responsive gels of hydrophobically modified poly(acrylic acid). Macromolecules 30:8278–8285
2. Torres-Lugo M, Peppas NA (1999) Molecular design and in vitro studies of novel pH-sensitive hydrogels for the oral delivery of calcitonin. Macromolecules 32:6646–6651
3. Peppas NA, Huang Y, Torres-Lugo M, Ward JH, Zhang J (2000) Physiochemical foundations and structural design of hydrogels in medicine and biology. Annu Rev Biomed Eng 2:9–29
4. Torres-Lugo M, Peppas NA (2000) Transmucosal delivery systems for calcitonin: a review. Biomaterials 21:1191–1196
5. Kost J, Langer R (2001) Responsive polymeric delivery systems. Adv Drug Deliv Rev 46 (1–3):125–148
6. Wang G, Wang X (2002) A novel hyperbranched polyester functionalised with azo chromophore: synthesis and photoresponsive properties. Polym Bull 49:1–8
7. Kavanagh CA, Rochev YA, Gallagher WM, Dawson KA, Keenan AK (2004) Local drug delivery in restenosis injury: thermoresponsive co-polymers as potential drug delivery systems. Pharmacol Ther 102(1):1–15
8. de las Heras Alarcon C, Pennadam S, Alexander C (2005) Stimuli responsive polymers for biomedical applications. Chem Soc Rev 34(3):276–285
9. Goodwin AP, Mynar JL, Ma Y, Fleming GR, Fréchet JMJ (2005) Synthetic micelle sensitive to IR light via a two-photon process. J Am Chem Soc 127(28):9952–9953
10. Kavanagh CA, Gorelova TA, Selezneva II, Rochev YA, Dawson KA, Gallagher WM, Gorelov AV, Keenan AK (2005) Poly(N-isopropy acrylamide) copolymer films as vehicles for the sustained delivery of proteins to vascular endothelial cells. J Biomed Mater Res A 72A (1):25–35
11. Miranda A, Millan M, Caraballo I (2006) Study of the critical points in lobenzarit disodium hydrophilic matrices for controlled drug delivery. Chem Pharm Bull 54(5):598–602
12. Matsuda N, Shimizu T, Yamato M, Okano T (2007) Tissue engineering based on cell sheet technology. Adv Mater 19(20):3089–3099
13. Ní Chearuil F, Corrigan OI (2009) Thermosensitivity and release from poly N-isopropylacrylamide-polylactide copolymers. Int J Pharm 366:21–30
14. Fundueanu G, Constantin M, Ascenzi P (2009) Poly (N-isopropylacrylamide-co-acrylamide) cross-linked thermoresponsive microspheres obtained from preformed polymers: influence of the physio-chemical characteristics of drugs on their release profiles. Acta Biomater 5:363–373
15. Galaev IY, Mattiasson B (1999) 'Smart' polymers and what they could do in biotechnology and medicine. Trends Biotechnol 17(8):335–340
16. Gil ES, Hudson SM (2004) Stimuli-responsive polymers and their bioconjugates. Prog Polym Sci 29(12):1173–1222
17. Bult H (2000) Restenosis: a challenge for pharmacology. Trends Pharmacol Sci 21(7):274–279
18. Lewis AL, Tolhurst LA, Stratford PW (2002) Analysis of a phosphorycholine- based polymer coating on a coronary stent pre- and post- implantation. Biomaterials 23(7):1697–1766

19. Szczupak B, Ryder AG, Togashi DM, Rochev YA, Gorelov AV, Glynn TJ (2009) Measuring the micro-polarity and hydrogen-bond donor/acceptor ability of thermoresponsive N-isopropylacrylamide/N-tert-butylacrylamide copolymer films using solvatochromic indicators. Appl Spectrosc 63(4):442–449
20. Smith AE, Xu XW, McCormick CL (2010) Stimuli-responsive amphiphilic (co)polymers via RAFT polymerization. Prog Polym Sci 35(1–2):45–93
21. Kumar A, Srivastava A, Galaev IY, Mattiasson B (2007) Smart polymers: physical forms and bioengineering applications. Prog Polym Sci 32(10):1205–1237
22. Qiu Y, Park K (2001) Environment-sensitive hydrogels for drug delivery. Adv Drug Deliv Rev 53(3):321–339
23. Jeong B, Gutowska A (2002) Lessons from nature: stimuli-responsive polymers and their biomedical applications. Trends Biotechnol 20(7):305–311
24. Okano T, Kikuchi A, Sakurai Y, Takei Y, Ogata N (1995) Temperature-responsive poly (N-isopropylacrylamide) as a modulator for alteration of hydrophilic/hydrophobic surface properties to control activation/inactivation of platelets. J Control Release 36(1–2):125–133
25. Kikuchi A, Okano T (2002) Intelligent thermoresponsive polymeric stationary phases for aqueous chromatography of biological compounds. Prog Polym Sci 27(6):1165–1193
26. Hoffman AS, Stayton PS, Bulmus V, Chen G, Chen J, Cheung C, Chilkoti A, Ding Z, Dong L, Fong R (2000) Really smart bioconjugates of smart polymers and receptor proteins. J Biomed Mater Res A 52(4):577–586
27. Schild HG (1992) Poly(N-isopropylacrylamide): experiment, theory and application. Prog Polym Sci 17:163–249
28. Zareie HM, Volga Bulmus E, Gunning AP, Hoffman AS, Piskin E, Morris VJ (2000) Investigation of a stimuli-responsive copolymer by atomic force microscopy. Polymer 41 (18):6723–6727
29. Nandivada H, Ross AM, Lahann J (2010) Stimuli-responsive monolayers for biotechnology. Prog Polym Sci 35(1–2):141–154
30. Aguilar MR, Elvira C, Gallardo A, Vázquez B, Román JS (2007) Smart polymers and their applications as biomaterials. Top in Tissue Eng 3:1–27
31. Zrinyi M, Barsi L, Szabo D, Kilian HG (1997) Direct observation of abrupt shape transition in ferrogels induced by nonuniform magnetic field. J Chem Phys 106(13):5685–5692
32. Zrinyi M (2000) Intelligent polymer gels controlled by magnetic fields. Colloid Polym Sci 278(2):98–103
33. Filipcsei G, Feher J, Zrinyi M (2000) Electric field sensitive neutral polymer gels. J Mol Struct 554(1):109–117
34. Maeda Y (2001) IR spectroscopic study on the hydration and the phase transition of poly (vinyl methyl ether) in water. Langmuir 17(5):1737–1742
35. Liu F, Urban MW (2008) Dual temperature and pH responsiveness of poly(2-(N, N-dimethylamino)ethyl methacrylate-co-n-butyl acrylate) colloidal dispersions and their films. Macromolecules 41(17):6531–6539
36. Liu F, Urban MW (2010) Recent advances and challenges in designing stimuli-responsive polymers. Prog Polym Sci 35(1–2):3–23
37. Roy D, Cambre JN, Sumerlin BS (2010) Future perspectives and recent advances in stimuli-responsive materials. Prog Polym Sci 35(1–2):278–301
38. Paricaud P, Galindo A, Jackson G (2003) Understanding liquid-liquid immiscibility and LCST behaviour in polymer solutions with a Wertheim TPT1 description. Mol Phys 101 (16):2575–2600
39. Klouda L, Mikos AG (2008) Thermoresponsive hydrogels in biomedical applications. Eur J Pharm Biopharm 68(1):34–45
40. Liu RX, Fraylich M, Saunders BR (2009) Thermoresponsive copolymers: from fundamental studies to applications. Colloid Polym Sci 287(6):627–643
41. Heskins M, Guillet JE (1968) Solution properties of poly (N-isopropylacrylamide). J Macromol Sci A 2(8):1441–1455

42. Schild HG, Muthukumar M, Tirrell A (1991) Cononsolvency in mixed aqueous solutions of poly(N-isopropylacrylamide). Macromolecules 24(4):948–952
43. Lutz JF, Akdemir Ö, Hoth A (2006) Point by point comparison of two thermosensitive polymers exhibiting a similar LCST: is the age of poly (NIPAM) over? J Am Chem Soc 128(40):13046–13047
44. Crespy D, Rossi RM (2007) Temperature responsive polymers with LCST in the physiological range and their applications in textiles. Polym Int 56(12):1461–1468
45. Graziano G (2000) On the temperature-induced coil to globule transition of poly-N-isopropylacrylamide in dilute aqueous solutions. Int J Biol Macromol 27(1):89–97
46. Lin CC, Metters AT (2006) Hydrogels in controlled release formulations: network design and mathematical modeling. Adv Drug Deliv Rev 58(12–13):1379–1408
47. Volpe CD, Cassinelli C, Morra M (1998) Wilhelmy plate measurements on poly(N-isopropylacrylamide)-grafted surfaces. Langmuir 14(16):4650–4656
48. Moran MT, Carroll WM, Selezneva I, Gorelov A, Rochev Y (2007) Cell growth and detachment from protein-coated PNIPAAm-based copolymers. J Biomed Mater Res A 81(4):870–876
49. Gras SL, Mahmud T, Rosengarten G, Mitchell A, Kalantar-zadeh K (2007) Intelligent control of surface hydrophobicity. ChemPhysChem 8(14):2036–2050
50. Crowther HM, Vincent B (1998) Swelling behavior of poly-N-isopropylacrylamide microgel particles in alcoholic solutions. Colloid Polym Sci 276(1):46–51
51. Yin ZZ, Zhang JJ, Jiang LP, Zhu JJ (2009) Thermosensitive behavior of poly (N-isopropylacrylamide) and release of incorporated hemoglobin. J Phys Chem C 113(36):16104–16109
52. Nash ME, Carroll WM, Nikoloskya N, Yang RB, Connell CO, Gorelov AV, Dockery P, Liptrot C, Lyng FM, Garcia A, Rochev YA (2011) Straightforward, one-step fabrication of ultrathin thermoresponsive films from commercially available pNIPAm for cell culture and recovery. ACS Appl Mater Interfaces 3(6):1980–1990
53. Rochev Y, O'Halloran D, Gorelova TA, Gilcreest V, Selezneva I, Gavrilyuk B, Gorelov A (2004) Rationalising the design of polymeric thermoresponsive biomaterials. J Mater Sci Mater Med 15(4):513–517
54. Rochev Y, Golubeva T, Gorelov A, Allen L, Gallagher WM, Selezneva I, Gavrilyuk B, Dawson KA (2001) Surface modification for controlled cell growth on copolymers of N-isopropylacrylamide. Prog Colloid Polym Sci 118:153–156
55. Doorty KB, Golubeva TA, Gorelov AV, Rochev YA, Allen LT, Dawson KA, Gallagher WM, Keenan AK (2003) Poly (N-isopropylacrylamide) co-polymer films as potential vehicles for delivery of an antimitotic agent to vascular smooth muscle cells. Cardiovasc Pathol 12(2):105–110
56. Han HD, Shin BC, Choi HS (2006) Doxorubicin-encapsulated thermosensitive liposomes modified with poly (N-isopropylacrylamide-co-acrylamide): drug release behavior and stability in the presence of serum. Eur J Pharm Biopharm 62(1):110–116
57. Demchenko AP (2002) The red-edge effects: 30 years of exploration. Luminescence 17(1):19–42
58. Bosch P, Catalina F, Corrales T, Peinado C (2005) Fluorescent probes for sensing processes in polymers. Chem-Eur J 11(15):4314–4325
59. Lakowicz JR (2006) Principles of fluorescence spectroscopy, 3rd edn. Springer, New York
60. Szczupak B (2009) Evaluation of polarity and hydrogen bonding ability of thermoresponsive N-isopropylacrylamide/N-tert-butylacrylamidecopolymer films using solvatochromic and fluorescence probes. Ph.D. Thesis, National University of Ireland, Galway
61. Morris C, Szczupak B, Klymchenko AS, Ryder AG (2010) Study of water adsorption in poly (N-isopropylacrylamide) thin films using fluorescence emission of 3-hydroxyflavone probes. Macromolecules 43(22):9488–9494
62. La Porte RJ (1997) Hydrophilic polymer coatings for medical devices structures/properties, development, manufacture and applications, Chapter 2. Technomic Pub, Lancaster

63. Ringsdorf H, Venzmer J, Winnik FM (1991) Fluorescence studies of hydrophobically modified poly(n-isopropylacrylamides). Macromolecules 24(7):1678–1686
64. Ringsdorf H, Simon J, Winnik FM (1992) Hydrophobically-modified poly(n-isopropylacrylamides) in water – a look by fluorescence techniques at the heat-induced phase-transition. Macromolecules 25(26):7306–7312
65. Ringsdorf H, Simon J, Winnik FM (1992) Hydrophobically-modified poly(n-isopropylacrylamides) in water – probing of the microdomain composition by nonradiative energy-transfer. Macromolecules 25(20):5353–5361
66. Kujawa P, Aseyev V, Tenhu H, Winnik FM (2006) Temperature-sensitive properties of poly (N-isopropylacrylamide) mesoglobules formed in dilute aqueous solutions heated above their demixing point. Macromolecules 39(22):7686–7693
67. Asano M, Winnik FM, Yamashita T, Horie K (1995) Fluorescence studies of Dansyl-labeled poly(N-isopropylacrylamide) gels and polymers in mixed water/methanol solutions. Macromolecules 28(17):5861–5866
68. Jones CD, McGrath JG, Lyon LA (2004) Characterization of cyanine dye-labeled poly (N-isopropylacrylamide) core/shell microgels using fluorescence resonance energy transfer. J Phys Chem B 108(34):12652–12657
69. Miller KE, Krueger RH, Torkelson JM (1995) Mobility-sensitive fluorescence probes for quantitative monitoring of water sorption and diffusion in polymer-coatings. J Polym Sci B-Polym Phys 33(17):2343–2349
70. Goodelle JP, Pearson RA, Santore MM (2002) Water-uptake kinetics in poly(methyl methacrylate) films with a fluorescent rotor probe. J Appl Polym Sci 86(10):2463–2471
71. Thijs HML, Remzi Becer C, Guerrero-Sanchez C, Fournier D, Hoogenboom R, Schubert US (2007) Water uptake of hydrophilic polymers determined by a thermal gravimetric analyzer with a controlled humidity chamber. J Mater Chem 17:4864–4871
72. Chou PT, Martinez ML, Clements JH (1993) Reversal of excitation behavior of proton-transfer vs. charge-transfer by dielectric perturbation of electronic manifolds. J Phys Chem 97(11):2618–2622
73. Klymchenko AS, Demchenko AP (2002) Electrochromic modulation of excited-state intramolecular proton transfer: the new principle in design of fluorescence sensors. J Am Chem Soc 124(41):12372–12379
74. Klymchenko AS, Demchenko AP (2003) Multiparametric probing of intermolecular interactions with fluorescent dye exhibiting excited state intramolecular proton transfer. Phys Chem Chem Phys 5(3):461–468
75. Létard JF, Delmond S, Lapouyade R, Braun D, Rettig W, Kreissler M (1995) New intrinsic fluoroionophores with dual fluorescence: DMABN Crown4 and DMABN Crown5. Recueil des Travaux Chimiques des Pays-Bas 114(11-12):517–527
76. Klymchenko AS, Ozturk T, Pivovarenko VG, Demchenko AP (2001) A 3-Hydroxychromone with dramatically improved fluorescence properties. Tetrathedron Lett 42(45):7967–7970
77. Klymchenko AS, Pivovarenko VG, Ozturk T, Demchenko AP (2003) Modulation of the solvent-dependant dual emission in 3-hydroxyflavones by substituents. New J Chem 27(9):1336–1343
78. Demchenko AP (2005) Optimization of fluorescence response in the design of molecular biosensors. Anal Biochem 343(1):1–22
79. Shynkar VV, Klymchenko AS, Duportail G, Demchenko AP, Mély Y (2005) Two-color fluorescent probes for imaging the dipole potential of cell plasma membranes. Biochim Biophys Acta (BBA)-Biomembr 1712(2):128–136
80. Turkmen Z, Klymchenko AS, Oncul S, Duportail G, Topcu G, Demchenko AP (2005) A triterpene oleanolic acid conjugate with 3-hydroxyflavone derivative as a new membrane probe with two-color ratiometric response. J Biochem Biophys Methods 64(1):1–18
81. Demchenko AP, Ercelen S, Roshal AD, Klymchenko AS (2002) Excited-state proton transfer reaction in a new benzofuryl 3-hydroxychromone derivative: the influence of low-polar solvents. Pol J Chem 76(9):1287–1299

82. Ercelen S, Klymchenko AS, Demchenko AP (2002) Ultrasensitive fluorescent probe for the hydrophobic range of solvent polarities. Anal Chim Acta 464(2):273–287
83. Klymchenko AS, Mély Y, Demchenko AP, Duportail G (2004) Simultaneous probing of hydration and polarity of lipid bilayers with 3-hydroxyflavone fluorescent dyes. Biochim Biophys Acta-Biomembr 1665(1–2):6–19
84. Klymchenko AS, Pivovarenko VG, Demchenko AP (2003) Elimination of the hydrogen bonding effect on the solvatochromism of 3-hydroxyflavones. J Phys Chem A 107:4211–4216
85. Reichardt C (1988) Solvents and solvent effects in organic chemistry. Federal Republic of Germany, VCH, Weinheim
86. Buncel E, Rajagopal S (1990) Solvatochromism and solvent polarity scales. Acc Chem Res 23(7):226–231
87. Valeur B (2001) Molecular fluorescence: principles and applications. Wiley-VCH Verlag GmbH, Weinheim (Federal Republic of Germany)
88. Fischer K, Spange S (2000) Empirical surface polarity parameters for native polysaccharides. Macromol Chem Phys 201(15):1922–1929
89. Kamlet MJ, Taft RW (1976) The solvatochromic comparison method. I. The. beta.-scale of solvent hydrogen-bond acceptor (HBA) basicities. J Am Chem Soc 98(2):377–383
90. Taft RW, Kamlet MJ (1976) The solvatochromic comparison method. 2. The. alpha.-scale of solvent hydrogen-bond donor (HBD) acidities. J Am Chem Soc 98(10):2886–2894
91. Kamlet MJ, Abboud JL, Taft RW (1977) The solvatochromic comparison method 6. The pi* scale of solvent polarities. J Am Chem Soc 99:6027–6038
92. Dong DC, Winnik MA (1984) The Py scale of solvent polarities. Can J Chem 62 (11):2560–2565
93. Reichardt C (1994) Solvatochromic dyes as solvent polarity indicators. Chem Rev 94:2319–2358
94. Abboud JLM, Notario R (1999) Critical compilation of scales of solvent parameters. Part I. Pure, non-hydrogen bond donor solvents. Pure Appl Chem 71(4):645–718
95. Katritzky AR, Fara DC, Yang H, Taemm K, Tamm T, Karelson M (2004) Quantitative measures of solvent polarity. Chem Rev 104(1):175–198
96. Laurence C, Nicolet P, Dalati MT, Abboud JLM, Notario R (1994) The empirical treatment of solvent-solute interactions: 15 years of. pi.*. J Phys Chem 98(23):5807–5816
97. Paley MS, McGill RA, Howard SC, Wallace SE, Harris JM (1990) Solvatochromism – a new method for polymer characterization. Macromolecules 23(21):4557–4564
98. McGill RA, Paley MS, Harris JM (1992) Solvatochromic characterization of polymers – effects of relative-humidity. Macromolecules 25(12):3015–3019
99. Matsuguchi M, Sadaoka Y, Mizuguchi H, Umeda K, Sakai Y (1997) Solvatochromic study of water sorption in polymer films. J Appl Polym Sci 63(12):1681–1691
100. Szczupak B, Ryder AG, Togashi DM, Rotchev YA, Klymchenko AS, Gorelov A, Glynn TJ (2010) Polarity assessment of thermoresponsive poly(NIPAM-co-NtBA) copolymer films using fluorescence methods. J Fluoresc 20(3):719–731
101. Kalyanasundaram K, Thomas JK (1977) Environmental effects on vibronic band intensities in pyrene monomer fluorescence and their application in studies of micellar systems. J Am Chem Soc 99(7):2039–2044
102. Karpovich DS, Blanchard GJ (1995) Relating the polarity-dependent fluorescence response of pyrene to vibronic coupling. Achieving a fundamental understanding of the py polarity scale. J Phys Chem 99(12):3951–3958
103. Barker IC, Cowie JMG, Huckerby TN, Shaw DA, Soutar I, Swanson L (2003) Studies of the "smart" thermoresponsive behaviour of copolymers of N-isopropylacrylamide and N, N-dimethylacrylamide in dilute aqueous solution. Macromolecules 36:7765–7770
104. Winnik FM (1990) Fluorescence studies of aqueous solutions of poly (N-isopropylacrylamide) below and above their LCST. Macromolecules 23(1):233–242

105. Shynkar VV, Mély Y, Duportail G, Piémont E, Klymchenko AS, Demchenko AP (2003) Picosecond time-resolved fluorescence studies are consistent with reversible excited-state intramolecular proton transfer in 4'-(Dialkylamino)-3-hydroxyflavones. J Phys Chem A 107:9522–9529
106. Prazeres TJV, Santos AM, Martinho JMG, Elaissari A, Pichot C (2004) Adsorption of oligonucleotides on PMMA/PNIPAM core-shell latexes: polarity of the PNIPAM shell probed by fluorescence. Langmuir 20(16):6834–6840
107. Matsumura Y, Katoh A (2008) Synthesis of 2,3-dimorpholino-6-aminoquinoxaline derivatives and application to a new intramolecular fluorescent probe. J Lumin 128(4):625–630
108. Bonne TB, Ludtke K, Jordan R, Stepanek P, Papadakis CM (2004) Aggregation behavior of amphiphilic poly(2-alkyl-2-oxazoline) diblock copolymers in aqueous solution studied by fluorescence correlation spectroscopy. Colloid Polym Sci 282(8):833–843
109. Bonne TB, Papadakis CM, Ludtke K, Jordan R (2007) Role of the tracer in characterizing the aggregation behavior of aqueous block copolymer solutions using fluorescence correlation spectroscopy. Colloid Polym Sci 285(5):491–497
110. Adelsberger J, Kulkarni A, Jain A, Wang WN, Bivigou-Koumba AM, Busch P, Pipich V, Holderer O, Hellweg T, Laschewsky A, Muller-Buschbaum P, Papadakis CM (2010) Thermoresponsive PS-b-PNIPAM-b-PS micelles: aggregation behavior, segmental dynamics, and thermal response. Macromolecules 43(5):2490–2501
111. Bonne TB, Ludtke K, Jordan R, Papadakis CM (2007) Effect of polymer architecture of amphiphilic poly(2-oxazoline) copolymers on the aggregation and aggregate structure. Macromol Chem Phys 208(13):1402–1408
112. Anikovsky MY, Petersen NO (2009) Photon counting histogram analysis as a tool for studying the nature of intermolecular interactions. J Phys Chem B 113(11):3404–3412
113. Wang W, Zhang CF, Wang SQ, Zhao J (2007) Diffusion of single polyelectrolytes on the surface of poly (N-isopropylacrylamide) brushes. Macromolecules 40(26):9564–9569
114. Wang SQ, Zhu YX (2010) Molecular diffusion on surface tethered polymer layers: coupling of molecular thermal fluctuation and polymer chain dynamics. Soft Matter 6(19):4661–4665
115. Glinel K, Sukhorukov GB, Mohwald H, Khrenov V, Tauer K (2003) Thermosensitive hollow capsules based on thermoresponsive polyelectrolytes. Macromol Chem Phys 204(14):1784–1790
116. Weber G, Laurence DJ (1954) Fluorescent indicators of adsorption in aqueous solution and on the solid phase. Biochem J 56(325th Meeting):xxxi
117. Slavík J (1982) Anilinonaphthalene sulfonate as a probe of membrane composition and function. Biochim Biophys Acta 694(1):1–25
118. Kane CD, Bernlohr DA (1996) A simple assay for intracellular lipid-binding proteins using displacement of 1-anilinonaphthalene 8-sulfonic acid. Anal Biochem 233(2):197–204
119. Uversky VN, Winter S, Lober G (1996) Use of fluorescence decay times of 8-ANS-protein complexes to study the conformational transitions in proteins which unfold through the molten globule state. Biophys Chem 60(3):79–88
120. Sirangelo I, Bismuto E, Tavassi S, Irace G (1998) Apomyoglobin folding intermediates characterized by the hydrophobic fluorescent probe 8-anilino-1-naphthalene sulfonate (ANS). Biochim Biophys Acta (BBA)-Protein Struct and Mol Enzymol 1385(1):69–77
121. Geuskens G, Soukrati A (2000) Investigation of polyacrylamide hydrogels using 1-anilinonaphthalene-8-sulfonate as fluorescent probe. Eur Polym J 36(8):1537–1546
122. Lis Thomas T, Mishra AK (2002) ANS fluorescence as a tool to monitor cross-linking polymerization of acrylamide. Eur Polym J 38(9):1805–1810
123. Kujawa P, Watanabe H, Tanaka F, Winnik FM (2005) Amphiphilic telechelic poly (N-isopropylacrylamide) in water: from micelles to gels. Eur Phys J E 17(2):129–137
124. Winnik FM (1993) Photophysics of preassociated pyrenes in aqueous polymer solutions and in other organized media. Chem Rev 93(2):587–614
125. Duhamel J (2006) Polymer chain dynamics in solution probed with a fluorescence blob model. Acc Chem Res 39(12):953–960

126. Yip J, Duhamel J, Qiu XP, Winnik FM (2011) Long-range polymer chain dynamics of pyrene-labeled poly(N-isopropylacrylamide)s studied by fluorescence. Macromolecules 44(13):5363–5372
127. Chee CK, Ghiggino KP, Smith TA, Rimmer S, Soutar I, Swanson L (2001) Time-resolved fluorescence studies of the interactions between the thermoresponsive polymer host, poly (N-isopropylacrylamide), and a hydrophobic guest, pyrene. Polymer 42(5):2235–2240
128. Chee CK, Rimmer S, Soutar I, Swanson L (2001) Fluorescence investigations of the thermally induced conformational transition of poly(N-isopropylacrylamide). Polymer 42(12):5079–5087
129. Fang Y, Qiang JC, Hu DD, Wang MZ, Cui YL (2001) Effect of urea on the conformational behavior of poly(N-isopropylacrylamide). Colloid Polym Sci 279(1):14–21
130. Tagit O, Tomczak N, Jafarpour A, Janczewski D, Han MY, Vancso GJ, Herek JL (2011) Influence of the length and grafting density of PNIPAM chains on the colloidal and optical properties of quantum dot/PNIPAM assemblies. Nanotechnology 22(26):265701
131. Shea KJ, Stoddard GJ, Shavelle DM, Wakui F, Choate RM (1990) Synthesis and characterization of highly cross-linked polyacrylamides and polymethylacrylamides – a new class of macroporous polyamides. Macromolecules 23(21):4497–4507
132. Yoshinari E, Furukawa H, Horie K (2005) Fluorescence study on the mechanism of rapid shrinking of grafted poly(N-isopropylacrylamide) gels and semi-IPN gels. Polymer 46 (18):7741–7748
133. Kurihara S, Sakamaki S, Mogi S, Ogata T, Nonaka T (1996) Crosslinking of poly(vinyl alcohol)-graft-N-isopropylacrylamide copolymer membranes with glutaraldehyde and permeation of solutes through the membranes. Polymer 37(7):1123–1128
134. Yin J, Li CH, Wang D, Liu SY (2010) FRET-derived ratiometric fluorescent K+ sensors fabricated from thermoresponsive poly(N-isopropylacrylamide) microgels labeled with crown ether moieties. J Phys Chem B 114(38):12213–12220
135. Serpe MJ, Jones CD, Lyon LA (2003) Layer-by-layer deposition of thermoresponsive microgel thin films. Langmuir 19(21):8759–8764
136. Wong JE, Muller CB, Laschewsky A, Richtering W (2007) Direct evidence of layer-by-layer assembly of polyelectrolyte multilayers on soft and porous temperature-sensitive PNiPAM microgel using fluorescence correlation spectroscopy. J Phys Chem B 111(29):8527–8531
137. Dertinger T, Pacheco V, von der Hocht I, Hartmann R, Gregor I, Enderlein J (2007) Two-focus fluorescence correlation spectroscopy: a new tool for accurate and absolute diffusion measurements. ChemPhysChem 8(3):433–443
138. Wong JE, Muller CB, Diez-Pascual AM, Richtering W (2009) Study of layer-by-layer films on thermoresponsive nanogels using temperature-controlled dual-focus fluorescence correlation spectroscopy. J Phys Chem B 113(49):15907–15913

Chapter 5
Principles of Fluorogenic Reagent Design for Forensics. Recent Progress Towards New Reagents to Develop Fingerprints in Blood and on Variable Surfaces

Martha Sibrian-Vazquez and Robert M. Strongin

Abstract The use of fluorogenic reagents for detecting and enhancing fingerprints in blood has been studied for many years. There are still several challenges. These include utility on dark and multi-colored substrates, loss of detail, print stability, and compatibility with acidic fixative agents. Herein, we describe progress in addressing each of these issues. The creation and evaluation of a promising class of leuco xanthenes dyes is highlighted.

Keywords Fingerprints • Blood • Leuco dye • Xanthenes • Peroxidase • Fluorone

5.1 Introduction

Fingerprints are the most common and useful physical evidence for the apprehension and conviction of crime perpetrators. The use of fluorogenic reagents for detecting and enhancing fingerprints in blood, however, has several associated challenges. For instance, the dyes used are generally unsuitable for dark and multi-colored substrates. Although examples such as luminol and fluorescin and other chemilumigens and fluorigens can function on dark and often multi-colored substrates, they are incompatible with acidic fixative reagents. The chemiluminescence of luminol and the fluorescence of fluorescein and other standard fluorophores are also quenched in acidic media. In addition, their oxidation products are not insoluble. They therefore may diffuse and degrade detail. The oxidation of luminol is a fast and irreversible reaction, so the user must act quickly to observe and capture an image or the print will be lost.

A goal of our research has been to develop fluorogenic compounds for detecting fingerprints in blood that have excellent sensitivity and stability, that enhance and preserve print details, and can be useful on dark and multi-colored substrates. Reagents and protocols were recently developed in our lab with invaluable critical

M. Sibrian-Vazquez • R.M. Strongin (✉)
Department of Chemistry, Portland State University, Portland, OR 97201, USA
e-mail: rmstrongin@gmail.com

© Springer International Publishing Switzerland 2016

C.D. Geddes (ed.), *Reviews in Fluorescence 2015*,
DOI 10.1007/978-3-319-24609-3_5

guidance and technical support from the Orange County Sherriff and Coroners Office. Desired solubility, fluorescence under acidic conditions and peroxidase-specific chemistry were achieved. It is envisioned that upon further optimization the materials described herein can be adaptable for use as part of a kit for crime scene response personnel.

5.2 Background: A Brief Survey of Classical Print Detection Reagents

Not all blood, and especially bloody prints, can be readily detected. Reagents are thus often employed for identification and/or enhancement. Many rely on the peroxidase-like activity inherent in red blood cells. It is generally accepted that the heme group of hemoglobin catalyzes the reduction of peroxide to water while a second redox-active species (the print reagent dye, Scheme 5.1) is concomitantly oxidized. The alteration in the redox or ionization state of the dye generates the optical readout allowing fingerprints to be observed in a desired matrix.

Benzidine and leuco dyes (e.g., phenolphthalin, leucocrystal violet), leucomalachite green [1] o-tolidine [2] diamino and tetramethyl-benzidine, [3] leucorhodamine 6G, [4] leucoeosin, [5] diaminobenzidine, and others [6, 7] function via the psudo-peroxidase activity of the heme moiety of hemoglobin with hydrogen peroxide or sodium perborate to detect blood and bloody prints.

Chemiluminogens and fluorogens have an extensive history. Luminol was first used in 1937 [8]. Its chemiluminescence decays rapidly unless the reaction is slowed by one of several techniques. In 2006 Young [9] compared the formulation known as BlueStar to luminol and flourescin. It was found to exhibit greater sensitivity and longer lasting chemiluminescence for the detection of blood, but not for bloody fingerprint enhancement. In general, chemiluminescence is indeed useful for blood detection in specific circumstances. However, in the case of bloody fingerprint enhancement it is challenging to determine how much reagent to use and to estimate the time available for capture and exposure.

Fluorescence allows for relatively better control of parameters via excitation intensity and wavelength and emission wavelength. Moreover, fluorescent lifetimes can be used to minimize background interference or unwanted autofluorescence [10, 11]. Fluorescin (reduced fluorescein; [12] and 2,2′-azino-di-[3-ethylbenzthiazolinesulfonate] (ABTSO, [13]) have been reported for the chemifluorescent detection of bloody fingerprints. Several other fluorogenic peroxidase substrates have been synthesized for use in biochemical and cytological studies. Quantiblue (Pierce Chemical, Rockford, IL), Amplex Red and Oregon Green (Invitrogen, Carlsbad, CA) are commercially available. However, none of these are compatible with strong protein denaturing (fixing) solutions. Many fluorophores are also subject to photo-bleaching by intense excitation light sources and susceptible to quenching by low pH.

Scheme 5.1 Peroxidase-mediated oxidation of pyrogallol to purpurogallin

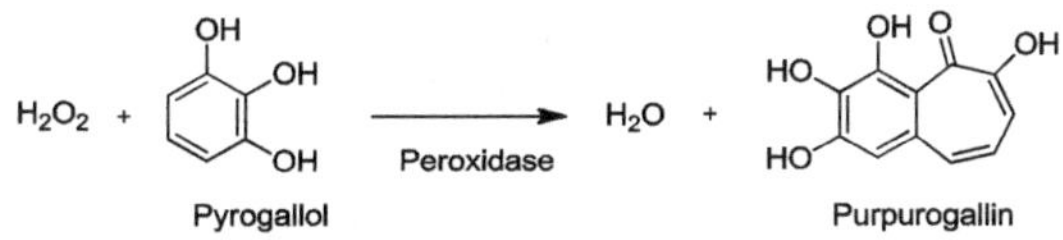

An issue with many reagents is that the relatively high protein content of blood may result in non-specific staining. Several groups (Jones 1982) [14], (Norkus 1986) [15], (Becraft 1987) [16] and Sears [17, 18] have studied protein stains for possible utility in bloody print development that include coomassie brilliant blue R and Crowle's double stain. General fluorescent stains for potential bloody fingerprint detection have also been reported, such as acid fuchsin (aka Hungarian Red, Theeuwen, [19]) and acid yellow 7 (Sears, [20]). Amino acid reagents such as ninhydrin and 1,8-diazafluoren-9-one (DFO), a fluorescent analog have also been screened. Other ninhydrin analogs including the 1-2-indandione family (Hauze, [21]) have been synthesized, but have not been screened for bloody fingerprint enhancement.

Apart from non-specific staining, there are other challenges associated with the available reagents. In particular, many leuco dyes are compatible with acidic protein fixative agents such as sulfosalicylic acid to preserve detail, but do not function well on dark and multi-colored substrates. In contrast, luminol and fluorescin can be used with dark and often multi-colored substrates, but cannot be used with fixatives. They are relatively less soluble in strong acids. Moreover, their chemiluminescence or fluorescence is quenched in acidic conditions. Furthermore, their oxidation products are not insoluble, and so tend to diffuse away, degrading detail. Benzidine and its homolog, TMB, benefited in this regard by the insolubility of their oxidized products. Krieg [22] has interestingly demonstrated that fluorogenic substrates can be designed to self-anchor for histological staining by covalent bonding. However, the majority of fluorogenic reagents require treatment with a separate fixing agent.

The oxidation of luminol is a fast and irreversible reaction, so the user must act quickly to observe and capture an image or it will be lost. Because of the inability to predict substrate color and the possibility of multi-colored substrates, a large range of emission wavelengths by a series of fluorophores would be useful, and a single fluorophore capable of wavelength shifts via a simple method such as pH change, or the addition or concentration change of some cation or anion, would be especially useful.

Although not studied in detail, the effectiveness of the previously reported dyes and stains depends very much on surface type (Sears, [20]). For instance, while Acid black 1 and Acid violet 17 are effective on all surfaces, Acid yellow 7 is the choice of reagent for nonporous surfaces. Similarly, DFO and ninhydrin can be employed only on porous surfaces. Hence, the study of dyes and stains on variable surfaces is of particular interest in developing more user-friendly and relatively simpler methods for bloody print detection.

In addition to detecting, enhancing and preserving bloody print detail, a relatively newer issue is the ability to successfully DNA type the residual blood content of the print Lee, [23], Laux, [24], Fregeau, [25], Budowle, [26], Martin, [27]. UV light and peroxides cause DNA damage. In general, most fingerprint enhancement treatments do not significantly degrade DNA. Nonetheless, any proposed new methods for detecting, fixing, or enhancing bloody prints should be validated for DNA testing.

The following, along with Table 5.1, summarizes the properties of many common reagents:

1. **Amido black and ninhydrin** form dark-colored dye complexes and have been used successfully on light colored transparent surfaces. They are not typically used on dark colored surfaces. Ninhydrin has a low background color but is unsuitable for porous surfaces since it runs off, and either distorts the print or fails to react before detail can be photographed. Amido black is very sensitive and works well on non-porous surfaces but its high background color (light to medium blue) compromises contrast on porous surfaces from which the stain can not be removed by rinsing.
2. **DFO (1,8-diazafluoren-9-one)** is a very sensitive dye that gives yellow fluorescence, but its use is limited, since for development, the specimen has to be heated at 100 °C. Treated porous items may become tainted with a yellow discoloration after time.
3. **Luminol** has had limited success since the only briefly appearing luminescence is weak. It is thus hard to photograph and fails to resolve fine ridge detail.
4. **Leucocrystal violet (LCV)** is used to develop prints mainly on light-colored backgrounds. Background development, however can occur under intense light due to photoionization products of the dye which may react with other substances not specific to blood. Porous materials can also strongly absorb the dye.
5. **ABTS** is a non-carcinogenic alternative to DAB. One must photograph the developed impressions within 2 weeks of development, as fading of the detail may occur. Some reports state a low recovery for DNA typing.
6. **Diaminobenzidine (DAB)** is highly toxic. Photographs must be taken shortly after development, due to light-induced photoionization.

5.3 Screening of Known Dyes for NBD- and Benzidine-Basic Dyes-Horseradish Peroxidase (HP) Activity

The purpose of this study was to screen the peroxidase-mediated reaction of NBD, and DAB with a known basic dyes series (compounds **1–12**, Fig. 5.1) in solution (Scheme 5.2). These dyes have been used in histological work; [28] however, the protocols for histological and fingerprint studies are significantly different. The aim was to investigate the formation of solid/polymeric products as well as changes in the spectroscopic properties of these dyes as a result of this reaction and their stability towards H_2O_2 and HP. The formation of solid/polymeric products can be desirable in order to preserve/enhance fingerprint details. Thus, the possibility of

Table 5.1 Common print reagents

Compound	Use	Type of stain	Color after development	Formula	Documentation	Shelf life
Amido black	Latent prints and to enhance visible blood prints on painted surfaces	Protein	Dark blue	HOAc/MeOH or	–	Indef.
Alcohol or water base				Citric Acid/ Kodak Photo Flo 600 solution		
Naphthol blue black						
Diaminobenzidine (DAB)	Latent prints and to enhance visible blood prints	Catalytic oxidation	Brown	Sulfosalicylic acid/H_2O_2	In the presence of light background development occurs due to dye photoionization	24 h at rt, 48 h if refrig.
ABTS	Latent prints and to enhance visible blood prints	Catalytic oxidation	Bright green	Water/H_2O_2	Fading of detail may occur within 2 weeks.	48 h refrig.
DFO (1,8-diazafluoren-9-one)	Latent prints on porous surfaces	Protein	Yellow fluorescence and red-pink ridge detail	MeOH/EtOAc/ HOAc	Treated porous items may become stained w/ a yellow discoloration after some time	6 months
Ninhydrin	Latent prints on porous surfaces	Protein	Purple	Acetone	Spraying, dipping or painting	1 year
				MeOH/ IsoprOH		
				Petroleum ether	75–80 % relative humidity is needed	
Comassie brilliant blue R(acid blue 83)	Latent prints and to enhance visible blood prints	Protein	Blue	HOAc/MeOH	–	Indef.
Crowle's double stain	Latent prints and to enhance visible blood prints	Protein	Blue	HOAc/TCAA	–	Indef.
Leucocrystal violet (LCV)	Latent prints, deposited blood on non-porous surfaces, visible blood prints	Catalytic oxidation	Purple	Sulfosalicylic acid	In the presence of light background development occurs due to dye photoionization	30 days
				H_2O_2		

(continued)

Table 5.1 (continued)

Compound	Use	Type of stain	Color after development	Formula	Documentation	Shelf life
Fluorescin	Reagent for latent bloodstains	Catalytic oxidation	Green fluorescence	EtOH/HOAc/ H_2O_2	Photographs should be taken within a few hours	Up to 24 h, recomm. to prep and use fresh
				Water/H_2O_2		
Luminol	Reagent for latent bloodstains	Catalytic oxidation	Blue glow	Water/H_2O_2, KOH	Photographs should be taken immediately	Solution has to be prepared and used immediately

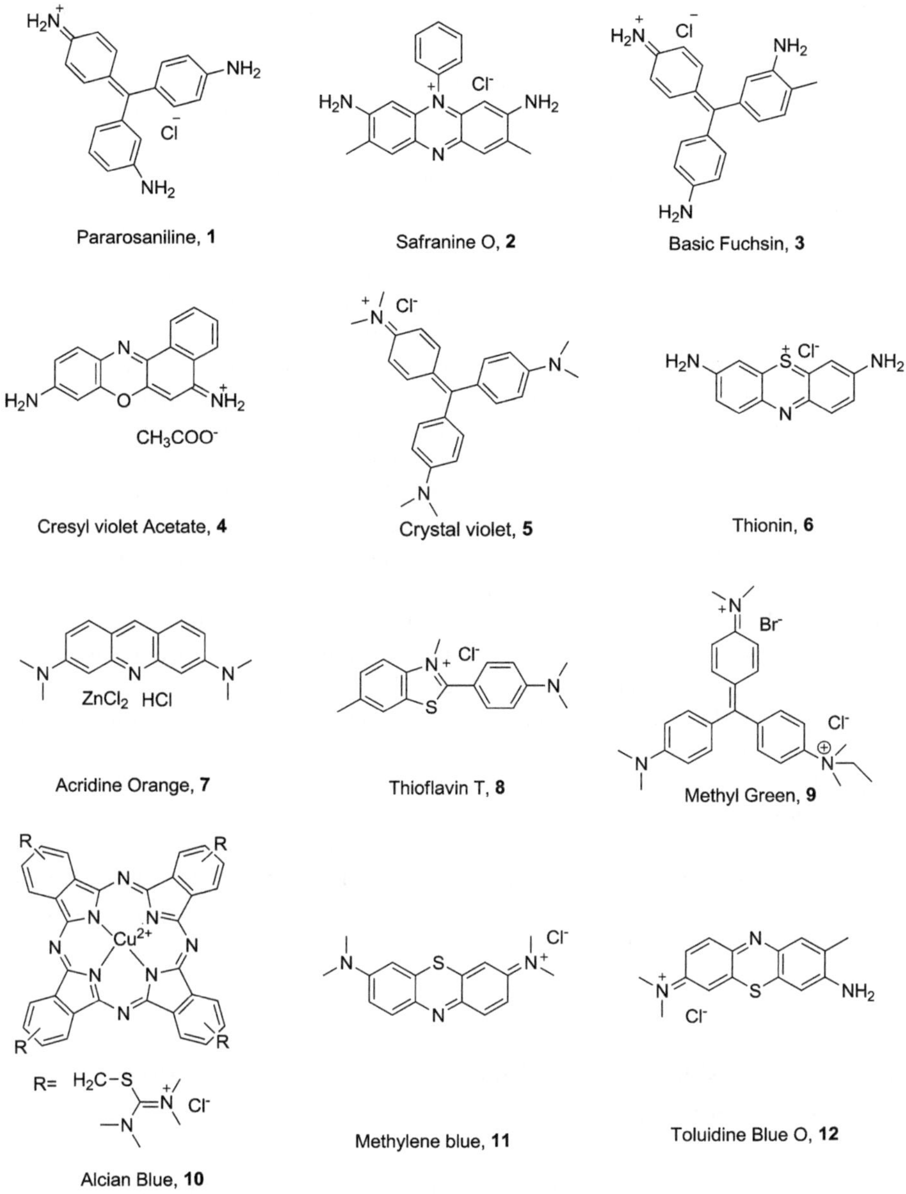

Fig. 5.1 Chemical structures of basic dyes used for screening

Scheme 5.2 Peroxidase mediated reaction of NBD and DAB

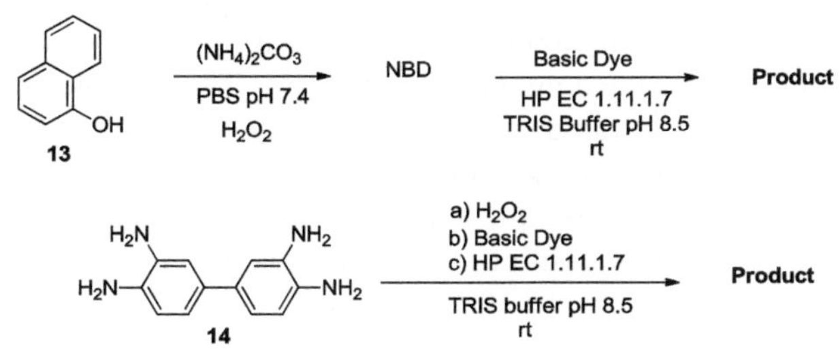

inducing this polymerization/solid formation via either covalent or non-covalent interactions with dyes or fluorophores sensitive to peroxidase was sought using NBD as a model system.

Changes observed in fluorescence after adding benzidine and NBD to each of the dye solutions are due to the emission of benzidine and its oxidation product, when an excitation wavelength at 280 nm is used. After addition of HP, the fluorescence of NBD and benzidine solutions is quenched (Table 5.2). The mixtures changed color and a precipitate was formed, (entries 1 and 2, Table 5.3). Although a precipitate was formed for the HP reaction with safranine O and toluidine blue, no changes in absorption or fluorescence were observed. No solid formation or changes in absorption or fluorescence were observed for azure A (entry 12, Table 5.3). After addition of HP, solid formation, changes in the absorption spectra and reduction or quenching of dye fluorescence was observed for dyes in entries 3, 5, 8, 9, 10, 13, 14, 15, and 16 when NBD was present. These changes may be attributed to the formation of precipitate that reduces the dye concentration in the mixture. When benzidine was present, in general, solid formation was not observed within 15 min; however, after 24 h at room temperature, a small amount of solid precipitated from each mixture. In the presence of NBD and after addition of HP, solid formation and a slight increase in the fluorescence of pyronine Y was observed. No major changes were observed when benzidine was used. In the presence of NBD or benzidine and after addition of HP, methyl green shows an increase in absorbance. Fluorescence emission is observed at 412 nm, (ex. 280 nm). A purple solid is formed in the presence of NBD and no precipitate is observed for the benzidine-methyl green mixture.

These preliminary results show that under the conditions tested, combinations of NBD-basic dyes that produce a precipitate, and enhance either absorption or fluorescence (for example pyronine Y and methyl green), represent potential candidates for further studies.

5.4 The Design of New Fluorogenic Reagents for Bloody Print Development

After our own screening and literature survey, it was evident that each known or potential reagent possesses inherent advantages and disadvantages. This prompted us to design new fluorophores.

We had previously reported a novel synthesis of xanthene dyes [29]. It involves the initial formation of methylated carbinol intermediates, followed by demethylation and concomitant condensation. This simple new method is inexpensive, efficient and rapid, and has led to the very first syntheses of benzo[a]- and [b] xanthene dye frameworks (Fig. 5.2) [30]. The benzofluorones are structurally related to fluorescein, a material already used in bloody fingerprint detection; however, they exhibit several advantages. They are able to absorb and emit at

Table 5.2 UV–vis and Fluorescence properties of basic dyes in the presence of NBD or benzidine

Entry	Compound	Absorption max. (nm)	Emission (nm) (exc. 280 nm)
1	NBD	291	465
2	Benzidine	–	410
3	Basic Fuchsin Flagella	540	–
4	Safranine O	519, 275, 251	–
5	Pararosaniline	540, 286, 243	–
6	Pyronin Y	546	558
7	Toluidine Blue	630, 288	648
8	Methyl Green	632, 255, 216	652
9	Crystal Violet	590, 303, 250	590
10	Thionine	600, 283	619
11	Methylene Blue	665, 290, 246	679
12	Azure A	635, 287	647
13	Giemsa Stain	–	410, 540, 678
14	Cresyl Violet	585, 269	321, 623
15	Acridine Orange	–	530
16	Thioflavine	–	530

Table 5.3 Summary of results for the reaction of mixtures of NBD and benzidine and basic dyes in the presence of HP

		NBD + HP		Benzidine + HP	
Entry	Compound	Solid	Solution	Solid	Solution
1	NBD	Purple	Purple		
2	Benzidine	–	–	Brown	Brown
3	Basic Fuchsin Flagella	Purple	Colorless	–	yellow
4	Safranine O	Dark red	Pink	Red	Red
5	Pararosaniline	Purple	Colorless	–	Light brown
6	Pyronin Y	Dark red	Red	–	Red
7	Toluidine Blue	Dark blue	Navy blue	–	Navy blue
8	Methyl Green	Purple	Colorless	–	Brown
9	Crystal Violet	Purple	Purple	–	Purple
10	Thionine	–	Purple	–	Purple
11	Methylene Blue	Dark blue	Dark blue	–	Green
12	Azure A	–	Dark blue	–	Dark blue
13	Giemsa Stain	Purple	Purple	–	Brown
14	Cresyl Violet	Purple	Purple	–	Brown
15	Acridine Orange	Brown	Brown	Brown	Brown
16	Thioflavine	Brown	Olive-green	–	Orange

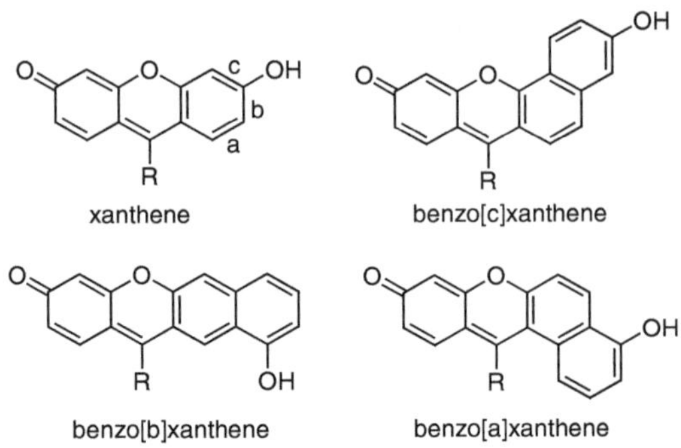

Fig. 5.2 The xanthene dye framework (*upper left*, typified by fluorescein) and three types of benzoxanthenes. Benzoxanthene types are each distinguished via the orientation of their naphthyl moieties. We have synthesized the first known benzo[b] and [c]xanthenes and have prepared the benzofluorones corresponding to types [a], [b], and [c] [31]. The benzofluorones, unlike xanthenes such as fluorescein, do not contain carboxylic acid/lactone moieties and thus are not quenched below pH = 5 (see Fig. 5.3)

relatively longer wavelengths, thereby reducing sample auto-fluorescence signals. Specific dyes in this new series exhibit multiple emission bands. They are pH sensitive and can absorb and emit over a very broad >400 nm range, spanning most of the visible (and part of the UV) region.

The compatibility of a prototypical dye in the benzofluorone series with commercial UV, blue and red imaging filter sets, and several dye laser excitation wavelengths, has already been demonstrated. This latter new compound was much more photostable compared to fluorescein and exhibited no cytotoxicity in imaging studies performed to date. Additionally, the benzofluorone fluorophores are less water soluble than fluorescein. We thus expected these materials to potentially preserve fingerprint detail better than fluorescein. The benzoxanthenes synthesized in our labs were also designed to be fluorescent over a broad solution pH range whereas fluorescein is essentially colorless and poorly fluorescent below pH 5 due to its carboxylic acid/lactone pH-dependent equilibrium (Fig. 5.3).

5.5 Facile One-Pot, Tri-Component Naphthoxanthene Syntheses

Multicolor-emitting benzofluorones were synthesized via one pot acid condensation.[32, 33] In addition and for the purpose of this study, the new benzorhodafluor **38** was synthesized in a two step process. Schemes 5.3 and 5.4 summarize the syntheses of these new fluorophores.

These fluorophores are reduced to the corresponding di-hydro derivatives (*leuco* forms) under either basic or acidic conditions [34, 35]. Deep colored solutions change to colorless or pale solutions once they are reduced. These new *leuco*

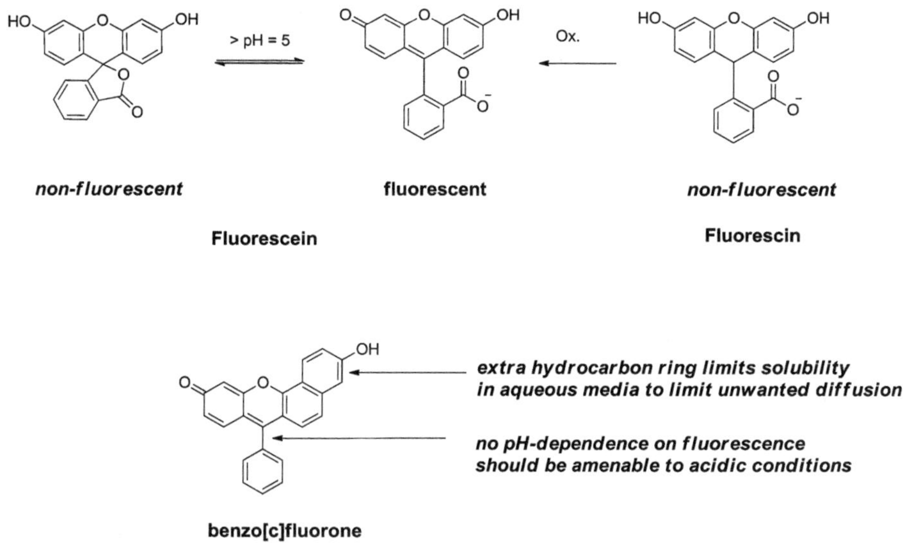

Fig. 5.3 Reagent design principles. The differences between fluoresceins (*TOP*) and the designed fluorones used in the authors' studies (*BOTTOM*) are (**i**) the carboxylate-lactone moiety and (**ii**) the degree of hydrophobicity

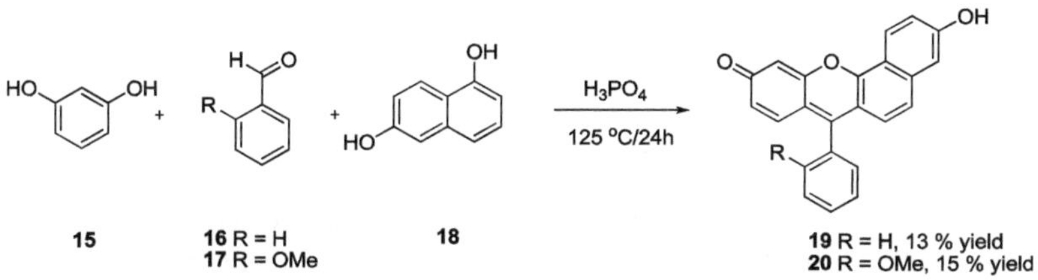

Scheme 5.3 One pot synthesis of benzofluorones type [c], (angular benzannulation)

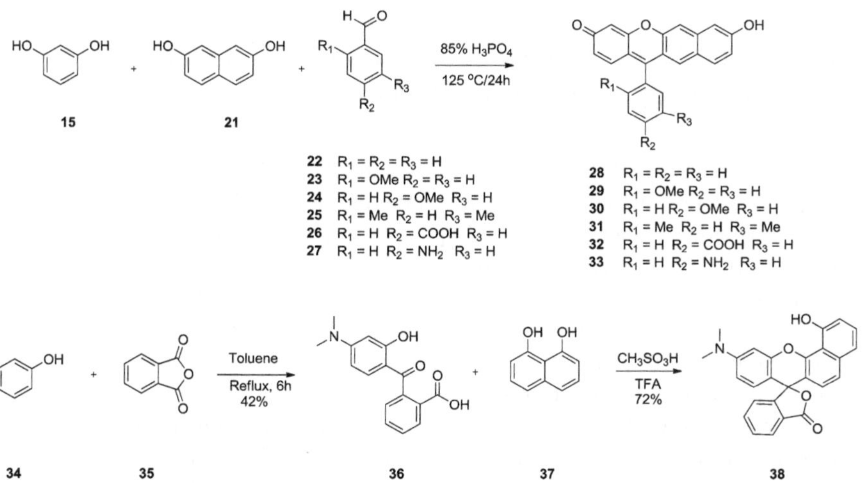

Scheme 5.4 *Top*: one pot synthesis of benzofluorones type [b], (linear benzannulation). *Bottom*: synthesis of benzorhodafluor **38**

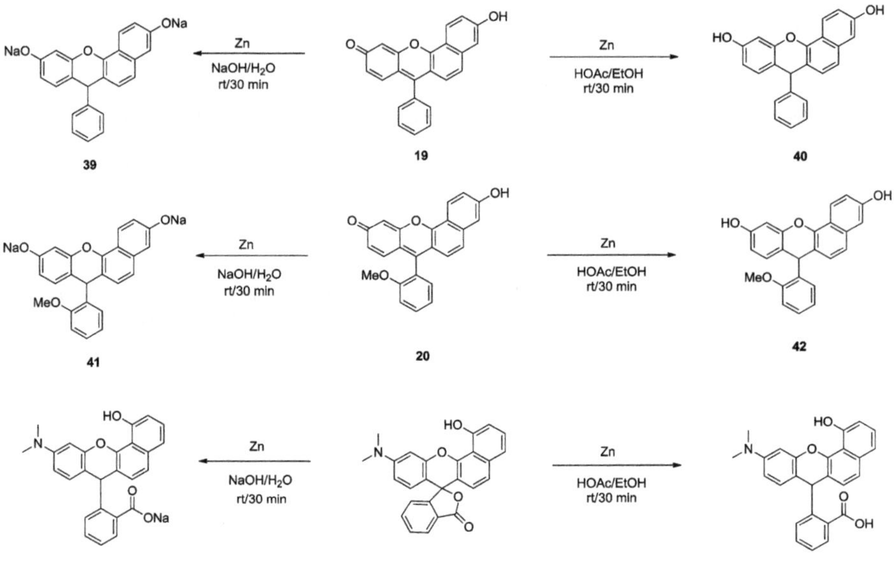

Scheme 5.5 Synthesis of the *leuco* forms of selected fluorophores

derivatives are colorless until exposed to the peroxidase activity of heme; at which time, the oxidized dyes became less soluble and visible. Scheme 5.5 shows representative examples of the synthesis of these *leuco* dyes.

5.6 Initial Key Findings Using Naphthoxanthene for Bloody Fingerprint Development

From the screening of the series of the new naphthoxanthene synthetic dyes, three compounds were selected for fingerprint development. This selection was based on the performance of their corresponding *leuco* bases when subjected to the oxidation mediated by HP (horseradish peroxidase)/H_2O_2 or commercial pig blood/H_2O_2 in solutions under varying pH conditions, as described below. It is important to recall that protocols for bioimaging (or biosensing) and fingerprint development studies can be significantly different. This investigation showed the feasibility of using the newly-developed reagents to respond to the peroxidase-type activity of blood.

In order to investigate the effect of pH on the oxidation reaction mediated by blood in solution, reactions were monitored for each of the selected dyes by UV–vis and fluorescence over a pH range from 4–9. A summary of results are shown in Figs. 5.4, 5.5, and 5.6. Compounds tested were the leuco dyes **39**, **41**, and **43**.

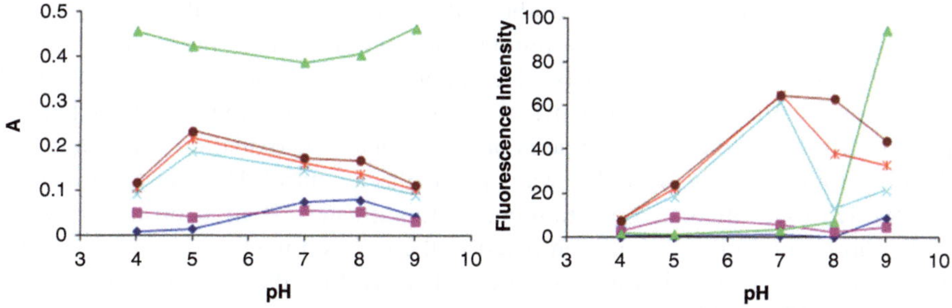

Fig. 5.4 pH profile for the oxidation reaction of compound **39**. (**a**) UV–vis. (**b**) fluorescence. Oxidation reaction mediated by: blood-H_2O_2; *black*, H_2O_2; *magenta*, H_2O_2-HP; *green*, 10 min pig blood- H_2O_2; *blue*, 20 min pig blood- H_2O_2; *red*, 30 min pig blood- H_2O_2, *brown*

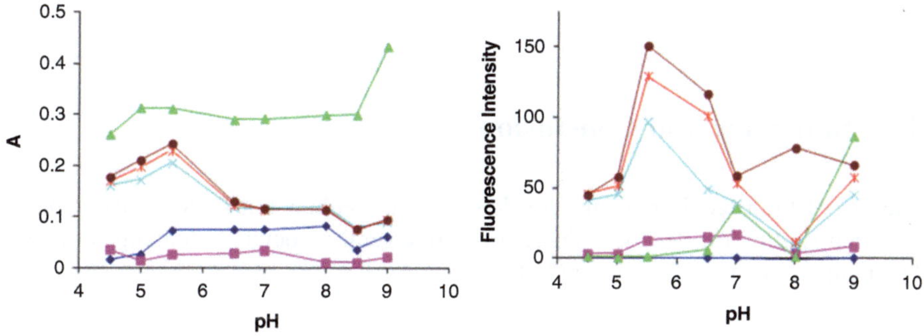

Fig. 5.5 pH profile for the oxidation reaction of compound **41**. (**a**) UV–vis. (**b**) fluorescence. Oxidation reaction mediated by: blood-H_2O_2; *black*, H_2O_2; *magenta*, H_2O_2-HP; *green*, 10 min pig blood- H_2O_2; *blue*, 20 min pig blood- H_2O_2; *red*, 30 min pig blood- H_2O_2, *brown*

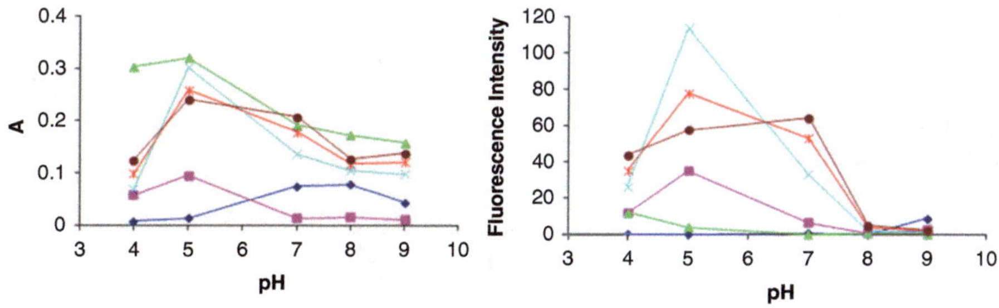

Fig. 5.6 pH profile for the oxidation reaction of compound **43**. (**a**) UV–vis. (**b**) fluorescence. Oxidation reaction mediated by: blood-H_2O_2; *black*, H_2O_2; *magenta*, H_2O_2-HP; *green*, 10 min pig blood- H_2O_2; *blue*, 20 min pig blood- H_2O_2; *red*, 30 min pig blood- H_2O_2, *brown*

Whereas the UV–vis spectra showed that the oxidation reaction proceed at higher rate under acidic pH with maxima between pH 4.5 and 5.6 for all compounds tested, the fluorescence spectra showed that optimal conditions for fluorescence emission from the oxidized products are slightly shifted to higher pH.

Importantly, these results also show that the signal from the blood at the excitation wavelengths used is minimum or negligible. It is interesting to note that while the oxidation products of compounds **39** and **41** show an increased fluorescence emission at higher pH, the fluorescence emission for the oxidation product from compound **43** decreases or is minimal. The oxidation reaction of **39**, **41** and **43** mediated by H_2O_2-HP was included as a control. UV–vis spectra show that the H_2O_2-HP mediated reaction occurs to a greater extent as observed by the higher absorption obtained; as compared to the reaction mediated by blood; however, fluorescence emission is only observed at pH 9 for the oxidation products of compounds **39** and **41**.

Bloody fingerprints were placed on non-porous and semi-porous surfaces using known methodologies [36, 37]. Fingerprints were processed with the corresponding *leuco* bases **39**, **41** and **43** obtained respectively from compounds **19**, **20**, and **38** as shown in Figs. 5.7, 5.8, 5.9, 5.10, and 5.11.

5.7 Summary and Conclusion

The use of fluorogenic reagents for detecting and enhancing fingerprints in blood has been studied for decades. There are still several associated challenges to date. The main issues to overcome include:

Fig. 5.7 Pig blood fingerprint on glass surface processed using reduced compound **39**. Print was illuminated with Laser 532 nm and visualized through an orange filter (Courtesy of Allison Murphy, Forensic Lab of the Orange County California Sheriff and Coroner's Office)

Fig. 5.8 Pig blood fingerprint on glass surface and processed using reduced compound **41** Print was illuminated with a Laser 532 nm and visualized through an orange filter (Courtesy of Allison Murphy, Forensic Lab of the Orange County California Sheriff and Coroner's Office)

Fig. 5.9 Pig blood fingerprint on glass surface and processed using reduced compound **43**. Print was illuminated with a Laser 532 nm and visualized through an orange filter (Courtesy of Allison Murphy, Forensic Lab of the Orange County California Sheriff and Coroner's Office)

Fig. 5.10 Pig blood fingerprint on drywall surface and processed using reduced compound **45**. Print was illuminated with a Laser 532 nm and visualized through an orange filter (Courtesy of Allison Murphy, Forensic Lab of the Orange County California Sheriff and Coroner's Office)

Fig. 5.11 Pig blood fingerprint on black cardboard surface and processed using reduced compound **39.** Pictures were taken after 30 min and 108 days of processing. Shown here after 108 days of storage without protection from air or light. The print was illuminated with an ALS 455 nm and visualized through an orange filter

(i) most common fingerprint reagents are unsuitable for dark and multi-colored substrates,
(ii) the visualizable oxidation products of luminol and fluorescein are not insoluble enough to prevent unwanted diffusion resulting in degradation of detail,
(iii) the oxidation of luminol is a fast and irreversible reaction, so the user must act quickly to observe and capture an image or it will be lost and
(iv) the chemiluminescence of luminol and the fluorescence of fluorescein is quenched in acidic fixing media.

Each of these issues has been addressed to some degree to date. Highlights include:

1. The oxidation reaction mediated by HP/H_2O_2 or pig blood/H_2O_2 indeed afforded fluorescent products in acidic media. This addresses issue **(iv)** above.
2. The use of NaOAc and PEG MW 35 000 as additives to the reaction mixture accelerated the oxidation reaction. It was not a goal to speed the development process. This is an observation that may be of value for future work.
3. The reduced form of the dyes is soluble at acidic pH (2–5) and is also compatible with strong protein denaturing solutions (i.e. sulfosalycilic acid, 20 g/L). This addresses an important aspect of solubility properties issue **(ii)** above in that the non-visualizable leuco dye forms should be soluble during application but turn into insoluble products when oxidized and visualized to enhance preservation and detail.
4. The oxidized forms have diminished solubility [issue **(ii)**]. Formation of colloidal type precipitates were observed for oxidations carried out in solution at acidic pH. Thus improvement in detail preservation may be possible.
5. Because of their broad absorption range, fingerprints developed using these dyes were visualized over the range of 400–570 nm [issue **(ii)**], using alternative light sources equivalent to several dye laser excitation wavelengths. Thus the use of UV-light for visualization which can damage DNA can be avoided-a promising area for further development.
6. The oxidation products are stable, allowing one to capture detailed images after long periods of time (at least 10 months) that do not differ from those taken immediately [(issue **(iii)**].
7. Improvement in the fluorescence emission is obtained by spraying the treated fingerprint with a solution of higher pH. Promising results were obtained on pig blood fingerprint trials on glass and colored paper board using the newly-created dyes.

In summary, the new fluorophores are useful on multicolor and dark substrates in a variety of fixative media, initial detail is not degraded even after many months. The dyes were found potentially useful on both porous and non-porous surfaces.

Acknowledgements The authors thank the generous support of the National Institute of Justice (Award 2007-DN-BX-K171) and the National Institutes of Health (R01 EB002044). We also very gratefully acknowledge the Forensic Lab of the Orange County California Sheriff and Coroner's

Office, in particular John Hartman, who initiated the project and guided our initial work in this area, and Allison Murphy, who obtained the prints shown in Figs. 5.7, 5.8, 5.9, and 5.10.

References

1. Adler O, Adler R (1904) Behavior of certain organic compounds towards blood, especially as regards the detection of blood. Zeit physiol Chem 41:59–67
2. Ruttan RF, Hardisty RHM (1912) A new reagent for detecting occult blood. Can Med Assoc J 2(11):995–998
3. Holland VR, Saunders BC, Rose FL, Walpole AL (1974) A safer substitute for benzidine in the detection of blood. Tetrahedron 30(18):3299–3302
4. Yapping L, Wang Y (2004) Bloody latent fingerprint detection using LeuR6G. J Forensic Identif 54(5):542–546
5. Wang Y, Weiping Z, Janping M (2007) Eosin Y detection of latent blood prints. J Forensic Identif 57(1):54–58
6. Gaensslen R (1983) Sourcebook in forensic serology, immunology, and biochemistry. National Institute of Justice, Department of Justice U.S. Government Printing Office, Washington, DC
7. Caldwell JP, Kim ND (2002) Extension of the color suite available for chemical enhancement of fingerprints in blood. J Forensic Sci 47(2):332–340
8. Specht W (1937) The chemiluminescence of hemin as a means of finding and recognizing blood traces of forensic importance. Angew Chem 50:155–157
9. Young T (2006) A photographic comparison of luminol, fluorescein, and Bluestar. J Forensic Identif 56(6):906–912
10. Ong SK, Seah LK, Murukeshan VM, Ong LS (2004) Enhancing fluorescence in time-resolved imaging of latent fingerprints. J Forensic Identif 54(3):281–295
11. Menzel ER, Menzel LW (2004) Ordinary and time-resolved photoluminescence field detection of traces of explosives and fingerprints. J Forensic Identif 54(50):560–571
12. Cheeseman R, DiMeo L (1995) Fluorescein as a field-worthy latent bloodstain detection system. J Forensic Identif 45:631–646
13. Caldwell JP, Henderson W, Kim ND (2000) ABTS: A safe alternative to DAB for the enhancement of blood fingerprints. J Forensic Identif 45(4):785–794
14. Jones RJ, Pounds CA (1982) The enhancement of fingerprints made in blood. Home Office Central Research Establishment (HOCRE), Aldermaston, England
15. Norkus P, Noppinger K (1986) New reagent for the enhancement of blood prints. Identification News, 355
16. Becraft M, Heintzman M (1987) Application of Crowle's stain in the enhancement of bloody fingerprints. Fingerprint World 12:65–66
17. Sears VG, Butcher CPG, Prizeman TM (2000) Enhancement of fingerprints in blood – part 2: protein dyes. J Forensic Identif 51(1):28–38
18. Sears VG, Prizeman TM (2000) Enhancement of fingerprints in blood – part 1: the optimization of amido black. J Forensic Identif 50(5):470–480
19. Theeuwen ABE, van Barneveld S, Drok JW, Keereweer I, Limborgh JCM, Naber WM, Velders T (1998) Enhancement of footwear impressions in blood. Forensic Sci Int 95 (2):133–151
20. Sears VG, Butcher CPG, Fitzgerald LA (2005) Enhancement of fingerprints in blood – part 3: reactive techniques, acid yellow 7, and process sequences. J Forensic Identif 55(6):741–763
21. Hauze DB, Petrovskaia O, Taylor B, Joullie MM, Ramotowski R, Cantu AA (1998) 1,2-indanediones: new reagents for visualizing the amino acid components of latent prints. J Forensic Sci 43(4):744–747

22. Krieg R, Halbhuber KJ (2004) Novel oxidative self-anchoring fluorescent substrates for the histochemical localization of endogenous and immunobound peroxidase activity. J Mol Histol 35(5):471–487
23. Lee HGR, Pagliaro E, Buman M (1989) The effect of presumptive test, latent fingerprint and some other reagents and materials on subsequent serological identification, genetic marker and DNA testing in bloodstains. J Forensic Identif 39:331–350
24. Laux DL (1991) Effects of luminol on the subsequent analysis of bloodstains. J Forensic Sci 36 (5):1512–1520
25. Fregeau CJ, Germain O, Fourney RM (2000) Fingerprint enhancement revisited and the effects of blood enhancement chemicals on subsequent Profiler Plus (TM) fluorescent short tandem repeat DNA analysis of fresh and aged bloody fingerprints. J Forensic Sci 45(2):354–380
26. Budowle B, Leggitt JL, Defenbaugh DA, Keys KM, Malkiewicz SF (2000) The presumptive reagent fluorescein for detection of dilute bloodstains and subsequent STR typing of recovered DNA. J Forensic Sci 45(5):1090–1092
27. Martin LA, Cahill CF (2004) Recovery of DNA from latent blood after identification by fluorescein. J Forensic Identif 54(6):660–667
28. Mauro A, Germano I, Giaccone G, Giordana MT, Schiffer D (1985) 1-Naphthol basic dye (1-NBD) – an alternative to diaminobenzidine (DAB) in immunoperoxidase techniques. Histochemistry 83(2):97–102
29. Yang YJ, Escobedo JO, Wong A, Schowalter CM, Touchy MC, Jiao LJ, Crowe WE, Fronczek FR, Strongin RM (2005) A convenient preparation of xanthene dyes. J Org Chem 70 (17):6907–6912
30. Yang YJ, Lowry M, Schowalter CM, Fakayode SO, Escobedo JO, Xu XY, Zhang HT, Jensen TJ, Fronczek FR, Warner IM, Strongin RM (2006) An organic white light-emitting fluorophore. J Am Chem Soc 128(43):14081–14092
31. Yang YJ, Lowry M, Xu XY, Escobedo JO, Sibrian-Vazquez M, Wong L, Schowalter CM, Jensen TJ, Fronczek FR, Warner IM, Strongin RM (2008) Seminaphthofluorones are a family of water-soluble, low molecular weight, NIR-emitting fluorophores. PNAS 105 (26):8829–8834
32. Hilderbrand SA, Weissleder R (2007) One-pot synthesis of new symmetric and asymmetric xanthene dyes. Tetrahedron Lett 48(25):4383–4385
33. Sibrian-Vazquez M, Strongin R (2009) Optimising the synthesis and red-green-blue emission of a simple organic dye. Supramol Chem 21(1–2):107–110
34. Trozzi TA, Schwartz RL, Hollars ML (2000) Processing guide for developing latent fingerprints U.S. Department of Justice. Federal Bureau of Investigation, Laboratory Division, Latent Print Unit.
35. Fluorescein detection of latent bloodstains. http://www.latent-prints.com/fluorecein2.htm
36. Jaret Y, Heriau M, Donche A (1997) Transfer of bloody fingerprints. J Forensic Identif 43:38–41
37. Langenburg G (2008) Deposition of bloody friction ridge impressions. J Forensic Identif 58:355–387

Chapter 6
Indicators for Ionic Copper in Biology

Richard B. Thompson and Hui Hui Zeng

Abstract Despite its status as a trace element in many organisms, copper is garnering increased interest for its biological functions and potential roles in many diseases. This review summarizes recent progress in the use of fluorescent indicators for determining copper ions in a variety of biological matrices. Following a brief summary of the chemistry and biology of Cu(I) and Cu(II), the review covers both organic fluorescent indicators as well as biologically derived fluorescent indicators or sensors. The future outlook for improved indicators and sensors is discussed.

We very much regret that it was not possible to discuss all the important developments in fluorescence-based copper sensors within this limited space, so we focused upon those that seemed of greatest interest to us. We ask forgiveness from the many creative investigators whose work could not be included.

Keywords Copper • Fluorescence sensor • Biosensor • Fluorescence lifetime • Fluorescence anisotropy • Biosensor

6.1 Basic Chemistry of Copper Ions

Copper is a transition element of Group 11 (IB) found in nature as a mixture of two isotopes, ^{63}Cu (69.09 %) and ^{65}Cu (30.91 %). In biology the element is nearly always found in one of two ionic forms, Cu(I) and Cu(II), and it is to these forms we will confine subsequent discussion. Like other transition metals with unfilled orbitals it exhibits an extensive coordination chemistry with a variety of ligands which accounts for many of its functions in biology. It frequently serves in a catalytic role in enzymatic reactions, especially reduction/oxidation reactions [1, 35]. The biological roles of copper ions are discussed in the next section, whereas a brief summary of its basic chemistry relevant to sensing is below:

The inorganic and bioinorganic chemistry of Cu(I) and Cu(II) have been reviewed [1, 8]. Cu(I) exhibits a $3d^{10}$ electronic structure, making it diamagnetic

R.B. Thompson (✉) • H.H. Zeng
Department of Biochemistry and Molecular Biology, University of Maryland School of Medicine, 108 N. Greene Street, Baltimore, MD 21201, USA
e-mail: RThompson@som.umaryland.edu

© Springer International Publishing Switzerland 2016

C.D. Geddes (ed.), *Reviews in Fluorescence 2015*,
DOI 10.1007/978-3-319-24609-3_6

and most of its complexes colorless in the visible wavelengths. However, it commonly exhibits metal to ligand charge transfer absorption bands in the ultraviolet, particularly with sulfur ligands. Much of Cu(I)'s chemistry can be attributed to the fact that Cu(I) is easily oxidized to Cu(II):

$$Cu^{+} + e^{-} \leftrightarrow Cu^{0} \qquad E^{0} = 0.52\ V$$

$$Cu^{2+} + e^{-} \leftrightarrow Cu^{+} \qquad E^{0} = 0.153\ V$$

The facile oxidation in water of Cu(I) by dissolved oxygen (lifetimes of the order of seconds) can be greatly slowed depending on the ligation state of the ion and the presence of high ionic strength; for instance, the CH_3CN complex is significantly more stable than the unligated ion but not very water soluble. Avoidance of oxidation *in vitro* typically requires the presence of reducing agents and anoxic conditions. Due to the reducing environment of the cell interior it is widely believed that what little copper is there is present as Cu(I). The most prevalent ligation state for Cu(I) is tetrahedral, but examples of others are known. Some Cu(I) complexes with proteins such as metallothionein exhibit very weak orange fluorescence when MLCT bands are excited with 250–300 nm light in the ultraviolet, but the low observed quantum yields seem weak for practical sensing [56].

By comparison, Cu(II) exhibits an electronic structure of $3d^9$, making it paramagnetic, and it exhibits a diverse coordination chemistry, preferring nitrogen and sulfur ligands in typically 4-, 5-, and 6-coordinate arrangements, often with square, distorted octahedral, or distorted tetrahedral stereochemistry. Many salts are water soluble, and complexes are frequently blue or green. In some cases there are charge transfer absorbance bands for some Cu(II) complexes, but these are usually found at shorter wavelengths. Cu(II) is a promiscuous fluorescence quencher due to its paramagnetism, the colored nature of many of its complexes, its ease of reduction, and its relatively high atomic number compared with most other biologically important elements. Some complexes (including with proteins [36]) exhibit weak d-d absorption bands ($\varepsilon \approx 10$–$50\ M^{-1}\ cm^{-1}$) in the red or near infrared; being much weaker than typical organic chromophore extinction coefficients, these bands have found little use analytically (except see below).

An aspect of copper chemistry that has aroused substantial concern in biological systems is Cu(I)'s propensity to form reactive oxygen species (such as hydroxyl radical) via a reaction with hydrogen peroxide similar to the Fenton reaction with ferrous ion:

$$Cu(I) + H_2O_2 \rightarrow Cu(II) + OH\cdot + OH^{-}$$

The hydroxyl radical is extremely reactive, capable of reacting with many compounds and having a lifetime of microseconds or less in aqueous solution. If means are available to re-reduce the copper ion *in situ* the metal catalyzes the production of a steady supply of the free radical, which can cleave DNA. In the laboratory Fe (II)-based reagents are used for DNA cleavage and exhibit varying levels of sequence specificity; *in vivo* the reaction is a source of strand breakage and hence mutagenesis. Thus it seems likely that, as a mutagenic catalyst, copper ions would not be permitted to "wander" free in the cell; and it has been proposed that there is little or no free copper ion in ordinary cells [50]. The discovery of copper "chaperones", proteins that carry copper ions in the cell from transporters on the cell surface to apoenzymes where they are needed, seems very logical in view of this consideration [47, 53].

6.2 Copper Ions in Biology

Compared with Fe, Zn, and Ca, copper species seem to play relatively few roles in biology, particularly in human biology. Copper is much less prevalent in humans (total body burden: 100 mg) than iron or zinc, and unlike zinc, copper is found in only a handful of enzymes in humans, all involved in redox reactions [1] (Table 6.1).

Copper is often found in the cell bound to metallothioneins, which are small, cysteine-rich proteins which are believed to play multiple roles in metal ion metabolism, distribution, sequestration, and detoxification [32, 51]. Other roles that copper may play apart from serving as an enzymatic cofactor are unclear, but we note that almost 50 % of copper in the body is found associated with the skeleton, and more than 25 % in muscle [35]. While lysyl oxidase plays a role a key role in collagen and elastin crosslinking in bone, and cytochrome *c* oxidase and superoxide dismutase are prevalent in tissues that consume (and produce) substantial amounts of energy such as muscle, the relatively large proportions of total copper suggest copper may also be present in other forms in those tissues. Recent reviews have summarized new roles for copper in cancer and the copper transporter *Ctr1* in cisplatin uptake [71], as well as the current level of understanding of copper transport [24]. At present, the picture of copper uptake from the diet of animals, its distribution to tissues, its excretion with the bile, and how these processes are controlled throughout the organism is far from complete.

Copper bound to proteins has been classified into four distinct groups. The extensively studied blue copper proteins from plants and microorganisms, such as plastocyanin, azurin, and stellacyanin, have Type I copper binding sites within a small polypeptide: a single Cu(II) ion ligated in a distorted trigonal geometry by (in the case of stellacyanin) two histidine imidazoles and a cysteine thiol, with a more distant methionine thioether ligated axially. They exhibit a characteristic blue absorbance around 610 nm (ε 2200–6000 M^{-1} cm^{-1}), an unusual EPR spectrum, and relatively high redox potentials which account for their function. Type 2 copper sites are found in "normal" copper enzymes such as superoxide dismutase, amine oxidases, and dopamine β-hydroxylase, all of which catalyze redox reactions. The Cu(II) they possess is bound in essentially a square planar site where the ligands are all histidinyl imidazoles. Type 3 copper comprises paired Cu(II) ions, which are

Table 6.1 Copper enzymes in humans and diseases arising from their deficiency

Enzyme	Function	Deficiency/Symptoms
Ceruloplasmin	Ferroxidase	Diabetes and basal ganglia degeneration resulting from iron accumulation
Superoxide dismutases	Destruction of toxic superoxide	Familial ALS (Lou Gehrig's disease)
Cytochrome *c* oxidase subunits	Electron transport/ ATP synthesis	Mitochondrial myopathies
Dopamine β-hydroxylase	Convert dopamine to norepinephrine	Hypotension, hypothermia
Peptidylglycine α-amidating monoxygenase	Peptide hormone amidation	Embryonic lethal in mice
Tyrosinase	Melanin synthesis	Absence of pigment in skin
Lysyl oxidase	Cross-linking in elastin and collagen	Osteolathyrism: laxity of skin and joints

Modified from [9]

EPR silent and function in oxidases such as tyrosinase. Finally, Type 4 comprises bound Cu(I), usually as the reduced form of Cu(II)-containing enzymes. These groupings are functionally useful, but as structural and functional details of copper proteins continue to be elucidated, more subtle differences are appearing and a more detailed classification may well be necessary [1].

6.3 Copper in Human Disease

Copper has been implicated to a greater or lesser degree in a range of human diseases. In some diseases it plays a central role, such as Wilson's Disease and Menke's Disease [9]. Wilson's disease [44] is a rare autosomal recessive genetic disorder that arises from defect(s) in a Cu-transporting ATPase (ATP7A) responsible for biliary excretion, resulting in copper buildup in the liver and other tissues. Menkes Disease also arises from an X-linked inherited defect in a different copper transporting ATPase (ATP7B), but in this case patients are copper-deficient, with symptoms arising from loss of function of copper-dependent enzymes; similar symptoms appear when the genes for these enzymes are mutated or knocked out, or chemically inhibited by compounds such as beta-aminopropionitrile [75]. Several reports suggest that Cu (and perhaps Zn and Fe) play roles in the development and progression of Alzheimer's Disease. In particular, some investigators report that metal ions promote the aggregation and precipitation of Aβ (or proteolysis products thereof) to form insoluble aggregates or fibrils that mimic the plaques that are a (postmortem) clinical hallmark and proposed cause of the disease [4, 21]. The concentrating effect on the metal ions of the protein is proposed to promote neuron death by oxidative reactions catalyzed by the copper, but this remains unclear. However, these ideas remain controversial, with some investigators claiming that Alzheimer's results from a shortage of copper, and others an excess that can be treated with chelation therapy [7, 49]; clear understanding of copper's role would likely suggest potential therapies. Copper has also been proposed as the source of pathology in some prion-mediated diseases such as bovine spongiform encephalopathy (BSE, "Mad Cow Disease"), Creutzfeld-Jakob Disease in humans, and scrapie in sheep. Prions (proteins that adopt different conformations, one conformer of which promotes aggregation of the same protein by a cooperative process) are known to bind copper [46], and have been hypothesized to produce free radicals via Fenton chemistry as described above.

6.4 Fluorescent Copper Indicators and Sensors: Figures of Merit

Fluorescent indicators and sensors have a number of properties which bear on their usefulness for particular applications or experiments. To begin with, we define a fluorescent indicator as a molecule or assembly of molecules which responds to the presence of a chemical analyte (copper ion, in this instance) by a change in a fluorescence observable; in favorable cases, this response may be correlated with a concentration, activity, or other property of the analyte. By comparison, fluorescence-based sensors represent a subset of indicators that furthermore permits continuous or quasi-continuous determination of the analyte over time, supplying information in near real time. Outside the narrow confines of biochemistry, sensors are often held to include the instrumentation and any computer or other device that converts the observable to a form (changes data domains, in Malmstadt's phrase) that provides usable information to the operator [39, 67]. Thus in the case of pH as an analyte, a pH electrode (and the associated voltmeter) would be a sensor, while the classical Fehling's test for reducing sugars is not a sensor since the analyte is irreversibly consumed in a single assay reaction. Of course, some sensors continuously monitor the analyte by continuously reacting it with some reagent in a flow system; the famous oceanographic nitrate sensor of Ken Johnson is an example [23]. Sensors that do not consume a reagent to produce a signal are termed "reagentless". We generally define a biosensor as a sensor whose recognition and/or transduction employ biologically derived or biomimetic molecules, not one whose cognate analyte is somehow biologically related.

The common figures of merit for sensors include sensitivity (sometimes conveniently expressed in terms of a minimum detection limit of the analyte), selectivity (ability to discriminate against potential interferents), accuracy (proximity of the measured analyte value to the true value), precision (variance in measured values), and dynamic range (range of analyte concentrations that can be reliably determined). From a chemical standpoint, important attributes include stability (particularly photostability), solubility, and compatibility; for quantitation, simple binding stoichiometry and rapid kinetics are preferred. As fluorescent indicators, it is typically desirable that they be photostable (and not sensitize photochemistry), have good absorbance at long wavelengths, exhibit a reasonable quantum yield, and in some cases have a long (or short) lifetime. For quantitative work, fluorescence sensors based on intensity ratios (wavelength ratiometric and anisotropy/polarization) or lifetime are almost indispensable. From the standpoint of biology, several additional factors are of importance: propensity to be taken up by (and excreted from) cells, tissues, and whole organisms (pharmacokinetics), lack of toxicity, and for protein-based sensors, expressibility in target cells or organelles. For microscopy, so-called "turn-on" indicators which exhibit an increase in quantum yield upon analyte binding are preferred but not essential. Of course, no indicator will possess all these attributes, and it is typically the case that a particular indicator will excel in some areas and not others, and thus be better suited for certain experiments than others.

Selectivity is a vital issue in sensors designed to operate in complex matrices such as natural waters or the interiors of cells. For the large proportion of chemical sensors that reversibly bind the analyte, interferents often include molecules that compete with the analyte for the binding site. For example, it is well known [16] that the widely used Ca indicators Fura-2 and Indo-1 also bind zinc ion, and in fact bind it with one hundred-fold higher affinity with similar fluorescence response. This is probably not an issue in many intracellular calcium determinations since intracellular free zinc levels are well below K_D for zinc binding to the Ca indicators. However, it is vital to recognize that an interferent may produce a substantial error even if it does not elicit the fluorescence response of the analyte, by simply competing for the binding site(s). This can be illustrated by the following example. The fluorescence sensors for Cu(II) we have devised are based on reversible binding to fluorescent-labeled variants of human apocarbonic anhydrase II (CA II) [62–64] resulting in changes in fluorescence lifetime of the fluorophores conjugated to the CA protein. These indicators are described in more detail below. In the case described below, binding of Cu(II) results in almost a three-fold reduction in lifetime (2.8–1.2 ns) and intensity, such that the proportion of fluorescence emitters with a reduced lifetime can be directly related to the concentration of Cu(II), and titration with Cu(II) causes a monotonic decrease in average lifetime. Zn(II) also binds to CA, with approximately forty-fold lower affinity [40], but has essentially no effect on the fluorescence intensity or lifetime. If we measure the Cu (II)-dependent average lifetime as a function of increasing [free Cu(II)] in the absence of Zn(II), we see a decline in the average lifetime as the Cu(II) is increased, with an inflection point near the Cu(II) binding constant Fig. 6.1. However, if the same experiment is performed in the presence of free zinc at a concentration about forty-fold higher than its binding constant (1.5 nM), the perturbation is dramatic: the apparent average lifetime measured at 10^{-11} molar Cu(II) with Zn(II) present is the same as that measured at 3×10^{-13} molar without, an error of thirty-fold. Thus

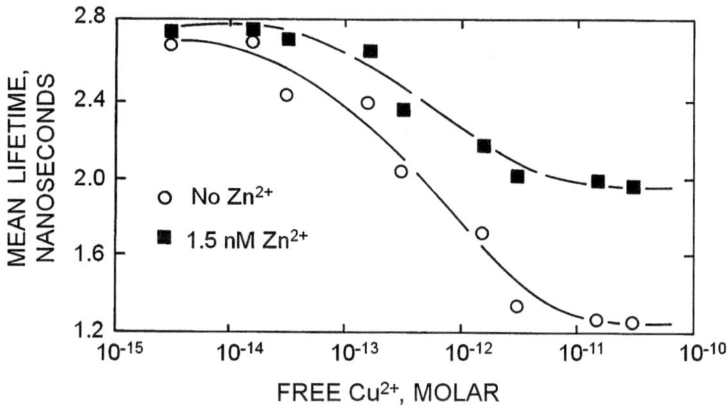

Fig. 6.1 Zinc interference in Cu(II) measurement. Fluorescent-labeled wild type apocarbonic anhydrase exhibits a substantial copper-dependent decline in average lifetime in the absence of zinc (*open circles*), but a much smaller decline in average lifetime when 1.5 nM free zinc (*filled squares*) is present, indicating substantial interference even though zinc binding does not affect the lifetime or intensity of the apoprotein

this particular sensor would likely be unsatisfactory for studying free Cu(II) in some cells which exhibit nanomolar free zinc under some conditions. It should be apparent that the effect will be similar with other indicators and fluorescence measurements. Many creators of fluorescent indicators seek to identify interferents by adding amounts of potential interferent to the indicator in the absence of the analyte and conclude there is no interference if no fluorescence change is observed; clearly that approach is unsatisfactory. In the case of CA-based sensors, sensors with altered selectivity were devised that gave the same lifetime response to copper in the presence and absence of 1.5 nM free zinc, so one could be confident of no interference from endogenous zinc [41].

Another source of error frequently overlooked is that the fluorescent indicator may bind to other things besides metal ions (such as proteins or membranes) in cells with an accompanying increase in fluorescence [3, 19, 26, 31], or simply bind to itself, forming colloidal aggregates [43]. This is indistinguishable by simple intensity measurements from binding the metal. For fluorophores with suitable lifetimes, binding to a typical protein or membrane will produce a substantial increase in fluorescence anisotropy (polarization) associated with binding to a more massive object with slower rotational diffusion [72]; this may be discerned by polarization microscopy. We note that "turn-on" indicators typically exhibit an increase or no change in their lifetime upon binding of metal ions and consequently should exhibit (if anything) a decrease in anisotropy on binding the proper analyte. In many cases it is likely that the indicator would exhibit a different lifetime bound to a protein compared to free in solution or with a metal ion bound; this may be determined in cells by fluorescence lifetime imaging microscopy (FLIM) [59].

6.5 Small Molecule Fluorescent Indicators/Sensors for Copper: Cu(II)

Scores of fluorescent Cu(II) indicators have been described over the last several decades (reviewed in [5, 14, 74]). The majority are organic heterocyclic fluorophores capable of coordinating with Cu(II), resulting in a complex whose fluorescence is reduced or abolished by quenching by the bound Cu(II). Of course Cu(II) may be determined by its ability to collisionally quench fluorophores without binding in solution, but the less specific interaction and relative insensitivity of the method in the absence of long-lived fluorophores make the method unattractive. More recently several workers have utilized long-lived fluorophores such as lanthanides to improve sensitivity into the nanomolar range [69]. A classic example of the binding approach is 8-hydroxy quinoline, which exhibits a substantial decrease in its fluorescence upon binding Cu(II). Unfortunately 8-HQ also binds to at least 25 other metals with measurable affinity [55], clearly raising the issue of selectivity in any realistic medium. Many similar indicators have been described, but are poorly suited for studying biological questions due to issues with selectivity, sensitivity, water solubility, background fluorescence, quantitation, and/or compatibility with the medium.

More recently, fluorescent indicators that were developed for metal ions such as Zn(II), Ca(II), and Fe in biological media have been shown to respond to Cu(II) and in some cases Cu(I) [18, 66]. For instance, we showed that Newport Green responds to much lower free Cu(II) than Zn(II), although binding of Zn(II) is not perturbed by physiological concentrations of Ca(II) and Mg(II) [66]. Zhao et al.'s recent, systematic study of the Zn(II) indicators FluoZin-3 and Newport Green determined apparent dissociation constants for these indicators with eight different metal ion species [81], and quantitatively demonstrated the propensity for interference. To the extent they are known, the affinities in most cases seem to follow the Irving-Williams series, so one might anticipate that the most likely interferent for Cu (II) measurements in living systems would be Zn(II), but depending upon circumstances Co(II), Cd(II), Hg(II), Ni(II), Fe(II), and Cr species might also be of concern. Newport Green's ten thousand-fold higher affinity for Cu(II) than Zn (II) [66, 81] is unsurprising in view of the typical affinity of its di-(2-picolyl) amine metal ion binding moiety and suggests that as a Cu(II) indicator it would be relatively free of Zn(II) interference. On the other hand, one might infer that the many Zn(II) indicators which also use di-(2-picolyl) amine as a metal ion binding moiety may also be quite sensitive to Cu(II). Evidently (Fig. 6.1) a metal ion can interfere if its concentration is high enough in comparison to its affinity, even if it causes no spectroscopic change. All the foregoing suggests that if one is using a metal ion indicator without foreknowledge of its affinity for likely interferents, one is really whistling past the graveyard.

6.6 Recent Cu(II) Indicators

An important recent advance has been the development of "turn on" fluorescent indicators for Cu(II) [48, 76]. These represent significant achievements because (as described above) Cu(II) is a potent quencher that can quench an excited state by several mechanisms, and "turn-on" indicators wherein binding of the analyte causes an increase in fluorescence are widely preferred, particularly for microscopy. Singhal et al. [54] describe a very clever turn-on Cu(II) indicator that employs a galactosamine moiety coupled to a naphthol moiety (Fig. 6.2a): the galactosamine not only supplies metal binding/recognition ligands (an hydroxyl and the amine) but also substantially augments the water solubility of the molecule. Binding of Cu (II) is relatively tight at about 22 μM and results in a seven-fold increase in quantum yield from 0.0005 to 0.0035; there are shifts in the absorbance spectra which might be useful for ratiometric determination. These authors made a careful study using mass spectrometry of the stoichiometry of metal-ligand complex speciation. Selectivity is more of an issue, with the apparent affinity of Zn^{2+} for the indicator differing only by 20 %, even though there is not the same quantum yield increase. Fe(II), Co(II), Ni(II), and Cd(II) affinities also are within a factor of two, with varying responses. This work represents a promising approach for highly soluble and potentially biocompatible probes.

It should be noted that the "turn-on" feature of many of these indicators is relative, and that even the Cu(II)-bound form of the indicator typically has a low (<5 %) quantum yield—it is just that the metal-free form of the indicator is even more efficiently quenched than the Cu-bound, and the binding of the metal is thus accompanied by an increase in fluorescence. These low quantum yields may be an issue in microscopy, particularly when the excitation is at shorter wavelengths, owing to the presence of interfering autofluorescence in cell and tissue samples. Moreover, the elevated excitation intensities required may prove toxic to some cell types.

Xie et al. [77] describe fascinating new sensors for Cu(II) based on bis-pyrenyl sugar macrolides (termed sugar aza-crowns (Fig. 6.2b)) which exhibit a ratiometric response. The advantages of ratiometric responses over simple intensity changes are well known and needn't be further addressed here. As is well known, pyrene molecules (or moieties) when kept in close proximity by high concentration or tethering can exhibit a broad excimer (excited state dimer) emission at longer wavelengths (480 nm) than the structured pyrene monomer emission bands around 380 nm [70]. When the indicators bind Cu2+ (with nanomolar range affinity), the emission from both monomer and excimer diminish, the latter preferentially such that the ratio of excimer to monomer emission drops about 7-fold as the binding saturates. The authors present evidence that the preferential quenching of the excimer occurs by electron transfer and calculated energy-minimized structures of the 1:1 Cu-indicator complexes. A ratiometric indicator is very desirable, but the indicators also exhibited very slow kinetics (hours) in view of the measured affinities, modest quantum yields (<8 % prior to Cu(II) binding), and required organic solvents, indicating that further development of this attractive concept will be necessary prior to its use in biological systems.

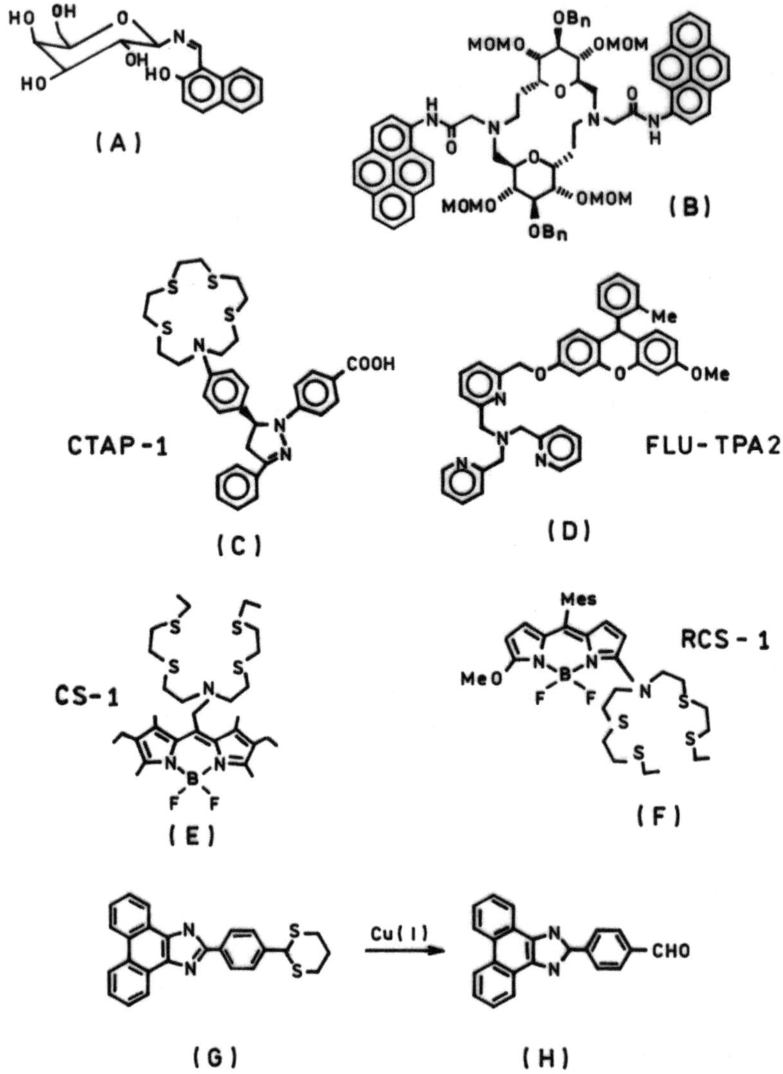

Fig. 6.2 Structures of fluorescent copper indicators

Another novel, ratiometric indicator for Cu(II) was devised by the Lin group of Hunan University [34]. They exploited the fact that Cu(II) rather specifically removes a protecting group called a dithiane (propane-1,3-dithiol) from an aryl aldehyde (dethioacetylization) (Fig. 6.2g, h); by attaching the aldehyde to a fluorescent phenanthroimidazole moiety the deprotection results in a 100 nm emission red shift as the dithiane is removed and the aldehyde is restored. Usually dithiane removal requires vigorous conditions ($CuCl_2$/CuO in refluxing acetone for 1 h), but in this case the deprotection took place within minutes in acetonitrile/$CuCl_2$ at room temperature. The selectivity of the approach is evidently high: other, similar metal ions capable of removing dithianes such as Ag(I), or Hg(II) do not interfere because they require different reaction conditions. Presumably metal ions such as Zn(II) or Hg(II) could interfere by competing for binding sites on the dithiane, but perhaps the affinity is low enough and the exchange rate fast enough to moderate any interference. The proposed reaction scheme does not indicate any reduction of the Cu2+, so presumably the metal ion acts as a catalyst and is not consumed in the reaction. If so, quantitation might be obtained by observing the rate of change of the ratio; unfortunately, sensitivity was not explored in much depth. Significant additional development will be necessary before this promising approach could be used in cells.

6.7 Cu(I) Indicators

The reducing environment found in most cells means that much of the un/weakly complexed copper to be found therein is in the form of Cu(I), not Cu(II). Of course, the oxygen tension (and thus propensity of weakly complexed Cu(I) to become oxidized) in various tissues in humans varies substantially with the degree of oxygen exposure and perfusion: the oxygen tension in an alveolar cell in the lung might be 2.5 fold-higher than that in the small intestinal lumen. Cu(I) in compartments with strong oxidants present (e.g., peroxisomes) might be very transient indeed. Thus much recent work has focused on Cu(I) (reviewed in [13]).

Chang's group at Berkeley [80] describe Coppersensor-1 (CS1), a tetrathio azo metal-binding moiety coupled to a BODIPY fluorophore with several attractive features for determining Cu(I) (Fig. 6.2e). Perhaps most important is its turn-on feature, wherein the quantum yield increases from 0.016 to 0.13 upon Cu (I) binding, but also the visible excitation wavelengths used are very desirable for microscopy to limit background fluorescence and phototoxicity. The selectivity of the indicator for Cu(I) appears good, but the caveats discussed above should be borne in mind. The affinity for Cu(I) is very high, $K_D \sim 40$ pM. The investigators demonstrated that the indicator would be taken up by HEK 293 cells, respond to added $CuCl_2$, and exhibit a corresponding decline in intensity following addition of the chelating moiety of the indicator. We were impressed by the cells' morphology following 7 h of exposure to 100 uM total $CuCl_2$ in view of its likely rapid (seconds) oxidation, but it is difficult to know the free metal ion concentration in the growth medium. Like the other turn-on indicators, this indicator may turn out to be a useful fluorescence lifetime-based indicator for quantitative studies. More recently, Domaille et al. [10] described an improved version of CS-1 termed RCS-1 (Fig. 6.2f), which exhibits an emission ratiometric response and reacts to increases in intracellular free Cu(I) levels induced by ascorbate reduction. This is an exciting development, as it offers the prospect of realistic quantitation in cells. A remarkable aspect of this fluorophore's photophysics is the appearance (in the free compound in water, with excitation at 480 nm) of fluorescence emission at shorter wavelengths

(505 nm) than the last absorption peak at 550 nm, suggesting emission from higher excited states such as is observed with compounds like azulene [70]. The modest quantum yields (0.05 in the Cu(I)-bound form, and twenty-fold lower in the free) suggest that further improvement will be necessary to measure the (likely) low concentrations of free Cu(I) in cells. A recent development from the Chang group is CS790, a heptamethine cyanine trithia azo indicator synthesized in ten steps that is excited at longer wavelengths (up to 800 nm) and provides a substantial turn-on; such long wavelengths permit in vivo imaging in small animals [20].

Rolinski, Birch, and their coworkers have developed a fluorescence lifetime-based group of Cu(I) indicators based on FRET within a Nafion matrix from a (Cu-insensitive) fluorophore such as perylene to a colored Cu(I)-bathocuproine (BCP = 2,9-dimethyl −4,7-diphenyl-1,10-phenanthroline) complex [52] Essentially the FRET-based quenching acts to reduce the lifetime of the donor, and the change in decay kinetics can be used to quantitate the analyte. The analytical advantages of transducing a change in analyte level as a change in fluorescence lifetime instead of simple changes in intensity are well known [30, 37], and discussed in greater detail below. These indicator systems are very sensitive because even a single colored BCP-Cu(I) complex can act as energy transfer acceptor for several nearby fluorophore donors in a manner precisely analogous to a single chromophore acting as a trap for exciton migration in a photosynthetic chloroplast. The perturbation to the time-resolved decay by the energy transfer process to the BCP-Cu(I) complexes depends in a complex way on the distribution of the complexes with respect to the donors and the translational and rotational diffusion of the donors and complexes; it is rather beyond the scope of this chapter. A simplified approach models the distribution of the molecules at a lower dimension than the 3-D distribution. There is a modest effect due to quenching of the perylene by Cu(II), but it requires millimolar levels which are unlikely to be encountered in biological systems.

Fahrni's group at Georgia Tech was among the first to focus on biologically compatible Cu(I) fluorescent indicators and their innovation continues. They described a disubstituted pyrazoline fluorescent indicator chosen for its redox properties for photoinduced electron transfer (PET), with a tetrathiaza crown ether as selective binding moiety **CTAP-1** (see Fig. 6.2c) [78]. The indicator has high affinity ($\log K = 10.4 \pm 0.1$) and evidently good selectivity against most divalent cations, with a substantial increase in intensity upon binding. Subsequently, this group has examined systematically the role of the electron transfer photophysics in the response of fluorescent indicators, and found (at least in the case of the triarylpyrazolines) that the PET efficiency could be optimized by changing the donor potential [6]. Their findings are not only of value for Cu (I) determination, but for other indicator systems that rely on PET, including some zinc indicators. They also recently described Cu(I) sensors much less prone to aggregate in solution [43].

Taki and colleagues at Kyoto University recently described a clever, catalytic approach to Cu(I) determination [60]. They conjugated the tetradentate ligand tris-[(2-pyridyl)methyl]amine to a fluorescein-like reduced xanthene moiety

(Fig. 6.2d); in the absence of Cu(I) the conjugate is non-fluorescent, but in its presence the reduced fluorescein is cleaved off and oxidized to form 3'-O-methylfluorescein, which of course emits strongly. The probe appears quite selective for Cu(I) due in part to the selectivity of the chelating moiety, as well as the catalysis of the cleavage. The investigators are currently seeking to understand the mechanism of the cleavage, as this is likely to affect quantitation: if the Cu(I) is oxidized during the cleavage then the intensity of the released xanthene dye will be proportional to the amount of Cu(I) (depending upon the rate at which it is re-reduced). However, if the Cu(I) is a true catalyst, then it would be sensible to measure the rate of appearance of the fluorescence. In a sense the need for the cellular environment to remain reducing to keep the copper reduced is a bit orthogonal to the need to oxidize the xanthene cleavage product to yield a fluorophore.

6.8 Macromolecule-Based Fluorescent Indicators/Sensors

Of course, it had been known for decades that polymers such as proteins bound metal ions such as copper, and in some cases the copper ions served as coenzymes (see above). Torrado et al. was among the first to utilize a polypeptide as the binding moiety for Cu(II) and incorporate it into a fluorescence sensor [68]. They utilized a pentapetide similar to the metal-binding ATCUN motif present in serum albumin [17] and conjugated a Dansyl (5-(dimethylamino)naphthalene-1-sulfonyl) moiety to the N-terminus of a β-amino alanine-β-alanyl-his-ser-ser peptide. The peptide binds Cu(II) with high affinity, which results in partial quenching of the Dansyl fluorophore. Importantly, Torrado demonstrated that relatively simple changes in the peptide (done by solid phase peptide synthesis) could produce improved quenching, and the ease and flexibility with which mutagenesis of polypeptides could improve the properties of fluorescent indicators has been extensively shown by ourselves and others.

Recently, the He group at Chicago have described a very promising expressible indicator for Cu(I) in cells [73] using a FRET-based approach akin to the Ca sensor of Miyawaki et al. [42]. They fused genes for Cyan Fluorescent Protein (CFP, FRET donor) and Yellow Fluorescent Protein (YFP, acceptor) on either end of the Cu(I)-binding domain of the copper-dependent regulator protein Amt1 from the yeast *Candida glabrata*. The domain (residues 36–110) contains eight cysteine residues which might be expected to adopt a different conformation upon binding metal ions such as Cu(I), thereby bringing the CFP and YFP domains into closer proximity, improving the efficiency of FRET from CFP to YFP. The protein binds four equivalents of Cu(I) with very high affinity (average $K_D \sim 2.5 \times 10^{-18}$ M), but only two equivalents of Zn^{2+}, with slightly different apparent affinities in the high nanomolar/low micromolar regime, which is excellent selectivity. While the signal change of this initial effort was modest (which is not unusual for such conformational change-based sensors, since few proteins exhibit two conformations as stable and distinct as calmodulin) [12], the results are very encouraging. We were puzzled that the affinity for Cu(II) was apparently not examined, inasmuch as the cell tests were done with this ion. This group subsequently reported sensors with improved response [38]. The Liang group of Nanjing University has also developed expressible Cu(I) sensors based on a different design concept; most recently, they described a red fluorescent protein with segments of the Amt1 copper binding motif subcloned into the structure. They observed an improved response over their earlier efforts and a range of affinities with subtle spectral changes. Importantly, the red emission of this construct will improve results in cell and tissue specimens due to reduced background emission and scattering [33].

6.9 Fluorescence Lifetime-Based Biosensing of Cu(II)

It is mentioned above that Cu(II) is a superlative quencher of fluorescence by several mechanisms, and in many cases the reduction in fluorescence intensity is accompanied by a change in fluorescence lifetime. Classically this is used to distinguish "static" quenching from "dynamic" quenching. In static quenching the quencher binds the fluorophore such that quenching is very rapid and efficient in the complex, no emission is observed, and thus no change in the lifetime is observed. By comparison, dynamic (or collisional) quenching occurs when the quencher collides with (or comes within a small distance from) the fluorophore. In dynamic quenching the fluorescence lifetime of the fluorophore declines since the longer an individual fluorophore remains in the excited state, the more likely it is to lose its energy by colliding with the quencher before emitting: essentially the quenching process selects for the slower emitters in the ensemble [28]. Our approach uses an intermediate case, wherein the Cu(II) quencher binds to a molecule to which a fluorescent label is attached (as in static quenching), but the fluorescence is only partially quenched, by a process that competes with emission, thereby lowering the lifetime. The process we chose is Förster resonance energy transfer (FRET), in part because Förster's theory [15] permits facile prediction of energy transfer efficiency from spectra, quantum yields, lifetimes, dielectric constants, and relative orientations that are known or can be estimated. In this case it was well-known that Cu (II) bound tightly to the enzyme CA II exhibited weak d-d absorbance bands in the red-infrared [36]. These bands are much weaker (($\varepsilon \approx 10$–$50\ M^{-1}\ cm^{-1}$) than those of regular organic chromophores, but serve perfectly well as energy transfer acceptors if the fluorescent donor is close enough (~10 Angstroms) Fig. 6.3.

For CA II the structure is known (and nearly the same for holo- and apo-carbonic anhydrase) so it is straightforward to position a unique cysteine close by using site-directed mutagenesis, and then selectively conjugate it with a thiol-reactive fluorescent label. Depending on the label and its proximity, other proximity-dependent processes may also contribute to the quenching and reduce the lifetime. Note that in this instance the Cu(II)-free and –bound species have discrete lifetimes; when the binding site is partly saturated, one observes (using time- or frequency-domain

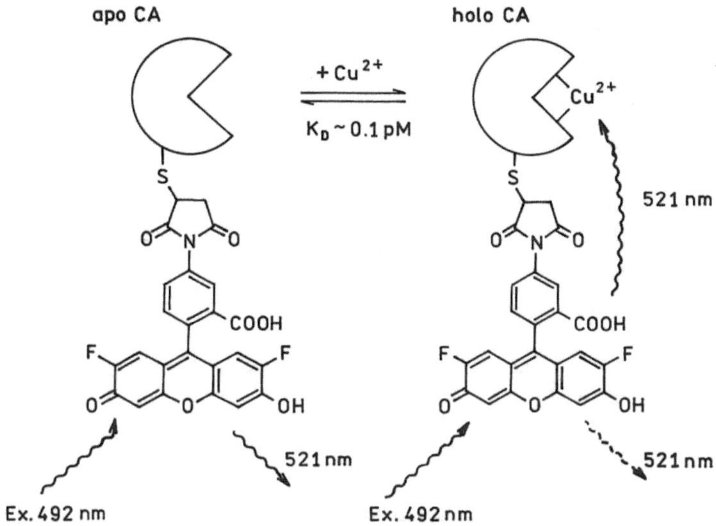

Fig. 6.3 Principle of fluorescence lifetime-based carbonic anhydrase II Cu(II) sensor. In the absence of Cu(II) the Oregon Green label on the protein does not transfer energy, exhibits unquenched lifetime. When Cu(II) binds to the protein the label is partly quenched by FRET to d-d absorbance bands of the bound Cu(II) and exhibits a reduced lifetime; the proportion of short and long lifetimes equals the proportions of bound and free, respectively, which are a simple function of the free Cu(II) concentration

fluorometry) not a single, intermediate fluorescence lifetime (as one does with collisional quenchers), but a mixture of the two lifetimes with their proportions (preexponential factors) equal to the free and bound forms. An example of this is apoCA II with a cysteine residue substituted at position 198 that has been labeled with Alexa Fluor 660 (abbreviated L198C-apoCAII-AF660). The cysteine is close to the metal binding site, so that the lifetime of the label drops from about 4 ns to about 600 ps upon copper binding, with a concomitant decrease in intensity Fig. 6.4. This large drop in lifetime is easily measured [58] in either the time or frequency domain. In frequency domain fluorometry, the measured phase angles and modulations at suitable frequencies smoothly vary with fractional binding site occupancy and thus Cu(II) concentration, such that the phase angle declines by 30° and modulation increases 30 %. One can uniquely relate the analyte concentration to the phase or modulation measurement [65]; e.g., Fig. 6.4 is a calibration curve. This is convenient for sensing applications because a precise phase and modulation measurement is easily obtained in seconds (or as little as microseconds [45]), whereas acquiring an entire multifrequency decay or time-resolved decay typically takes longer. We used the high sensitivity and selectivity of this approach to measure free picomolar Cu(II) concentration changes in real time in the ocean remotely through a fiber optic [79].

As has been pointed out above [30, 37], lifetime measurements are relatively insensitive to variations in excitation intensity and fluorophore concentration, in a

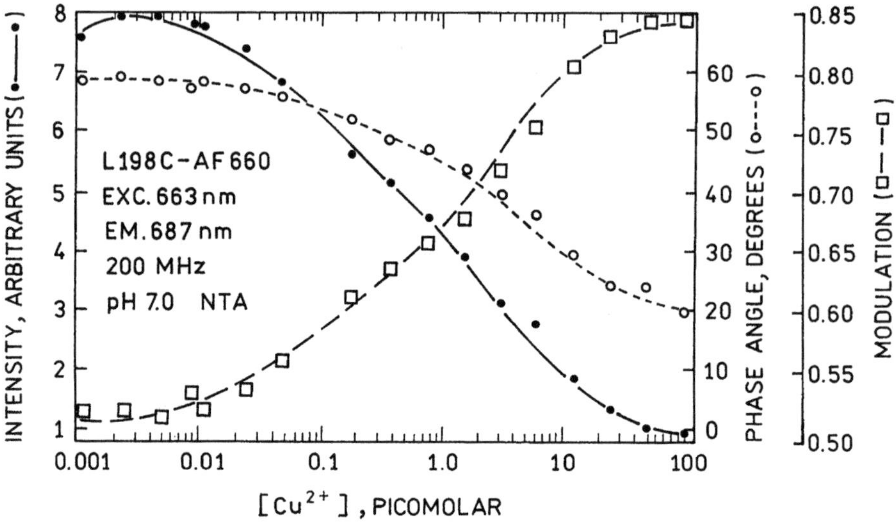

Fig. 6.4 Free Cu(II)concentration-dependent intensities (*filled circles*), phase angles (*open circles*), and modulations (*squares*) at 200 MHz for L198C-Alexa Fluor 660-labeled apocarbonic anhydrase II (Reproduced from Zeng et al. [79] with permission)

manner analogous to fluorescence ratiometric methods. This is particularly useful in microscopy and fiber optic sensing applications, where simple intensity-based sensors are notoriously difficult to calibrate. The commercial availability of fluorescence lifetime-based ("FLIM") microscopes for FRET-based measurements has made lifetime-based imaging and sensing more widespread [61]. Szmacinski et al. showed that if analyte binding also induced a wavelength shift as well as a change in lifetime, expansion of the sensor dynamic range beyond five orders of magnitude could be achieved [57].

We used this approach to attempt to image free Cu(II) in PC-12 cells using frequency-domain fluorescence lifetime imaging microscopy (FLIM), wherein the contrast in the image arises from differences in fluorescence lifetime within the visual field [29, 59]. Previously, Rae et al. had famously predicted that (on the average) no free copper ion would be found in mammalian cells, that essentially all of it would be bound in relatively stable complexes ([50]). A key issue in measuring Cu(II) in low levels in cells would be the potential for endogenous free zinc in cells at picomolar levels interfering with Cu(II) measurements at femtomolar levels. Fortunately, Professor Fierke and her colleagues had previously created a variant of wild type carbonic anhydrase II (Q92A) with substantially better selectivity for Cu (II) than the wild type [25]: the wild type exhibits 40-fold higher affinity for Cu (II) than Zn(II) (100 f. vs 4 pM), whereas Q92A is more than 25-fold more selective, such that even 1.5 nM Zn(II) does not interfere [41]. By *in situ* calibration we found we were able to discern Cu(II) levels in the femtomolar range in cells but were unable to detect any difference in resting cells between that and normal Cu (II) levels: that is, the ordinary free Cu(II) level in cells is in the femtomolar range, such that Rae, et al., are correct that on the average there is no free Cu(II) in these cells at least. It will be interesting to use this technique on cells that might be expected to have differing free Cu(II) levels.

Fluorescence Polarization (Anisotropy)-Based Cu(II) Sensing Although fluorescence lifetime instruments (including microscopes) are becoming more common, the instrumentation remains significantly more costly and complex compared with steady state. Thus we also sought a steady state, ratiometric measurement of Cu(II). The effect of changes in the fluorescence lifetime on the measured anisotropy (polarization) was worked out more than fifty years ago by Perrin in his famous equation:

$$r_0/r = 1 + (\tau/\theta_c)$$

where r_0 is the limiting anisotropy at a given excitation wavelength, r is the measured anisotropy, θ_c is the rotational correlation time (inverse of the rotational rate), and τ is the fluorescence lifetime. The terms "fluorescence polarization" and "fluorescence anisotropy" are synonymous in that they describe the same phenomenon and are measured the same way, but normalized differently; we prefer anisotropy because theory expressed in terms of anisotropy is simpler; see Lakowicz for a thorough treatment [28]. Essentially, the anisotropy depends on the ratio of the lifetime to rotational correlation time, such that the anisotropy increases if the lifetime decreases, or the rotational correlation time increases due to lower temperature, increased viscosity, or increased molecular size. For suitable values of τ and θ_c, a process causing a decline in lifetime can be detected by an increase in the anisotropy. Anisotropy (like the similar polarization) measurements are ratios of steady state fluorescence intensity measurements and therefore enjoy the relative freedom from artifact and facile calibration which have spurred the growth of wavelength ratiometric fluorescence indicators like the Fura and Indo families of calcium indicators. Being intensity measurements they are also simpler and cheaper than lifetime measurements.

We can predict the response of such an indicator from Perrin's equation with a few (testable) assumptions. Suppose we attach a fluorescent label to CAII such that it rotates essentially rigidly with the protein, reflecting the whole protein's rotational correlation time of approximately 15 ns and a lifetime of 15 ns and a normalized quantum yield of 1.0. Under these conditions the fluorophore will exhibit an anisotropy of 0.2 when excited at the $S_0 \rightarrow S_1$ transition. If binding the metal ion reduces the lifetime and quantum yield by 1/2, the anisotropy of the bound form increases to 0.30, and the observed anisotropy r_{OBS} at any metal ion concentration is the average of the free (F) and bound (B) forms, weighted by their respective quantum yields:

$$r_{OBS} = (r_B \times f_B \times QY_B) + (r_F \times f_F \times QY_F)$$

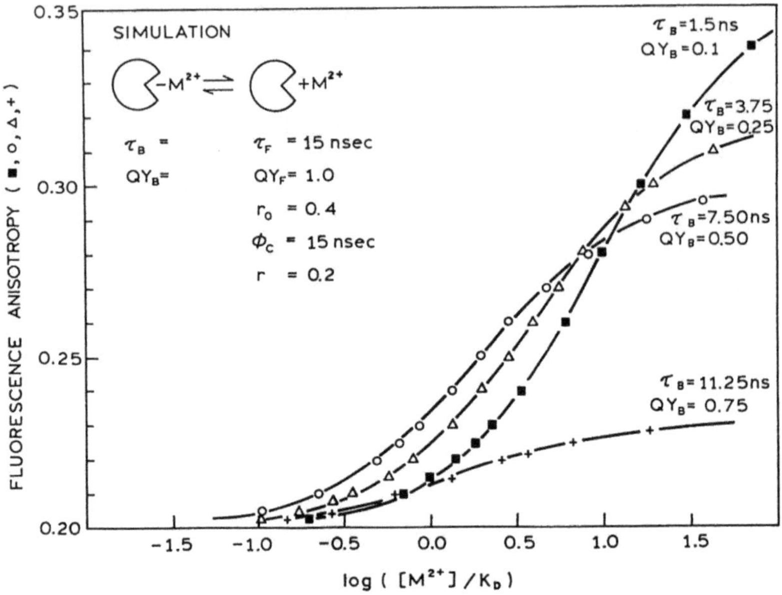

Fig. 6.5 Simulated metal-ion dependent fluorescence anisotropies for a metal ion that partially quenches and reduces the lifetime of the label on a macromolecule to which it binds. Results are depicted for a macromolecule with a 15 ns rotational correlation time, labeled with a 15 ns lifetime (unquenched) fluorophore when binding reduces the lifetime and quantum yield by 25 % (*crosses*), 50 % (*circles*), 75 % (*triangles*), and 90 % (*filled squares*) (Reproduced from Thompson, et al., with permission)

where r, f, and QY are the anisotropies, fractions, and quantum yields of the respective forms. Of course, the fractions of free and bound are determined by the concentration of the metal ion and its K_D. Simulated metal-dependent anisotropies are depicted in Fig. 6.5 for different degrees of quenching. Evidently 50 % quenching provides a useful increase in anisotropy from 0.2 to 0.3 as the binding site becomes saturated with the metal since the typical accuracy and precision of anisotropy measurements is ± 0.002. If metal ion binding is accompanied by a larger decrease in lifetime (say 75 or 90 %), there is a bigger increase in anisotropy, but note that the apparent binding constant is shifted to higher concentration because the free form contributes a larger proportion of the observed emission. Note that if the lifetime is much bigger or smaller than the rotational correlation time, the anisotropy change upon binding is reduced.

The Cu(II)-dependent response of a fluorescent-labeled carbonic anhydrase is depicted in Fig. 6.6 below. In this example the fluorophore ABD-T is conjugated to a cysteinyl residue inserted into the CAII sequence to replace the phenylalanine at position 131, which is about 9 Angstroms from the metal ion binding site. Binding of Cu(II) results in approximately a threefold decrease in intensity and lifetime, and about a 33 % increase in anisotropy, which is easily measured [63].

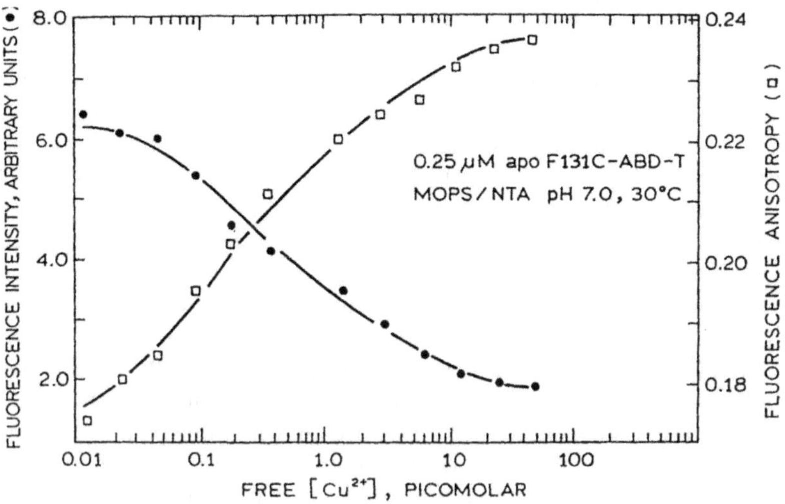

Fig. 6.6 Cu(II)-dependent fluorescence intensities (*filled circles*) and anisotropies (*open squares*) for apo-F131C-ABD-T (Reproduced from Thompson et al. [63], with permission)

6.10 Outlook and Potential Future Directions

A potential issue in measuring intracellular copper (compared with Zn(II) or Ca(II), for instance) is the lower total amounts present in most cells may not offer the same opportunity for buffering the free metal ion concentration. For instance, the vast majority of intracellular Zn(II) is bound to proteins and small molecules within the cell, such that these buffer changes in free Zn(II) caused by additions of zinc or zinc chelators (such as zinc indicators). Consequently, indicators present at nanomolar levels can still accurately measure the picomolar free levels of free zinc present [2, 11, 27] since the buffering opposes the otherwise dramatic reduction in free zinc concentration caused by the thorough scavenging by the high affinity indicator. In the case of copper ions, the total cellular copper ion is somewhat lower than total zinc in heart, kidney, liver, and brain, and much lower in other tissues [35]; moreover, there is reason to believe that only a very small proportion is not bound very tightly to the few intracellular enzymes that utilize it [50]. Thus the free levels of copper may not only be very low, but may require very low levels of sensor binding sites to avoid scavenging all the available copper. The recent developments in single molecule fluorescence measurements may enable one to limit the numbers of sensor molecule binding sites present. It should be noted, however, that single molecule fluorescence studies mainly are the province of fluorophores with the highest absorption, quantum yields, and photostability. With a limited ensemble of binding sites measuring free analyte levels from fractional occupancy levels of many sites to obtain adequate precision would appear challenging, but the fractional occupancy in time should be a valid measure, as it is for ligand binding to (numerous) receptor channels measured by patch clamping methods.

Time is also a factor in measuring low levels of free metal ions in all kinds of matrices. The slow dissociation rate constants (off-rates) required for high receptor affinity even with association rate constants as fast as diffusion-controlled (probably $>10^8$ M^{-1} s^{-1} at room temperature in water) raises the issue of how long an indicator takes to equilibrate. In the case of intracellular free zinc the equilibration happens much more rapidly than one would expect based on the known kinetics of the indicator [2], which we attribute to catalysis by small molecule ligands. It is not at all clear that this will occur with either Cu(I) or Cu (II) in the cell. CA II variants we have used exhibit diffusion-controlled on-rates with Cu(II) [22], so we would expect them to equilibrate in cells in the presence of picomolar copper concentrations in tens of minutes. The kinetics of few of the other copper indicator systems described above are known at this point, but some high affinity ligands have notably slow kinetics. It remains to be seen what free levels of copper are (or are not) present, but it seems likely that rapid concentration changes will be difficult to measure.

Acknowledgments We would like to thank all of our co-workers who did the work described herein, as well as the National Institutes of Health, National Science Foundation, Office of Naval Research, and National Oceanic and Atmospheric Administration for support.

References

1. Bertini I, Sigel A et al (eds) (2001) Handbook on metalloproteins. Marcel Dekker, New York
2. Bozym RA, Thompson RB et al (2006) Measuring picomolar intracellular exchangeable zinc in PC-12 cells using a ratiometric fluorescence biosensor. ACS Chem Biol 1(2):103–111
3. Burdette SC, Walkup GK et al (2001) Fluorescent sensors for Zn(2+) based on a fluorescein platform: synthesis, properties and intracellular distribution. J Am Chem Soc 123 (32):7831–7841
4. Bush AI, Pettingell WH et al (1994) Rapid induction of Alzheimer AB amyloid formation by zinc. Science 265:1464–1467
5. Carter KP, Young AM et al (2014) Fluorescent sensors for measuring metal ions in living systems. Chem Rev 114(8):4564–4601
6. Chaudry AF, Verma M et al (2010) Kinetically controlled photoinduced electron transfer switching in Cu(I)-responsive fluorescent probes. J Am Chem Soc 132:737–747
7. Cherny RA, Atwood CS et al (2001) Treatment with a copper-zinc chelator markedly and rapidly inhibits beta-amyloid accumulation in Alzheimer's disease transgenic mice. Neuron 30:665–676
8. Cotton FA, Wilkinson G (1988) Advanced inorganic chemistry. Wiley-Interscience, New York
9. Culotta VC, Gitlin JD (2001) Disorders of copper transport. In: Scriver CR, Beaudet AL, Sly WS, Valle D (eds) The metabolic and molecular bases of inherited disease, vol 2, 8th edn. McGraw-Hill, New York, pp 3105–3126
10. Domaille DW, Zeng L et al (2010) Visualizing ascorbate-triggered release of labile copper within living cells using a ratiometric fluorescence sensor. J Am Chem Soc 132:1194–1195

11. Dongen EMWMv, Evers TH et al (2007) Variation of linker length in ratiometric fluorescent sensor proteins allows rational tuning of Zn(II) affinity in the picomolar to femtomolar range. J Am Chem Soc 129:3494–3495
12. Eis PS, Lakowicz JR (1993) Time-resolved energy transfer measurements of donor-acceptor distance distributions and intramolecular flexibility of a CCHH zinc finger peptide. Biochemistry 32:7981–7993
13. Fahrni CJ (2013) Synthetic fluorescent probes for monovalent copper. Curr Opin Chem Biol 17 (4):656–662
14. Fernandez-Gutierrez A, Munoz de la Pena A (1985) Determinations of inorganic substances by luminescence methods. In: Schulman SG (ed) Molecular luminescence spectroscopy, Part I: Methods and applications, vol 77. Wiley-Interscience, New York, pp 371–546
15. Forster T (1948) Intermolecular energy migration and fluorescence (Ger.). Ann Phys 2:55–75
16. Grynkiewicz G, Poenie M et al (1985) A new generation of calcium indicators with greatly improved fluorescence properties. J Biol Chem 260(6):3440–3450
17. Harford C, Sarkar B (1997) Amino-terminal Cu(II) and Ni(II)-binding (ATCUN) motif of proteins and peptides: metal binding, DNA cleavage, and other properties. Acc Chem Res 30:123–130
18. Haugland RP (2005) The handbook: a guide to fluorescent probes and labeling technologies. Invitrogen Corp, Carlsbad
19. Hawkins BE, Frederickson CJ et al (2012) Fluorophilia: Fluorophore-containing compounds adhere non-specifically to injured neurons. Brain Res 1432:28–35
20. Hirayama T, Van de Bittner GC et al (2012) Near-infrared fluorescent sensor for in vivo copper imaging in a murine Wilson disease model. Proc Natl Acad Sci 109(7):2228–2233
21. Hung YH, Bush AI et al (2010) Copper in the brain and Alzheimer's disease. J Biol Inorg Chem 15:61–76
22. Hunt JA, Ahmed M et al (1999) Metal binding specificity in carbonic anhydrase is influenced by conserved hydrophobic amino acids. Biochemistry 38:9054–9060
23. Johnson KS, Elrod VA et al (2000) Continuous flow techniques for on site and in situ measurements of metals and nutrients in sea water. In: Buffle J, Horvai G (eds) In situ monitoring of aquatic systems. Wiley, Chichester, pp 223–252
24. Kaplan JH, Lutsenko S (2009) Copper transport in mammalian cells: special care for a metal with special needs. J Biol Chem 284(38):25461–25465
25. Kiefer LL, Paterno SA et al (1995) Second shell hydrogen bonds to histidine ligands enhance zinc affinity and catalytic efficiency. J Am Chem Soc 117:6831–6837
26. Kolb DA, Weber G (1975) Cooperativity of binding of anilinonaphthalenesulfonate to serum albumin induced by a second ligand. Biochemistry 14(20):4476–4481
27. Krezel A, Maret W (2006) Zinc-buffering capacity of a eukaryotic cell at physiological pZn. J Biol Inorg Chem 11:1049–1062
28. Lakowicz JR (1999) Principles of fluorescence spectroscopy. Kluwer Academic/Plenum Publishers, New York
29. Lakowicz JR, Szmacinski H et al (1992) Fluorescence lifetime imaging. Anal Biochem 202:316–330
30. Lakowicz JR, Szmacinski H et al (1993) Fluorescence lifetime-based sensing: applications to clinical chemistry and cellular imaging. In: SPIE Conference on Ultrasensitive Laboratory Diagnostics, Los Angeles, SPIE
31. Landero Figueroa JA, Subramanian Vignesh K et al (2014) Selectivity and specificity of small molecule fluorescent dyes/probes used for the detection of Zn2+ and Ca2+ in cells. Metallomics 6(2):301–315
32. Li Y, Maret W (2008) Human metallothionein metallomics. J Anal At Spectrosc 23:1055–1062
33. Liang J, Guo L et al (2014) Genetically encoded red fluorescent copper(I) sensors for cellular copper(I) imaging. Biochem Biophys Res Commun 443(3):894–898

34. Lin W, Yuan L et al (2009) Construction of fluorescent probes via protection/deprotection of functional groups: a ratiometric fluorescent probe for Cu2+. Chem Eur J 15:1030–1035
35. Linder MC (1991) Biochemistry of Copper. Plenum, New York
36. Lindskog S, Nyman PO (1964) Metal-binding properties of human erythrocyte carbonic anhydrases. Biochim Biophys Acta 85:462–474
37. Lippitsch ME, Pusterhofer J et al (1988) Fibre-optic oxygen sensor with the fluorescence decay time as the information carrier. Anal Chim Acta 205:1–6
38. Liu J, Karpus J et al (2013) Genetically encoded Copper(I) reporters with improved response for use in imaging. J Am Chem Soc 135(8):3144–3149
39. Malmstadt HV, Enke CG et al (1974) Electronic measurements for scientists. W.A. Benjamin, Inc., Menlo Park
40. McCall KA (2000) Metal Ion specificity and avidity in carbonic anhydrase variants. Doctoral, Duke University
41. McCranor BJ, Szmacinski H et al (2014) Fluorescence lifetime imaging of physiological free Cu(II) levels in live cells with a Cu(II)-selective carbonic anhydrase-based biosensor. Metallomics 6(5):1034–1042
42. Miyawaki A, Llopis J et al (1997) Fluorescent indicators for Ca2+ based on green fluorescent proteins and calmodulin. Nature 388:882–887
43. Morgan MT, Bagchi P et al (2011) Designed to dissolve: suppression of colloidal aggregation of Cu(I)-selective fluorescent probes in aqueous buffer and in-gel detection of a metallochaperone. J Am Chem Soc 133(40):15906–15909
44. Pfeiffer RF (2007) Wilson's disease. Semin Neurol 27:123–132
45. Pinsky BG, Ladasky JJ et al (1993) Phase-resolved fluorescence lifetime measurements for flow cytometry. Cytometry 14(2):123–135
46. Prince RC, Gunson DE (1998) Prions are copper-binding proteins. Trends Biochem Sci 23 (6):197–198
47. Pufahl RA, Singer CP et al (1997) Metal ion chaperone function of the soluble Cu(I) receptor Atx I. Science 278:853–856
48. Qi X, Jun EJ et al (2006) New BODIPY derivatives as OFF-ON fluorescent chemosensor and fluorescent chemodosimeter for Cu2+: cooperative selectivity enhancement toward Cu2+. J Org Chem 71:2881–2884
49. Quinn JF, Crane S et al (2009) Copper in ALzheimer's disease: too much or too little? Expert Rev Neurother 9(5):631–637
50. Rae TD, Schmidt PJ et al (1999) Undetectable intracellular free copper: the requirement of a copper chaperone for superoxide dismutase. Science 284:805–808
51. Riordan JF, Vallee BL (eds) (1991) Metallobiochemistry Part B: Metallothionein and related molecules, Methods in enzymology. Academic, San Diego
52. Rolinski OJ, Birch DJS (1999) A fluorescence lifetime sensor for Cu(I) ions. Meas Sci Technol 10:127–136
53. Rosenzweig AC (2002) Metallochaperones: bind and deliver. Chem Biol 9:673–677
54. Singhal NK, Ramanujam B et al (2006) Carbohydrate-based switch-on molecular sensor for Cu(II) in buffer: absorption and fluorescence study of the selective recognition of Cu(II) ions by galactosyl derivatives in HEPES buffer. Org Lett 8(16):3525–3528
55. Stevens HM (1959) The effect of the electronic structure of the cation upon fluorescence in metal-8-hydroxyquinoline complexes. Anal Chim Acta 20:389–396
56. Stillman MJ, Gasyna Z (1991) Luminescence spectroscopy of metallothioneins. In: Riordan JF, Vallee BL (eds) Metallobiochemistry Part B: Metallothionein and related proteins, vol 205, Methods in Enzymology. Academic, San Diego, pp 540–555
57. Szmacinski H, Lakowicz JR (1993) Optical measurements of pH using fluorescence lifetimes and phase-modulation fluorometry. Anal Chem 65:1668–1674
58. Szmacinski H, Lakowicz JR (1994) Lifetime-based sensing. In: Lakowicz JR (ed) Topics in fluorescence spectroscopy, Vol. 4: Probe design and chemical sensing. Plenum, New York, pp 295–334

59. Szmacinski H, Lakowicz JR et al (1994) Fluorescence lifetime imaging microscopy: homodyne technique using high-speed gated image intensifier. In: Johnson ML, Brand L (eds) Numerical computer methods, vol 240. Academic, New York, pp 723–748
60. Taki M, Iyoshi S et al (2010) Development of highly sensitive fluorescent probes for detection of intracellualr copper (I) in living systems. J Am Chem Soc 132:5938–5939
61. Thompson RB (2008) Fluorescence lifetime biosensing: entering the mainstream. In: Ligler FS, Taitt CR (eds) Optical biosensors: today and tomorrow. Elsevier, Amsterdam, pp 287–315
62. Thompson RB, Ge Z et al (1996) Fiber optic biosensor for Co(II) and Cu(II) based on fluorescence energy transfer with an enzyme transducer. Biosens Bioelectron 11(6):557–564
63. Thompson RB, Maliwal BP et al (1998) Determination of picomolar concentrations of metal ions using fluorescence anisotropy: biosensing with a "reagentless" enzyme transducer. Anal Chem 70(22):4717–4723
64. Thompson RB, Maliwal BP et al (1999) Selectivity and sensitivity of fluorescence lifetime-based metal ion biosensing using a carbonic anhydrase transducer. Anal Biochem 267:185–195
65. Thompson RB, Patchan MW (1995) Lifetime-based fluorescence energy transfer biosensing of zinc. Anal Biochem 227:123–128
66. Thompson RB, Peterson D et al (2002) Fluorescent zinc indicators for neurobiology. J Neurosci Methods 118:63–75
67. Thompson RB, Walt DR (1994) Emerging strategies for molecular biosensors. Naval Res Rev 46(3):19–29
68. Torrado A, Walkup GK et al (1998) Exploiting polypeptide motifs for the design of selective Cu(II) ion chemosensors. J Am Chem Soc 120:609–610
69. Turel M, Duerkop A et al (2009) Detection of nanomolar concentrations of copper(II) with a Tb-quinoline-2-one probe using luminescence quenching or luminescence decay time. Anal Chim Acta 644:53–60
70. Turro NJ (1978) Modern molecular photochemistry. Benjamin/Cummings Publishing Co., Menlo Park
71. Turski ML, Thiele DJ (2009) New roles for copper metabolism in cell proliferation, signaling, and disease. J Biol Chem 284(2):717–721
72. Weber G (1953) Rotational Brownian motion and polarization of the fluorescence of solutions. Adv Protein Chem 8:415–459
73. Wegner SV, Arslan H et al (2010) Dynamic copper(I) imaging in mammalian cells with a genetically encoded fluorescent copper(I) sensor. J Am Chem Soc 132:2567–2569
74. White CE, Argauer RJ (1970) Fluorescence analysis: a practical approach. Marcel Dekker, New York
75. Wilmarth KR, Froines JR (1992) In vitro and in vivo inhibition of lysyl oxidase by aminopropionitriles. J Toxicol Environ Health 37(3):411–423
76. Xiang Y, Tong A et al (2006) New fluorescent rhodamine hydrazone chemosensor for Cu (II) with high selectivity and sensitivity. Org Lett 8(13):2863–2866
77. Xie J, Menand M et al (2007) Synthesis of bispyrenyl sugar-aza-crown ethers as new fluorescent molecular sensors for Cu(II). J Org Chem 72:5980–5985
78. Yang L, McRae R et al (2005) Imaging of the intracellular topography of copper with a fluorescent sensor and by synchrotron x-ray fluorescence microscopy. Proc Natl Acad Sci U S A 102(32):11179–11184
79. Zeng HH, Thompson RB et al (2003) Real-time determination of picomolar free Cu(II) in seawater using a fluorescence-based fiber optic biosensor. Anal Chem 75(24):6807–6812
80. Zeng L, Miller EW et al (2006) A selective turn-on fluorescent sensor for imaging copper in living cells. J Am Chem Soc 128:10–11
81. Zhao J, Bertoglio BA et al (2009) The interaction of biological and noxious transition metals with the zinc probes FluoZin-3 and Newport Green. Anal Biochem 384:34–41

Chapter 7
Heterogeneous Lipid Distributions in Membranes as Revealed by Electronic Energy Transfer

Radek Šachl and Lennart B.-Å. Johansson

Abstract The techniques achieving the highest resolution only can characterize membrane heterogeneities on the lowest molecular level. When Förster resonance energy transfer (FRET) is combined with Monte-Carlo (MC) simulations and is applied to the measurement of nanodomain/pore sizes it reaches an unbeatable resolution of 2–50 nm. While other techniques start being less efficient at such short distances FRET is most efficient in this region. Here, usefulness of MC-FRET is demonstrated on three different systems that contain heterogeneously distributed lipids: a nanoscopically phase separated bilayer, a bilayer containing pores and finally on bicelles consisting of highly curved and flat regions. Moreover, this paper gives the reader information on how a FRET experiment should be designed to achieve the highest FRET resolution but also which experimental conditions should be avoided. The theory describing FRET between randomly distributed donors and acceptors in a lipid bilayer is also described in this paper as well as reasons are explained why for heterogeneous probe distribution MC simulations should rather be used.

Keywords FRET • Domains • Pores • Monte Carlo simulations • Lipid bilayer

Abbreviations

A_r	Area fraction of the L_o domains
B-F	Baumann-Fayer
BODIPY-FL	4,4-difluoro-4-bora-3a,4a-diaza-5,7-dimethyl-*s*-indacenyl
BODIPY 564/570	4,4-difluoro-4-bora-3a,4a-diaza-5-styryl-dimethyl-*s*-indacenyl

R. Šachl
Department of Biophysical Chemistry, J. Heyrovský Institute of Physical Chemistry of the Academy of Sciences of the Czech Republic, Dolejškova 2155/3, Prague 8, 182 23, Czech Republic

L.B.-Å. Johansson (✉)
Department of Chemistry: Biophysical Chemistry, Umeå University, S-901 87, Umeå, Sweden
e-mail: lennart.johansson@chem.umu.se

© Springer International Publishing Switzerland 2016

C.D. Geddes (ed.), *Reviews in Fluorescence 2015*,
DOI 10.1007/978-3-319-24609-3_7

C_2	Reduced surface concentration
C_{2A}	Reduced surface concentration in phase A
C_{2B}	Reduced surface concentration in phase B
CTxB	Cholera toxin
DAET	Donor-acceptor energy transfer
DDEM	donor-donor energy migration
DiD	1,1′-Dioctadecyl-3,3,3′,3′-Tetramethylindodicarbocyanine-5,5′-Disulfonic Acid ($DiIC_{18}(5)$-DS)
FRET	Förster resonance energy transfer
$G^s(t)$	the excitation probability of the initially excited donor
κ	Angular dependence of dipole-dipole coupling
τ_D	Fluorescence lifetime of the donor
MC	Monte Carlo
R	Radius of a lipid domain/raft
R_0	Förster radius
r_0	Limiting anisotropy
TCSPC	Time-correlated single-photon counting
L_d	Liquid disordered phase
L_o	Liquid ordered phase
K_i (i = D or A)	Partition coefficient of donors (i = D) or acceptors (i = A)
R	Domain radius
R_{plane}	Radius of the bicellar planes (*cf.* Picture 7.3)
R_{pore}	Radius of the pores
SS	Steady-state
Sph	Sphingomyelin
TR	Time-resolved

7.1 Introduction

Methods based on fluorescence and NMR spectroscopy [1, 2] are frequently applied for exploring lipid bilayer structure and its connection to the functioning of membranes. In the recent years, particular interest has been focussed on lateral lipid phase separation, usually referred to as lipid domains or rafts [3]. These structures (>200 nm) can be detected by microscopy techniques, whereas methods based on optical spectroscopy and electronic energy transport are required for revealing smaller domains [4–6]. Actually, despite the advances in super-resolution spectroscopy [4] the domain sizes comparable to the Förster radius (R_0), or even smaller, have hitherto only been determined by experiments based on electronic energy transfer.

Plasma/model membranes may contain heterogeneities which cause heterogeneous fluorophore distributions in bilayers. For instance, the formation of L_o domains in bilayers results in a preferential delocalisation of most fluorescence probes to the liquid-disordered L_d phase. Few exceptions are, *e.g.* cholera toxin or

NBD-DPPE [7–9], which exhibit a modest affinity to L_o phases. A bicelle resembles a disc that consists of flat regions and a highly curved rim. Bicelles represent interesting systems, where constituting molecules are heterogeneously distributed. For instance, bulky DSPE-PEG2000 molecules are predominantly localised on the rim, whereas cylindrically shaped DSPE molecules can form bicellar planes [10, 11]. Perforated bilayers contain regions with high and low curvature which imply the presence heterogeneous distribution of molecules [12]. The pores can be built up by *e.g.* short acyl chain DHPC molecules, whereas the flat regions could constitute DMPC molecules with two longer acyl chains [13]. In torroidal pores, the highly curved regions are usually formed by antibacterial peptides and lipid molecules, which preferentially pack into regions of high curvature [14].

Irreversible electronic energy transfer from excited donors to acceptor molecules [15–18] (FRET or DAET) depends on the spatial distribution of the interacting chromophores. Therefore, FRET studies are very useful for revealing heterogeneous lipid distributions in membranes. Hitherto, electronic energy migration between fluorescent molecules of the same kind, so called donor-donor energy migration (DDEM) or homotransfer [19–22], is less frequently applied, although DDEM contains the same information about spatial distributions. In practise, DAET can be monitored by the efficiency of energy transfer between donors and acceptors, or preferably by lifetime measurements of donors and acceptors in the presence and absence of ET. However, when utilising the DDEM approach, fluorescence depolarisation experiments are required for monitoring the electronic energy transport.

The present work on the advances in fluorescence spectroscopy aims at showing how heterogeneous lipid distributions can be explored in phase separated model lipid bilayers, pore containing bilayers, as well as in bicelles. In particular, the time-correlated single-photon counting technique has been applied and the obtained data have been analysed by Monte Carlo (MC) simulations. Of particular interest is determination of lipid domain sizes that are beyond the resolution of optical microscopy.

7.2 Modelling Energy Migration/Transfer in Lipid Structures

Homogenous distribution of donor- and acceptor-labelled lipids. Quite often, lipid molecules are labelled with fluorescent groups which aim at probing membranes at different lateral positions [23]. A possible case is pictured in Picture 7.1, where the fluorescent label is covalently attached to the lipid head-group. If the labels are donor groups, DDEM can occur within and between the bilayer leaflets, *i.e.* as intra- and interlayer migration, respectively. This is also the case if one studies mixtures of donor- and acceptor-labelled lipids. Within a two-particle approximation, the fluorescence relaxation and depolarisation of the donor in DAET and DDEM

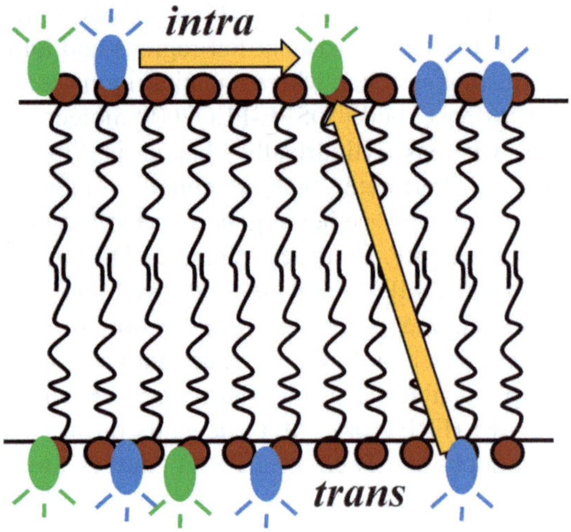

Picture 7.1 Energy transfer/migration can occur between fluorophores within one bilayer leaflet (*i.e.* an ***intra*** layer process), and between fluorophores located in opposite leaflets (*i.e.* an ***inter*** layer process). The interlayer energy transport occurs between layers separated at the distance *d*

experiments is described by a function $G^s(t)$, which stands for the probability that initially excited donor is still excited at a time t later [24]. Models for the probabilities of the intra- and interlayer processes have previously been derived [24];

$$\ln G^s_{\text{intra}}(t) = -C_2\left(\frac{15}{8\lambda}\right)^{1/3}\Gamma\left(\frac{2}{3}\right)\left(\frac{t}{\tau}\right)^{1/3} \tag{7.1}$$

$$\ln G^s_{\text{inter}}(\mu) = -\frac{\lambda C_2}{6\nu^2}\int_0^1\left\{1 - exp\left[-\mu\left(\frac{5}{4}s - 3s^{4/3} + \frac{9}{4}s^{5/3}\right)\right]\right\}s^{-4/3}ds \tag{7.2}$$

Here C_2 denotes the reduced concentration, which stands for the number of electronically interacting molecules within the area of a circle determined by the Förster radius, R_0. The reduced concentration can be calculated from $C_2 = \pi\rho R_0^2$ with knowledge of the surface acceptor density (ρ). The parameter λ is a number equal to 1 or 2 for DAET and DDEM, respectively. In Eq. 7.2 $\mu = (3\lambda/2)t\nu^6\tau^{-1}$, $s = \cos^6\theta_r$and $\nu = R_0 d^{-1}$. Here, d denotes the distance between the monolayers, which is approximately equal to the thickness of a lipid bilayer (*cf.* Picture 7.1). The angle θ_r is between the surface normal and a line connecting the centres of mass of each interacting dipole. For the energy transfer/migration within and between the two-dimensional planes, the electronic transition dipoles are approximated to be either randomly or in-plane oriented [24].

The total excitation probability,$G^s(t)$, that accounts for both intra- and interlayer energy transfer/migration is given by the joint probability

$$G^{s}(t) = G^{s}_{\text{intra}}(t)\prod_{j} G^{s}_{j,\text{inter}}(t) \tag{7.3}$$

In Eq. 7.3, the multiplication accounts for the general case of energy transfer/ migration between several planes of interacting donors/ acceptors. In the case when there is no intralayer energy transport$G^{s}_{\text{intra}}(t) = 1$.

For a donor-acceptor system, the fluorescence relaxation of the donor is given by

$$F(t) = G^{s}(t)F_{D}(t) \tag{7.4}$$

To monitor the rate of energy migration among donors one needs to measure the fluorescence depolarisation, which is conveniently expressed by the time-resolved anisotropy, *r(t)*, according to [25]:

$$r(t) = r_0\{\rho(t) - S^2\}G^{s}(t) + r_0S^2 \tag{7.5}$$

Here *S* describes the order of transition dipole *S* with respect to the bilayer normal. The reorientation of the excited donors is described by

$$\rho(t) = \sum \rho_j \exp(-t/\phi_j) \tag{7.6}$$

where θ_j are rotational correlation times which describe local motions of the donors in a lipid bilayer.

For lipids exhibiting inherent mutual affinity within lipid mixtures, the assumption of random distribution is no longer valid and consequently the above Baumann-Fayer (B-F) approach is not valid. When applying the B-F model one then finds an apparent higher reduced concentration of the labelled lipids than that calculated for a random distribution [21].

Lipid domains: Heterogeneous distributions of donor- and acceptor-labelled lipids. The complexities involved in describing these systems limit the derivation and application of analytical models. The difficulties concern various lipid phase geometries, as well as non-random lipid distributions. A recent review discusses these models [26]. The main complication is to analytically account for the nontrivial radial density distribution of acceptors surrounding each donor [27].

In order to estimate domain sizes in model lipid bilayers a majority of studies are based on assuming an infinite phase separation [26, 28]. Briefly, the acceptor surface concentrations C_{2A} and C_{2B} in phases A and B are estimated by means of the B-F model (*i.e.* by assuming the distribution of the labelled lipids within the segregated areas as homogenous) and are then recalculated to the partition coefficient K_A. Consequently, FRET that occurs across the phase separation boundary is neglected, which causes errors in the estimation of K_A. The true value of K_A (which can be determined by other methods) will deviate from that estimated with decreasing domain size. A comparison of both K_A values allows one to estimate the domain size.

An attempt to account for domain sizes comparable to R_0 has been presented by Brown et al. [29]. Their model includes the following assumptions: (i) The donors are distributed between the L_d and L_o phase according to K_D, while the acceptors are completely excluded from the domains; (ii) The donors localized in the domains reside on a circle with the radius $R_{circle} = 0.67\,R$, which corresponds to the average distance of donors from the domain centre; (iii) FRET within the same phase, as well as FRET across the phase boundary occur among homogeneously distributed acceptors, but the integration boundary differs [29]. These assumptions can be circumvented by using Monte Carlo simulations. Hereby, different random energy transfer processes can be mimicked [22, 30, 31]. This approach has been applied to describe DAET/FRET in lipid bilayers containing circular nanodomains [6, 9], torroidal pores, as well as in lipid bicelles [32]. In general, studies that involve MC simulations of domain size are very limited. The simulations have mainly been applied to test various analytical models [27, 33, 34].

The simulation procedure is carried out in a step-wise manner [6, 9, 32]: (i) For pore/nanodomain containing membranes, a certain number of pores/domains are generated (*cf.* Picture 7.2), which corresponds to a pre-defined peptide to lipid ratio for a pore system, and to the area fraction of domains present in lipid bilayers. (ii) The donors and acceptors were generated for definite probabilities of localisation inside and outside pores, in domains or on bicellar rims. These distributions are dictated by an equilibrium constant $K_{D,A}$ (= [probes inside]/[probes outside]). (iii) It is assumed that a donor is randomly excited and that its excitation energy is transferred to an acceptor. The time for this event to occur depends on the overall energy transfer rate Ω_i according to [30]

$$\Delta t_i = \ln\alpha / \Omega_i \tag{7.7}$$

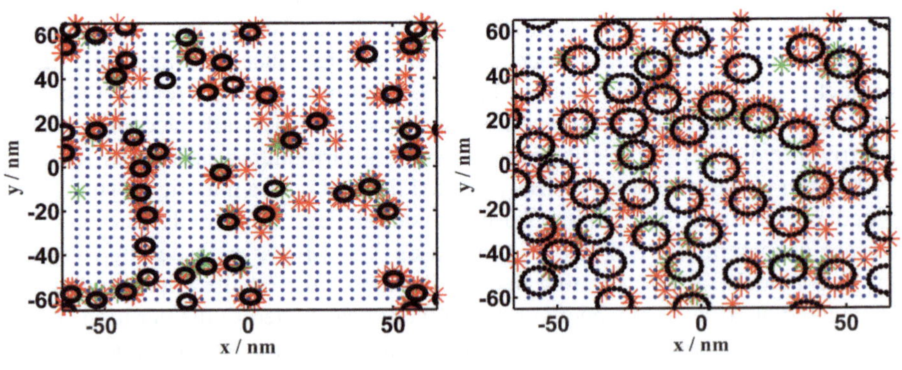

Picture 7.2 Schematic picture of a lipid bilayer with cylindrical pores (top view). The pore radii were 3 nm (on the *left*) and 6 nm (on the *right*). On the average, 5.5 magainin-2 molecules occupy each pore. At a peptide to lipid ratio of 1:200 this corresponds to *ca*. 39 pores distributed over the bilayer surface with dimensions of $20R_0 \times 20\ R_0$, $R_0 = 6$ nm. The partition coefficient of donors and acceptors distributed between the pores and remaining bilayer was 100

In Eq. 7.7 α denotes a random number between 0 and 1. The total rate of energy transfer is given by the sum of energy transfer rates from the excited donor i to all acceptors. Acceptors that are beyond the cut-off distance $10R_0$ are included via the continuum approach [30]

$$\Omega_{\mathrm{i}} = \sum 3/2\kappa_{\mathrm{ij}}^2 \left(R_0/R_{\mathrm{ij}}\right)^6 \tau_{\mathrm{D}}^{-1} + C_2 \left(R_0/R_{\mathrm{c}}\right)^4 \tau_{\mathrm{D}}^{-1} \tag{7.8}$$

In Eq. 7.8 the subscript j $(1 \leq j \leq N)$ refers to the number of acceptors (N) found within a cut-off distance R_{c}. R_{ij} is the distance between the i-th donor and j-th acceptor. The second term in Eq. (7.8) accounts for rates between the excited donor and the continuum of acceptors for two parallel planes (*i.e.* for a model lipid bilayer). The simulations correspond to energy transfer taking place under the dynamic limit condition and among isotropically oriented donors and acceptors. Therefore, the orienting factor $<\kappa_{\mathrm{ij}}^2> = 2/3$. New configurations were generated approximately 3000 times, and each generated configuration set was used 100 times in the calculation step. Periodic boundary conditions are used in order to mimic an infinitely large membrane containing pores/domains. The size of a replicated box is taken to be $20R_0 \times 20R_0$. The simulation yields $G^s(t)$-functions, which are related to the fluorescence decay according to

$$F(t) = G_{\mathrm{DA}}^s(t) F_{\mathrm{D}}(t) \tag{7.9}$$

Here $F_{\mathrm{D}}(t)$ denotes the donor fluorescence relaxation in the absence of acceptors. Translational diffusion can be neglected since DAET/FRET occurs typically within a few nanoseconds.

7.3 Specific Modelling Achievements

The modelling and analyses of lipid domain size in previous studies involve a large number of free parameters, which need to be optimized [26]. Recent studies show that these challenges can be circumvented by using the MC simulations [6, 9]. In our approach, the following parameters are considered and they can be either fixed or free: The donor fluorescence decay [$F(t)$], the Förster radius, the bilayer thickness (d), the partition coefficients of donors (K_{D}) and acceptors (K_{A}), the reduced surface concentration of acceptors (C_{A}), the domain radius (R) as well as their domains area in the bilayer (A_{r}). Unlike previous studies, all parameters were fixed except for R and A_{r}. The parameters K_{D} and K_{A} values could be kept constant because they were estimated at compositions being close to those investigated, and for which the bilayer was microscopically separated. This enabled an estimation of K by using the weighted fluorescence signal originating from the L_{d} or L_{o} phase.

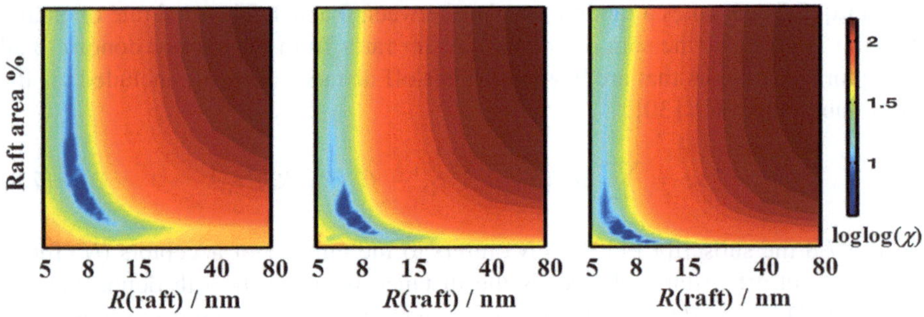

Fig. 7.1 Best fitting χ^2-parameter as a function of the domain radii (R) and the area domains occupied in the bilayers (A_r). The figures display the influence of the donor partition coefficient (K_D) in L_o and L_d phases on the position of the χ^2minimum. The figures correspond to $K_D = 3$ (*left*), $= 6$ (*middle*) and 11 (*right*), respectively. The partition coefficient of DiD acceptors was $K_A = 0.01$, *i.e.* practically all acceptors are excluded from the L_o domains

One could argue that this is not strictly true because a partition coefficient is constant solely along a tie-line of a phase diagram, whereas it changes along other lines. Nevertheless, changes of the partition coefficient (*cf.* Fig. 7.1) within ±50 % have a minor influence on the estimated bilayer area, and practically no impact on the raft size. The above deviations are large as compared to the error expected when K_D is estimated at different points in the phase diagram. The small influence on R is explained by an increase of K_D that would only cause an accumulation of more donors in the L_o phase. Similar results are obtained upon increasing the domain area, at a constant domain radius.

The reduced acceptor concentration was also kept constant in our analysis. This parameter can be independently obtained by applying the B-F model for compositions that correspond to a homogenous bilayer. Minor errors are expected as compared to non-homogenous lipid bilayers, but modification of C_A within ±10 % led to a small influence on the output parameters.

7.4 FRET in Bilayers Containing Liquid-Ordered Domains

In FRET experiments, the fluorescent probes (*i.e.* donors and acceptors) must exhibit different affinity to liquid-ordered L_o domains and the remaining bilayer. In order to observe small domains which are comparable to R_0, the probes need to be predominately excluded from one of the phases [9]. However, such ideal probes are rare [7, 35]. For instance, fluorophore-labelled cholera toxin is known for its affinity to the L_o domains, but still the partition coefficient is not high ($K \approx 6$) [6, 7]. Probes that prefer the L_d phase are more common with K-values as low as 0.01–0.005.

Domains are frequently detected as changes in the steady-state (SS) or time-resolved (TR) fluorescence intensity of donors that undergo FRET to acceptors before and after the formation of domains, *i.e.* for the homogenous and heterogeneous distribution of fluorescent probes. Three cases are of particular interest, namely; (1) The donors and acceptors are excluded from the domains, (2) the donors and acceptors are preferentially localized within the domains, and (3) the donors and acceptors prefer a different phase. The first two cases cause an increased FRET efficiency (= shorter fluorescence lifetimes), since the average D-A distance decreases. The last case results in the decrease of FRET efficiency (= longer fluorescence lifetimes) due the formation of domains, which segregate donors from acceptors.

The impact of the partition coefficient on the resolution of FRET in the domain size estimation is clearly illustrated in Fig. 7.2. Here, the steady-state intensity ratio between the homogenous (*e.g.* in the absence of domains) and heterogeneous (*e.g.* in the presence of domains) probe distribution is plotted versus the domain radius and the area fraction of domains in the bilayer. The data correspond to D:s and A:s, which are localized at the lipid/water interface of the bilayer. In our experience, the domains are beyond the resolution of TR-FRET and SS-FRET for the *ratio* I_D/I_D(uni) (case 3) or I_D(uni)/I_D (case 1 & 2) < 1.05 (red). The domains are close to the detection limit of TR-FRET when 1.05 < *ratio* < 1.1 (orange). For the range 1.1 < *ratio* < 1.2 (yellow) the domains can be resolved by TR-FRET and are

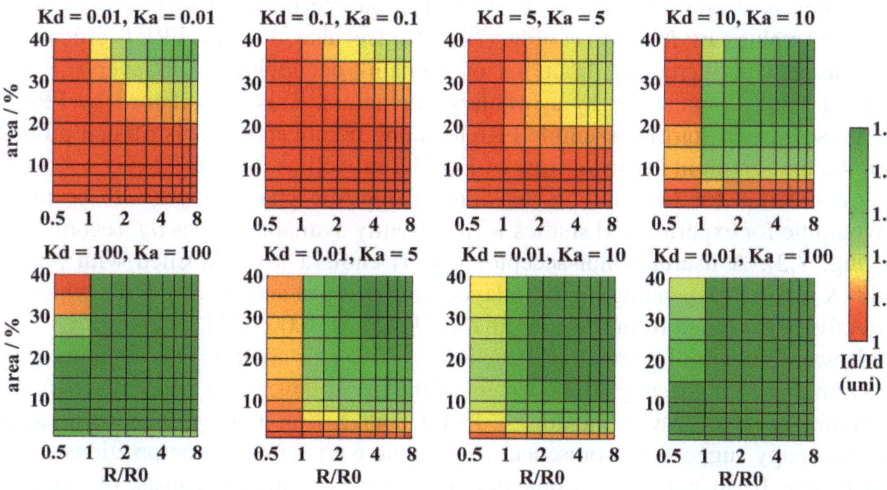

Fig. 7.2 The resolution of FRET as a function of the domain area and the relative domain radius (*i.e.* R/R_0). The FRET resolution is given by the ratio I_D/I_D(uni) for D:s and A:s preferring different phases, or by the ratio I_D(uni)/I_D for D:s and A:s preferring the same phase. I_D denotes the steady-state intensity of donors for probe distributions determined by K_D and K_A, whereas I_D(uni) denotes the steady-state intensity for probes which are homogeneously distributed over the entire bilayer surface. The D-lifetime was 6 ns in the L_o and L_d phase and the probes were localized at the lipid water interface. The *red*, *orange*, *yellow* and *green* colour corresponds to the conditions 1–4, respectively. Values exceeding 1.5 are displayed with the same colour as the limiting value

Table 7.1 The following conditions (1–4) were used in the discussion on FRET limits in the determination of nanodomain sizes. F_{uni} and F denote steady-state intensities that correspond to a homogenous and heterogeneous probe distribution, respectively

Conditions	TRFM	Steady-state range	Colour
1	Domains are beyond the resolution	$F/F_{uni} < 1.05$	Red
2	Domains are close to the resolution limit	$1.05 < F/F_{uni} < 1.1$	Orange
3	Domains are detectable	$1.1 < F/F_{uni} < 1.2$	Yellow
4	Domains are safely resolvable	$F/F_{uni} > 1.2$	Green

close to the detection limit of SS-FRET. Finally for *ratio* > 1.2 (green, cf. Table 7.1) the domains are detectable by TR-FRET and SS-FRET.

Case 1, *i.e.* when D:s and A:s are excluded from the domains, is not convenient for the detection of domains comparable to R_0, except for at low values of K (<0.001). First two upper graphs in Fig. 7.2 demonstrate that probes with $K > 0.01$ cannot resolve domains within a broad range of $R \in (0; R_0)$ and this experimental setup is convenient in cases of domains that occupy large bilayer areas. This is because the probes are localized in a small area of the L_d phase (the remaining part of the bilayer is occupied by the large domains), which makes FRET much more effective as compared to homogenous probe distributions.

Case 2, *i.e.* D:s and A:s exhibiting an increased affinity to the L_o domains, appears more sensitive to the presence of small domains, although the FRET efficiency is strongly influenced by the partition coefficient. Useful probes need to show values of $K > 10$ (*cf.* Fig. 7.2). Notice also that the FRET efficiency decreases with increasing fractions of domains occupying the bilayer (*i.e.* by several small domains, or by a few large domains). This is explained by FRET processes which start to resemble FRET among homogenously distributed probes (*cf.* the first graph in the lower row of Fig. 7.2 as well as Picture 7.2).

Case 3, *i.e.* for D:s and A:s residing in opposite phases appears to be most favourable for experimental studies with currently available probes (*cf.* second row in Fig. 7.2). A useful donor-acceptor pair is cholera toxin labelled with Alexa 488 ($K = 6$–11, depending on the bilayer composition) and the dye DiD, which is effectively excluded from the L_o domains ($K = 0.01$–0.005) [7].

It was shown in the previous work [6] that this pair could reveal domains with $R \approx 5$ nm and that occupy a few percent of the total bilayer area. More specifically, a combination of MC simulations, FRET and z-scan fluorescence correlation spectroscopy suggests the presence of domains with different compositions (type I and type II) in model lipid bilayers. The two types mainly consist of variable amounts of DOPC/Sph, constant levels of cholesterol, and low as well as high contents of cholera toxin [6]. The protein cholera toxin is responsible for the cholera disease and it is known to induce microscopically detectable domains in lipid bilayers [36, 37]. The properties of type I domains are influenced by the concentration of cholera toxin (Table 7.2): the domain radii were found to increase from 5 to 8 nm, and the diffusion time of labelled cholera toxin across the focal beam

Table 7.2 Estimated domain sizes obtained by MC simulations and FRET studies. The model lipid bilayer was composed of 25 mol% of cholesterol, different amounts of Sphingomyelin (Sph) and DOPC, as well as at low and high content of cholera toxin. In addition, the bilayer contained 5 mol % of DOPG and 2 mol % of G_{M1}

Sph/DOPC CT×B	0/0.68 low	0.19/0.49 low	0.19/0.49 high	0.24/0.44 low/high
Domain sizes	Homogenous	$R = 5$ nm	$R = 8$ nm	$R = 24$ nm
		area = 9 %	area = 6 %	area = 3 %
Transition time	6.6 ms	5.9 ms	46.5 ms	13.3 ms

increased substantially from approximately 6 to 47 ms. On the other hand, the type II domains (which contain more Sph) became much larger ($R \approx 24$ nm), whereas the diffusion time decreased to *ca.* 13 ms. It has been hypothesized that increased levels of Sph stabilize the type II domains, since further addition of CT×B has no influence. The shorter diffusion time of CT×B indicates that type II domains, although larger, are more permeable for CT×B and also bind less tightly to these. Thus, it seems that CT×B is less important for the existence of the type II domains.

7.5 FRET in Bilayers Containing Pores

Several similarities exist between heterogeneously distributed fluorescent molecules in a domain containing bilayer, and a lipid bilayer that contains antibacterial torroidal pores [11, 14, 38–40]. Both peptides and lipids constitute the surface curvature of torroidal pores. A typical torroidal pore has a radius of *ca.* 3 nm [14]. In FRET simulations, the pore shape can be approximated by a cylinder with probe molecules distributed on the cylindrical surface. Actually, the data in Fig. 7.2 illustrate the limitations of FRET in the size-estimation of L_o domains, which is related to FRET on the pore-containing bilayers.

A molecule which has certain affinities to the curved and non-curved regions (and is for FRET studies labelled by either a D or an A) is unevenly distributed between the pores and the remaining flat bilayer, leading to a heterogeneous distribution. FRET is with the exception of a few cases affected by redistribution of probes on the pore-containing bilayer, as is shown in this section. It is important to notice that FRET cannot distinguish between fluorescent molecules which are excluded from pores or exhibiting a modest affinity to pores, $K \in \langle -\infty; 10 \rangle$ (*cf.* Fig. 7.3) In all shown cases the fluorescence decays coincide with a homogenous distribution of probes for $K = 1$ [32].

Furthermore, if the peptide to lipid ratio is too high, FRET cannot report on the localisation of probes in pores or in the remaining bilayer surface, *i.e.* when the relative area of the bilayer occupied by the pores approaches 30 % [32]. For instance, in the case of magainin-2 the FRET efficiency seems to be invariant to K at the peptide to lipid ratio of 1:50. When considering that one pore constitutes approximately 5.5 magainin-2 molecules [14] this ratio would correspond to bilayers which are to 30 % covered by 3 nm sized pores. The ratio 1:50 appears

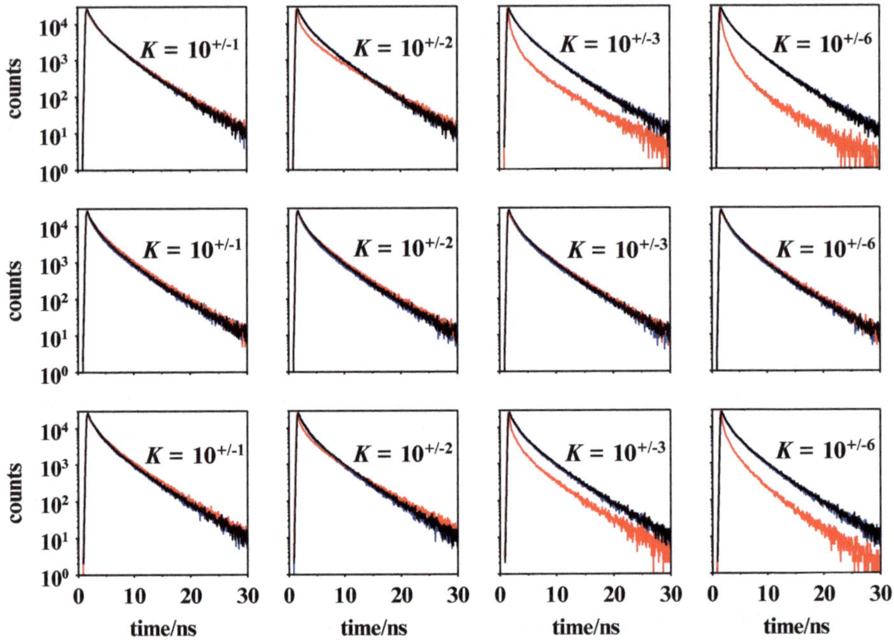

Fig. 7.3 Time-resolved decay curves of BODIPY-FL donors which transfer excitation energy to BODIPY 564/570 acceptors in lipid bilayers containing antibacterial pores. The probes were distributed on the lipid/water interface between the highly curved torroidal pores and the remaining flat regions according to the partition coefficient K. $K = 1$ (*black*), $K = 10^{+n}$ (*red*) or $K = 10^{-n}$ (*blue*), where $n = 1, 2, 3$ or 6. In the most upper row, the pore radius was 3 nm and the peptide to lipid ratio was 1:200. In the middle row, the pore radius was 6 nm, and the peptide to lipid ratio was 1:200. In the lowest row, the pore radius was 6 nm and the peptide to lipid ratio was 1:400

biologically relevant, since antibacterial pores are usually formed at peptide to lipid ratios higher than 1:300 [39]. Regarding the FRET resolution, a reasonable compromise is a peptide concentration when most peptides form pores, which is at the ratio 1:200. At this ratio the FRET decay (for $K \in \langle 10; \infty)$ is clearly distinguished from that for a random distribution (*cf.* the upper row in Fig. 7.3), unless for pore radii > 5 nm. At a peptide to lipid ratio of 1:200 and $R_{\mathrm{pore}} = 6$ nm the relative pore area will namely reach the critical occupancy of 30 %. This means that the distribution of probes starts to resemble a random probe distribution, even at very high affinity to the pores (*cf.* the middle row in Fig. 7.3 and Picture 7.2). A decrease of the peptide to lipid ratio (=1:400) improves the resolution even for $R_{\mathrm{pore}} = 6$ nm (*cf.* the lowest row in Fig. 7.3).

The results of this section can also be used in considerations related to the distribution of molecules that constitute the pore itself (as *e.g.* the antibacterial peptides). For this case $K_{\mathrm{D,A}} \to \infty$. As is shown in the last column of Fig. 7.3, pore formation could be very well monitored by FRET unless one of the above restricting conditions is fulfilled.

7.6 FRET in Bicelles

A bicelle consists of a flat plane and a highly curved rim, *i.e.* regions with variable curvature (*cf.* Picture 7.3). Therefore, bicelles might be useful for exploring the probe affinity to regions of different curvature. As is shown in Fig. 7.4, FRET actually enables the localisation of probes on bicellar surfaces. Here, the following three cases were considered. The probes are randomly distributed over the bicellar surface (indicated in red), only reside on the rims (indicated in blue), and, only reside in the planar regions (indicated in green). The FRET efficiency also depends on the bicellar size (Fig. 7.4). Bicelles formed by DMPC and DHPC have a radius of the planes, $R_{plane} \approx 3$ nm [13]. These are not suitable for studies of probe localisation because R_{plane} and R_0 are comparable and donors can couple with all acceptors. This corresponds to almost identical decay for the three distributions (*cf.* A in Fig. 7.4).

Bicelles spontaneously formed by DSPE and DSPE-PEG2000 (75/25 wt %) appear much better suited for affinity studies [10, 11]. An average radius of $R_{plane} \approx 8$ nm will suppress the FRET efficiency, and a negligible transfer should occur to the most distant acceptors. Present study strongly suggests that the partition coefficient K (= the probe localisation) can be obtained from the analysis of TR-fluorescence decays over the entire range, *i.e.* $K \in (-\infty; \infty)$ and for$R_{plane} \in (8, \infty)$ (tested region).

For $R_{plane} \approx 8$ nm one should obtain faster donor decay for DA pairs that prefer the planar region, and a slower if they preferentially are localized at the rims. Such a qualitative study has been performed with two DA pairs of differently shaped phosphatidylethanolamines (PE) [32]. The first DA pair had a cylindrical shape, whereas the other pair had a conical like shape. Because the influence of FRET on

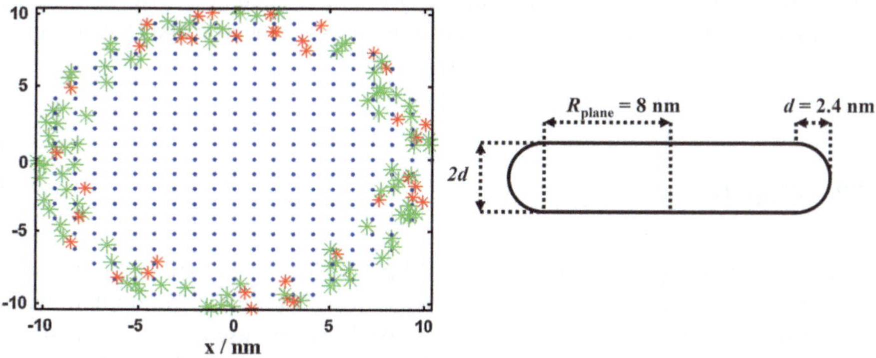

Picture 7.3 (*Left*) A top view of a bicelle with donors and acceptors distributed over the rim. (*Right*) A side view of a bicelle. The radius of the flat planes R_{plane} was 8 nm and the bilayer thickness $2d = 4.8$ nm. These dimensions correspond to bicelles formed by DPPC/DSPE-PEG2000 (75/25 mol %) lipid derivatives

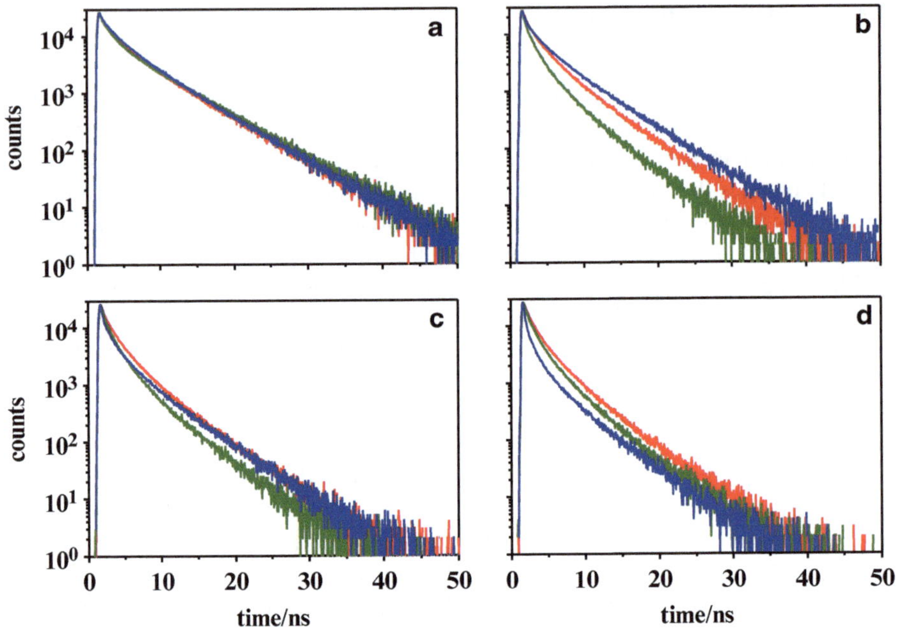

Fig. 7.4 Time-resolved fluorescence decays for BODIPY-FL which are acting as FRET donors to the acceptors BODIPY 564/570. The following conditions were mimicked: The probes were distributed homogenously over the entire bicellar surface (*red*); solely over the bicellar edges (*blue*) or over the bicellar planes (*olive*). The radii of the flat planes R_{plane} (*cf.* Picture 7.3) were (**a**) 2.0 nm, (**b**) 8.0 nm, (**c**) 16.0 nm, (**d**) 24.0 nm. The bilayer thickness was 4.8 nm

the time-resolved decays was very similar for both pairs, it was concluded that these pairs exhibit a similar affinity to the curved rims and the bicellar planes.

By increasing R_{plane}, the planar area grows much faster than the area of the rim. This causes a more efficient FRET for the DA pairs located on the rim as compared to DA pairs which are excluded from this region, since the average D-A distance is shorter among the former (*cf.* A–D in Fig. 7.4).

7.7 Conclusions

In studies of heterogeneous distribution of lipid molecules in various lipid systems, Förster resonance energy transfer combined with Monte Carlo simulations is a powerful technique. For instance, this approach can be used to estimate sizes of liquid-ordered nano-sized domains, *i.e.* domains which are far beyond the resolution of a light microscope. However, the FRET resolution strongly depends on probe affinities to the domains as well as the remaining bilayer. Currently available donors and acceptors that efficiently segregate into the different phases resolve domain sizes comparable to R_0, contrary to donors and acceptors which reside in the same phase. Furthermore, the approach described enables studies of the distribution of labelled lipids between small antibacterial pores and the remaining bilayer. The limitations concern high pore densities, or for modest probe affinities to the pores. Bicelles appear to be very suitable for similar localisation studies, because the FRET efficiency strongly depends on whether the probes are distributed over the entire bicellar surface, are only residing on the rims, or, are only residing in the planar regions.

Acknowledgements This work was financially supported by the Grant Agency of the Czech Republic via the grant 14-03141 (RŠ) and the Swedish Research Council (LBÅJ).

References

1. Lindblom G, Orädd G (1994) NMR-studies of translational diffusion in lyotropic liquid-crystals and lipid-membranes. Prog Nucl Magn Reson Spectrosc 26:483–515
2. Filippov A, Orädd G, Lindblom G (2007) Domain formation in model membranes studied by pulsed-field gradient-NMR: the role of lipid polyunsaturation. Biophys J 93:3182–3190
3. Lingwood D, Simons K (2010) Lipid rafts as a membrane-organizing principle. Science 327:46–50
4. Eggeling C, Ringemann C, Medda R, Schwarzmann G, Sandhoff K, Polyakova S, Belov VN, Hein B, von Middendorff C, Schonle A, Hell SW (2009) Direct observation of the nanoscale dynamics of membrane lipids in a living cell. Nature 457:1159–U1121
5. Heberle FA, Wu J, Goh SL, Petruzielo RS, Feigenson GW (2010) Comparison of three ternary lipid bilayer mixtures: FRET and ESR reveal nanodomains. Biophys J 99:3309–3318
6. Štefl M, Šachl R, Humpolíčková J, Cebecauer M, Macháň R, Johansson LB-Å, Hof M (2012) Dynamics and size of crosslinking-induced lipid nanodomains in model membranes. Biophys J 102(9):2104–2113
7. Chiantia S, Ries J, Kahya N, Schwille P (2006) Combined AFM and two-focus SFCS study of raft-exhibiting model membranes. ChemPhysChem 7:2409–2418
8. de Almeida RFM, Loura LMS, Fedorov A, Prieto M (2005) Lipid rafts have different sizes depending on membrane composition: a time-resolved fluorescence resonance energy transfer study. J Mol Biol 346:1109–1120
9. Šachl R, Humpolíčková J, Štefl M, Johansson LB-Å, Hof M (2011) Limitations of energy transfer in the determination of lipid nanodomain sizes. Biophys J 101:L60–L62
10. Sandström MC, Johansson E, Edwards K (2008) Influence of preparation path on the formation of discs and threadlike micelles in DSPE-PEG(2000)/lipid systems. Biophys Chem 132:97–103
11. Lundquist A, Wessman P, Rennie AR, Edwards K (2008) Melittin-lipid interaction: a comparative study using liposomes, micelles and bilayer disks. Biochim Et Biophy Acta-Biomembr 1778:2210–2216
12. Almgren M (2011) Stomatosomes: perforated bilayer structures. Soft Matter 6:1383–1390
13. van Dam L, Karlsson G, Edwards K (2004) Direct observation and characterization of DMPC/DHPC aggregates under conditions relevant for biological solution NMR. Biochim et Biophys Acta-Biomembr 1664:241–256
14. Ludtke SJ, He K, Heller WT, Harroun TA, Yang L, Huang HW (1996) Membrane pores induced by magainin. Biochemistry 35:13723–13728
15. Van der Meer BW, Coker G III, Chen S-YS (1994) Resonance energy transfer: theory and data. VCH Publishers Inc, New York

16. Valeur B (ed) (2001) Molecular fluorescence principles and applications. Wiley-VCH Verlag GmbH, Weinheim
17. Lakowicz JR (2006) Principles of fluorescence spectroscopy. Springer, Singapore
18. Vekshin NL (1997) Energy transfer in macromolecules. Spie Press, Bellingham/Washington, DC
19. Kalinin S, Johansson LB-Å (2004) Utility and considerations of donor-donor energy migration as a fluorescence method for exploring protein structure-function. J Fluoresc 14:681–691
20. Mikhalyov I, Bogen ST, Johansson LB-Å (2001) Donor-donor energy migration (DDEM) as a tool for studying aggregation in lipid phases. Spectrochim Acta a-Mol Biomol Spectrosc 57:1839–1845
21. Marushchak D, Gretskaya N, Mikhalyov I, Johansson LB-Å (2007) Self-aggregation – an intrinsic property of G(M1) in lipid bilayers. Mol Membr Biol 24:102–112
22. Marushchak D, Grenklo S, Johansson T, Karlsson R, Johansson LB-Å (2007) Fluorescence depolarization studies of filamentous actin analyzed with a genetic algorithm. Biophys J 93:3291–3299
23. Šachl R, Boldyrev I, Johansson LB-Å (2010) Localisation of BODIPY-labelled phosphatidylcholines in lipid bilayers. Phys Chem Chem Phys 12:6027–6034
24. Baumann J, Fayer MD (1986) Excitation transfer in disordered two-dimensional and anisotropic 3-dimensional systems – effects of spatial geometry on time-resolved observables. J Chem Phys 85:4087–4107
25. Medhage B, Mukhtar E, Kalman B, Johansson LB-Å, Molotkovsky JG (1992) Electronic-energy transfer in anisotropic systems. 5. Rhodamine lipid derivatives in model membranes. J Chem Soc-Faraday Trans 88:2845–2851
26. Loura LMS, Fernandes F, Prieto M (2010) Membrane microheterogeneity: Forster resonance energy transfer characterization of lateral membrane domains. Eur Biophys J Biophys Lett 39:589–607
27. Towles KB, Brown AC, Wrenn SP, Dan N (2007) Effect of membrane microheterogeneity and domain size on fluorescence resonance energy transfer. Biophys J 93:655–667
28. Loura LMS, Fedorov A, Prieto M (2001) Fluid-fluid membrane microheterogeneity: a fluorescence resonance energy transfer study. Biophys J 80:776–788
29. Brown AC, Towles KB, Wrenn SP (2007) Measuring raft size as a function of membrane composition in PC-based systems: part 1 – binary systems. Langmuir 23:11180–11187
30. Johansson LB-Å, Engström S, Lindberg M (1992) Electronic-energy transfer in anisotropic systems. 3. Monte-Carlo simulations of energy migration in membranes. J Chem Phys 96:3844–3856
31. Engström S, Lindberg M, Johansson LB-Å (1988) Monte-Carlo simulations of electronic-energy transfer in 3-dimensional systems – a comparison with analytical theories. J Chem Phys 89:204–213
32. Šachl R, Mikhalyov I, Gretskaya N, Olzynska A, Hof M, Johansson LB-Å (2011) Distribution of BODIPY-labelled phosphatidylethanolamines in lipid bilayers exhibiting different curvatures. Phys Chem Chem Phys 13:11694–11701
33. Kiskowski MA, Kenworthy AK (2007) In silico characterization of resonance energy transfer for disk-shaped membrane domains. Biophys J 92:3040–3051
34. Troup GM, Tulenko TN, Lee SP, Wrenn SP (2004) Estimating the size of laterally phase separated cholesterol domains in model membranes with Forster resonance energy transfer: a simulation study. Colloids Surf B-Biointerfaces 33:57–65
35. de Almeida RFM, Loura LMS, Prieto M (2009) Membrane lipid domains and rafts: current applications of fluorescence lifetime spectroscopy and imaging. Chem Phys Lipids 157:61–77
36. Hammond AT, Heberle FA, Baumgart T, Holowka D, Baird B, Feigenson GW (2005) Crosslinking a lipid raft component triggers liquid ordered-liquid disordered phase separation in model plasma membranes. Proc Natl Acad Sci U S A 102:6320–6325

37. Lingwood D, Ries J, Schwille P, Simons K (2008) Plasma membranes are poised for activation of raft phase coalescence at physiological temperature. Proc Natl Acad Sci U S A 105:10005–10010
38. Matsuzaki K, Sugishita K, Ishibe N, Ueha M, Nakata S, Miyajima K, Epand RM (1998) Relationship of membrane curvature to the formation of pores by magainin 2. Biochemistry 37:11856–11863
39. Bechinger B (1999) The structure, dynamics and orientation of antimicrobial peptides in membranes by multidimensional solid-state NMR spectroscopy. Biochim et Biophys Acta-Biomembr 1462:157–183
40. Brogden KA (2005) Antimicrobial peptides: pore formers or metabolic inhibitors in bacteria? Nat Rev Microbiol 3:238–250

Chapter 8
Fluoimaging Determination of Poisons, Pollutants, Narcotics and Drugs with No Added Reagents

Natalia V. Strashnikova, Shlomo Mark, Yehoshua Kalisky, and Abraham H. Parola

Abstract In this article we present optical photoluminescence spectroscopic method for the qualitative and quantitative detection of impurities, hazardous materials, pesticides, and pollutants in water, which lays the grounds for its future potential application to the detection of drugs in body fluids without added reagents. The method is based on synchronous fluorescence spectroscopy (SFS) of organic aromatic compounds or poly-aromatic hydrocarbons (PAH) and is carried out by following simultaneously their excitation and emission spectra. The full excitation emission matrix (EEM) generated in this way provides a 2-D and 3-D fluorescence map of the tested sample and the diagonals through the axes origin provide the synchronous fluorescence spectra at a constant wavelengths differences between the emission and excitation wavelengths, thus enabling simultaneous, multitude components identification. This map contains all the relevant spectroscopic information of the tested sample, and serves as a unique "fingerprint" with a very specific and accurate identification. When compared with pre-determined spectra and calibration curves from a database, there is a one-to-one correspondence between the image and the specific compound, and it can be identified accurately both qualitatively and quantitatively. This method offers several significant

---

N.V. Strashnikova
Faculty of Art and Science, New York University-Shanghai (NYUSH), Shanghai, 200122, P.R. China

S. Mark
Department of Software Engineering, Shamoon College of Engineering, Beer Sheva 84100, Israel

Y. Kalisky (✉)
Research and Development Division, Nuclear Research Center Negev, 9001, Beer Sheva 84190, Israel
e-mail: kalisky@netvision.net.il

A.H. Parola
Faculty of Art and Science, New York University-Shanghai (NYUSH), Shanghai, 200122, P.R. China

Present Affiliation: Structural Biology program, Memorial Sloan Kettering Cancer Center, New York, NY 10021, USA

© Springer International Publishing Switzerland 2016

C.D. Geddes (ed.), *Reviews in Fluorescence 2015*,
DOI 10.1007/978-3-319-24609-3_8

advantages, and it provides a sensitive (ppm detection level), accurate and simple spectroscopic tool to monitor impurities and pollutants in water. The design and performance of the spectrofluorimeter, i.e., fluoimager prototype, the software development and analysis of chemical organic compounds and mixtures in water as well as preliminary studies of drugs in body fluids, will be discussed in this paper.

Keywords Synchronous fluorescence spectroscopy • Fluorescence imaging • Detection of water impurities • Concentration calibration curves • 3D fluorescence • Drugs in body fluids • Neuronal networks • Pattern recognition software

8.1 Introduction

The qualitative and quantitative detection of either drugs in body fluids, poisons and pollutants in water resources and even street drugs in a powder form have the following common challenges that are addressed and tackled in this review chapter: high sensitivity, quick and simple determination, utilization by unskilled personnel, overcoming the use of sophisticated expensive equipment and on site determination with no need for sample transportation. For quick and simple determination, one would prefer to avoid uncalled for steps of sample purification as well as the elimination of use of specific reagents, e.g. antibodies. Moreover, the ability to detect and identify simultaneously multi component samples is a definite advantage. The method of choice is the utilization of 3D fluoimaging method, which is based on synchronous fluorescence spectral analysis. This method requires the development of an archetype of portable and continuous flow fluoimager and the preparation of data bank which is than scanned by advanced pattern recognition programs, which enable the identification of the unknown compound. We shall now expand on the various issues raised above and clarify one at a time.

The presently available procedures for the qualitative and quantitative determination of drugs in body fluids are based on the use of specific antibodies (Ab). The method requires a separation step between free and Ab bound drug, unless fluorescence polarization method is used (which suffers from a relatively reduced sensitivity) usually utilizing competition experiments of fluorescently labeled Ag with unlabeled Ag for the same Ab. It is time consuming and requires very advanced equipment, which calls for very highly qualified personnel to maintain. Those are usually done in hospitals and regional laboratories. Not only does one need to ship the samples to these center laboratories, often by air cargo, but it keeps us away from the most called for “lab on a chip” approach, where bed side testing are introduced. The fluoroimaging method brings us closer to the situation of a bed side testing, eliminates the need of shipping and instead calls for the spreading of relatively inexpensive fluoimagers worldwide, having the samples measured on site with no requirement for highly qualified personnel, with no step of either purification or addition of any Ab. Instead, the fluoimager, reads the unknown sample and sends the data though the internet to a central data bank which is able to analyze the data and ship though e-mail the answer; alternatively, dedicated fluoimagers may have their own data bank on a disk, utilized directly for their specific analysis e.g. poisons in water resources, or street drugs.

The detection of poisons and pollutants in water resources has recently become not only an environmental issue, but even more so, a homeland security problem. This requires the constant monitoring of those water supplies worldwide, utilizing the same fluoimagers with the advantage of quick, simple and reliable determinations.

Testing for street samples of narcotics, utilizing presently employed equipment and procedures, while very precise, are time consuming, require sophisticated expensive bench top devices, highly trained personnel with strong scientific background. Such street samples are multicomponent in nature, containing a mixture of drugs with adulterants and diluents at variable ratios, require a sample preparation step as well as the occasional addition of chemicals. As a result, legal actions and court procedures based on those analyses are frequently dismissed.

The method we describe here should be reliable, inexpensive, and can be extended towards various types of poisons, pollutants and impurities, with potential application to the detection of drugs in body fluids, e.g. saliva, serum and further in the future, even urine.

Analytical spectroscopic methods based on absorption and emission techniques are well established technologies. Emission spectroscopy provides more detection sensitivity and more information (included by the absorption and emission wavelengths), but it is less specific than the absorption spectroscopy when the emitting species possess broadband and overlapping emission bands.

An improvement of these classical detection methods is to use synchronous fluorescence spectroscopy (SFS) [1] where both the emission and excitation wavelengths are scanned simultaneously while maintaining a constant interval (Stokes shift) between the emission and excitation wavelengths. Practically, this method is a combination of both absorption and emission techniques, and it provides a very accurate and ***sharp*** emission spectrum of the chemical substance. This method is more useful in terms of chemical selectivity than the conventional fluorimetric techniques, particularly in the case of mixtures with spectral overlap. The synchronous fluorescence intensity, I_s, is given by [2]:

$$\mathrm{I_s}(\lambda_{\mathrm{ex}}, \lambda_{\mathrm{em}}) = Kcd\mathrm{E_{ex}}(\lambda_{\mathrm{ex}}) \times \mathrm{E_{em}}(\lambda_{\mathrm{ex}} + \Delta\lambda) \tag{8.1}$$

where E_{ex} and E_{em} are the excitation and luminescence spectral functions, respectively, and $\Delta\lambda$ is the difference between the emission (λ_{em}) and excitation (λ_{ex}) wavelengths. The other parameters are the instrumental geometrical constant K, c is the sample concentration, and d is the optical path. From Eq. (1) it is observed that the synchronous luminescence intensity is an explicit function of two parameters, λ_{ex} and λ_{em} and this enhances the sensitivity and selectivity of the detection by providing more information about the system. The value of $\Delta\lambda$ is an additional degree of freedom to improve the selectivity and sensitivity, especially when it is around Stokes shift.

Further improvement of the selectivity of the synchronous technique is to perform series of luminescence by repetitive emission scans over different excitation wavelengths range, hence creating a full excitation/emission matrix, which is represented by a 3-D or a 2-D contour map. This 3-D or 2-D image contains all the relevant spectral information including both the absorption and the emission spectra of a specific chemical compound, and is thus termed as the "total luminescence spectrum". It provides a unique "fingerprint" of a chemical compound, as well as more selectivity that enables the detection of individual components in the case of multi-component mixtures. It should be emphasized that this method provides a general picture while the synchronous fluorescence is equivalent to the information included in any diagonal line along the 3-D contour map, which represents simultaneous scan of emission and excitation wavelengths with a constant wavelength separation.

The synchronous fluorescence spectroscopy method has been applied first into qualitative and quantitative detection of various luminescent organic-aromatic compounds [3] and later has been applied towards trace analysis [4], human serum [5] and body fluid drug-analysis [6], water samples [7], as well as into studying photochemical and photophysical decay mechanisms of excited aromatic compounds [8].

In this context, our goal is to develop a compact, robust, deployable, and easy-to-use analytical tool for fast, accurate, and continuous analytical measurements. The proposed method scans the samples without pre-treatment or other time-consuming procedures of sample preparation. By using a pre-determined "data bank" and calibration curves of known concentrations and pH, it is possible to identify the compounds both quantitatively and qualitatively.

8.2 Experimental

8.2.1 The Fluoimager

A PC controlled multichannel spectrofluorimeters, model M50 & M50C have been designed and built according to our specification by SKALAR-LDI Ltd (Fig. 8.1) This apparatus automatically scans over 240–360 nm excitation and 265–580 nm emission spectral range, measuring simultaneously both fluorescence excitation and emission spectra in liquid samples. Model M50C enables also front surface measurements when the tested sample is not optically transparent.

A Continuous 75/150 W quartz Xe-lamp is used as a light source to excite the fluorescence. The spectral resolution is 3.5 nm for 200 μm slit width or 1 nm for a 50 μm slit. The ventilator on top of the lamp housing-device provides cooling of the lamp and is switched ON automatically. The Xe-lamp lights through the condenser the entrance slit excitation monochromator. The power supply is intended to provide the ignition and work of Xe-lamp in the linear regime. The Xe-lamp is fixed in a metallic cylindrical housing. The lamp position in the housing is adjusted for optimal illumination of the optical unit. The excitation monochromator is based on a holographic diffraction grating in the holder, rotating by a step motor. Through the exit slit the monochromatic light induces the fluorescence of the sample in the optical cell. The cell unit consists of two objectives, cell camber and cell holder

Fig. 8.1 The layout of the Fluoimager system used for the excitation/emission image and synchronous spectra

with a quartz cell. The first objective is used to illuminate the cell by monochromatic light, the second one collects the fluorescence signal and delivers it to the registration monograph. The cell holder is removable and is fixed on the front panel of M50C. The excitation light is transferred via a dispersive element into the sample placed in a 1 cm^2 quartz cuvette. The registration monograph with the holographic diffraction grating provides flat-field optical design. The fluorescence is collected at 90 deg to the excitation light, then analyzed by a monochromator (driven by a step motor), coupled with holographic diffraction grating and an intensified linear CCD-detector, (500 channels). The detector unit consists of image intensifier and linear CCD camera with 500 channels. The CCD camera includes digital and analogue controllers. Digital controller provides the CCD camera with necessary level control signals. The CCD output video signal is processed by analogue controller, digitized and transferred to the Fluoimager micro controller. The compact detector unit contains image intensifier (type DEP-XX1450 Delft Electronics) with PSU, fiber-optical tablet in optical contact (immersion liquid) with CCD – array at the video controller plate. The detector is assembled in the housing and fixed by the holder on the outlet frame of the registration monograph. The fixing and adjustment screws at the holder are used for positioning the detector in the horizontal and vertical directions. The video controller plate is connected by plane cable to the digital controller plate. At this plate are the following couplings: parallel data transfer to the microcontroller, reference and photometer cells. The camera operates steadily, while data collection and video signal transformation and processing taking place. The CCD detector is operated by two controllers. The

Table 8.1 Technical parameters

Component	Description
Light source	XE-lamp, continuous 75/150 W
Spectral range of excitation	240–360 nm
Spectral range of registration	250–570 nm
Spectral resolution	3.5 nm (200 μm slit), 1 nm (50 μm slit)
Detector type	Intensified CCD-array, 500 channels
Measurement cycle duration	Minimal 25 s
PC-connection	RS232 protocol via Serial Port
Dimensions B × H × T	482.6 × 266.7 × 500 mm
Weight	30 kg
Mains voltage	110/220 V, 60/50 Hz
Power consumption	250 W (when 150 W lamp operates)

sample geometry is a classical transmission arrangement, with short light path, hence low losses and high sensitivity. For SFS measurements, the constant interval between the excitation/emission wavelengths ($\Delta\lambda$) was kept in the range of $\Delta\lambda \approx 10$–30 nm. A microcontroller provides the control of measurement cycle, storage of data, preliminary data processing and data transfer to the PC via a standard Serial Port. Figure 8.1 presents the layout of the Fluoimager. The technical parameters of the Fluoimager are given in Table 8.1.

The Fluoimager controlling software provides the measurement and analysis of two-dimensional fluorescent spectra as the matrix of fluorescence intensity in coordinates of excitation and emission wavelengths. It operates on MS Windows platform and requires a standard home computer.

Lifetime measurements were performed using phase modulation method as was described briefly in Ref. [9].

8.2.2 The Identification Software Package

The identification software was developed in order to automate and facilitate the analysis. The idea beyond the developing of the software is to build an automatic, objective, sensitive and rapid analyzing tool that would be designated as a complementary tool for the embedded real time package Fluoimager M50C. The Fluoimager M50C package belongs to the family of real time embedded tool as well as Scientific software that require operating numerical methods and database management capability, while ensuring efficiency and reliability. The overall package has been developed as a three iteration process with three versions each: alpha, beta and release. The first iteration module serves as an independent confirmation tool. After a Software Test Plan, Description and Report the program moves to JAVA under NetBeans 7.0, and every module passed all the testing phases (requirements, design, program, installation and regression testing). User feedback, suggestions and accuracy assessments lead us, especially in order to facilitate and expedite the development process, to move the package to C# under Visual Studio . NET 2008 and again, every module, passed all the testing phases as well as user feedback; suggestions and accuracy assessments were incorporated into the package which was subjected to regression testing to make sure none of the improvements cause earlier passed tests to fail. The software package consists of four independent units: the embedded unit, the database unit, the pattern recognition and analysis unit and the GUI (graphic user interface) unit. In fact the application package is divided into six function oriented modules. Each module is responsible for specific functionality. The modules are: database connection module, graphic module, florescent imaging device connection module, math module, identification module and GUI module.

After the fluorescent imaging device finishes scanning the substance (e.g. the embedded unit), the database unit that was developed under MySQL (Open Source Database), saves for reference the measured data in a local MySQL database, where the data is presented in 3D and eight map graphs.

The pattern recognition and analysis unit – SPECTRALYZER, was created specifically for analyzing the spectra. The idea beyond this independent unit is that it is not enough to use one method to find the various substances under various conditions [10, 11].

This unit was designed to have the ability to readily use multiple methods for analyzing spectra, improving the chances of accurately detecting the important markers. The SPECTRALYZER was developed as neuronal network-pattern recognition software in order to identify and measure the concentration of the substances during analysis. The identification is made by comparing the obtained spectrum with the preliminary pre-gathered bank of spectra that contains the 3D images and the quantitative calibration data as well. This data bank is stored in the PC memory and can be easily exchanged with a remote central computer. Thus, the Fluoimager may easily be integrated into a large system of analysis, e.g., a national service bank. Spectra may be manipulated by this program in many different ways: subtracted, shifted, scaled and "smoothed", as well as peak matching. The user may determine which kind of preferred background spectrum should be subtracted for better quantitative results. The identification process which is based on several algorithms makes some manipulation such as smoothing the recorded signal, filtering the noise and identifying its peak values. The identification process is selected as the hierarchical pyramid of identification. Each step in the pyramid consists of several parallel and complementary detection methods; different numerical methods with similar complexity. In the first base of the pyramid we search peaks (verbose, Range peaks identification), at the second base we move from 3D graph to 2D and use the adjusted synchronous spectroscopy analysis, at the third base we use the principal-component analysis (PCA) and linear-discriminant analysis (LDA) as spectra comparison methods. From any linear function that describes the diagonal line slicing the 3-D image, it is possible to construct 2-D spectra which is also the synchronous spectra of the tested compound. The slope and the intercept can serve as analytical parameters to identify the compound both in a single or multicomponent analysis. At the top of the hierarchical pyramid we use the neuronal network-pattern pattern recognition analysis.

As was mentioned before this identification software package belongs to the family of scientific software, the narrow realm of Scientific Programming within the broader discipline of Software development, which refers to the development of algorithms and codes to solve problems in science and engineering. Due to the complexity of scientific calculations, testing scientific software is much more difficult as compared to other types of software because faults may arise from a variety of sources, including: lack of precise understanding of the process, flaws in the mathematical model as an abstraction of the process, difficulties in translating the mathematical model into computable form, faults due to program error, logical flaws, insufficient precision in computer arithmetic functions, etc.

8.2.3 *Materials*

The chemicals investigated in this research were purchased from Sigma-Aldrich LTD, Riedel-de Haen or Merk. Insecticides were obtained from either Machteshim LTD or alternatively, purchased from Aldrich and Fluka.

Samples were dissolved in distilled water. The initial concentration of the dissolved chemicals was adjusted to the level of ~10^{-4} M/L. The samples were further diluted into 10^{-6}–10^{-8} M/L concentration range. For fluorescence measurements, solutions at concentrations in the range of μg/ml were prepared.

Drugs to be tested were obtained from the pharmacy at the Soroka Medical Center, at Ben Gurion University. Others were purchased from Sigma-Aldrich LTD, Riedel-de Haen or Merk. Insecticides were obtained from either Machteshim LTD or purchased from Aldrich.

Samples where either dissolved in doubly distilled water (DDW) or in 1:100 plasma or serum in DDW. Plasma was purchased from the blood bank of Soroka Hospital and serum was separated from blood samples donated by healthy donors. Typically, for a compound in powder, 5 mg were dissolved in 5 ml of the examined solvent. For bottled drugs or those in ampoules, the initial concentration was registered on them. Compounds with reduced solubility were vigorously stirred at elevated temperature, below 50 °C. For fluorescence measurements, solutions at concentrations ranging from 2.0×10^{-2} to 200.0 μg/ml or saturated solutions (their exact concentration was determined spectrophotometrically) have been measure at room temperature.

For absorption studies, the HP, UV/VIS spectrophotometer, model 8452A and Jasco, model V-550 were used. The linear range of absorption obtained at variable concentrations was sought. Unless otherwise stated, the pH was kept in the range of 6–7, and was measured always at room temperature. Control solvent as well as the diluted plasma was subtracted from the sample fluorescence or absorption spectrum.

8.2.4 Instrument Calibration

The Fluoimager exhibits different sensitivity towards various fluorophores, depending on their fluorescence quantum yield. Therefore we have pre-calibrated the Fluoimager with each of the studied fluorophores at several concentrations. The 3D spectra obtained during this calibration procedure were inserted into a specific catalogue which is than treated by the software. This software defines the calibration constants from the absolute maxima of the 3D spectra.

8.2.5 Measurements

Prior to the more complicated measurements, *e.g.* in body fluids, we have checked the reliability and accuracy of the Fluoimager by analyzing several fluorophores in aqueous solutions. We prepared solutions of several known concentrations of each fluorophore, measured their 3D spectra which were inserted into the Fluoimager data bank. We then prepared similar "unknown" solutions to be qualitatively identified and quantitatively determined.

Following the proof of concept in aqueous solutions, we initiated drug analysis in diluted plasma. Different from aqueous solutions, human plasma has several complicating properties for fluorescence analysis. Probable drug modification by body fluids, variability in pH affecting the emission spectra and relatively high absorption in the UV region are among these complicating properties. We intend to overcome these problems. For instance, drug modification by body fluids, when occurs, would require the enlargement of the data bank and would be considered as new substances. In cases where a pH dependence is anticipated (based on preliminary screening), one can include a pH-meter signal with the automatic measurement cycle as well as with the software, which will be applied only to pH dependent drugs. Presently, in order to overcome the problem of high optical density of the plasma we study 1:100 diluted plasma. Its fluorescence background is subtracted from the 3D spectra of the added drug. The problem of the background rising from diluted body liquids of healthy, untreated individuals would be solved in the future after careful learning of the spectral characteristics of the diluted plasma. One possibility is decoding of the nature of each background peak and resolving it by adequate software.

An additional approach is the evaluation of the use of front surface spectroscopy specifically for opaque body fluids. Saliva might be the least complicated body fluid to study while urine samples could require additional pharmakokinetics studies to be added to the data bank.

8.3 Results and Discussion

8.3.1 *Impurities in Water*

We have investigated several aromatic organic compounds that have potential health risk due to their hazardous nature, following the list of such compounds provided by the Pharmaceutical Services Division Ministry of Health [12]. These compounds include poisons, pesticides, or drugs, which have a potential risk to civilians when dissolved in drinking water resources. The sensitivity of the technique (in the range of ppm-ppb detection limit), as well as its accuracy (in the range of ±1.8 % or less), provides an efficient tool to detect qualitatively and quantitatively the tested chemicals and consequently, prevents possible health damage. In order to optimize the detection method, the fluorescence data base and the calibration, we have to characterize the excitation and emission wavelengths of the various compounds. This is performed by measuring the "total luminescence spectrum" of each sample. This spectrum is represented in the form of excitation/emission matrix (EEM).

Figure 8.2 presents a typical 3-D and 2-D (EEM) of a commonly used insecticide (1,2 Naphtoquinone-4-Sulfonic acid) dissolved in water. This image provides a unique "fingerprint" of this material and hence, an accurate identification of this material. The EEM was obtained by scanning the excitation and emission wavelengths in the spectral range of 270–360 nm and 400–580 nm, respectively. This EEM spectral image provides the optimum spectral conditions that characterize specific chemical compound, in our case the Naphtoquinone-Sulfonic acid for example. This is because the data provides excitation/emission maxima 335 nm and 480 nm, respectively, which corresponds only to Naphtoquinone-Sulfonic acid.

Figure 8.3 presents a typical EEM of a two-component drug mixture diazepam and digoxin in water. Since the EEM contains all the excitation/emission data of the individual compounds, it is possible to get a well-resolved fluorescence 3-D image or fingerprint of each component that belongs to the mixture. Here we obtain from a single scan simultaneous and detailed analysis of two compounds. There is also a one-to-one correspondence between each chemical compound in the mixture and the wavelengths of excitation/emission maxima, as can be depicted from the fluorescence map in Fig. 8.3.

The information contained in the total luminescence spectrum (represented also by EEM), is specific to the tested compound. Particularly, excitation emission wavelengths at maxima can serve as an indicative tool for chemical analysis. Our results indicate that the pair of excitation/emission wavelengths at maxima ($\lambda_{ex}{}^{max}$, $\lambda_{em}{}^{max}$) is enough to analyze the fluorescent aromatic compounds accurately. In other words, there is one-to-one correspondence between the values of ($\lambda,{}_{ex}^{max}, \lambda,{}_{em}^{max}$) and the type of the analyzed sample. This is demonstrated in Fig. 8.4 for several chemical reagents including typical insecticides, pollutants and drugs dissolved in water. It should be noted here that the excitation/emission matrix (EEM) is a sensitive and accurate tool to detect small photochemical

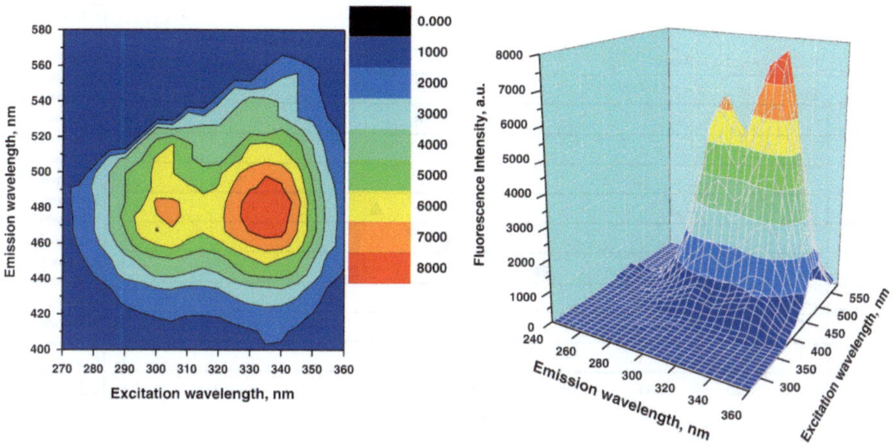

Fig. 8.2 A 2-D (*left*) and 3-D (*right*) excitation/emission matrix (fluorescence image) of 1,2 Naphtoquinone-4-Sulfonic acid insecticide

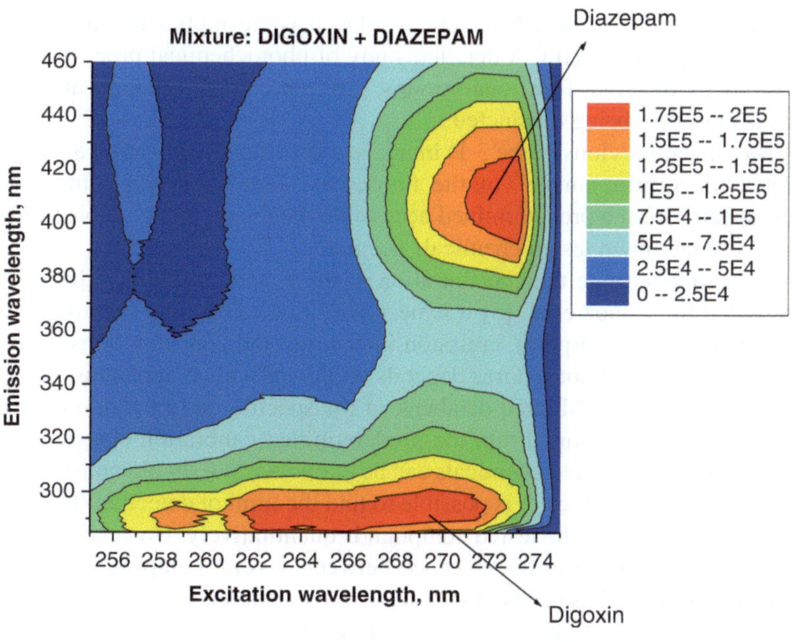

Fig. 8.3 The emission/excitation spectral matrix of a drug mixture (digoxin and diazepam) dissolved in water. The different colors indicate different fluorescence intensities

changes and spectral shift of the wavelengths pair ($\lambda,_{ex}^{max}$, $\lambda,_{em}^{max}$) in chemical compounds with similar excitation/emission characteristics. Substituting the phenolic aromatic ring with electronegative atoms (chlorine or fluorine) or with electron donors molecules, results in significant spectral shift, from

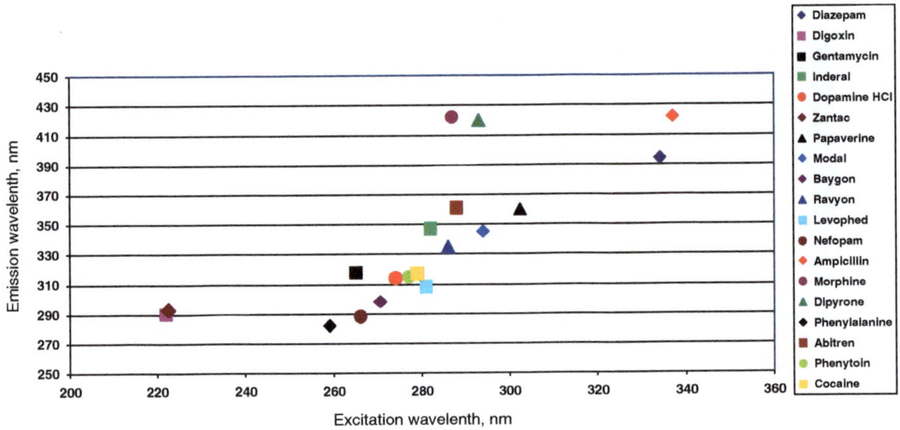

Fig. 8.4 Excitation/emission maximum wavelengths of several chemical compounds dissolved in water. The data is based on the 3-D fluorescence emission

($\lambda,_{ex}^{max} = 270\,\text{nm}$, ($\lambda,_{ex}^{max} = 310\,\text{nm}$) (in phenol) to ($\lambda,_{ex}^{max} = 305\,\text{nm}$, $\lambda,_{ex}^{max} = 395\,\text{nm}$) in aminophenol. Hence, the EEM technique has feasible application in the case of a slightly overlapping spectra. A detailed study of photochemical processes in compounds having different functional groups, electronic configurations, and pH conditions is currently being conducted.

As was stated previously, the 3-D image is a specific identification and provides a unique "fingerprint image" of the fluorescent sample. When compared with various spectra from a predetermined accessible databank, it is possible to identify qualitatively unknown tested chemicals accurately.

In order to determine the concentration of the tested sample, similarly to what was reported by Munoz et al. [5] it is necessary to establish the functional dependence between the absorption, emission intensities and concentrations. Since, the sensitivity of various fluorophores depends on the individual apparatus, it is essential to generate a pre-calibrated databank of the spectral matrix at several concentrations for each apparatus. From the predetermined calibration curves, unknown concentrations of the tested chemicals can be extracted. We should note here that by using the calibration curves, it is possible to find the limit of detection (LOD) of our system and characterize the tested compounds quantitatively. However, we have to avoid areas of overlapping fluorescence between several components in a multi-component mixture. This can be done owing to the detailed technique of excitation/emission matrix spectral analysis presented in [7]. Figure 8.5a presents the excitation/emission matrix of a commonly used pesticide Indole-3-Acetic acid, (1.0 ug/ml), and the excellent correlation ($R^2 = 0.995$, R is the correlation factor) between the measured and pre-determined concentrations is observed-see Fig. 8.5b.

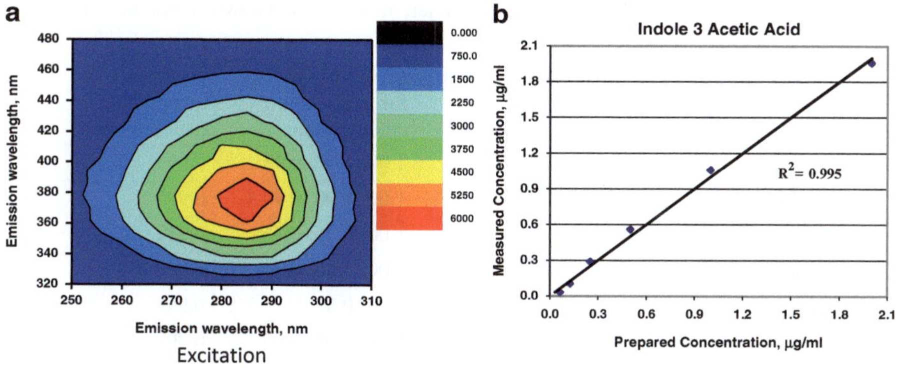

Fig. 8.5 (**a**) The fluorescence image of excitation/emission matrix (1.0 μg/ml) in water (**b**) Quantitative determination of indole-3-acetic acid dissolved in water

8.3.2 Street Narcotics and Drugs Determination in Body Fluids

The results of the Fluoimager, for testing of fluorescent drugs in aqueous solution (usually found in body fluids) is presented in Table 8.2. The high reliability of substance identification is evident. Figure 8.6 depicts an example of the quantitative determination one of these substances-phenol. The results of the analysis of drugs in plasma are given in Table 8.3. A three- component mixture was correctly analyzed both qualitatively and quantitatively. Note that the concentrations in the table are given as measured, i.e., in diluted plasma.

8.4 Lifetime Measurements

The values of the lifetime of the fluorescent emitting state is an indication as to the chemical dynamics and photophysical processes of the organic compounds under various conditions such as concentration, PH or temperature. The measured lifetimes of various drugs in water are presented in Table 8.4. The Table includes also the excitation wavelength and the concentration. It was found that the fluorescence lifetime of various drugs dissolved in water were short, in the range of ns temporal regime. These short lifetime results from the $S_1 \rightarrow S_0$ allowed transitions. It exhibited in some cases multi-exponential decay curve which indicates complex photophysical processes. Similar behavior was observed by using the same compounds in plasma. However, more research is planned to explore the chemical dynamics and PH dependence of narcotics and various drugs involved.

Table 8.2 Testing the fluoimaging reliability in determining of unknown compounds in water

No	Solution in water[a]	Prepared/Average Measured Concentrations[b]	Reliability factor for qualitative determination[c]	Correlation coefficient for quantitative determination[d]
1	Aminocarb	20.0/20.4 2.5/2.2 0.6/0.8	1.00; 1.00; 0.95	0.995
2	Aniline	22.0/22.6 2.7/2.6 0.3/0.2	1.00; 1.00; 0.90	0.995
3	Asulam	24.0/23.5 3.0/3.1 0.2/0.4	1.00; 0.95; 0.95	0.917
4	Bendiocarb	26.0/25.7 6.5/7.0 3.0/2.8	1.00; 1.00; 0.90	0.999
5	Desmedipham	3.0/3.3 0.7/1.0	1.00; 0.95	0.995
6	3,5-Dimethylphenol	20.0/18.8 1.2/1.0 0.1/0.05	1.00; 1.00; 0.95	0.999
7	Ethoxyquin	3.0/2.7 0.4/0.3 0.05/0.02	1.00; 0.95; 0.95	0.999
8	Baygon	30.0/28.6 2.0/1.5 0.1/0.06	1.00; 0.95; 0.90	0.999
9	Carbofuran	50.0/50.5 6.2/6.0 1.5/1.9	1.00; 0.95; 0.90	0.999
10	Methyl cotnion (Azinphos-methyl)	10.0/10.4 5.0/4.8 2.5/2.1	1.00; 1.00; 1.00	0.946
11	Naphthalene	15.0/15.3 3.7/4.0 0.1/0.2	1.00; 0.95; 0.90	0.985
12	Phenol	6.0/5.9 1.0/1.0 0.06/0.03	1.00; 1.00; 0.95	0.999
13	Acenaphthenequinone	10.0/11.0 5.0/5.4 1.0/0.7	1.00; 1.00; 1.00	0.995
14	1,2-Naphthoquinone-4-sulfonic acid sodium salt	20.0/19.2 2.5/2.1 0.3/0.2	1.00; 0.95; 0.90	0.995
15	Indole 3 Acetic Acid	2.0/1.9 0.5/0.6 0.06/0.07	1.00; 1.00; 0.90	0.995
16	Estriol	100.0/88.2 12.5/12.0 1.5/1.3	1.00; 1.00; 0.95	0.999
17	17 Estradiol	100.0/93.6 25.0/24.1 6.2/5.5	0.95; 0.95; 0.90	0.985
18	Fluorene	1.0/1.1 0.5/0.4 0.1/0.07	1.00; 1.00; 1.00	0.995
19	2-Cyanophenol	10.0/9.4 5.0/4.5 1.2/1.0	0.95; 1.00; 0.90	0.957
20	Azoxystrobin	20.0/20.6 10.0/10.3 2.5/2.6	1.00; 1.00; 1.00	0.999
21	Fenpropidin	66.0/65.0 13.5/12.8	1.00; 0.90	0.995
22	1-Naphthol-3,6-disulfonicacid, disodium salt hydrate	12.5/12.7 3.1/3.4 0.8/1.3	1.00; 1.00; 0.95	0.999
23	4-Fluorophenol	10.0/10.1 2.5/2.7 0.3/0.3	1.00; 1.00; 0.95	0.995
24	4-Chlorophenol	12.0/11.5 3.0/2.7 0.4/0.3	1.00; 0.95; 0.95	0.968
25	1-Naphthol	12.0/12.6 1.5/1.8 0.2/0.3	1.00; 1.00; 1.00	0.995
26	Toluene	86.6/90.0 11.0/10.8	0.95; 0.95	0.814

[a]Tapped water at neutral pH and room temperature

[b]Unknown solution prepared at the relevant soluble concentration (μg/ml), nominator, were prepared and their concentration was determined (in μg/ml) by the fluoimager, denominator, based on the pre-prepared data bank and the concentration calibration curves

[c]Defined as the ratio of correct vs. total number of repeat measurements (between 10 and 20). In most cases of the substance identification dismissing it was found operator's mistake

[d]The correlation coefficients of the linear curves obtained by plotting the measured against the predetermined concentrations

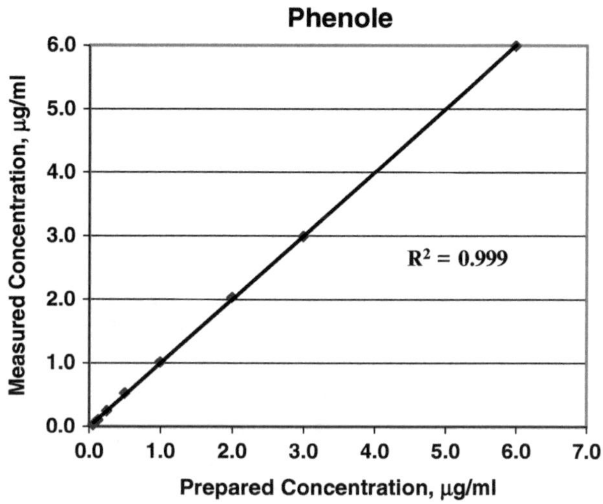

Fig. 8.6 Quantitative determination of phenol in water

8.5 Conclusion

Our method is useful in the detection of impurities in water, aimed at homeland security applications. Moreover, it shows the potential of application to drug and narcotics diagnostics, as well.

Four street narcotics, determined by the fluoimaging method (based on a preprepared data bank containing as many as 150 fluoimages of known narcotics from various sources and their known impurities), were compared with the classical tests run at the Israel National Forensic Laboratories at Tel-Hashomer Hospital by the most trained technicians, on conventional, expensive and time consuming instruments. The results obtained by our novel approach for the detection of cocaine, heroin, exstasy and marijuana, were confirmed both qualitatively and quantitatively [13].

The inherent compactness, reliability and wide functional flexibility of the designed Fluoimager prototype promise high perspectives for its suitability to therapeutic drug monitoring (TDM).

The displayed results on the screen shows the chemical nature of the detected substance, its concentration and further shows the overall 3D view of the spectrum.

The pre-prepared mixtures of three drugs, sodium salicylate at the ng/ml range, dobutamine at the µg/ml range and quinidine at the ng/ml range, all at physiologically relevant concentrations, were tested in human plasma (1:100 diluted by DDW) by our fluoimaging method and revealed the expected concentration values. It is conceivable that further studies addressing the above mentioned difficulties should result in the successful implementation of this approach to TDM.

The advantages of the method make it unique: there is no need for pre-treatment of the sample and no use of antibodies. The method is specific with highly resolved

Table 8.3 Quantitative determination of individual fluorophores in mixed samples: determining of drugs in diluted human plasma

Solution of the 3 components in diluted plasma		Prepared vs. Average measured Concentrations						
Na-salicylate, ng/ml	Prepared	48.0;	96.0;	96.0;	192.0;	192.0;	192.0;	48.0
	Measured	48.4;	94.3;	96.3;	192.3;	192.3;	191.8;	48.8
		±0.8	±1.7	±0.3	±0.2	±0.2	±0.1	±1.8
Dobutamine, μg/ml	Prepared	5.0;	10.0;	20.0;	50.0;	50.0	5.0;	50.0
	Measured	5.0;	10.0;	20.0;	50.0;	50.0	5.0;	50.0
		±0.1	±0.1	±0.1	±0.1	±0.1	±0.1	±0.1
Quinidine, ng/ml	Prepared	78.0;	78.0;	156.0;	312.0;	78.0;	312.0;	312.0
	Measured	78.0;	76.7;	153.1;	312.0;	78.0;	312.7;	311.8
		±1.2	±1.8	±1.8	±0.3	±0.6	±0.2	±0.6

Table 8.4 Lifetime data of various drugs dissolved in water measured with the K2/ISS Multifrequency Phase Modulation Spectrofluorimeter

No	Sample	Solvent	Concentration, μg/ml	Excitation wavelength, nm (Filter)[a]	Lifetime, ns	Fraction
1	Atropine Sulfate	Water	181.8	230 (B1)	Not enough fluorescence	-
			470.6	258 (UG1)		
2	Diazepam	Water	100	340 (B2)	0.57 ± 0.010.0.000 ± 0.003	0.585 ± 0.004
						0.344 ± 0.005
3	Digoxin[b]	Water	250	250 (UG1)	3.45 ± 0.47	0.82 ± 0.02
4	Luminal (Phenobarbital)	Water	5–300	242 (B5(1))	not enough fluorescence	–
5	Gentamycin	Water	40, 92.3,	256 (B5(1))	not enough fluorescence	–
			184.6	280 (UG1)		
6	Aminophylline	Water	–	272 (UG1)	not enough fluorescence	–
7	Esracaine (Lignocaine HCl)	Water	–	264 (UG1)	not enough fluorescence	–
8	Pramine (Metoclorpra-mide)	Water	12.5, 25, 125	280 (UG1)	very short	–
				300, 310 (B3)		
9	Procainamide HCl	Water	6.2	300 (B3, B6, B365)	very short	–
		PBS	6.2	300 (B365)	very short	–
10	Inderal (Propranolol HCl)	Water	0.3, 0.6	300 (B365)	9.90 ± 0.53	0.93 ± 0.03
		Plasma 1/100	1.0	290 (UG1)	6.26 ± 1.07	0.55 ± 0.03
					0.71 ± 0.11	0.35 ± 0.02
11	Dopamine HCl	Water	20	280 (UG1)	0.84 ± 0.05	0.60 ± 0.1
					0.37 ± 0.07	0.32 ± 0.1
12	Zantac (Ranitidine HCl)	Water	–	314 (C6)	not enough fluorescence	–
13	Verapamil (Ikakor)	Water	125, 250	278 (UG1)	1.78 ± 0.01	0.84 ± 0.1
		Plasma 1/100	125	278 (UG1)	5.91 ± 0.09	0.51 ± 0.01
					1.68 ± 0.05	0.43 ± 0.01
14	Cerubidin (Daunorubicin HCl)	Water		500 (Y1)	1.09 ± 0.03	0.88 ± 0.07

(continued)

Table 8.4 (continued)

No	Sample	Solvent	Concentration, μg/ml	Excitation wavelength, nm (Filter)[a]	Lifetime, ns	Fraction
			322, 75, 150, 300, 600			
15	Papaverine HCl	Water	5	327 (B6)	<0.3	–
16	Tagamet (cimetidine)	Water	–	218(UG1)	–	–
17	Modal (Sulpiride)	Water	25, 50	298 (B3)	0.476 ± 0.004	0.972 ± 0.005
				300 (B365)		
18	Baygon	Water	25	278 (UG1)	1.086 ± 0.004	0.921 ± 0.002
			50, 100	278 (UG1)	1.88 ± 0.16	0.37 ± 0.03
					0.85 ± 0.01	0.63 ± 0.04
19	Ravyon	Ethanol	5.6, 11.3, 22.5, 45, 90	282 (UG1)	14.2 ± 0.2	0.88 ± 0.01
20	Carbofuran	Water	18.8, 75	280 (UG1)	very short	–
21	Anginine	Water	31.2	280 (UG1)	not enough fluorescence	–
22	Phenylpropyl carbamate	Ethanol	300	270 (UG1)	10.7 ± 0.8	0.79 ± 0.03
23	Cardoxine (Dipyridamole)	Water	3.1, 6.2, 125	404 (G1)	17.70 ± 0.45	0.931 ± 0.005
		Plasma	12.5	404 (G1)	17.63 ± 0.05	0.935 ± 0.002
24	Adenosine (Adenocor)	Water	32.5	360 (C6)	0.156 ± 0.002	0.838 ± 0.001
			16.2	284 (UG1)	5.28 ± 0.07	0.28 ± 0.02
					0.109 ± 0.03	0.526 ± 0.002
25	Dobutamine HCl (Butamine)	Water	7.8, 15.6, 31.2, 62.5, 125	280 (UG1)	1.13 ± 0.14	0.54 ± 0.13
					0.49 ± 0.07	0.42 ± 0.14
26	Levophed (Norepinephri-ne)	Water	3.1, 125, 25	280 (UG1)	1.27 ± 0.01	0.80 ± 0.06
					0.42 ± 0.06	0.194 ± 0.05
27	Isoket (Isosorbide dinitrate)	Water	–	208 sh	–	–

28	Nitroglycerin	Water	–	202 sh	–	–
29	Neo Synephrine (Phenylephrine HCl)	Water	3.1, 6.2, 12.5, 25, 50	280 (UG1)	2.45 ± 0.06	0.94 ± 0.06
30	Regitine (Phentolamine methanesulfo-nate)	Water	12.5	280 (UG1)	<0.1	0.707 ± 0.004
31	Nefopam	Water	72.5, 145	268 (UG1)	2.55 ± 0.06	0.87 ± 0.02
32	Esmolol (Brevibloc)	Water	25, 50	275 (UG1)	3.83 ± 0.15	0.18 ± 0.02
					0.97 ± 0.02	0.73 ± 0.03
33	Bretylate (Bretylium tosylate)	Water	250	264 (UG1)	3.78 ± 0.04	
					0.00	
34	Zovirax (Acyclovir sodium salt)	Water	–	265 (UG1)	–	
35	Fungizone (Amphoteracin B)	Water	–	328 (B2)	11.25 ± 0.38	
					0.39 ± 0.15	
36	Amikacin Sulfate (Amikin)	–	–	–	–	–
37	Nebcin (Tobramycin sulfate)	–	–	–	–	–
38	Mycomyst (Acetylcys-teine)	–	–	–	–	–
39	Metronidazole	–	–	–	–	–
40	Ampicillin	Water	15.0–25.0	344 (C6)	3.84 ± 0.40	0.24 ± 0.01
					0.19 ± 0.03	0.65 ± 0.01
41	Morphine	Water	10.0, 20.0	284 (UG1)	0.554 ± 0.002	0.97 ± 0.03
42	Dipyrone (optalgin)	–	–	–	–	Very low intensity
43	Meperidine	Water	100.0, 200.0, 500.0	230,258 (B1, UG1)	–	Very low intensity
44	Phenylalanine	Water	100.0, 150.0, 200.0	266 (B1 + WG280)	6.033 ± 0.097	0.802 ± 0.045

(continued)

Table 8.4 (continued)

No	Sample	Solvent	Concentration, μg/ml	Excitation wavelength, nm (Filter)[a]	Lifetime, ns	Fraction
45	Rafazocine (pentazocine)	Water	4.5, 9.0, 18.0	280 (UG1)	3.162 ± 0.037	0.910 ± 0.008
46	Abitren (diclofenac sodium)	Water	2.5	289 (B365)	–	Very low intensity
47	Triclonam	–	–	–	–	Non band in absorption spectrum
48	Qunidine sulfate	Water	2.0, 3.0, 4.0	332 (B6)	15.59 ± 0.79	0.253 ± 0.012
					4.15 ± 0.10	0.657 ± 0.025
					1.41 ± 0.15	0.090 ± 0.017
49	Sodium salycilate	Water	9.4, 4.7, 2.4	297 (B6)	4.022 ± 0.014	1.0
50	Lomidal	Water	5.0, 12.5, 20.0	328 (B6)	0.488 ± 0.022	0.775 ± 0.011
					0.000	0.131 ± 0.019
51	Phenytoin	Water	1000.0, 2000,0, 3000.0	290 (B1 + G1)	1.64 ± 0. 12	0.522 ± 0.024
					0.10 ± 0.08	0.364 ± 0.033
52	Promethazine hydrochloride	Water	62.5, 125,0, 250.0	340 (B6)	1.791 ± 0.107	0.330 ± 0.090
					0.300 ± 0.060	0.632 ± 0.088
53	Zofran (Ondansetron hydro-chloride dihydrate)	Water	4.0	323 (B6)		Very short lifetime
54	Furosemide	Water	6.3, 12.5	346 (C6)	4.685 ± 0.125	0.512 ± 0.078
					0.266 ± 0.011	0.440 ± 0.098
55	Haloperidol Decanoate	Water	500.0, 1000.0	250 (UG1)	3.941 ± 0.086	0.708 ± 0.032
					0.352 ± 0.048	0.165 ± 0.023
56	Taroctyl (Chlorpromazine hydrochloride)	–	–	–	–	–
57	Kemadrin (procyclidine hydrochloride)	Water	500.0, 750.0	250 (UG1)	3.108 ± 0.032	0.764 ± 0.040

58	Adolan (Methadon)	Water	1.56, 3.25	250 (UG1)	–	Signal very small for good measuring

[a]Transmittance of Filters, nm:

WG280	>250
B1	230–420
UG1	280–418
B5 (1)	290–410
B365	300–420
B3	308–400
B6	340–560
C6	356–454
B2	400–500
G1	440–560
Y1	>500

[b]Large scattering of points.

capabilities, instantaneous, highly sensitive and enabling both quantitative and qualitative identification of fluorescent unknowns. It provides simultaneous identification of multitude components and enables real time data transfer.

As for homeland security applications, the spectral technique presented here is an additional layer of countermeasures currently being used to fight global terrorism.

We have developed specific software that controls the system's operation, data acquisition and mathematical analysis, automatically. By using pre-determined databank (we acquired more than 50 aromatic compounds for such databank) and calibration curves, it was possible to identify the tested compounds both quantitatively and qualitatively.

The Apparatus has several clear advantages: it is inexpensive, compact and user-friendly and does not require any special operation know-how. It can be set for continuous flow analysis as well as field work. The following implementation is anticipated: the device will be freely distributed and reimbursement will be made upon delivery of the analysis results at much reduced rates than presently charged. It will be based on the continuous dynamic expansion of the data bank and results will be delivered through the internet rather than depend on air shipment of test samples.

The anticipated success of the proposed determination of drugs in body fluids will allow use not only in hospital laboratories, but will also include bed-side testing and MD's office diagnostics. Furthermore, since it is quick and simple it will allow the screening of large populations for drug abuse (e.g., high school students, army and industrial personnel, drivers, etc.).

The technology has several limitations since it is limited to chemical compounds with efficient fluorescence and with significant Stokes shift. The issue of detecting a mixture of several chemicals requires additional studies. The detection of impurities in water is more feasible than the detection of drugs in body fluids due to emission and scattering from background chemicals, in the latter. Moreover, the application to TDM will require the expansion into pharmacokinetic which will require additional fluoimages of drug metabolites to be included in the data bank.

Acknowledgements Work was supported by The Israeli Water Authority, contract No. 4500196511 and by the Technological Incubators Program, Office of the Chief Scientist (OCS) of the Israel Ministry of Industry, Trade and Labor.

We are grateful to Dr. Vitaly Sominsky Ph.D., and Dr. Shlomo Almog Ph.D., for enabling the calibration tests of the narcotics at the Sheba Medical Center and to the Israeli Police for providing us with them. We thank for the fruitful discussions with Professor Yona Amitai, MD, and Mr. Razi Efron.

References

1. Lloyed JBF (1971) Synchronized excitation fluorescence emission spectra. Nat Phys Sci 231:64–65
2. Vo-Dinh T (1978) Multicomponent analysis by synchronous luminescence spectroscopy. Anal Chem 50(3):396–401

3. Lloyed JBF, Evett IW (1977) Prediction of peak wavelengths and intensities in synchronously excited fluorescence emission spectra. Anal Chem 49(12):1710–1715
4. Rubio S, Gomez-Hens A, Valcarcel M (1986) Analytical applications of synchronous fluorescence spectroscopy. Talanta 33:633–640
5. Munoz de la Pena A, Moreno MD, Duran-Meras I, Salinas F (1996) Synchronous fluorimetric determination of salicylic acid and diffusional serum using partial least-squares calibration. Talanta 43:1349–1356
6. Strashnikova NV, Gershanik AP, Papiashvili N, Khankin D, Luria RC, Mark S, Kalisky Y, Parola A (2010) Towards instantaneous quantitative fluoroimaging drugs determination in body fluids with no added reagents. Spectroscopy 24:317–324
7. Eiroa A, Blanco EV, Mahia PL, Lorenzo SM, Rodriguez DP (2000) Resolution of benzo[*a*] pyrene in complex mixtures of other polycyclic aromatic hydrocarbons. Comparison of two spectrofluorimetric methods applied to water samples. Analyst 125:1321–1326
8. Nevin A, Comelli D, Valentini G, Cubeddu R (2009) Total synchronous fluorescence spectroscopy combined with multivariate analysis: method for the classification of selected resins, oils, and protein based media used in paintings. Anal Chem 81:1784–1791
9. Zamai M, Hariharan C, Pines D, Safran M, Yayon A, Caiolfa VR, Cohen-Luria R, Pines E, Parola AH (2002) Nature of interaction between basic fibroblast growth factor and the antiangiogenic drug 7,7-(carbonyl-bis[imino-N-methyl-4,2-pyrrolecarbonylimino[N-methyl-4,2-pyrrole]-carbonylimino])-bis-(1,3-naphtalene disulfonate): 2. Removal of polar interactions affects protein folding. Biophys J 82:2652–2664
10. Mark S, Ronen Y (2010) The Milne and Numorov methods as independent approaches for solving the Schroedinger equation: test case for the deuteron model for odd – odd N = Z nuclei. Phys Scr 82(4):1–8
11. Dubi A, Yaar I, Mark S (2012) Two independent approaches used for estimating 2d contamination distribution on the ground level- based on air monitoring information, mathematics in engineering. Sci Aerosp 3(2):151–158
12. Available from: http://www.pharmacy.gov.my/index.cfm?menuid=66&parentid=9. Accessed 15 Aug 2011
13. Sominsky V (2005) A portable device and method for on-site detection and quantification drugs. Patent No. WO 2005/111586 A1, 24 Nov 2005

Chapter 9
Intramolecular Mechanisms for the Occurrence of Fluorescence from Upper Excited States of Aromatic Molecules and Linear Polyenes

Takao Itoh

Abstract Fluorescence from upper excited states is surveyed as a milestone for future works. The mechanisms of the occurrence of such anomalous fluorescence are classified. The chemical species treated in the present review are aromatic molecules and linear polyenes in the papers reported up to 2013. Information on the fluorescence properties reported so far in numerous papers allows generalizations of the mechanism of the appearance of fluorescence from upper excited states. There are three intramolecular mechanisms for the occurrence of fluorescence from upper excited states for aromatic molecules and liner polyenes. That is, the fluorescence from the upper state occurs; (i) through thermal population from the lower excited state, S_1; (ii) through reverse internal conversion from the lower singlet state under collision-free conditions (in this case the upper and lower singlet states are mixing); (iii) directly from the upper singlet state without involvement of the fluorescence component via reverse internal conversion from the lower singlet state, i. e., prompt fluorescence.

Keywords S_2 fluorescence • S_3 fluorescence • Aromatic molecules • Linear polyenes • Intramolecular mechanisms

9.1 Introduction

Fluorescence from organic molecules originates normally only from the lowest excited singlet state (S_1), irrespective of the photon energy used for the excitation. The generalization of this observation is referred to as Kasha's rule "*The emitting level of a given multiplicity is the lowest excited level of that multiplicity*" [1]. This rule has been found true for a number of organic molecules that exhibit detectable

T. Itoh (✉)
Graduate School of Integrated Arts and Sciences, Hiroshima University, 1-7-1 Kagamiyama, Higashi-Hiroshima City 739-8521, Japan
e-mail: titoh@hiroshima-u.ac.jp

© Springer International Publishing Switzerland 2016

C.D. Geddes (ed.), *Reviews in Fluorescence 2015*,
DOI 10.1007/978-3-319-24609-3_9

fluorescence. Although Kasha's rule applies to most of the fluorescent organic molecules, there are exceptions to this rule. It is known that some of organic molecules show fluorescence from upper excited states such as S_2, S_3 or S_4. Such an anomalous fluorescence has been of theoretical and experimental interest. Further, it is of importance to investigate fluorescence of this kind in conjunction with the mechanisms of intramolecular electronic relaxation processes.

There have already been several partial reviews on the present subject. In 1978 Turro et al. presented a review on organic photoreactions, where they surveyed fluorescence from upper electronic states [2]. A brief review on $S_2 \rightarrow S_0$ fluorescence (S_2 fluorescence) is given also in standard textbooks such as the one by Birks first edited in 1973 [3–6]. In 1993 Maciejewski and Steer presented a review on the photochemistry and photophysics of thiocarbonyl compounds which typically exhibit S_2 fluorescence [7]. Ermolaev also presented a review in 2001 on ultrafast nonradiative transitions between upper excited states in organic molecules, where he summarized the results of the determination of nonradiative rate constants using a range of methods for various molecules [8]. Although molecules such as azulenes and thiones are known to be typical systems which exhibit fluorescence from upper excited states, there are still a number of other molecules which show various types of such anomalous emission [9]. During the past three or four decades spectroscopic data on such molecules have been accumulated extensively. Information on the fluorescence properties reported so far in numerous papers allows generalizations of the mechanism of the appearance of such emission. In this review, this specific subject is surveyed as a milestone for future works, and the mechanisms of the occurrence of the anomalous fluorescence are classified. The chemical species treated in the present review are aromatic molecules and linear polyenes reported up to 2013.

9.2 Overview of Fluorescence from Upper Excited States of Aromatic Molecules and Linear Polyenes

9.2.1 Aromatic Molecules

Fluorescence from upper excited states has been reported for a number of aromatic molecules of medium to large size such as pyrene, coronene, 1,12-benzoperylene, 3,4-benzopyrene, 1,2-benzanthracene, 3,4-benzotetraphene (benzochrysene), ovalene, 1,2-:3,4-dibenzanthracene, picene and chrysene as well as their derivatives in a variety of media (*vide infra*). In addition, extremely weak fluorescence from upper excited states was reported in the 1970s for smaller molecules such as benzene, mesitylene (1,3,5-trimethylbenzene), naphthalene and *p*-xylene, toluene in solution and in the vapor phase [10–12]. Although these smaller molecules are commercially available and some are commonly used as solvents, detailed information on fluorescence from upper excited states has not been provided recently. However, it was reported later that the anthracene crystal exhibits weak S_2

fluorescence when excited through a two-step process via the lowest excited state, direct excitation or via annihilation of singlet excitons [13–15]. Further, naphthalene vapor excited by means of glow discharge is reported to show $T_k \rightarrow T_1$ fluorescence around 400 nm [16].

The excited states of most medium or large aromatic molecules are characterized by comparatively small $\Delta E(S_1 - S_2)$ values (normally less than about 4000 cm^{-1}) and much larger oscillator strength of S_2 than that of S_1. In solution or in matrices, S_2 florescence is reported to occur through the thermal population from the S_1 state for 3,4-benzopyrene, 1,12-benzoperylene, 1,2-benzanthracene and 2′-methyl-1,2-benzanthracene, 3-methylpyrene, 3,4-benzotetraphene and ovalene [17–24].

The emission properties of medium or large molecules in the vapor phase are of particular interest, because the intrinsic photophysical property can be observed in the absence of environmental effects, and because the appearance of the S_2 fluorescence is much more significant compared to smaller aromatic molecules such as naphthalene. Hoytink et al. observed S_2 fluorescence from pyrene (0.05 Torr at 170 °C), and 3,4-benzpyrene (0.1 Torr at 260 °C) in the static vapor phase [25]. At almost the same time, the emission properties of some of the aromatic molecules such as pyrene and 3,4-benzpyrene were investigated by a number of researchers in the static vapor phase at low pressure [25–27]. In particular, the fluorescence properties of pyrene and its derivatives were investigated in detail in the static vapor phase by Baba et al. [28–33]. Although pyrene vapor is known to show weak S_2 fluorescence in addition to strong S_1 fluorescence at high total pressure, this S_2 emission was assigned to delayed fluorescence occurring through the thermal activation of the S_1 state [25]. However, the observation of the S_2 fluorescence under collision-free conditions demonstrated that the occurrence of the S_2 fluorescence is not due to thermal population alone [25, 28]. Fluorescence and excitation spectra of pyrene and benzanthracene vapor at low pressure are displayed in Figs. 9.1 and 9.2, respectively. These molecules exhibit similar fluorescence properties in the vapor phase. Under low pressure condition, the relative intensity of the S_2 fluorescence of pyrene vapor increases to a certain extent compared to that in solution or under high pressure condition (~100 Torr) in the presence of added buffer gas [30]. When the excitation energy is increased, the S_2 fluorescence shifts to the red in almost the same way as does the S_1-fluorescence [30]. These observations indicate that both the S_1 and S_2 fluorescence originate from the unrelaxed upper vibronic levels of the excited electronic states. Further, the S_2/S_1 fluorescence quantum yield ratio, $\Phi_F(S_2)/\Phi_F(S_1)$, increases almost exponentially with increasing excitation energy [30]. Similar fluorescence properties were described for pyrene-d_{10}, 1-methylpyrene and 4-methylpyrene vapors [31]. Analyses of the fluorescence spectral data of pyrene vapor revealed that the $\Phi_F(S_2)/\Phi_F(S_1)$ value can be related to the ratio $\rho(S_1)/\rho(S_2)$, with $\rho(S_1)$ and $\rho(S_2)$ denoting the vibrational-state densities of S_1 and S_2 states at the energy of excitation, respectively [30]. The pressure dependence of the $\Phi_F(S_2)$ and $\Phi_F(S_1)$ values also were investigated by changing the pressure of added buffer gases [33]. It was shown that $\Phi_F(S_2)$ decreases, but $\Phi_F(S_1)$ increases with increasing the buffer gas pressure [33]. These observations have been interpreted kinetically in terms of reversible internal conversion between the

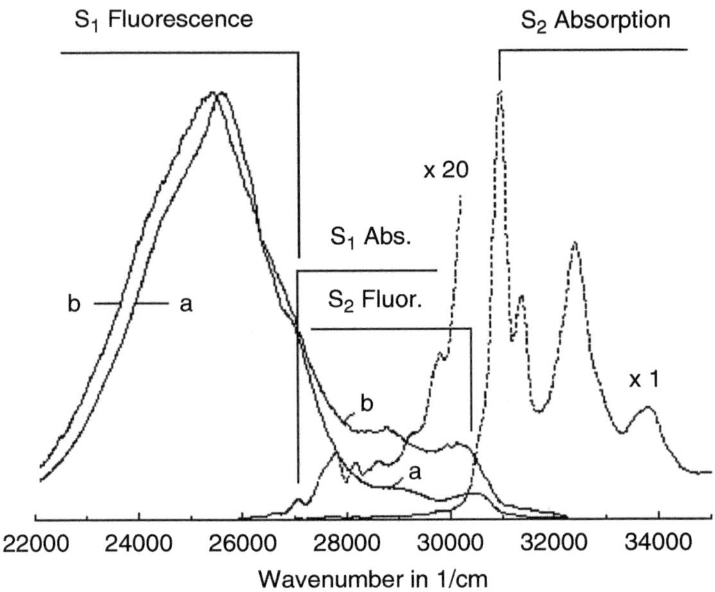

Fig. 9.1 Corrected fluorescence and excitation spectra of pure pyrene vapor measured at 140 °C. (**a**) Fluorescence spectrum obtained by excitation into the S_3 origin, (**b**) that obtained by excitation into the S_4 origin

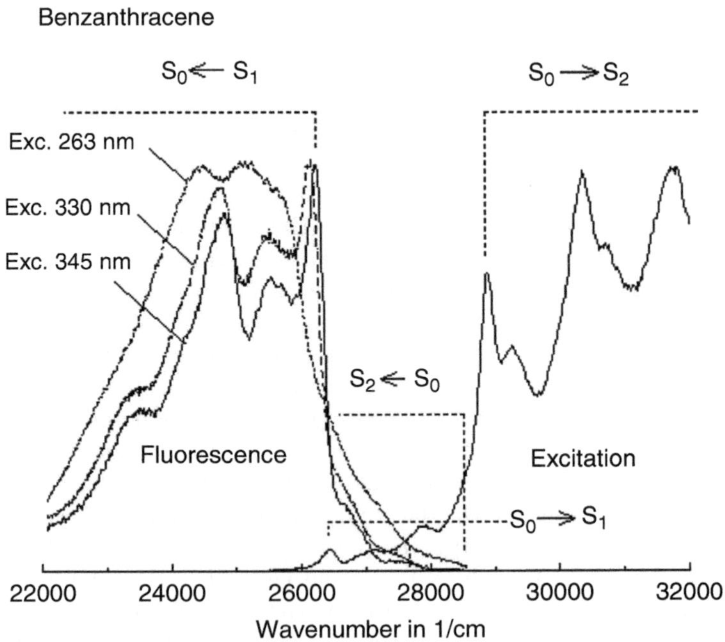

Fig. 9.2 Corrected fluorescence spectra of pure benzanthracene vapor obtained by excitation at different wavelengths and the corrected excitation spectrum

S_1 and S_2 states. Through the investigation of the effect of added buffer gas, including the quenching effects of S_2 fluorescence by molecular oxygen, the $\Phi_F(S_2)$ value was shown to consist of fast and slow components [32, 33]. The fast component was considered to correspond to the S_2 fluorescence directly emitted from S_2, and the slow component to the S_2 fluorescence which is emitted from S_2 formed from S_1 through reverse $S_1 \rightarrow S_2$ internal conversion. The decrease of $\Phi_F(S_2)$ upon increasing the buffer gas pressure was interpreted in terms of the decrease of the slow component of S_2 fluorescence. In the case of reversible internal conversion between S_2 and S_1, the rate constants of the forward $S_2 \rightarrow S_1$ (k_{21}) and reverse $S_1 \rightarrow S_2$ internal conversions (k_{12}) can be formulated as $k_{21} = (2\pi/\hbar)V^2\rho(S_1)$ and $k_{12} = (2\pi/\hbar)V^2\rho(S_2)$, respectively, utilizing Fermi's golden rule, where V is the coupling constant between S_1 and S_2 and $\rho(S_i)$ is the density of the vibronic states in S_i at the energy of excitation. It was demonstrated that the excitation-energy dependence of the ratio, k_{21}/k_{12}, agrees qualitatively with that calculated based on Fermi's golden rule [30]. Further, it was shown for pyrene-h_{10} and -d_{10} vapors through the investigation of the effect of added buffer gas on the fluorescence that the k_{12} value increases almost linearly with increasing excitation energy from 33,600 to 42,700 cm^{-1}, but k_{21} is almost constant regardless of excitation energy [32].

The fluorescence and excitation spectra of pyrene vapor were also measured in a jet [34–37]. The excitation spectrum in a jet for the $S_0 \rightarrow S_2$ region shows a complicated structure due to the interaction between the S_2 and S_1 states [35]. The dispersed fluorescence spectrum in a jet obtained by excitation into the S_4 state is similar to that in the static vapor phase, but the S_2/S_1 fluorescence intensity ratio is intensified by a factor of 1.7, and is weakened for pyrene-d_{10} compared to pyrene-h_{10} [36]. These observations indicate that the contribution of the thermally-activated S_2 fluorescence is small under collision-free conditions and that the S_2/S_1 fluorescence intensity ratio is related to the vibrational-level density.

Nakajima measured the fluorescence and excitation spectra of chrysene, 1,2-benzanthracene, 20-methylcholanthrene, coronene, 1,12-benzperylene in the static vapor phase at low pressure [29]. He compared the fluorescence quantum yield ratios, $\Phi_F(S_2)/\Phi_F(S_1)$, with those calculated based on Fermi's golden rule [29, 38]. The values for $\Phi_F(S_2)/\Phi_F(S_1)$ with the excitation into the S_3 state are 0.037 (0.038) and 0.037(1.0), respectively, for 1,2-benzanthracene and 20-methylcholanthrene with the calculated values shown in parentheses [29]. The reported difference between experimental and theoretical quantum yield ratio, 0.037 and 1.0, is large for 20-methylcholantrene, which may arise due to underestimation of the ρ values. With coronene vapor, the value for $[\Phi_F(S_3) + \Phi_F(S_2)]/\Phi_F(S_1)$ obtained by the excitation into the S_4 state is 0.11(0.079) [29]. However, recent investigation based on careful purification of the coronene sample revealed that the S_2 and S_3 fluorescence reported by Nakajima likely originate from impurities and that the real S_2 and S_3 fluorescence are much weaker than those reported by Nakajima [39]. Further, the S_2 and S_3 fluorescence of coronene vapor were analyzed as prompt fluorescence which does not involve the fluorescence occurring through reverse internal conversion [39]. In general, the purification of coronene is tremendously difficult [40]. In any case, the occurrence of the S_2 fluorescence for most of the medium to large-size aromatic molecules such as pyrene, benzpyrene, benzanthracene and benzperylene in the vapor phase at low pressure can be interpreted by the relaxation model involving the reversible internal conversion between S_1 and S_2. The excitation-energy dependence of the relative S_2/S_1 fluorescence quantum yield ratio can be interpreted at least qualitatively using Fermi's golden rule. Recently, picene is reported to show weak S_2 fluorescence in the vapor phase [41].

9.2.2 Linear Polyenes

Linear polyenes have the chemical structure, R_1-(CH=CH)$_N$-R_2, with $N \geq 2$ and R_1 or R_2 being hydrogen, alkyl, aryl or other groups. These molecules are prototypes of π-electron conjugated systems for which a number of spectroscopic and quantum mechanical studies have been carried out since the 1930s [42, 43]. Most *all*-s-trans conformer of symmetrically substituted polyenes (R_1=R_2.) such as α, ω-diphenylpolyenes belong to the C_{2h} point group. *All*-s-trans conformer of linear polyenes exhibit an intense absorption band that corresponds to an allowed electronic transition, 1^1A_g (S_0) → 1^1B_u*(π, π*), in the UV-vis region. The excitation energy of the ^1B_u*(π, π*) state is known to decrease systematically with increasing polyene chain-length. The chain-length dependence of the excitation energy and oscillator strength of the strong 1^1A_g (S_0) → 1^1B_u*(π, π*) absorption bands have been interpreted at least qualitatively with simple molecular orbital theory such as Hückel or PPP methods [43, 44]. The 1^1B_u*(π, π*) state had been considered to be the lowest excited singlet state (S_1) for a long time, but in 1972 the presence of a forbidden excited singlet state, 2^1A_g, located below the allowed 1^1B_u state was pointed out by Hudson and Kohler [45, 46]. That is, the 1^1B_u state is not the first singlet excited state, S_1, but the second, S_2. Schulten and Karplus provided a theoretical rationalization for the forbidden 2^1A_g, state which is doubly excited in nature, and described it with extensive configuration interaction [47]. Since those findings, a number of spectroscopic studies of polyenes have been carried out concerning the forbidden low-lying $2^1A_g(S_1)$ state [48, 49].

At present, there is a general consensus that for unsubstituted *all*-s-trans linear polyenes (H-(CH=CH)$_N$-H) with the number of the polyene double bonds (N) over 3, the S_1 state is not 1^1B_u but 2^1A_g [49, 50]. Excited-state energy level scheme of linear polyenes is presented in Fig. 9.3. Although the $S_1(2^1A_g)$ state is one-photon forbidden, it is two-photon allowed through the excitation of an imaginary B_u state. The forbidden 2^1A_g state is normally not observable in room temperature absorption spectra and is considered to obtain its one-photon transition intensity mainly through vibronic coupling with the strongly allowed $S_2(1^1B_u)$ state. The reasons for the absence of the $S_1(2^1A_g)$ state in absorption spectra at room temperature are that its origin band is forbidden and that it is masked by the onset of the strong $S_2(1^1B_u)$ absorption band [46]. In this sense, the feature of the forbidden $S_1(2^1A_g)$ state of

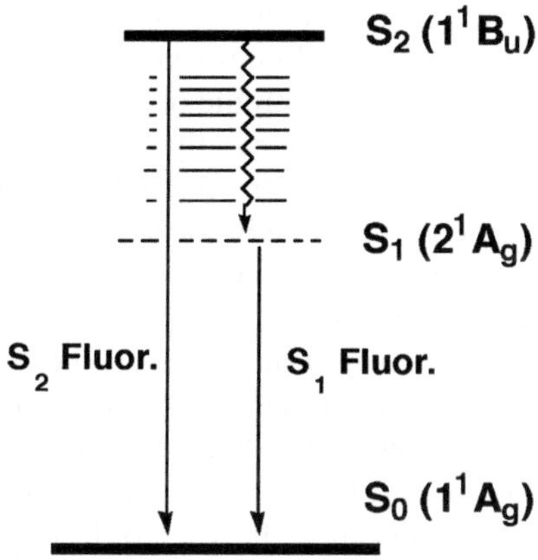

Fig. 9.3 Energy level scheme and emitting states of typical linear polyenes

polyenes is different from that of the $S_1(n, \pi^*)$ state of many (=C=O)- or (–N=)-containing molecules for which the band origin of the forbidden $S_0 \rightarrow S_1(n, \pi^*)$ absorption appears distinctly. It is known that the energy level of the 1^1Bu state of polyenes decreases significantly with increasing solvent polarizability, (n2 – 1)/(n2 + 2) with n being the refractive index of the solvent, while that of the 21Ag state is almost unchanged [48]. Therefore, the ΔE(S2(11Bu) – S1(21Ag)) value of polyenes can be changed systematically by changing solvent polarizability. The trend of recent research on the electronic states of polyenes has seemingly shifted from the determination of the location and nature of their 2^1A_g states to their photophysics and dynamical behavior. The spectroscopy and excited electronic states of octatetraene and other polyenes were reviewed by Hudson et al. in 1982 [48], and by Kohler in 1993 [49].

Prototypical polyenes would be R-(CH=CH)$_N$-R with R=H or alkyl group. For *all*-s-trans conformers of unsubstituted polyenes (H-(CH = CH)$_N$-H) with $N = 3 \sim 8$, the S_1 and S_2 states have been assigned as 2^1Ag and 1^1B_u, respectively [50–55]. In rigid matrices at low temperature and in room temperature solution, unsubstituted *all*-s-trans polyene with $N = 4$ (octatetraene) shows only $S_1(2^1A_g)$ fluorescence, but in the vapor phase, including in a jet, it shows S_2 fluorescence in addition to weak S_1 fluorescence [51]. In a rigid matrix at low temperature, unsubstituted polyenes with $N = 4 \sim 6$ show only S_1 fluorescence, but that with $N = 7$ shows both S_1 and S_2 fluorescence [54]. In room temperature EPA solution, dimethyl-substituted polyenes (H_3C-(CH=CH)$_N$-CH_3) with $N = 4$ and 5 show only S_1 fluorescence, but the ones with $N = 6$ and 7 show both S_1 and S_2 fluorescence [52]. Since the research history of the emission properties of octatetraene (H-(CH=CH)$_4$-H) is somewhat complicated, it is described briefly below. Gavin et al. measured the absorption and emission spectra of octatetraene along with the fluorescence quantum yield and lifetime in solution and in the static vapor phase [56]. They observed fluorescence from the $S_2(1^1B_u)$ and $S_1(2^1A_g)$ states in solution, but observed only the S_2 fluorescence in the vapor phase. It seems that they did not recognize the 1^1B_u fluorescence to be S_2 emission [56]. In 1979, Granville et al. reported highly resolved one- and two-photon $S_0(1^1A_g) \rightarrow S_1$ excitation and $2^1A_g(S_1) \rightarrow S_0$ fluorescence spectra of octatetraene in octane at 4.2 K, where the vibrationless electronic origin was observed in the two-photon excitation spectrum between the vibronically-induced one-photon excitation and fluorescence origins [57]. This measurement confirmed the assignment of the $S_1(2^1A_g)$ fluorescence of octatetraene. Heimbrook et al. measured the emission and excitation spectra of octatetraene in a supersonic jet and assigned the emission as the fluorescence from the $S_2(1^1B_u)$ state [58].

Bouwman et al. observed the structured $S_2(1^1B_u)$ fluorescence and the broad $S_1(2^1A_g)$ fluorescence in the emission spectra of *all*-s-trans octatetraene and decatetraene (1,8-dimethyl-substituted octatetraene) in the static vapor phase [59]. Later, Petek et al. reported similar emission for decatetraene vapor in supersonic molecular beams [60]. In order to reveal the relaxation processes of decatetraene, the pressure and excitation-energy dependence of the S_1 and S_2 fluorescence were investigated in the vapor phase [61]. The pressure dependence of the fluorescence yields of decatetraene vapor was interpreted in terms of a relaxation model involving reversible internal conversion between S_1 and S_2 and the vibrational relaxation in the S_1 manifold. The reverse $S_1 \rightarrow S_2$ internal conversion rate was shown to be significantly slower compared to the forward $S_2 \rightarrow S_1$ internal conversion rate [61]. That is, the S_2 fluorescence can be regarded practically as prompt fluorescence without being accompanied by the reverse $S_1 \rightarrow S_2$ internal conversion for decatetraene. Fluorescence spectra of *all*-s-trans decatetraene vapor following the excitation at different wavelengths are displayed in Fig. 9.4.

All-s-*trans*-diphenylpolyenes are typical C_{2h} model polyenes for which a number of spectroscopic data have been accumulated as described below. The emission properties of diphenylpolyenes with different polyene chain-lengths are informative in considering the mechanism for occurrence of fluorescence from upper excited states. Figure 9.5 shows fluorescence and absorption spectra of diphenylpolyenes with N ranging from 3 to 7 in CCl_4 at room temperature. These diphenylpolyenes exhibit dual fluorescence from the $S_1(2^1A_g)$ and $S_2(1^1B_u)$ states in room temperature solution [62], but the mechanism for the occurrence of the S_2 fluorescence differs depending on N [63]. With shorter diphenylpolyenes ($N = 3$ and 4), the S_2 fluorescence occurs as the result of thermal population from the S_1 state due to small $S_1 - S_2$ energy separations, $\Delta E(S_1 - S_2)$, (~1000–2000 cm^{-1}) [64–69]. Thus, the relative intensity of S_2 fluorescence increases with increasing temperature for these shorter diphenylpolyenes. Although it was shown later that the fluorescence spectrum of diphenylpolyene with $N = 3$ includes a fluorescence contribution from the s-cis,s-trans conformer, which increases at higher excitation energies and temperatures [70, 71], it did not influence the analyses of the S_2/S_1 fluorescence-intensity ratio of this molecule significantly [65]. On the other hand, the S_2 fluorescence of the longer diphenylpolyenes with $N = 6$ and 7 occurs as the prompt emission due to the comparatively large $\Delta E(S_1 - S_2)$ values (~3500–5000 cm^{-1}). Thus, the relative S_1/S_2 fluorescence intensity ratio is almost independent of temperature for diphenylpolyenes with $N = 6$ and 7 [63]. For the

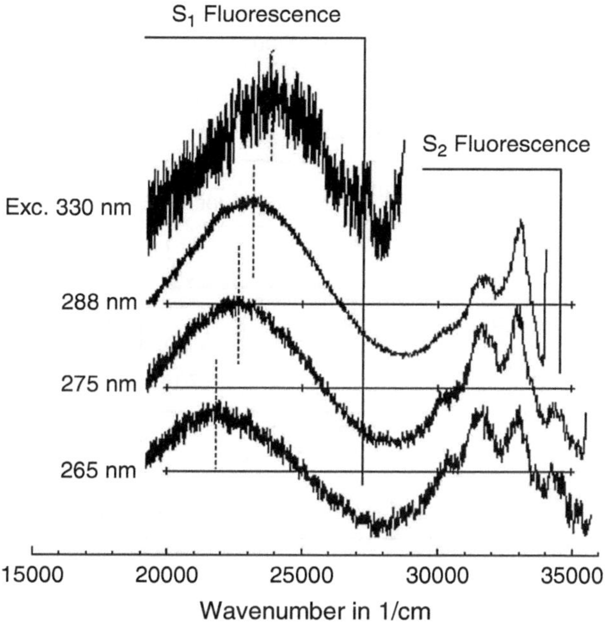

Fig. 9.4 Fluorescence spectra of pure 2,4,6,8-decatetraene [$CH_3(CH{=}CH)_4CH_3$] vapor at 35 °C obtained by excitation at different wavelengths [61]

diphenylpolyene with $N = 5$, both of the thermally activated and prompt S_2 fluorescence emissions were observed depending on the conditions such as solvent polarizability and temperature [63]. Therefore, the S_2/S_1 fluorescence quantum yield ratio, $\Phi_F(S_2)/\Phi_F(S_1)$, decreases with increasing N for diphenylpolyenes with $N = 3 \sim 5$, but increases for diphenylpolyenes with $N = 5 \sim 7$, when the temperature is kept constant (Fig. 9.6) [63, 72]. The $\Delta E(S_1 - S_2)$ dependence of the relative S_2 fluorescence intensity was investigated in detail for the diphenylpolyene with $N = 7$ [73]. As previously mentioned, the $\Delta E(S_1 - S_2)$ value of linear polyenes depends significantly on the solvent polarizability [74, 75]. The S_2/S_1 fluorescence intensity ratio of the diphenylpolyene with $N = 7$ was shown to increase with increasing ΔE $(S_1 - S_2)$ values which are varied by changing the solvent polarizability. This observation was interpreted in terms of "intensity borrowing mechanism" by S_1 from S_2 [73]. Hirata et al. have measured the $S_2 \rightarrow S_1$ internal conversion rate constants for diphenylpolyenes with $N = 3 \sim 8$ in solution [76]. They have shown that the internal conversion rate does not depend significantly on N, indicating that the energy gap law for the $S_2 \rightarrow S_1$ internal conversion does not apply for diphenylpolyenes [76]. The diphenylpolyene with $N = 2$ (diphenylbutadiene) shows fluorescence from $S_2(1^1B_u)$ in the static vapor phase at high total pressure in the presence of buffer gas, but shows only the $S_1(2^1A_g)$ fluorescence in a jet [77]. Since the $\Delta E(S_1 - S_2)$ value of the diphenylpolyene with $N = 2$ is only $1100\ cm^{-1}$ in the vapor phase, the weak S_1 fluorescence is considered to be masked by the strong S_2 fluorescence in the static vapor phase at high total pressure. Further, it is suggested that the S_1 state of the diphenylpolyene with $N = 2$ is probably assigned to 2^1A_g in low polarizable solvents such as perfluoropentane at

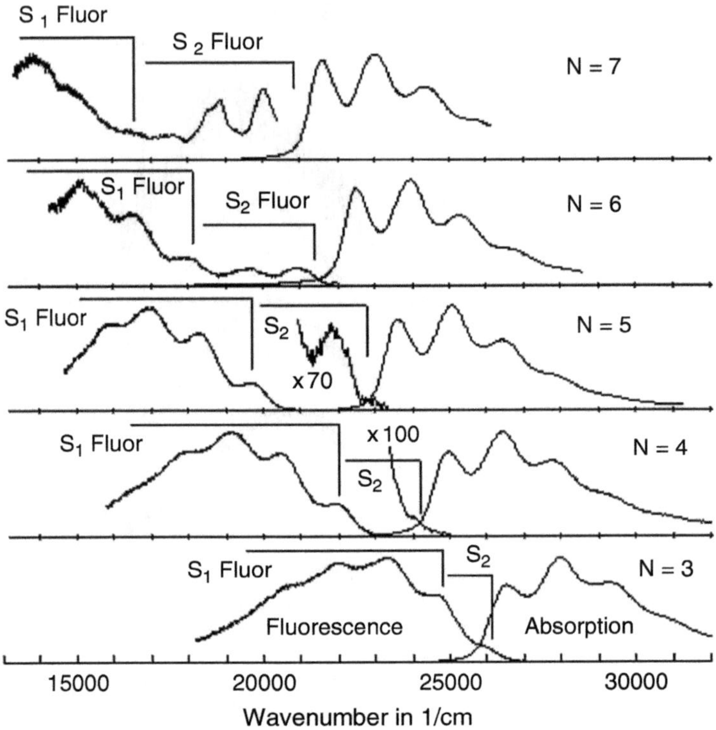

Fig. 9.5 Fluorescence and absorption spectra of α, ω-diphenylpolyenes with different polyene double bond numbers N in CCl_4 at room temperature [62a]. All spectra are normalized to a common magnitude

Fig. 9.6 S_2/S_1 fluorescence quantum yield ratios of α,ω-diphenylpolyenes with different polyene double bond number in CCl_4 and CS_2 at room temperature [62a]

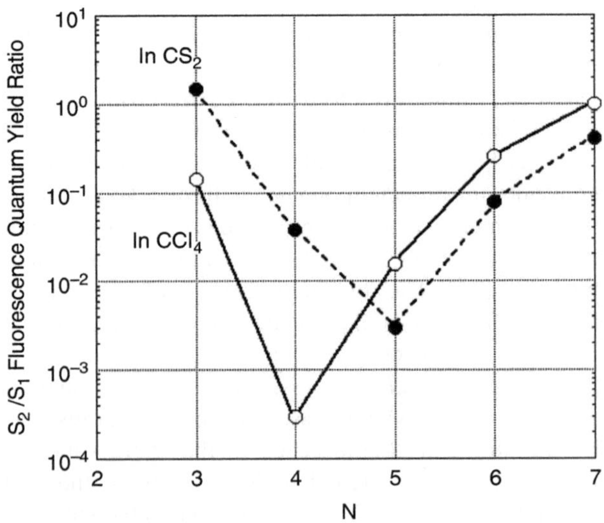

room temperature and that the fluorescence occurs mainly from $S_2(1^1B_u)$ in such solvents [78, 79]. The 1^1B_u fluorescence obtained by the two-photon excitation into the 2^1A_g state also was observed for the diphenylpoyene with $N=2$ in a low polarizability solvent [80]. Diphenylpoyene with $N=3$ shows only $S_1(2^1A_g)$ fluorescence both in the static vapor phase and in a jet, while diphenylpoyenes with $N=4$ and 5 show weak $S_2(1^1B_u)$ fluorescence and stronger $S_1(2^1A_g)$ fluorescence in the static vapor phase [63].

It is possible to invert the 2^1A_g and 1^1B_u levels for some diphenypolyenes and to observe the full 1^1B_u fluorescence in sufficiently polarizable solvents (in this case, the 1^1B_u state is S_1). The full 1^1B_u fluorescence was observed for the diphenylpoyene with $N=3$ in a highly polarizable solvent at 77 K, although the emission consists of dual fluorescence from the $S_1(2^1A_g)$ and $S_2(1^1B_u)$ states in commonly used solvents such as hexane or benzene at temperatures near room temperature, [81]. Fluorescence properties similar to those of diphenylpolyenes were reported recently for unsymmetrically substituted polyenes, α-methyl-ω-phenylpolyenes with $N=3$ and 4 (Fig. 9.7) [82].

As the case of a rigid analogue of the diphenylpoyene with $N=3$, *all*-s-trans 1,4-diindanylidenyl-2-butene exhibits weak thermally activated S_2 fluorescence only in low polarizability solvents along with strong S_1 fluorescence [83, 84]. The $\Delta E(S_1 - S_2)$ value of this molecule is smaller than that of diphenylhexatriene, and the $S_1(2^1A_g)$ and $S_2(1^1B_u)$ energy levels are inverted when solvent polarizability is increased.[61] Weak thermally activated S_2 (1^1B_u) fluorescence has also been observed for derivatives of diphenylhexatriene and 1,3,5-heptatrienylbenzene along with S_1 fluorescence [64, 82, 85, 86]. In addition, 1-phenyl-4-(1′-pyrenyl)-1,3-butadiene and isomers of 1-(1′-naphthyl)-6-phenylhexatriene are reported to show thermally-activated S_2 fluorescence in addition to stronger S_1 fluorescence in room temperature solutions [87, 88].

All-s-trans dithienylpolyenes with $N=2$ and 3 are also reported to show thermally-activated S_2 fluorescence in room temperature solution and in the static vapor phase, although these polyenes possess rotational isomers depending on the orientation of the two thienyl groups on the two edges of the polyene chain [89, 90]. These rotational isomers were identified and separated by the emission, excitation and hole-burning spectral measurements in a jet [91, 92].

Carotenoids are organic pigments that occur naturally in plants as well as in some other light-harvesting organisms in photosynthesis. There are over 623 known carotenoids. These can be split into two classes, xanthophylls which contain oxygen and carotenes which are purely hydrocarbons containing no oxygen [93, 94]. Some of these carotenoids can be extracted from natural sources such as carrot or spinach. A number of carotenoids possess polyene structures with long polyene chain-lengths. Emission properties of carotenoids have been intensively investigated, mainly in conjunction with photosynthesis. The strong absorption band in the visible region of carotenoids is caused by the allowed $S_0 \rightarrow S_2(1^1B_u^+)$ transition, the energy of which decreases as N increases.

β-Carotene is a typical carotenoid on which numerous spectroscopic studies have been carried out. Emission from β-carotene consists mainly of fluorescence

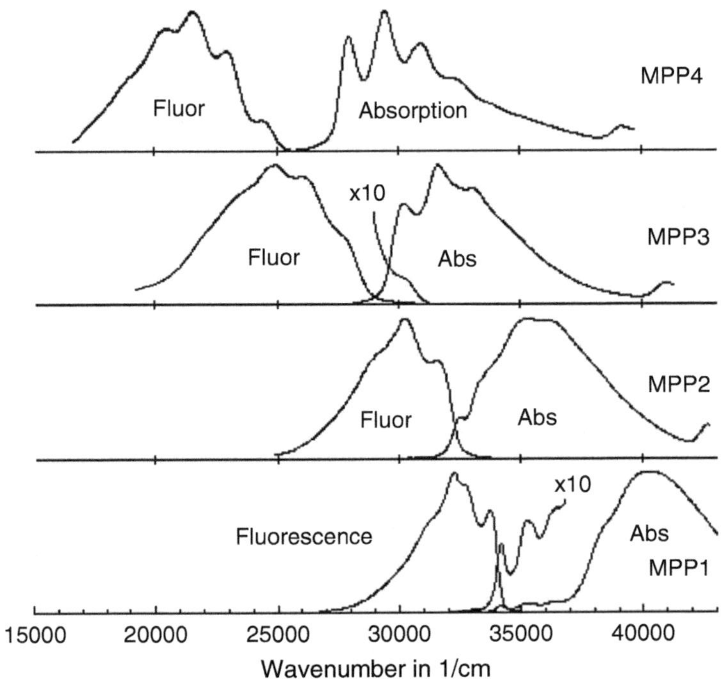

Fig. 9.7 Fluorescence and absorption spectra of MPPn with n = 1 ~ 4 in hexane at room temperature: MPP1; β-methylstyrene, MPP2; α-methyl-ω-phenylbutadiene, MPP3; α-methyl-ω-phenylhexatriene, MPP4; α-methyl-ω-phenyloctatetraene. MPP3 exhibits weak delayed S_2 fluorescence [82]

from the $S_2(1^1B_u^+)$ state, which was once thought to be the S_1 state. With carotenoids possessing a long polyene chain such as β-carotene, it seems to be the S_1 fluorescence that is anomalous, rather than the S_2 fluorescence. The S_1 fluorescence of β-carotene was reported first by Bondarev and Knyukshto to appear only weakly at the far red side of the S_2 fluorescence in highly polarizable solvents [95, 96]. Soon after, Anderson et al. reported the fluorescence spectrum from the S_1 (2^1A_g) state of β-carotene appearing at about 13,000 cm^{-1} [97]. Careful analyses of the emission spectrum indicated that the S_1 state of β-carotene is located at 14,100 cm^{-1} [98]. Ultrafast dynamics of the excited states of carotenoids were reviewed by Polivka and Sundstroem in 2004 [99]. Extremely rapid decay times of the S_2 state ranging from 50 to 300 fs were reported for β-carotene as well as for other carotenoids [99–105]. Recently, α-carotene was also shown to emit dual fluorescence from the S_1 and S_2 states in highly polarizable solvents [106]. Christensen et al. measured the absorption and fluorescence spectra of a series of apo-carotenes with polyene double bonds N ranging from 5 to 11 in EPA at 77 K [107]. Apo- and diapo-carotenes with N ranging from 7 to 10 exhibit both S_1 and S_2 fluorescence, for which the relative S_2 fluorescence intensity tends to increase with increasing N [107]. In general, the S_2/S_1 fluorescence intensity ratio tends to increase with increasing N or with increasing $\Delta E(S_2 - S_1)$ energy gap, which is achieved by changing the solvent polarizability. This tendency can be seen for a number of different carotenoids [108–112]. Among a number of fluorescent linear polyenes, the molecules possessing the longest polyene chain-length may be decapreno-β-carotene ($N = 15$) and dodecapreno-β-carotene ($N = 19$) which show only weak S_2 fluorescence with the quantum yields of 10^{-4}–10^{-5} [97]. However, S_2 fluorescence spectral data are available for the $N = 13$ polyene (1′,2′-dihydro-3′,4′,7′,8′-tetradehydrospheroidene) as the molecule having the longest polyene chain-length [113].

It was predicted theoretically that there is a forbidden singlet excited state, $1^1B_u^-$, in addition to the forbidden $2^1A_g^-(S_1)$ state, located below the allowed $1^1B_u^+$ state for polyenes with long polyene chain length ($N > 9$) [114, 115]. Later, the presence of the $1^1B_u^-$ was indicated through the measurements of the resonance Raman excitation profile and fluorescence spectra of carotenoids [116–118]. The observed fluorescence was interpreted as a superposition of the three kinds of fluorescence from $S_1(2^1A_g^-)$, $S_2(1^1B_u^-)$ and $S_3(1^1B_u^+)$, although the measured fluorescence spectra are extremely weak and noisy [118]. More recently, Kosumi et al. measured femto-second time-resolved absorption spectra of β-carotene homologs with $N = 7 \sim 15$ [119, 120]. These spectra showed that the energy gap law for the $S_2 \rightarrow S_1$ internal conversion applies for carotenoids with N less than 9, but the reverse energy gap law applies for those with N greater than 11. The observed reverse energy gap law was interpreted in terms of the presence of an excited state located between the $2^1A_g^-$, and $1^1B_u^+$ states for carotenoids with N greater than 11 [120]. Frank et al. measured the absorption, fluorescence, excitation and time-resolved absorption spectra of a series of spheroidenes with N ranging from 7 to 13 and showed that the S_2 fluorescence quantum yield shows a maximum for spheroidenes with $N = 9$, which then decreases with increasing N [113]. In any case, carotenoids with large N value can be regarded as unique molecules in the sense that these may exhibit triple fluorescence from S_1, S_2 and S_3. The electronic states of carotenoids were reviewed by Christensen in 1999 [121].

Although there is general agreement that several longer *all*-s-trans polyenes show both S_1 and S_2 fluorescence, the situation of polyene spectroscopy is not as simple as has been explained heretofore. Recently, it was suggested by Christensen et al. that the reported fluorescence from the forbidden S_1 state of carotenoids and longer *all*-s-trans linear polyenes likely originates in part from the s-cis-conformer which is possibly produced photochemically and that the S_1 fluorescence of *all*-s-trans polyenes is somewhat weak compared to that of s-cis conformer [122]. The results presented by Christensen et al. may require a reinterpretation of the fluorescence obtained previously for *all*-s-trans conformers of long polyenes and carotenoids [122]. Further, the complexity of the spectroscopy of carotenoids also arises from the observation that the S_1 fluorescence spectra of *all*-s-trans conformers resemble those of the s-cis conformers [122]. However, even in such a situation, the observations that many of the longer polyenes, irrespective of which regioisomer is present, show dual fluorescence from S_1 and S_2 seem to be still valid at present.

There are two distinct mechanisms for the occurrence of the S_2 fluorescence for *all*-s-trans linear polyenes. One is the thermally activated delayed S_2 fluorescence which occurs as a result of thermal activation from the S_1 state and is observable when the $\Delta E(S_1 - S_2)$ value is comparatively small ($1000 \sim 2000\ cm^{-1}$). The other is the prompt S_2 fluorescence which occurs as a result of the fast radiative process from S_2 competing with the fast $S_2 \rightarrow S_1$ internal conversion and is observable when the $\Delta E(S_1 - S_2)$ value is large (over about $3000\ cm^{-1}$). In the case of the prompt S_2 emission, there is a tendency for increasing S_2/S_1 fluorescence intensity ratio with increasing N or $\Delta E(S_1 - S_2)$. This observation was interpreted in terms of intensity borrowing by S_1 from S_2 and/or the energy gap law for the $S_2 \rightarrow S_1$ internal conversion rate [73, 109, 113, 123, 124]. In the case of the intensity-borrowing mechanism, the S_2/S_1 fluorescence quantum yield ratio, $\Phi_F(S_2)/\Phi_F(S_1)$, increases with increasing $\Delta E(S_1\text{-}S_2)$ for a fixed molecular species, since the S_1 fluorescence intensity is expressed approximately by the form, $f(S_2) \times V/\Delta E(S_1\text{-}S_2)$, where $f(S_2)$ is the oscillator strength for the $S_0 \rightarrow S_2$ transition and V is the coupling constant between S_1 and S_2 [73, 123, 124]. Further, the energy gap law can also explain the variation of $\Phi_F(S_2)/\Phi_F(S_1)$ with increasing $\Delta E(S_1\text{-}S_2)$, since the relative $S_2 \rightarrow S_1$ internal conversion rate tends to decrease with increasing $\Delta E(S_1\text{-}S_2)$ [113, 120]. At present, a number of polyenes are known to show prompt S_2 fluorescence emission that was once thought to be the conventional S_1 fluorescence. As mentioned above, the S_2/S_1 fluorescence intensity ratio tends to increase with increasing polyene chain length. This tendency can be seen clearly when N is changed successively for a series of polyenes, R-$(CH{=}CH)_n$–R [52, 97]. This observation has been interpreted in terms of the energy gap law for the $S_2 \rightarrow S_1$ internal conversion in some cases [109, 113], but it cannot be explained simply by the energy gap law [63, 125].

9.3 Mechanisms for the Appearance of Fluorescence from Upper Excited States of Aromatic Molecules and Linear Polyenes

Regarding to fluorescence from upper excited singlet states of aromatic molecules and polyenes, there seem to be at least three intramolecular mechanisms for the occurrence of the emission (Fig. 9.8). The fluorescence from the upper state occurs; (i) through thermal population from the lower excited state, S_1; (ii) through reverse internal conversion from the lower singlet state under collision-free conditions (in this case the upper and lower singlet states are mixing); (iii) directly from the upper singlet state without involvement of the fluorescence component via reverse internal conversion from the lower singlet state, i. e., prompt fluorescence. Mechanisms (i) and (ii) resemble each other, but in case (ii) the fluorescence originates from unrelaxed vibronic levels of the excited state, e. g. S_2*, and it disappears in condensed phases or under high-pressure conditions where the collisional

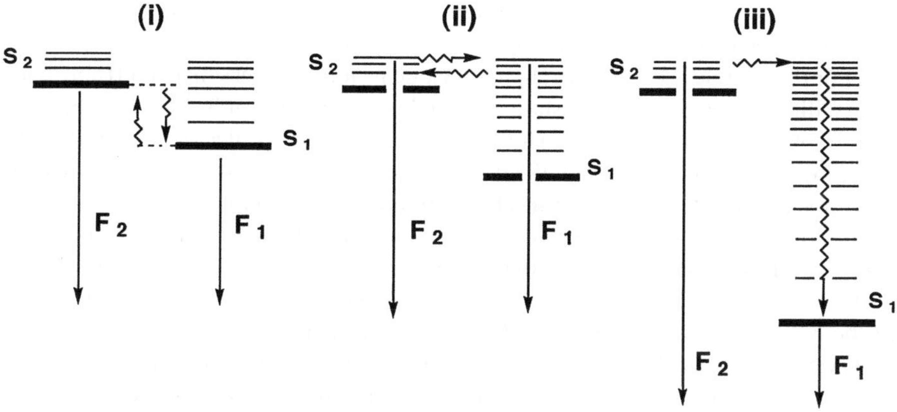

Fig. 9.8 Schemes showing the three intramolecular mechanisms for occurrence of S_2 fluorescence. F_1 and F_2 represent S_1 and S_2 fluorescence, respectively

deactivation occurs effectively. Actually, when the rate of the forward internal conversion $S_2^* \rightarrow S_1^*$ is much faster than that of the reverse internal conversion $S_2^* \leftarrow S_1^*$, case (ii) corresponds to case (iii). In case (i) normally the relative quantum yield of the S_2 fluorescence tends to increase with decreasing the $\Delta E(S_1 - S_2)$ value, while in case (iii) the relative quantum yield of the S_2 fluorescence tends to decrease with decreasing $\Delta E(S_1 - S_2)$ due to the energy gap law for the $S_2 \rightarrow S_1$ internal conversion. Further, in case (ii) the S_2 fluorescence quantum yield in the vapor phase tends to increase with decreasing pressure, while in case (iii) it is almost invariant against pressure.

The S_2 fluorescence of diphenylhexatriene, diphenyloctatetraene and 1,3,5-heptatrienylbenzene as well as pyrene in solution corresponds to case (i). The $\Delta E(S_1 - S_2)$ values of these molecules are normally less than 2000 cm^{-1} and the oscillator strength of the S_2 state is much larger than that of the S_1 state. In case (i), whether or not the S_2 fluorescence is observable can be predicted using the approximate relationship, $\Phi_F(S_2)/\Phi_F(S_1) = k_{F2}/k_{F1} \times \exp[-\Delta E(S_1 - S_2)/k_B T]$, where k_{F2} and k_{F1} are the intrinsic radiative rate constants of the upper (S_2) and lower states (S_1), respectively, k_B is the Boltzmann constant and T is the absolute temperature. If the quantum yield ratio, $\Phi_F(S_2)/\Phi_F(S_1)$, is lower than 10^{-2}, it is normally difficult to observe the fluorescence from the upper state. When $k_{F2}/k_{F1} = 200$, $\Delta E(S_1\text{-}S_2) = 1500$ cm^{-1} and $T = 300$ K, for example, the $\Phi_F(S_2)/\Phi_F(S_1)$ value is evaluated to be about 0.16. Thus, in such a case fluorescence from the upper state can be detected. Diphenylbutadiene vapor exhibits S_2 fluorescence at high total pressure in the presence of buffer gas, but shows only S_1 fluorescence in a jet. The $\Delta E(S_1 - S_2)$ value of diphenylbutadiene vapor is about 1100 cm^{-1} with the k_{F2}/k_{F1} value being estimated to be over 100. Therefore, the observation that diphenylbutadiene exhibits only S_2 fluorescence in the static vapor phase at 100 °C seems to be reasonable, since the $\Phi_F(S_2)/\Phi_F(S_1)$ value at this temperature is evaluated to be over 1.5.

S_2 fluorescence of some of aromatic molecules of medium to large size such as pyrene and benzanthracene in the vapor phase at low pressure corresponds to case (ii). For these molecules, the $\Delta E(S_1 - S_2)$ value is normally less than 4000 cm^{-1} and the oscillator strength of the S_2 state is much larger than that of S_1. The excited-state feature of case (ii) molecules is similar to that of the molecules of case (i), but the mechanism of the occurrence of the fluorescence differs from (i), since the emission from upper excited states is observable only under collision free conditions.

S_2 fluorescence of polyenes with longer polyene chain lengths corresponds to case (iii). For these molecules, the $\Delta E(S_1 - S_2)$ value is normally in the range 5000–10,000 cm^{-1} and the oscillator strength of the S_2 state is much larger than that of the S_1 state. For example, the $\Delta E(S_1 - S_2)$ value of hexadecaheptaene is about 5500 cm^{-1}. With azulene, mainly the S_2 fluorescence has been observed, while the quantum yields of the S_1 and S_2 fluorescence are almost the same for hexadecaheptaene in solution [124]. Spectroscopic studies of polyenes and carotenoids demonstrated that the systematic increase in conjugation lengths results in the S_2 fluorescence replacing the S_1 fluorescence observed for short polyenes. The increase of the S_2/S_1 fluorescence intensity ratio with increasing conjugation length can be accounted for by the energy gap law and/or by the intensity borrowing mechanism.

Even if a particular molecule satisfied the conditions that the emission from the upper state can be observable, as mentioned above, it cannot be always said that the molecule actually shows the emission from upper excited states. Whether or not the emission from upper excited states is observable depends also on other factors such as the stability of the molecule with respect to photon irradiation and relative rates of the radiative and nonradiative processes. For example, even if a molecule possesses a large $\Delta E(S_1 - S_2)$ value, it may be difficult to detect the S_2 fluorescence when the $S_2 \rightarrow S_1$ internal conversion rate is much faster than the radiative rate of S_2 by a factor of over 10^5. In this sense, the emission from upper excited states can be considered as anomalous or rare and provide an opportunity to reveal the dynamic behavior of the molecules in the excited states.

References

1. Kasha M (1950) Characterization of electronic transitions in complex molecules. Discuss Faraday Soc 9:14–19
2. Turro NJ, Ramamurthy V, Cherry W, Farneth W (1978) The effect of wavelength on organic photoreaction in solution. Reaction from upper excited states. Chem Rev 78:125–145
3. Birks JB (1973) Organic molecular photophysics. Wiley, New York
4. Klan P, Wirz J (2009) Photochemistry of organic compounds. Wiley, New York
5. Turro NJ, Ramamurthy V, Scaiano JC (2010) Modern molecular photochemistry of organic molecules. University Science, Sausalito
6. Turro NJ (1991) Modern molecular photochemistry. University Science, Sausalito
7. Maciejewski A, Steer RP (1993) The photophysics, photochemistry and spectroscopy of thiocarbonyls. Chem Rev 93:67–98

8. Ermolaev VL (2001) Ultrafast nonradiative transitions between higher excited states in organic molecules. Russ Chem Rev 70:471–490
9. Itoh T (1995) Low-lying electronic states, spectroscopy and photophysics of linear para acenequinones. Chem Rev 95:2351–2367
10. Wannier P, Rentzepis PM, Jortner J (1971) Resonance fluorescence from the second excited state of naphthalene. Chem Phys Lett 10(2):193–196
11. Gregory TA, Hirayama F, Lipsky S (1973) Fluorescence from highly excited states of some aromatic molecules in the vapor phase. J Chem Phys 58(10):4697–4699
12. (a) Hirayama F, Gregory TA, Lipsky S (1973) Fluorescence from highly excited states of some aromatic molecules in solution. J Chem Phys 58(10):4696–4697. (b) Gregory TA, Lipsky S (1976) Studies in the mechanism of radiationless conversion of electronic energy. I, *p*-xylene in solution. J Chem Phys 65(1):296–305
13. Katoh R, Kotani M (1999) Fluorescence from the second excited state of an anthracene crystal observed by two-step excitation. Chem Phys Lett 300(5–6):734–738
14. Katoh R, Fujiyoshi S, Kotani M (1998) Observation of weak fluorescence from the second excited state in an anthracene crystal. Chem Phys Lett 292(4–6):621–624
15. Katoh R, Kotani M (1993) Observation of fluorescence from higher excited states in an anthracene crystal. Chem Phys Lett 201(1–4):141–144
16. Schueler H, Arnold G (1961) Beobachtung der emission eines triplett-triplett-ueberganges beim naphthalin in der glimmentladung. Z Naturforsch A 16:1091–1093 (in German)
17. (a) Easterly CE, Christophorou LG, Blaunstein RP, Carter JG (1970) Fluorescence from the second excited pai-singlet state of 1.2-benzanthracene and 3.4-benzopyrene in solution. Chem Phys Lett 6(6):579–582. (b) Easterly CE, Christophorou LG (1974) Fluorescence emission from the first- and second-excited pai-singlet states of aromatic hydrocarbons in solution, and their temperature dependence. J Chem Soc Faraday Trans 2 70(2):267–273
18. Johnson PC, Offen HW (1972) Two-level fluorescence of ovalene. J Chem Phys 57 (1):336–338
19. Kropp JL, Stanley CC (1971) Temperature dependence of ovalene fluorescence. Chem Phys Lett 9(6):534–538
20. Carter JG, Christophorou LG, Easterly CE (1973) Fluorescence from the second pai-singlet state of aromatic hydrocarbons in solution. J Chem Soc Faraday Trans 2 69(3):471–483
21. Hoytink GJ (1973) The anomalous fluorescence of 1, 12-benzperylene in n-heptane. Chem Phys Lett 22(1):10–12
22. Birks JB, Easterly CE, Christophorou LG (1977) Stokes and anti-stokes fluorescence of 1,12-benzoperylene in solution. J Chem Phys 66(9):4231–4236
23. (a) Lyubimtsev VA, Ermolaev VL (1986) Spectra and quantum yields of fluorescence from upper excited singlet states of aromatic molecules in solutions: cascade relaxation. Optika i Spektroskopiya 61(3):511–517 (In Russian). (b) Tolkachev VA, Tugbaev VA (1975) Anomalous fluorescence of 3,4-benzopyrene vapors. Optika i Spektroskopiya 38(5):897–903 (In Russian)
24. Dawson WR, Kropp JL (1969) Radiationless deactivation and anomalous fluorescence of singlet 1.12-benzperylene. J Phys Chem 73(6):1752–1758
25. Geldof PA, Rettschnick RPH, Hoytink GJ (1969) Fluorescence from the second excited singlets of pyrene and 3,4-benzpyrene. Chem Phys Lett 4(2):59–61
26. Langelaar J, Leeuw WM, Van Voorst JDW, Rettschnick RPH (1979) Dual fluorescence (S_1, S_2) in chrysene and 3,4-benzophenanthrene. Chem Phys Lett 62(1):14–18
27. Deinum T, Werkhoven CJ, Langelaar J, Rettschnick RPH, Van Voorst JDW (1974) Sequence congestion and temperature effects on the fluorescence of isolated pyrene molecules. Chem Phys Lett 27(2):206–209
28. Nakajima A, Baba H (1970) Fluorescence spectrum of pyrene vapor: emission from the second excited singlet state. Bull Chem Soc Jpn 43:967
29. Nakajima A (1972) Fluorescence emission in the gas phase of several aromatic hydrocarbons. Bull Chem Soc Jpn 45:1687–1695

30. Baba H, Nakajima A, Aoi M, Chihara K (1971) Fluorescence from the second excited singlet state and radiationless processes in pyrene vapor. J Chem Phys 55(3):2432–2438
31. Baba H, Aoi M (1973) Vapor-phase fluorescence spectra from the second excited singlet state of pyrene and its derivatives. J Mol Spectrosc 46:214–222
32. Chihara K, Baba H (1977) Quenching of dual fluorescence of pyrene vapor by high pressure oxygen or nitric oxide. Chem Phys 25:299–306
33. Chihara K, Baba H (1975) Effects of foreign gases on dual fluorescence of pyrene vapor. Bull Chem Soc Jpn 48(11):3093–3100
34. Steidl H, Nowak D (1975) Die Verwendung eines Molekuelstrahls zur Messung der S_2-Fluoreszenz des Pyrens. Z Phys Chem 94(1–3):95–99 (in German)
35. Ohta N, Baba H, Marconi G (1987) Vibrational coupling and intramolecular dynamics of pyrene as revealed by the $S_0 \rightarrow S_2$ excitation spectrum in a supersonic jet. Chem Phys Lett 133(3):222–229
36. Suzuka I, Suzuki Y, Numata Y (1995) 11th Annual meeting of chemical reaction, Japan, 1P2, 13 (In Japanese)
37. Amirav A, Even U, Jortner J (1980) Intermediate level structure in the S_2 state of the isolated ultracold ovalene molecule. Chem Phys Lett 69(1):14–17
38. Nakajima A (1973) Short-wavelength fluorescence of 1,12-benzperylene. Chem Phys Lett 21 (1):200–204
39. Itoh T (2008) Multiple fluorescence and the electronic relaxation processes of coronene vapor: the fluorescence from the S_1, S_2, and S_3 states. J Mol Spectrosc 252:115–120
40. Iwashima S, Ohno K, Kajiwara T, Aoki J (1969) Synthesis and purification of coronene. Nihon Kagaku Zasshi 90(9):804–888 (in Japanese)
41. Itoh T, Yamaji M, Okamoto H (2013) S_2 fluorescence from picene vapor. Chem Phys Lett 570:26–28
42. Lennard-Jones JE (1937) The electronic structure of some polyenes and aromatic molecules. I, The nature of the links by method of molecular orbitals. Proc R Soc Lond 158:280–296.
43. Mulliken RS (1939) Intensities of electronic transitions in molecular spectra VII. Conjugated polyenes and carotenoids. J Chem Phys 7:364–373
44. Jorgensen P, Oddershede J (1972) Self-consistent polarization propagator calculations in the Pariser-Parr-Pople model. A modified random phase method. J Chem Phys 57(1):277–285
45. Hudson BS, Kohler BE (1972) A low-lying weak tranition in the polyene alfa, omega-diphenyloctatetraene. Chem Phys Lett 14(3):299–304
46. Hudson BS, Kohler BE (1973) Polyene spectroscopy: the lowest energy excited state of diphenyloctatetraene and other linear polyenes. J Chem Phys 59(9):4984–5002
47. Schulten K, Karplus M (1972) On the origin of a low-lying forbidden transition in polyenes and related molecules. Chem Phys Lett 14(3):305–309
48. Hudson BS, Kohler BE, Schulten K (1982) Linear polyene electronic structure and potential surfaces. Excited States 6:1–95
49. Kohler BE (1993) Octatetraene photoisimarization. Chem Rev 93(1):41–54
50. Buma WJ, Kohler BE, Song K (1990) Location of the 2^1A_g state in hexatriene. J Chem Phys 92(7):4622–4623
51. Kohler BE, Spangler C, Westerfield C (1988) The 2^1A_g state in the linear polyene 2,4,6,8,10,12,14,16-octadecaoctaene. J Chem Phys 89(9):5422–5426
52. Christensen RL, Galinato MGI, Chu EF, Howard JN, Broene RD, Frank HA (2008) Energies of low-lying excited states of linear polyenes. J Phys Chem A 112(49):12629–12636
53. Simpson JH, McLaughlin L, Smith DS, Christensen RL (1987) Vibronic coupling in polyenes: high resolution optical spectroscopy of all-*trans*-2,4,6.8.10,12,14-hexadecaheptaene. J Chem Phys 87(6):3360–3365
54. Snyder R, Arvidson E, Foote C, Harrigan L, Christensen RL (1985) Electronic energy levels in long polyenes: $S_2 \rightarrow S_0$ emission in all-*trans*-1,3,5,7,9,11,13-tetradecaheptaene. J Am Chem Soc 107(14):4117–4122

55. D'Amico KL, Manos C, Christensen RL (1980) Electronic energy levels in a homologous series of unsubstituted linear polyenes. J Am Chem Soc 102(6):1777–1782
56. Gavin RM Jr, Weisman C, McVey JK, Rice SA (1978) Spectroscopic properties of polyenes. III, 1,3,5,7-octatetraene. J Chem Phys 68(2):522–529
57. Granville MF, Holtom GR, Kohler BE, Christensen RL, D'Amico KL (1979) Experimental confirmation of the dipole forbidden character of the lowest excited singlet state in 1,3,5,7-octatetraene. J Chem Phys 70(1):593–594
58. Heimbrook LA, Kohler BE, Levy IJ (1984) Fluorescence from 11Bu state of trans, trans-1.3.5.7-octatetraene in a free jet. J Chem Phys 81(4):1592–1597
59. Bouwman W, Jones A, Phillips D, Thibodeau P, Friel C, Christensen R (1990) Fluorescence of gaseous tetraene and pentaene. J Phys Chem 94(19):7429–7434
60. Petek H, Bell A, Yoshihara K, Christensen R (1991) Spectroscopic and dynamical studies of the S_1 and S_2 states of decatetraene in supersonic molecular beams. J Chem Phys 95 (7):4739–4750
61. Itoh T (2009) Electronic relaxation processes of *all-trans*-2,4,6,8-decatetraene vapor revealed by the measurements of the S_1 and S_2 fluorescence. Mol Phys 107:1705–1711
62. (a) Allen MT, Whitten DG (1989) The photophysics and photochemistry of alpha, omega-diphenylpolyene singlet states. Chem Rev 89(8):1691–1702 and references cited therein. (b) Bachilo SM, Spangler CW, Gillbro T (1998) Excited state energies and internal conversion in diphenylpolyenes: from diphenylbutadiene to diphenyltetradecaheptaene. Chem Phys Lett 283(3–4):235–242
63. (a) Itoh T (2004) Evidence for the coexistence of two different mechanisms for the occurrence of anti-Kasha S_2 (1^1B_u) fluorescence from α,ω-diphenylpolyenes. J Chem Phys 121:6956–6960. (b) Itoh T (2007) Photophysics of α,ω-diphenyloctatetraene in the vapor phase. J Phys Chem A 111:3502–3506. (c) Itoh T (2007) S_2 (1^1Bu) fluorescence of diphenyl-decapentaene vapor. Chem Phys Lett 444:226–228
64. Alford PC, Palmer TF (1982) Fluorescence of DPH derivatives. Evidence for emission from S_2 and S_1 excited states. Chem Phys Lett 86(3):248–253
65. Itoh T, Kohler BE (1987) Dual fluorescence of diphenylpolyenes. J Phys Chem 91 (7):1760–1764
66. Jones GR, Cundall RB (1986) The effectt of moderately high pressure on the fluorescence spectra of 1,6-diphenylhexatriene, 1,6-[4,4′-dicyanophenyl]hexatriene and 1,8-diphenyloctatetraene. Chem Phys Lett 126(2):129–133
67. Bachilo BM, Bachilo EV, Gillbro T (1998) Spectral shapes of diphenylpolyene fluorescence and mixing of the S_1 and S_2 states. Chem Phys 229(1):75–91
68. Itoh T (1989) Evaluation of the coupling constants between the 2^1Ag(S_1) and 1^1Bu(S_2) states for diphenylhexatriene and diphenyloctatetraene. Chem Phys Lett 159:263–266
69. Itoh T (2002) Fluorescence spectral patterns of α, ω-diphenylpolyenes near the S1(21Ag) and S2(11Bu) level crossing point in solution. Bull Chem Soc Jpn 75:1973–1976
70. Turek AM, Krishnamoorthy G, Sears DF Jr, Garcia I, Dmitrenko O, Saltiel J (2005) Resolution of three fluorescence components in the spectra of all-*trans*-1,6-diphenyl-1,3,5-hexatriene under isopolarizability conditions. J Phys Chem A 109(2):293–303
71. Saltiel J, Sears DF Jr, Sun Y-P, Choi JO (1992) Evidence for ground-state s-cis conformers in the fluorescence spectra of all-trans-1,6-diphenyl-1,3,5-hexatriene. J Am Chem Soc 114 (10):3607–3612
72. Itoh T, Kohler BE, Spangler CW (1994) Fluorescence quantum yields of α,-ω-diphenylpolyenes. Spectrochim Acta A 50(13):2261–2263
73. (a) Itoh T (2003) Solvent-polarizability dependence of the relative 2^1Ag (S_1)- and 1^1Bu (S_2)-fluorescence intensities of 1, 14-diphenyl-1,3,5,7,9,11,13-tetradecaheptaene. J Chem Phys 119:4516–4521. (b) Itoh T (2003) Fluorescence spectrum of α,ω-diphenyltetradecaheptaene in a rigid matrix at 77 K. Chem Phys Lett 377:577–581
74. Sklar LA, Hudson BS, Petersen M, Diamond J (1977) Conjugated polyene fatty acids as fluorescent probes: spectroscopic characterization. Biochemistry 16(5):813–819

75. Catalan J, Hopf H, Klein D, Martus M (2008) On the photophysics of polyenes 1. Bathochromic shifts in their $1A_g - 1B_u$ electronic transitions caused by the polarizability of the medium. J Phys Chem A 112(25):5653–5657
76. Hirata H, Mashima K, Fukumoto H, Tani K, Okada T (1999) Energy gap dependence of the S2–S1 internal conversion of alpha, omega-diphenylpolyenes (N = 3–8) in solution phase. Chem Phys Lett 308:167–180
77. Itoh T, Kohler BE (1988) Fluorescence from diphenylpolyene vapors. J Phys Chem 92 (7):1807–1813
78. Itoh T (2001) Fluorescence spectra and the 2^1Ag level of 1,4-diphenylbutadiene in perfluorohexane and perfluoropentane. Chem Phys Lett 342:550–554
79. Dahl K, Biswas R, Maroncelli M (2003) The photophysics and dynamics of diphenylbutadiene in alkane and perfluoroalkane solvents. J Phys Chem B 107(31):7838–7853
80. Itoh T, Numata Y, Suzuka I (2007) Fluorescence spectra of all trans 1,4-diphenylbutadiene obtained by the two-photon excitation into the 2^1Ag state. Chem Phys Lett 445:179–182
81. Kohler BE, Itoh T (1988) Fluorescence from the 1^1B_u state of diphenylhexatriene: inversion of the 1^1B_u and 2^1A_g levels in CS_2. J Phys Chem 92(18):5120–5122
82. Itoh T (2013) Excited electronic states and spectroscopy of unsymmetrically substituted polyenes. J Chem Phys 139(9):094304
83. Saltiel J, Wang S (1995) Absorption and fluorescence spectra of a rigid analogue of all-trans-1,6-diphenyl-1,3,5-hexatriene. Solvent-controlled order inversion of 2^1A_g and 1^1B_u energy levels. J Am Chem Soc 117(43):10761–10762
84. Saltiel J, Wang S (2006) The photochemistry and photophysics of *all-trans*-1,4-diindanylidenyl-2-butene, a rigid analogue of *all-trans*-1,6-diphenyl-1,3,5-hexatriene. Photochem Photobiol Sci 5:883–895
85. Alford PC, Palmer TF (1983) Photophysics of derivatives of all-trans-1,6-diphenyl-1,3,5-hexatriene (DPH). Part 1, Model involving fluorescence from S_2 and S_1 excited states. J Chem Soc Faraday Trans 2 79(3):433–447
86. Sonoda Y, Goto M, Tsuzuki S, Tamaoki N (2007) Fluorinated diphenylpolyenes: crystal structures and emission properties. J Phys Chem A 111(51):13441–13451
87. Marri E, Galiazzo G, Mazzucato U, Spalletti A (2000) Effect of solvent polarizability on dual fluorescence of EE-1-phenyl,4-(1′-pyrenyl)-1,3-butadiene. Chem Phys 260(3):383–390
88. Marri E, Galiazzo G, Masetti F, Mazzucato U, Zuccaccia C, Spalletti A (2005) Effect of the chain length on the excited state properties of alpha-naphthyl, omega-phenyl-polyenes: photobehaviour of hexatrienes. J Photochem Photobiol, A 174(3):181–186
89. Bartocci G, Spalletti RS, Becker RS, Elisei F, Floridi S, Mazzucato U (1999) Excited-state behavior of some all-trans-α, ω-dithienylpolyenes. J Am Chem Soc 121(5):1065–1075
90. Itoh T, Yamaji M (2008) 1^1Bu (S_2) and 2^1Ag (S_1) fluorescence and the 2^1Ag-state of α, ω-dithienylbutadiene and α, ω-dithienylethylene. J Phys Chem A 112:13413–13418
91. Itoh T, Numata Y (2009) Spectroscopy and photophysics of low-lying excited singlet states of α, ω-dithienylbutadiene and α, ω-dithienylethylene vapors. J Phys Chem A 113 (38):10160–10166
92. Numata Y, Itoh T, Omokawa T, Okuyama K (2010) Hole-burning spectroscopy of jet-cooled dithienylethylene and dithienylbutadiene vapors. Chem Phys Lett 488:126–129
93. Armstrong GA, Hearst JE (1996) Carotenoids 2: genetics and molecular biology of carotenoid pigment biosynthesis. FASEB J 10:228–237
94. Britton G, Liaaen-Jensen S, Pfander H (2004) Carotenoids handbook. Birkhauser, Basel/Boston
95. Bondarev SL, Knyukshto VN (1994) Fluorescence from the $S_1(2^1A_g)$ state of all-trans-beta-carotene. Chem Phys Lett 225:346–350
96. Bondarev SL, Knyukshto VN (1994) β-Carotene S1–S0 fluorescence. Opt Spektrosk 76:591–592

97. Andersson PO, Bachilo SM, Chen RL, Gillbro T (1995) Solvent and temperature effects on dual fluorescence in a series of caritenes. Energy gap dependence of the internal conversion rate. J Phys Chem 99(44):16199–16209
98. Chynwat V, Frank HA (1995) The application of the energy gap law to the S1 energies and dynamics of carotenoids. Chem Phys 194:237–244
99. Polivka T, Sundstroem V (2004) Ultrafast dynamics of carotenoid excited states – From solution to natural and artificial systems. Chem Rev 104(4):2021 and references cited therein
100. Lustres JLP, Dobryakov AL, Holzwarth A, Veiga M (2007) S2–S1 internal conversion in beta-carotene: strong vibronic coupling from amplitude oscillations of transient absorption bands. Angew Chem Int Ed 46:3758–3761
101. Kosumi D, Yanagi K, Nishio T, Hashimoto H, Yoshizawa M (2005) Excitation energy dependence of excited states dynamics in all-*trans*-carotenes determined by femtosecond absorption and fluorescence spectroscopy. Chem Phys Lett 408:89–95
102. Shreve AP, Trautman JK, Owens TG, Albrecht AC (1991) Determination of the S_2 lifetime of β-carotene. Chem Phys Lett 178(1):89–96
103. Akimoto S, Yamazaki I, Takaichi S, Mimuro M (1999) Excitation relaxation of carotenoids within the S2 state probed by the femtosecond fluorescence up-conversion method. Chem Phys Lett 313:63–68
104. Kandori H, Sasabe H, Mimuro M (1994) Direct determination of a lifetime of the S_2 state of β-carotene by femtosecond time-resolved fluorescence spectroscopy. J Am Chem Soc 116 (6):2671–2672
105. Campillo AJ, Hyer RC, Kollman VH, Shapiro SL, Sutphin HD (1975) Fluorescence lifetimes of α- and β-carotenes. Biochim Biophys Acta 387:533–535
106. Itoh T (2011) Lowest singlet excited state and spectroscopy of α-carotene. Chem Phys Lett 505:96–99
107. Christensen RL, Goyette M, Gallagher L, Duncan J, DeCoster B, Lugtenburg J, Jansen FJ, van der Hoef I (1999) S_1 and S_2 states of apo- and diapocarotenes. J Phys Chem A 103 (14):2399–2407
108. Andersson PO, Gillbro T, Asato AE, Liu RSH (1992) Dual singlet state emission in a series of mini-carotenes. J Lumin 51(1–3):11–20
109. Mimuro M, Nagashima U, Nagaoka S, Nishimura Y, Takaichi S, Katoh T, Yamazaki I (1992) Quantitative analysis of the solvent effect on the relaxation processes of carotenoids showing dual emissive characteristics. Chem Phys Lett 191(3, 4):219–224
110. Mimuro M, Nishimura Y, Takaichi S, Yamano Y, Ito M, Nagaoka S, Yamazaki I, Katoh T, Nagashima U (1993) The effect of molecular structure on the relaxation processes of carotenoids containing a carbonyl group. Chem Phys Lett 213(5, 6):576–580
111. Macpherson AN, Gillbro T (1998) Solvent dependence of the ultrafast S2–S1 internal conversion rate of β-carotene. J Phys Chem A 102(26):5049–5058
112. Ehlers F, Wild DA, Lenzer T, Oum K (2007) Investigation of the S_1/ICT – S_0 internal conversion lifetime of 4′-apo-β-caroten-4′-al and 8′-apo-β-caroten-8′-al: dependence on conjugation length and solvent polarity. J Phys Chem A 111(12):2257–2265
113. Frank HA, Desamero RZB, Chynwat V, Gebhard R, van der Hoef I, Jansen FJ, Lugtenburg J, Gosztola D, Wasielewski MR (1997) Spectroscopic properties of spheroidene analogs having different extents of pai-electron conjugation. J Phys Chem A 101(2):149–157
114. Tavan P, Schulten K (1986) The low-lying electronic excitations in long polyenes: a PPP-MRD-CI study. J Chem Phys 85(11):6602–6609
115. Tavan P, Schulten K (1987) Electronic excitations in finite and infinite polyenes. Phys Rev B 36:4337–4358
116. Sashima T, Nagae H, Kuki M, Koyama Y (1999) A new singlet-excited state of all-trans-spheroidene as detected by resonance-Raman excitation profiles. Chem Phys Lett 299 (2):187–194
117. Sashima T, Koyama Y, Yamada T, Hashimoto H (2000) The $1B_u^+$, $1B_u^-$, and $2A_g^-$ energies of crystalline lycopene, β-carotene, and mini-9-β-carotene as determined by resonance-

Raman excitation profiles: dependence of the $1B_u^-$ state energies on the conjugation length. J Phys Chem B 104(20):5011–5019
118. Fujii R, Ishikawa T, Koyama Y, Taguchi M, Isobe Y, Nagae H, Watanabe Y (2001) Fluorescence spectroscopy of all-*trans*-anhydrorhodovibrin and spirilloxanthin: detection of the $1B_u$- fluorescence. J Phys Chem A 105(22):5348–5355
119. Kosumi D, Yanagi K, Fujii R, Hashimoto H, Yoshizawa M (2006) Conjugation length dependence of relaxation kinetics in b-carotene homologs probed by femtosecond Kerr-gate fluorescence spectroscopy. Chem Phys Lett 425:66–70
120. Kosumi D, Fujiwara M, Fujii R, Cogdell RJ, Hashimoto H, Yoshizawa M (2009) The dependence of the ultrafast relaxation kinetics of the S2 and S1 states in b-carotene homologs and lycopene on conjugation length studied by femtosecond time-resolved absorption and Kerr-gate fluorescence spectroscopies. J Chem Phys 130(21):214506
121. Christensen RL (1999) The electronic states of carotenoids. Adv Photosynth 8:137–159
122. Christensen RL, Galinato MGI, Chu EF, Fujii R, Hashimoto H, Frank HA (2007) Symmetry control of radiative decay in linear polyenes: low barriers for isomerization in the S_1 state of hexadecaheptaene. J Am Chem Soc 129(6):1769–1775
123. Birks JB, Tripathi GNR, Lumb MD (1978) The fluorescence of all-*trans* diphenylpolyenes. Chem Phys 33(2):185–194
124. Andrews JR, Hudson BS (1978) Environmental effects on radiative rate constants with application to linear polyenes. J Chem Phys 68(10):4587–4594
125. Cosgrove SA, Guite MA, Burnell TB, Christensen RL (1990) Electronic relaxation in long polyenes. J Phys Chem 94(21):8118–8124

Chapter 10
Developments in the Photonic Theory of Fluorescence

Jamie M. Leeder, David S. Bradshaw, Mathew D. Williams, and David L. Andrews

Abstract Conventional fluorescence commonly arises when excited molecules relax to their ground electronic state, and most of the surplus energy dissipates in the form of photon emission. The consolidation and full development of theory based on this concept has paved the way for the discovery of several mechanistic variants that can come into play with the involvement of laser input – most notably the phenomenon of multiphoton-induced fluorescence. However, other effects can become apparent when off-resonant laser input is applied during the lifetime of the initial excited state. Examples include a recently identified scheme for laser-controlled fluorescence. Other systems of interest are those in which fluorescence is emitted from a set of two or more coupled nanoemitters. This chapter develops a quantum theoretical outlook to identify and describe these processes, leading to a discussion of potential applications ranging from all-optical switching to the generation of optical vortices.

Keywords Multiphoton process • Anisotropy • Nonlinear optics • Optical vortex • All-optical switch

10.1 Introduction

Fluorescence is a form of luminescence whose measurement is widely employed in optical devices, microscopy imaging, biology and medical research. The basic theory describing fluorescence emission from individual molecules is extremely well-established. It centres upon the release of a photon from an excited molecule as it relaxes in a transition that is spin-allowed, and usually electric dipole-allowed, to a lower electronic state; this follows an initial electronic excitation, and usually some intervening vibrational relaxation. Delving more deeply into this model, the underlying quantum theory offers us additional physical insights into single- and multi-photon fluorescence. It also enables the prediction of several other novel, fluorescence-related processes, potentially leading to the production of useful

J.M. Leeder • D.S. Bradshaw • M.D. Williams • D.L. Andrews (✉)
School of Chemistry, University of East Anglia, Norwich NR4 7TJ, UK
e-mail: d.l.andrews@uea.ac.uk

© Springer International Publishing Switzerland 2016

C.D. Geddes (ed.), *Reviews in Fluorescence 2015*,
DOI 10.1007/978-3-319-24609-3_10

devices and applications across the sciences. This chapter offers a look at the latest developments in the photonic theory of fluorescence.

The structure of this work is as follows. Section 10.2 first affords a brief outline of the fundamental formalism utilized throughout this chapter, working up from quantum amplitudes to expressions for measurable rates of fluorescence, cast in terms that can duly elicit photonic attributes of the processes they describe. This section provides a basis for understanding the connection and common ground between the most familiar form of fluorescence, and the newer processes. It also includes a summary of the way in which the theory can be developed as a two-state model, whenever optical response is dominated by just the excited and ground electronic levels. Successive sections address recent research on specific applications. Section 10.3 discusses advances in the theory of multiphoton fluorescence, casting expressions for the output signals in terms of the associated electric polarization and molecular transition moment properties. Results established by means of an isotropic orientational average determine the fluorescence response of a fully disordered molecular environment – a complete system, or micro-domains within a complete system – also revealing one means by which multiphoton imaging can be further developed to monitor and quantify variations in chromophore orientation. Section 10.4 explores the development of 'laser-controlled fluorescence', a process whereby the rate of fluorescent emission is modified by an off-resonant probe beam of sufficient intensity. Associated changes in fluorescence behaviour afford new, chemically-specific information and a potential for novel technological applications through all-optical switching. Finally, in Sect. 10.5, theoretical developments in the field of multi-emitter fluorescence are described. Following a focus on the quantum mechanisms that operate between a pair of electromagnetically coupled nanoantenna emitters, attention is then given to designer systems based on an arrangement of molecular nanoemitters – which can be used as a basis for the generation of optical vortex radiation. Simple illustrations of the topics to be discussed in Sects. 10.3, 10.4 and 10.5 are shown in Fig. 10.1 The chapter concludes in Sect. 10.6 with a Discussion.

10.2 Photonic Theory of Fluorescence

In any molecular system that exhibits fluorescence, the primary stage – associated with the absorption of input radiation – is the electronic excitation of individual chromophores. Typically, ultrafast intramolecular vibrational redistribution produces a degree of immediate relaxation that results in a partial degradation of the acquired energy, with subsequent fluorescence occurring from the lowest level of the electronic excited state. For present purposes, assuming the validity of a Born-Oppenheimer separation of wavefunctions, we can focus on the character of these electronic transitions, since it is these that primarily determine the energetics and selection rules. The corresponding vibrational energies are generally small compared to the difference in electronic energy states: their impact on the fluorescence

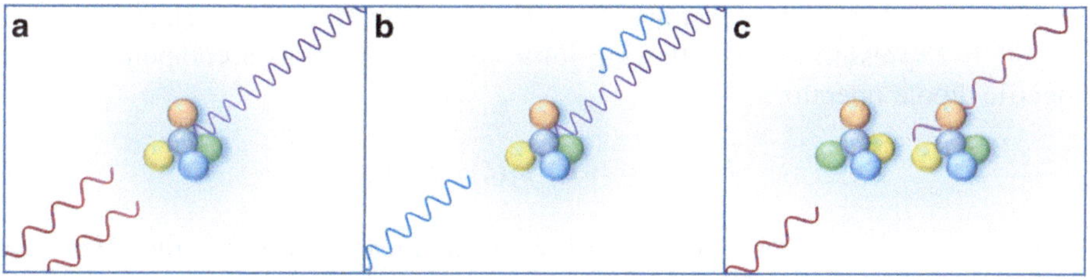

Fig. 10.1 Variations on molecular fluorescence: (**a**) Multiphoton fluorescence; (**b**) Laser-controlled fluorescence; (**c**) Fluorescence from coupled nanoemitters

transition, although important, principally features in the linewidth, determined by Franck-Condon factors.

With these considerations, the following representation of theory can now be built on the basis of parameters delivered by a quantum framework for both the radiation and the matter. Specifically, these will essentially be the quantum amplitudes (strictly 'matrix elements', M_{FI}, as they are in principle derivable for any specified pair of states) for the initial excitation and for the fluorescent decay, duly representing the input and output photons as quanta of the radiation field. Assuming that the energy associated with the strength of coupling between matter and radiation is far less than any molecular bond energy, such matrix elements which characterize the transition between initial and final system states, $|I\rangle$ and $|F\rangle$ respectively, are typically derived through time-ordered perturbation theory [1]. Such a perturbation is formally cast as an infinite, converging series, *i.e.*:

$$M_{FI}(\xi) = \sum_{n=1}^{\infty} \langle F|H_{\text{int}}(\xi)(T_0 H_{\text{int}}(\xi))^{n-1}|I\rangle, \tag{10.1}$$

where ξ represents a molecule or chromophore label, $T_0 = (E_I - H_0)^{-1}$ (in which E_I is the energy of the initial state and H_0 is the Hamiltonian for an unperturbed system), and $H_{\text{int}}(\xi)$ is the interaction Hamiltonian whose operation defines the system perturbation. The development of Eq. (10.1) usually involves implementation of the completeness relation $\sum_R |R\rangle\langle R| = 1$ etc., so that the expression becomes;

$$\begin{aligned} M_{FI} =& \langle F|H_{\text{int}}|I\rangle + \sum_{R} \frac{\langle F|H_{\text{int}}|R\rangle\langle R|H_{\text{int}}|I\rangle}{(E_I - E_R)} + \sum_{R,S} \frac{\langle F|H_{\text{int}}|S\rangle\langle S|H_{\text{int}}|R\rangle\langle R|H_{\text{int}}|I\rangle}{(E_I - E_R)(E_I - E_S)} \\ &+ \sum_{R,S,T} \frac{\langle F|H_{\text{int}}|T\rangle\langle T|H_{\text{int}}|S\rangle\langle S|H_{\text{int}}|R\rangle\langle R|H_{\text{int}}|I\rangle}{(E_I - E_R)(E_I - E_S)(E_I - E_T)} + \dots , \end{aligned} \tag{10.2}$$

where the intermediate system states are given by $|R\rangle$, $|S\rangle$, $|T\rangle\ldots$ and E_N is the energy of the state denoted by its subscript; the leading non-zero term for a process involving n photons is generally the nth term. The interaction Hamiltonian is explicitly expressed in the following form, featuring $\mu_i(\xi)$ as a component of the electric dipole operator:

$$H_{\text{int}}(\xi) = -\varepsilon_0^{-1}\mu_i(\xi) \cdot d_i^{\perp}(\mathbf{R}_\xi), \tag{10.3}$$

using the convention of summation over repeated Cartesian subscripts. Here, the contributions of magnetic and higher-order transition moments are legitimately ignored; the contribution from both are typically insignificant under conditions in which the molecular dimensions are significantly smaller than the optical wavelength. The transverse electric displacement field operator $d_i^{\perp}(\mathbf{R}_\xi)$ at position $\mathbf{R}_\xi$ acts upon the radiation system states within the arbitrary quantization volume V as:

$$d_i^{\perp}(\mathbf{R}_\xi) = \sum_{\mathbf{p},\eta}\left(\frac{\hbar c p \varepsilon_0}{2V}\right)^{\frac{1}{2}} i\left[e_i^{(\eta)}(\mathbf{p})a^{(\eta)}(\mathbf{p})\exp(i\mathbf{p}.\mathbf{R}_\xi) - \bar{e}_i^{(\eta)}(\mathbf{p})a^{\dagger(\eta)}(\mathbf{p})\exp(-i\mathbf{p}.\mathbf{R}_\xi)\right], \tag{10.4}$$

where $e_i^{(\eta)}$ is the unit electric polarization vector, with an overbar denoting its complex conjugate. The electric field operator is linear in both a and $a^\dagger$, which are the optical mode annihilation and creation operators, respectively, for a mode ($\mathbf{p}$, η); hence each operation of $d_i^{\perp}(\mathbf{R}_\xi)$ is responsible for either the creation or annihilation of a single photon. The parameter n in Eq. (10.1) defines the order of the matrix element with respect to $H_{\text{int}}(\xi)$, therefore effectively being determined by the number of matter-radiation interactions that occur within a given optical process. In order to exact results amenable to practical verification, it is common practice to report results in a form based on a measurable experimental observable. Throughout this review a commonly deployed methodology is utilized through application of Fermi's Golden Rule:

$$\Gamma = \frac{2\pi\rho_F}{\hbar}\left\langle |M_{FI}|^2 \right\rangle. \tag{10.5}$$

In this expression, Γ is the rate observable, proportional to the modulus square of the relevant matrix element, while on the right-hand side ρ_F represents a density of final system states defined as the number of molecular levels per unit energy associated with $|F\rangle$; the angular brackets here denote an orientational average to be effected for a system of randomly oriented molecules, as in the liquid phase [2]. Moreover, when the initial excitation is the rate-determining step (as is usually the case), then an effective rate can be cast in terms of an average for the product of matrix element quadratic terms for the excitation and fluorescent emission.

Equation (10.2) is often modified by the introduction of a simplifying assumption, to describe the optical response of a model system with just two electronic states (although, when applied to some other kinds of optical interaction, this common simplification can produce significantly misleading predictions) [3]. In the context of fluorescence, it is in most cases entirely defensible to consider only the ground and lowest excited states, *i.e.* employ the two-state model [4–18], since Kasha's rule states that fluorescence only occurs in appreciable yield from the lowest electronic excited state – although it cannot be presumed that the state from which the fluorescence decay occurs is necessarily the same as the state initially populated by photoexcitation. Upon application of such a two-state strategy, the quantum completeness identity becomes;

$$1 = \sum_{R} |R\rangle\langle R| \equiv \sum_{\rho,r} |\rho_{\mathrm{rad}}\rangle|r_{\mathrm{mol}}\rangle\langle r_{\mathrm{mol}}|\langle\rho_{\mathrm{rad}}| = \mathbf{1}_{\mathrm{rad}} \times (|0\rangle\langle 0| + |\alpha\rangle\langle\alpha|). \quad (10.6)$$

where the system state is decomposed into radiation and molecular states, the latter involving only a ground state $|0\rangle$ and a first excited state $|\alpha\rangle$. Limiting any intermediate molecular states to just $|0\rangle$ and $|\alpha\rangle$ restricts the number of transition sequences from the excited to ground molecular states. In complex interactions, each sequence generates a combined sequence of transition electric dipole moments, such as $\boldsymbol{\mu}^{0\alpha}$ and $\boldsymbol{\mu}^{\alpha 0}$, in combination with the static dipole moments of the ground and excited energy levels, $\boldsymbol{\mu}^{00}$ and $\boldsymbol{\mu}^{\alpha\alpha}$ respectively. It can be legitimately assumed that the former transition electric moments are real (as is always possible, given a suitable choice of basis set for the molecular wavefunctions) and therefore equal, by virtue of the Hermiticity of the dipole operator. Detailed analysis reveals that the dependence on static moments emerges only in terms of their vector difference. With the benefit of an algorithmic method, the following prescription, $\boldsymbol{\mu}^{\alpha\alpha} \to \boldsymbol{\mu}^{\alpha\alpha} - \boldsymbol{\mu}^{00} = \mathbf{d}$; $\boldsymbol{\mu}^{00} \to 0$ can be adopted [19], whose general validity has been proven to rest on a canonical transformation of the quantum interaction operator [20]. Applying this protocol requires application of an associated rule: any transitional mechanism that connects the initial and final system states (here, for the emission process) through a ground state static dipole is discarded.

10.3 Multiphoton Fluorescence

In laser-based studies of fluorescence, it is well-known that polarization features of the emission convey rich information on structural details of the sample, particularly in condensed phase molecular media. For example, detailed information can be secured on the degree of chromophore orientational order through polarization-resolved measurements [21, 22]. Numerous studies have focused on confined, highly ordered materials where the chromophores are held in crystalline structures [23–25], or else samples such as cell membranes, molecular films or fibers, where they are less rigidly bound to a physical matrix [26, 27]. In such instances, the rotational freedom of the targeted species is commonly restricted, enforcing a degree of orientational order relative to the external structure. Whereas polarization-derived information is often restricted to two spatial dimensions, the determination of three-dimensional orientation can also be explored [28]. Further investigations have extended the scope of such studies into the single-molecule regime, to elucidate information that is obscured in ensemble studies [29–31].

This section assesses the output signal resulting from *multiphoton* induced fluorescence, the application of which is highly prevalent in modern research owing primarily to the technique's unparalleled ability to deliver high-resolution, three dimensional imaging of heterogeneous samples. In general terms, the capture of high quality images aids the investigation of chemically specific information, since fluorescence intensity distributions allow the determination of the relative location, concentration and structure of specific molecular species *in situ* [32–34]. However, the attendant advantages offered by multiphoton methods include further features that have as yet received surprisingly little attention. In this respect it shall be shown that multiphoton imaging has a potential for further development as a diagnostic tool, to selectively discriminate micro-domains within a sample that exhibit a degree of orientational correlation. Any such technique could then equally monitor dynamical changes in this localized order, perhaps resulting from a chemical interaction, or acting in response to an externally applied stimulus.

The theory that follows duly provides a means of interrogating the extent of correlation between the transition moments associated with the process of fluorescence, namely those responsible for photon absorption and emission. Specific attention is given to the extent to which fluorescence retains a directionality of polarization from the initial excitation. To approach such issues involved in multiphoton processes, it is appropriate to begin with a representation of the optical process in its entirety, subsuming both the multi-photon absorption of laser input and the emission of fluorescent radiation. The output optical signal, $I_{\text{flu}}^{(n)}(\phi)$, is thus introduced as a function of the experimentally controllable angle between the polarization vector of the incident light and the resolved polarization of the emission, ϕ:

$$I_{\text{flu}}^{(n)}(\phi) = K^{(n)} \sum_{\xi} \left\langle \left| M_{\nu 0}^{(n)}(\xi) \right|^2 \left| M_{0\alpha}(\xi) \right|^2 \right\rangle. \tag{10.7}$$

The signal separates matrix elements for nth order multiphoton absorption and single-photon emission, $M_{\nu 0}^{(n)}(\xi)$ and $M_{0\alpha}(\xi)$ respectively. The possibility for excited state processes such as internal conversion, hindered rotation, rotational diffusion, intramolecular energy transfer etc. are accommodated through the adoption of labels 0 and ν to denote the molecular ground and initially excited energy levels, and α for the level from which emission occurs. The fluorescence signal in Eq. (10.7) is thus portrayed in terms of the physically separable efficiencies of the absorption and emission processes; the constant of proportionality $K^{(n)}$ is itself dependent on experimental parameters including the nth power of the mean irradiance delivered by the input laser beam, and also the corresponding degree of nth order coherence [35]. Angular brackets once again denote implementation of an orientation average, providing for the likely case in which the transition moments associated with multiphoton absorption and single photon emission are randomly oriented relative to the propagation vector of the input. If rotational diffusion during the excited state lifetime is significant, then the orientational average itself decouples into separate averages for the excitation and decay processes. To determine the results for one-, two- and three-photon induced fluorescence, the form of all associated matrix elements is required. Each is derived by standard methods; the underlying principles are introduced in a detailed description of single-photon induced fluorescence that directly follows.

10.3.1 One-Photon Induced Fluorescence

As indicated above, the theory for the process of single-photon induced fluorescence is characterized by the development of two distinct matter-radiation interactions: the first describes the optical excitation of a chromophore by single-photon absorption; the second entails molecular relaxation and photon emission that returns the chromophore to its ground electronic state. Addressing first the former process, the matrix element for single photon absorption is derived by substitution of Eq. (10.3) into (10.1) where $n = 1$:

$$M_{\nu 0}^{(1)}(\xi) = -i\left(\frac{q\hbar cp}{2\varepsilon_0 V}\right)^{\frac{1}{2}} e_i{}^{(\eta)}\mu_i{}^{\nu 0}\exp(i\mathbf{p}\cdot\mathbf{R}_\xi). \tag{10.8}$$

The level of intensity of the input mode is such it conveys q photons within a quantization volume V that is assumed to enclose the absorbing chromophore. By comparison, the matrix element for the process of photon emission, which engages electronic decay of the excited chromophore and the creation of a single photon into the vacuum radiation field, is expressed as:

$$M_{0\alpha}(\xi) = i\left(\frac{\hbar cp'}{2\varepsilon_0 V}\right)^{\frac{1}{2}} \overline{e}_i{}^{(\eta')}\mu_i{}^{0\alpha}\exp\left(-i\mathbf{p}'\cdot\mathbf{R}_\xi\right). \tag{10.9}$$

Here, prime labels have been utilized to distinguish the wave-vector and polarization of the output fluorescence from corresponding properties of the input beam. By substituting the matrix elements for both absorption and emission into Eq. (10.7), a complete expression for the signal output emerges:

$$I_{\text{flu}}^{(1)}(\phi) = \sum_{\xi} K^{(1)}\left\langle S_{ij}\overline{S}_{kl}T_{ij}\overline{T}_{kl}\right\rangle, \tag{10.10}$$

where the square modulus of Eqs. (10.8) and (10.9) have been employed, and the products of scalar parameters within the parentheses of each matrix element are incorporated into the proportionality constant $K^{(1)}$. For ease of notation, the orientation-dependent products of the unit electric polarization vectors, and those of the molecular transition moments, are each incorporated into second rank tensors where specifically S_{ij} and $\overline{S}_{ij}$ denote $e_i^{(\eta)}\overline{e}_j^{(\eta')}$ and $\overline{e}_i^{(\eta)}e_j^{(\eta')}$. Likewise, the molecular transition moment products described by T_{ij} and $\overline{T}_{ij}$ correspond to $\mu_i^{\nu 0}\mu_j^{0\alpha}$ and $\overline{\mu}_i^{\nu 0}\overline{\mu}_j^{0\alpha}$. Here, and in all subsequent applications of this notation, the last index in the electric polarization and molecular transition tensors relates to the photon emission. Equation (10.10) thus expresses a result that embraces the angular disposition of the chromophore transition moments with respect to the input and output polarization vectors. In a rigidly oriented system, forgoing the orientational average, the result would thus exhibit a dependence on $\cos^2\eta\cos^2\theta$, where η is the angle between the absorption moment and the input polarization, and θ is that between the emission moment and the fluorescence polarization.

10.3.2 Multiphoton Induced Fluorescence

Having derived the matrix element for one-photon emission, the fluorescence signal for multiphoton processes now requires expressions that account for the concerted absorption of two or more photons. First addressing the specific case of *two-photon* absorption, the associated matrix element entails a progression through an intermediate system state in which one photon is annihilated and the chromophore, lacking a resonant level to match the photon energy, is accordingly in a transient superposition of virtual molecular states. Any such energy non-conserving state can be sustained only as long as is allowed by the time-energy uncertainty principle – necessary summation is made over all possible intermediate states, as required by quantum principles. The full matrix element is thus developed by substitution of Eqs. (10.3) and (10.4) into Eq. (10.1) where $n = 2$ such that:

$$M_{\nu 0}^{(2)}(\xi) = \left(\frac{q_2^{1/2}\hbar c p}{2\varepsilon_0 V}\right) e_i^{(\eta)} e_j^{(\eta)} \alpha_{(ij)}^{\nu 0}. \tag{10.11}$$

Here, the quantization volume initially contains the chromophore and two photons of the incident radiation; the factor of $q_2^{1/2} \equiv [q(q-1)]^{1/2}$ correspondingly arises from the successive operations of the photon annihilation operator. The above expression exploits the symmetry of the electric polarisation terms $e_i^{(\eta)}e_j^{(\eta)}$ with respect to exchange of the indices i and j. The second rank molecular response tensor $\alpha_{(ij)}^{\nu 0}$ is duly defined as:

$$\alpha_{(ij)}^{\nu 0} = -\frac{1}{2}\sum_r (E_{r0} - \hbar c p)^{-1}\left(\mu_i^{\nu r}\mu_j^{r0} + \mu_j^{\nu r}\mu_i^{r0}\right), \tag{10.12}$$

where $E_{rs} = E_r - E_s$ is an energy difference between molecular states. The two dipole product contributions in the above expression relate to each of the possible time-orderings in which the two, indistinguishable input photons can be annihilated; the factor of $\frac{1}{2}$ is introduced to preclude over-counting, and bracketed subscripts denote symmetry with respect to interchange of the enclosed indices. In cases where the electronic level accessed by two-photon absorption equates to that from which subsequent radiative decay occurs, i.e. $\nu = \alpha$, it is prudent to allow the two-level approximation for the two-photon absorbing chromophore [36]. The tensor that determines the two-photon absorption properties of such a system then reduces to a form that features both static and transition dipoles, the former expressed as a shift in dipole moment that accompanies the transition:

$$\alpha_{(ij)}^{\alpha 0(TLA)} = -\frac{1}{2}(E_{\alpha 0} - \hbar c p)^{-1}\left(d_i \mu_j^{\alpha 0} + d_j \mu_i^{\alpha 0}\right). \tag{10.13}$$

By combining Eqs. (10.9) and (10.11), the following expression represents the output signal resulting from two-photon induced fluorescence:

$$I_{\text{flu}}^{(2)}(\phi) = \sum_{\xi} K^{(2)} \left\langle S_{(ij)k} \overline{S}_{(lm)n} T_{(ij)k} \overline{T}_{(lm)n} \right\rangle, \tag{10.14}$$

Here, the electric vector and molecular transition moment products are expressed as third rank tensors such that $S_{(ij)k}$ and $\overline{S}_{(ij)k}$ correspond to $e_i^{(\eta)} e_j^{(\eta)} \overline{e}_k^{(\eta')}$ and $\overline{e}_i^{(\eta)} \overline{e}_j^{(\eta)} e_k^{(\eta')}$, whilst $T_{(ij)k}$ and $\overline{T}_{(ij)k}$ signify $\alpha_{(ij)}^{\nu 0} \mu_k^{0\alpha}$ and $\overline{\alpha}_{(ij)}^{\nu 0} \overline{\mu}_k^{0\alpha}$ respectively. In this case, for an oriented sample, the dependence on emission angle is again $\cos^2\theta$. However the dependence on input polarization is considerably more intricate, being determined by a weighted combination of $\cos^2$ functions for each angle between the input polarization vector and one of a number of transition moments, i.e. $\boldsymbol{\mu}^{\nu r}$, $\boldsymbol{\mu}^{r0}$, for each level r.

For *three-photon* induced fluorescence, the transition between the ground and excited state of the chromophore, which requires the concerted absorption of an additional photon, progresses through two distinct virtual intermediate states. Substitution of Eqs. (10.3) and (10.4) into (10.1) where $n = 3$ yields the following matrix element cast in terms of a third rank molecular response tensor $\beta_{(ijk)}^{\nu 0}$:

$$M_{\nu 0}^{(3)}(\xi) = -q_3^{1/2} i \left(\frac{\hbar c p}{2\varepsilon_0 V}\right)^{\frac{3}{2}} e_i^{(\eta)} e_j^{(\eta)} e_k^{(\eta)} \beta_{(ijk)}^{\nu 0}. \tag{10.15}$$

As with the case of second rank tensor used to describe two-photon absorption, $\beta_{(ijk)}^{\nu 0}$ features a sum of dipole product contributions that account for all possible time-orderings of the identical input photons. Including a factor of $\frac{1}{6}$, again to offset over-counting, this third-rank molecular response tensor is defined thus:

$$\beta^{\nu 0}_{(ijk)} = \frac{1}{6}\sum_{r,s} [(E_{r0} - \hbar c p)(E_{s0} - 2\hbar c p)]^{-1} \left(\mu_i^{\nu s}\mu_j^{sr}\mu_k^{r0} + \mu_i^{\nu s}\mu_k^{sr}\mu_j^{r0} + \mu_j^{\nu s}\mu_i^{sr}\mu_k^{r0} + \mu_j^{\nu s}\mu_k^{sr}\mu_i^{r0} + \mu_k^{\nu s}\mu_i^{sr}\mu_j^{r0} + \mu_k^{\nu s}\mu_j^{sr}\mu_i^{r0}\right). \tag{10.16}$$

As with two-photon absorption, it is again expedient to re-express this general three-photon tensor in a more specific two-level form:

$$\beta^{\alpha 0}_{(ijk)} = \frac{1}{3}(E_{\alpha 0} - \hbar c p)^{-1}\left[(E_{\alpha 0} - 2\hbar c p)^{-1}\left(\mu_i^{\alpha 0} d_j d_k + \mu_j^{\alpha 0} d_k d_j + \mu_k^{\alpha 0} d_i d_k\right) - (2\hbar c p)^{-1}\left(\mu_i^{\alpha 0}\mu_j^{0\alpha}\mu_k^{\alpha 0} + \mu_i^{\alpha 0}\mu_k^{0\alpha}\mu_j^{\alpha 0} + \mu_j^{\alpha 0}\mu_i^{0\alpha}\mu_k^{\alpha 0}\right)\right]. \tag{10.17}$$

The fluorescence signal due to three-photon excitation can now be represented as:

$$I^{(3)}_{\text{flu}}(\phi) = \sum_{\xi} K^{(3)}\left\langle S_{(ijk)l}\overline{S}_{(mno)p} T_{(ijk)l}\overline{T}_{(mno)p}\right\rangle, \tag{10.18}$$

The electric polarization and molecular transition moments are described in terms of fourth rank tensors, where $S_{(ijk)l}$ and $\overline{S}_{(ijk)l}$ respectively represent $e_i^{(\eta)} e_j^{(\eta)} e_k^{(\eta)} \overline{e}_l^{(\eta')}$ and $\overline{e}_i^{(\eta)}\overline{e}_j^{(\eta)}\overline{e}_k^{(\eta)} e_l^{(\eta')}$, whilst $T_{(ijk)l}$ and $\overline{T}_{(ijk)l}$ correspond to $\beta^{\nu 0}_{(ijk)}\mu_l^{0\alpha}$ and $\overline{\beta}^{\nu 0}_{(ijk)}\overline{\mu}_l^{0\alpha}$, the final index of each again being associated with the one-photon emission. The orientation relative to the input polarization again depends on a multitude of angles, corresponding in this case to the orientations of the transition moments $\boldsymbol{\mu}^{\nu s}$, $\boldsymbol{\mu}^{sr}$, $\boldsymbol{\mu}^{r0}$, summed over states r and s.

10.3.3 Freely Tumbling Molecules

The general results presented so far for the fluorescence output in one-, two- and three-photon induced systems are applicable to systems in which the responsible chromophores have arbitrary orientations with respect to experimentally determined input and detection configurations. As such, the derived expressions are already directly applicable to all ordered samples in which individual chromophores are held in a fixed orientation. To address disordered systems it is expedient to secure corresponding results for an opposite extreme – systems of completely random dipole orientation, which represents a set of freely tumbling molecules. To this end, the above results can be subjected to an established, integration free, orientational averaging protocol utilizing isotropic tensors [37–39].

Beginning with the signal for *one-photon* induced fluorescence, Eq. (10.10) exhibits a sum over four separate Cartesian indices. By first uncoupling the molecular and radiation components of the system – achieved by assigning the former to a molecule-fixed frame of reference and the latter to a laboratory-fixed equivalent – and then performing a fourth-rank average, tensor contractions are effected. All of the ensuing results are then expressible in terms of scalar products between input and output polarization components. In the commonly utilized deployment of plane-polarized input laser light, the polarization vectors are real and the scalar product of any two is concisely summarized by:

$$\overset{(-)(\eta)}{\mathbf{e}} \cdot \overset{(-)(\eta')}{\mathbf{e}} = \delta_{\eta\eta'} + \left(1 - \delta_{\eta\eta'}\right) \cos\phi \,, \tag{10.19}$$

where ϕ is the angle between the input and output polarization vectors. The final result for the orientationally averaged fluorescence output emerges in terms of ϕ as;

$$I_{\text{flu}}^{(1)}(\phi) = \frac{K^{(1)}}{30}\left[\left(T_{\lambda\lambda}\overline{T}_{\mu\mu} + T_{\lambda\mu}\overline{T}_{\mu\lambda}\right)\left(3\cos^2\phi - 1\right) - \left(2T_{\lambda\mu}\overline{T}_{\lambda\mu}\right)\left(\cos^2\phi - 2\right)\right], \tag{10.20}$$

involving three molecular invariants, $T_{\lambda\lambda}\overline{T}_{\mu\mu}$, $T_{\lambda\mu}\overline{T}_{\mu\lambda}$ and $T_{\lambda\mu}\overline{T}_{\lambda\mu}$ – whose form and means of characterization, using linear and circular polarizations, were first identified in pioneering work by McClain [40, 41]. For this case of one-photon induced fluorescence, it is further possible to express the molecular tensors in Eq. (10.20) relative to the magnitude of the molecular transition moments $\boldsymbol{\mu}^{\nu 0}$ and $\boldsymbol{\mu}^{0\alpha}$, and the angle between them, β, such that:

$$I_{\text{flu}}^{(1)}([\phi],[\beta]) = \frac{K^{(1)}\left|\boldsymbol{\mu}^{\nu 0}\right|^2\left|\boldsymbol{\mu}^{0\alpha}\right|^2}{30}\left[\left(3\cos^2\phi - 1\right)\left(2\cos^2\beta\right) - 2\left(\cos^2\phi - 2\right)\right], \tag{10.21}$$

where the identities $T_{\lambda\lambda}\overline{T}_{\mu\mu} = T_{\lambda\mu}\overline{T}_{\mu\lambda} = \left|\boldsymbol{\mu}^{\nu 0}\right|^2\left|\boldsymbol{\mu}^{0\alpha}\right|^2\cos^2\beta$, and $T_{\lambda\mu}\overline{T}_{\lambda\mu} = \left|\boldsymbol{\mu}^{\nu 0}\right|^2 \left|\boldsymbol{\mu}^{0\alpha}\right|^2$ apply. Resolving Eq. (10.21) for fluorescence components parallel or perpendicular to the input polarization leads to the familiar degree of *fluorescence anisotropy* for a randomly oriented sample [21, 22].

To resolve the corresponding anisotropy equations for two- and three-photon induced fluorescence invokes identical methods based on sixth- and eighth-rank averages [42], respectively. From a detailed analysis of the results, it emerges that the equations determining the multiphoton fluorescence response prove expressible in a relatively simple, generic form:

$$I_{\text{flu}}^{(n)}(\phi) = K^{(n)}\left[\Lambda^{(n)}\left(3\cos^2\phi - 1\right) - \Upsilon^{(n)}\left(n\cos^2\phi - (n+1)\right)\right], \tag{10.22}$$

with both $\Lambda^{(n)}$ and $\Upsilon^{(n)}$ each representable as a sum of distinct molecular invariants. While the detailed form of such invariants is reported elsewhere [43], it should be noted that an inherent summation in each over accessible virtual intermediate states precludes the possibility of further simplifying the above result by attempting to factorize out the absorption and emission transition moments, as was achieved for Eq. (10.21), without introducing further assumptions that would compromise the generality of the outcome.

These generic results represent tools that can be applied in the analysis of polarization-determined features in two- and three-photon fluorescence from samples of considerable molecular complexity. By determining how either type of multiphoton-induced fluorescence signal responds to the orientation of a polarizer, it is in principle possible to distinguish and quantify any departure from local orientational order or disorder within a bulk sample. Key to this discrimination is the difference in angular disposition of the fluorescence polarization. In samples whose chromophores are rigidly oriented, the fluorescence signal from an ensemble with common orientation takes the form of a $\cos^2$ distribution with respect to the angle θ between the emission moment and the resolved polarization. On rotation of the polarizer through 180° there will be an angle at which the signal is extinguished – both for single- and multi-photon induced fluorescence. However, as will be shown, the behaviour from a randomly oriented sample is in general distinctively different.

To proceed, it is helpful to cast the general result that determines the multiphoton fluorescence signal, in the following form:

$$I_{\text{flu}}^{(n)}(\phi) = K^{'(n)}\left[n + 1 - y + (3y - n)\cos^2\phi\right], \tag{10.23}$$

where $K^{'(n)} = K^{(n)}\Upsilon^{(n)}$, $y = \Lambda^{(n)}/\Upsilon^{(n)}$. The latter parameter is a scalar that characterizes the relative values of the molecular invariant groupings in Eq. (10.22). Although the precise value of y will depend on the component values of the transition tensors, it can be shown that it is positive and limited to an upper bound of $(n+1)$. Figure 10.2 exhibits the functional form of the fluorescence

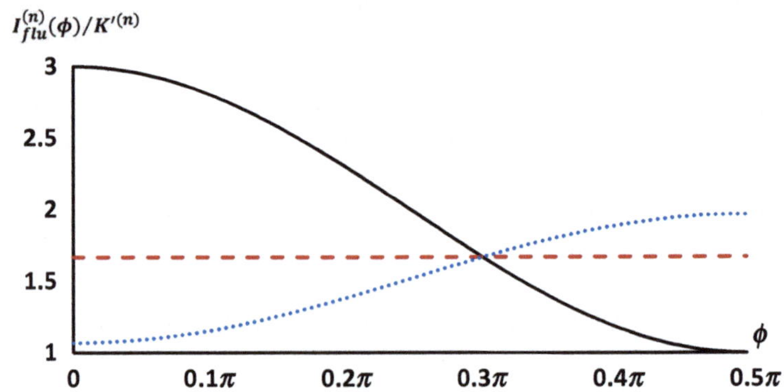

Fig. 10.2 Angular disposition of polarization in fluorescence produced by single-photon absorption ($n = 1$): *blue* (*dotted*) curve $3y/n = 0.1$; *red* (*dashed*) curve $3y/n = 1$; *black* (*solid*) curve $3y/n = 3$

polarization, for single-photon induced fluorescence, over the range $(0 \leq \phi \leq \pi/2)$ – results of identical form but different scale have also been recorded for the multiphoton processes. Each graph shows the behavior for different values of y; the "magic angle condition" represented as the point at which the curves for all different values of y intersect. The curve for $3y/n = 0.1$, for example, represents an extreme condition, $\Lambda^{(n)} \ll \Upsilon^{(n)}$, characterized by strongly depolarized emission. On the other hand the case $3y/n = 1.0$ is of special interest because the fluorescence proves to be independent of the resolving polarization, thus representing a condition under which the fluorescence produced through the concerted absorption of any number of photons becomes completely unpolarized.

Results for $3y/n = 3.0$ are perhaps the most interesting, being indicative of the statistically most likely outcome. This condition arises when, within the general result, all featured molecular invariants are of approximately equal value. It is remarkable that this condition leads in every case to $\Lambda^{(n)}/\Upsilon^{(n)} = y = n$. Here, there is a strong retention of polarization, the corresponding emission anisotropies $r = (I_{\parallel} - I_{\perp})/(I_{\parallel} + 2I_{\perp})$ proving to conform to the simple formula $r = 2n/(2n+3)$ and yielding the following specific values: (i) $n = 1$; $r = 2/5 = 0.4$, the familiar one-photon result [21]; also (ii) for two-photon excitation, $n = 2$; $r = 4/7 = 0.57$; (iii) for the three-photon case $n = 3$; $r = 6/9 = 0.67$. These limiting results are in precise agreement with the values that arise specifically when all transition moments are considered parallel, a special case originally considered and reported by Lakowicz et al. [44]

The correlation serves to verify a limiting case of the present, more general results – but it is also notable that the conditions under which such behavior arises are not only associated with parallel transition moments. The same observations will result, for example, if all of the molecular transition tensor elements have similar values. In conclusion, the considered cases all satisfy the condition that the ratio of maximum and minimum fluorescence intensities $I^{(n)}_{\min}/I^{(n)}_{\max}$ lies in the interval $[0,\ 1/(2n+1)]$. It is worth recalling that rotation of the resolving polarizer can entirely extinguish the fluorescence from an orientationally perfectly ordered sample or domain. This suggests that in a general case the measured value of $I^{(n)}_{\min}/I^{(n)}_{\max}$ registered against the scale $[0,\ 1/(2n+1)]$ should represent a robust, easily determined single-value indicator of the degree of disorder in fluorescence produced by n-photon excitation.

10.4 Laser-Controlled Fluorescence

Using the same quantum formalism, a completely novel development in relation to fluorescence is now to be discussed. Its background is the well-known fact that the throughput of a laser beam into a photo-activated system may produce stimulated emission when the laser frequency matches the fluorescence energy – a phenomenon that has found analytical applications, for example, in the recently Nobel Prize winning technique of stimulated emission depletion spectroscopy [45–51]. However, outside of stimulated emission, it has further emerged that a moderately intense, *off*-resonant laser beam may significantly alter the rate and intensity of fluorescence [52–55]. Under these circumstances, the probe laser essentially confers optical nonlinearity onto the fluorescent emission – and consequently, each excited-state lifetime is appreciably modified. One may draw analogies with the well-known enhancement of optical emission through its coupling with a plasmonic surface [56–65]. However, this novel development modifies spontaneous fluorescent emission through direct interaction with the oscillating electric field of throughput radiation, without the presence of any surface. With initial estimates suggesting that conventional fluorescence lifetimes could be reduced by 10 % or more [53], for input beam intensities in the 10^{15} W m^{-2} range, such a prospect is readily amenable to measurement with modulation-based instrumentation – and it also affords a new means of exerting control over the fluorescence process.

In the following summary of theory, the effects of laser-controlled emission on fluorescence anisotropy are determined for a system of randomly oriented chromophores. It is also shown that a two-level formulation of theory leads to relatively tractable expressions with a broad validity extending from quantum dots [11, 12, 14] to fluorescent proteins [13, 15–18] – indeed any material whose emission spectrum is dominated by one excited electronic state. A limiting case is then considered in which fluorescence arises solely through activation by the off-resonant input. First, outside of these two-level considerations, the mechanism of laser-controlled fluorescence is to be more fully described.

10.4.1 The Mechanism

We first return to the well-established tenet that the theory of single-photon emission from any individual chromophore, since it involves a single matter-radiation interaction, is cast in terms of first-order time-dependent perturbation theory. In cases where no other light is present – as is the case in normal experiments, once the radiation responsible for initial electronic excitation has traversed the system – then higher order (odd-rank) perturbation terms are insignificant, and only denote self-energy corrections. However, higher-order interactions of much greater significance can occur on the application of an off-resonant probe laser, *i.e.* a beam whose wavelength lies within a transparent region of the chromophore. Although no net absorption or stimulated emission of the beam then occurs, elastic forward-scattering events are present – namely, photons are annihilated and created into the same radiation mode (the latter photon emerging as if unchanged from the former). Through nonlinear coupling, such events may directly engage with the fluorescent emission in a mechanism comprising *three* concerted matter-radiation

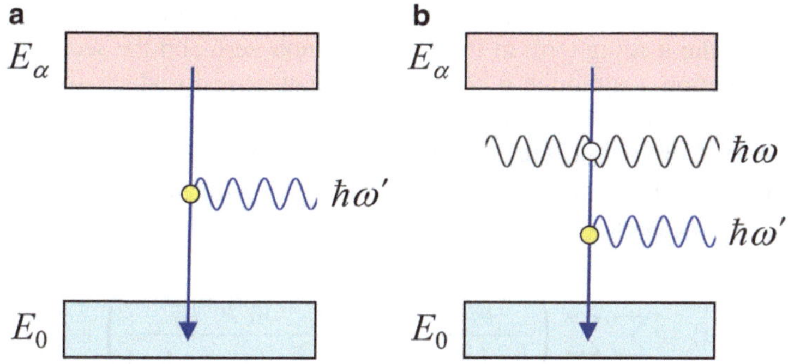

Fig. 10.3 Energy level representation for: (**a**) spontaneous and (**b**) nonlinear coupling mechanisms for fluorescence. Electronic states (and their vibrational manifolds) are signified by boxes, where E_0 and E_α are correspondingly the ground and excited molecular states. Wavy lines indicate photon propagation and the vertical arrow is a decay transition due to the emission. The emitted fluorescence has energy defined by $\hbar\omega'$, while photons of the off-resonant laser beam are of energy $\hbar\omega$. The yellow dot symbolizes a single matter-radiation interaction and the white dot represents two such interactions (i.e. elastic forward-scattering)

interactions (Fig. 10.3), *i.e.* a process that has to be described using third-order perturbation theory. In passing, it is noteworthy that the off-resonant probe beam produces similar effects in connection with resonance energy transfer [66–68].

The intensity of fluorescence $I'(\Omega')$ (or power per unit solid angle) is determined by multiplying the Fermi Rule of Eq. (10.5) by the energy of a fluorescent photon, $\hbar\omega' \equiv \hbar ck'$ [69, 70], and it signifies the single-molecule fluorescence signal that follows relaxation from the relevant excited state. By including laser-controlled fluorescence, the net intensity is found from $I'(\Omega')\mathrm{d}\Omega' = 2\pi\rho ck'\left|M_{\mathrm{flu}}^{(1)} + M_{\mathrm{flu}}^{(3)}\right|^2$, where $M_{\mathrm{flu}}^{(1)}$ and $M_{\mathrm{flu}}^{(3)}$ are the quantum amplitudes for first- and third-order fluorescent processes, respectively, and the density of radiation states is $\rho = \left(k'^2 V/8\pi^3\hbar c\right)\mathrm{d}\Omega'$. As determined elsewhere [53], the following general result is derived from this Fermi-related expression;

$$I'(\Omega') = \left(\frac{ck'^4}{8\pi^2\varepsilon_0}\right)\Big[e'_i e'_j \mu_i^{0\alpha}\bar{\mu}_j^{0\alpha} + (I/c\varepsilon_0)e_i e_j e'_k e'_l \chi_{ijk}^{0\alpha}(\omega'; -\omega, \omega)\bar{\mu}_l^{0\alpha} + (I^2/4c^2\varepsilon_0^2)e_i e_j e'_k e_l e_m e'_n \chi_{ijk}^{0\alpha}(\omega'; -\omega, \omega)\bar{\chi}_{lmn}^{0\alpha}(\omega'; -\omega, \omega)\Big], \tag{10.24}$$

where I is the irradiance of the laser probe, and **e** now represents the polarization vector of the probe, off-resonant photons with energy $\hbar ck$. The first term corresponds to spontaneous emission, intrinsic to the system and independent of the probe laser beam, while the last term signifies a coupling of the elastically forward-scattered probe beam with the fluorescence emission. The middle term, linear in I, signifies a quantum interference of these two concurrent processes. In general, it is assumed that the leading term in Eq. (10.24) is non-zero and the second one is the leading correction – although a circumstance can arise in which solely the third term exists, *i.e.* when the first and second terms are null (this is discussed in Sect. 10.4.3). Continuing, the sum-over-states form of the third-order nonlinear optical 'transition hyperpolarizability' tensor $\chi_{ijk}^{0\alpha}(\omega';-\omega,\omega)$, explicitly exhibiting the frequency dispersion, is as follows;

$$\begin{aligned}\chi_{ijk}^{0\alpha}(\omega';-\omega,\omega) &= \sum_{r}\sum_{s\neq\alpha}\left(\frac{\mu_k^{0s}\mu_j^{sr}\mu_i^{r\alpha}}{\tilde{E}_{s\alpha}\left(\tilde{E}_{r\alpha}-\hbar\omega\right)}+\frac{\mu_k^{0s}\mu_i^{sr}\mu_j^{r\alpha}}{\tilde{E}_{s\alpha}\left(\tilde{E}_{r\alpha}+\hbar\omega\right)}\right)\\ &+\sum_{r}\sum_{s}\left(\frac{\mu_j^{0s}\mu_k^{sr}\mu_i^{r\alpha}}{\left(\tilde{E}_{s\alpha}-\hbar\omega+\hbar\omega'\right)\left(\tilde{E}_{r\alpha}-\hbar\omega\right)}+\frac{\mu_i^{0s}\mu_k^{sr}\mu_j^{r\alpha}}{\left(\tilde{E}_{s\alpha}+\hbar\omega+\hbar\omega'\right)\left(\tilde{E}_{r\alpha}+\hbar\omega\right)}\right)\\ &+\sum_{r\neq 0}\sum_{s}\left(\frac{\mu_j^{0s}\mu_i^{sr}\mu_k^{r\alpha}}{\left(\tilde{E}_{s\alpha}-\hbar\omega+\hbar\omega'\right)\left(\tilde{E}_{r\alpha}+\hbar\omega'\right)}+\frac{\mu_i^{0s}\mu_j^{sr}\mu_k^{r\alpha}}{\left(\tilde{E}_{s\alpha}+\hbar\omega+\hbar\omega'\right)\left(\tilde{E}_{r\alpha}+\hbar\omega'\right)}\right).\end{aligned} \tag{10.25}$$

The tildes serve as a reminder to add to the excited state energies, in the case of near-resonance conditions, imaginary terms to accommodate line-shape and damping. With reference to later comments, it is worth noting here that there is no assumption of Kleinman symmetry [71] at this stage – this being a simplifying device, commonly made for calculational expediency that would impose complete index symmetry for such a tensor.

Considering the dependence of the fluorescence signal on the optical frequency of the probe, it is evident that the denominators within the transition hyperpolarizability tensor of Eq. (10.25) have an important role in determining any degree of enhancement or suppression of the fluorescence emission. These factors are dependent on the relative spacing of the chromophore energy levels relative to the magnitude of the probe photon energy. It is convenient to assume that the probe light is delivered in the form of a tunable beam with optical frequency $\omega < \omega'$, a condition that specifically precludes single-photon excitation of ground-state molecules. It will also be assumed that the chosen range of probe frequencies cannot produce multiphoton excitation.

The main challenge in evaluating the nonlinear response characterized by the transition tensors within Eq. (10.24) now lies with implementing the required sum over intermediate states. As mentioned earlier, it is fully justifiable to consider only states in which the majority of the optical transitions occur, *i.e.* to employ a two-state model. Limiting the intermediate states of Eq. (10.25) to just $|0\rangle$ and $|\alpha\rangle$ restricts the transition sequences from excited to ground states that progress through r and s. Applying the prescription described in Sect. 10.2 to the six terms of

Eq. (10.25) produces a two-level hyperpolarizability tensor that is generally expressible as a sum of twelve separate contributions. Further simplification ensues because a number of these terms, for which $r = 0$ and/or $s = \alpha$, are precluded by the conditions of perturbation theory, namely the exclusion of virtual states that equate to the initial or final state. The two-state form of $\chi_{ijk}^{0\alpha}(\omega'; -\omega, \omega)$ thus re-emerges as: [55]

$$\chi_{ijk}^{0\alpha}\left(\omega'; -\omega, \omega\right) = \frac{2}{\hbar^2} \frac{\mu_j^{0\alpha}\mu_k^{0\alpha}\mu_i^{0\alpha}}{\left(\omega^2 - \omega'^2\right)} + \frac{\mu_j^{0\alpha} d_i d_k}{\hbar^2 \omega\omega'} - \frac{\mu_i^{0\alpha} d_j d_k}{\hbar^2 \omega\omega'} . \tag{10.26}$$

It may be observed that the second and third terms on the right in Eq. (10.26) exhibit an antisymmetry with respect to interchange of the indices *i* and *j*. However, in the physical observable delivered by Eq. (10.24), this tensor is index-contracted with a *i,j*-symmetric product of polarization vectors. Consequently, since only the *i,j*-symmetric part of Eq. (10.26) can contribute to the fluorescence signal, it is expedient to replace $\chi_{ijk}^{0\alpha}(\omega'; -\omega, \omega)$, without further approximation, by an index-symmetrised form, $\chi_{(ij)k}^{0\alpha}(\omega'; -\omega, \omega)$ that is defined as follows:

$$\chi_{(ij)k}^{0\alpha}\left(\omega'; -\omega, \omega\right) \equiv \frac{1}{2}\left(\chi_{ijk}^{0\alpha}\left(\omega'; -\omega, \omega\right) + \chi_{jik}^{0\alpha}\left(\omega'; -\omega, \omega\right)\right) = \frac{2}{\hbar^2} \frac{\mu_i^{0\alpha}\mu_j^{0\alpha}\mu_k^{0\alpha}}{\left(\omega^2 - \omega'^2\right)} . \tag{10.27}$$

It is notable that the expression on the right is, in fact, fully index-symmetric, therefore the two-level model delivers a result that is consistent with the adoption of Kleinman symmetry – even though the latter condition (a simplification that is often effected in other realms of nonlinear optics) has not been artificially imposed. Furthermore, there is a significant physical consequence; it emerges that the mechanism for the laser-controlled emission depends only on the transition dipole, and not on the static moments.

10.4.2 Effects of the Probe Beam on the Fluorescence Anisotropy

As discussed earlier, a great deal of information that is highly relevant to speciation and structure determination can be derived from fluorescence anisotropy. Specifically, the anisotropy parameters normally signify the degree to which fluorescence retains a directionality of polarization from the initial excitation – see for example chapter 7 of the classic text by Valeur [22]. The associated experimental measurements can also inform on excited state photophysical processes such as internal conversion, hindered rotation, rotational diffusion, intramolecular energy transfer etc. Each of these processes represents one means by which the fluorescent emission can exhibit properties quite different from the preceding absorption – quite apart from the Stokes shift in wavelength that is normally apparent. The former processes all provide situations in which the emission dipole moment need not be parallel to the absorption moment. To accommodate such features in the present theory, the initial absorption must again be incorporated into our analysis. Since the probe beam is only delivered to the system after the initial excitation, we have:

$$\left\langle I'\left(\Omega'\right)\right\rangle \sim \left\langle \left|M_{\mathrm{abs}}^{(1)}\right|^2 \left|M_{\mathrm{flu}}^{(1)} + M_{\mathrm{flu}}^{(3)}\right|^2 \right\rangle, \tag{10.28}$$

an expression established under identical conditions, regarding the separation of excitation and emission processes, as was employed in the derivation of the fluorescent signal presented earlier as Eq. (10.7). The anisotropy is now determined from the general expression $r' = \left(\left\langle I'_{\parallel}\right\rangle - \left\langle I'_{\perp}\right\rangle\right) / \left(\left\langle I'_{\parallel}\right\rangle + 2\left\langle I'_{\perp}\right\rangle\right)$, where $\left\langle I'_{\parallel}\right\rangle$ and $\left\langle I'_{\perp}\right\rangle$ are the components of fluorescence intensity polarized parallel and perpendicular, respectively, to the electric polarization vector of the initial excitation beam – the initial absorption is proportional to $\mathbf{e}_0 \cdot \boldsymbol{\mu}^{\nu 0}$, where $\mathbf{e}_0$ represents the input polarization vector aligned by definition to the z-direction.

For initial purposes it is assumed, as will often be the case, that the third contribution to the fluorescence signal in Eq. (10.24) is negligibly small. Duly considering the first two terms, the rotationally averaged fluorescence output from a two-level molecular system is determined and can be expressed as follows, explicitly cast in terms of the three distinct angles between each pair of polarization vectors, for the incident, off-resonant probe and emitted light: $\theta = \cos^{-1}(\mathbf{e}_0 \cdot \mathbf{e})$, $\varphi = \cos^{-1}\left(\mathbf{e} \cdot \mathbf{e}'\right)$ and $\phi = \cos^{-1}\left(\mathbf{e}_0 \cdot \mathbf{e}'\right)$:

$$\begin{aligned}\left\langle I'\left(\Omega'\right)\right\rangle = \; & K^{(1)}\big[T_{ii}\overline{T}_{jj}(3\cos^2\phi - 1) + T_{ij}\overline{T}_{ij}(-\cos^2\phi + 2) \\ & + \frac{I}{7c\varepsilon_0}\big(T_{i(ijj)}\overline{T}_{kk}(6\cos\theta\cos\varphi\cos\phi - 2\cos^2\theta - 2\cos^2\varphi + 5\cos^2\phi - 1) \\ & + T_{i(ijk)}\overline{T}_{jk}(6\cos\theta\cos\varphi\cos\phi + 5\cos^2\theta - 2\cos^2\varphi - 2\cos^2\phi - 1) \\ & + T_{i(jjk)}\overline{T}_{ik}(-4\cos\theta\cos\varphi\cos\phi - \cos^2\theta + 6\cos^2\varphi - \cos^2\phi + 3)\big)\big].\end{aligned} \tag{10.29}$$

In this expression, the first two terms signifying the expected response have a form identical to Eq. (10.20) under the condition that $T_{\lambda\lambda}\overline{T}_{\mu\mu} = T_{\lambda\mu}\overline{T}_{\mu\lambda}$, which is always true under the standard assumption that the inherent electric dipole moments are real.

The higher order contributions in the above Eq. (10.29) represent the lead corrections produced by the probe laser. The expression continues the established shorthand notation that represents transition moment products in terms of second- and fourth-rank molecular tensors, within each of which the first index is associated with the initial molecular excitation. In deriving specific results for independent polarization components, further simplification can be achieved by writing each of the above molecular tensors explicitly in terms of components of the two transition dipole moments, the photo-selected $\boldsymbol{\mu}^{\upsilon 0}$ and the emission $\boldsymbol{\mu}^{0\alpha}$. Following the introduction of β as the angle between these two dipoles, the fluorescence is readily resolved for polarizations $\mathbf{e}'$ in the z- and x-directions, respectively. The results are given as follows for $\phi = 0$, $\varphi = \pi/2$, $\theta = \pi/2$;

$$\left\langle I'_{\parallel}\left(\Omega'\right)\right\rangle = K^{(1)}\left|\boldsymbol{\mu}^{0\alpha}\right|^2\left|\boldsymbol{\mu}^{\upsilon 0}\right|^2\left[2\cos^2\beta + 1 + \frac{I\left|\boldsymbol{\mu}^{0\alpha}\right|^2\left(\cos^2\beta + 2\right)}{7\varepsilon_0\hbar^2 c\left(\omega^2 - \omega'^2\right)}\right], \qquad (10.30)$$

and for $\phi = \pi/2$, $\varphi = 0$, $\theta = \pi/2$:

$$\left\langle I'_{\perp}\left(\Omega'\right)\right\rangle = K^{(1)}\left|\boldsymbol{\mu}^{0\alpha}\right|^2\left|\boldsymbol{\mu}^{\upsilon 0}\right|^2\left[2 - \cos^2\beta + \frac{3I\left|\boldsymbol{\mu}^{0\alpha}\right|^2\left(3 - 2\cos^2\beta\right)}{7\varepsilon_0\hbar^2 c\left(\omega^2 - \omega'^2\right)}\right]. \qquad (10.31)$$

Hence, upon substitution of Eqs. (10.30) and (10.31) into the general anisotropy expression, it is found that:

$$r' = \frac{3\cos^2\beta - 1 + KI\left|\boldsymbol{\mu}^{0\alpha}\right|^2\left(\cos^2\beta - 1\right)}{5 + KI\left|\boldsymbol{\mu}^{0\alpha}\right|^2\left(20 - 11\cos^2\beta\right)/7}, \qquad (10.32)$$

where $K = \left(\varepsilon_0\hbar^2 c\left(\omega^2 - \omega'^2\right)\right)^{-1}$. In the limiting case $I = 0$ the well-known expression [22] $r' = (1/5)(3\cos^2\beta - 1)$ for conventional fluorescence is recovered. Generally, however, a change in fluorescence anisotropy will be apparent due to the interaction with the probe beam – even though the radiation state of the probe laser beam is unaltered.

10.4.3 *Configuration for higher order effects*

Up until now, the nonlinear contribution to the fluorescence output (the I^2 quadratic term) has not been considered in detail. Nevertheless, there are circumstances in which such a term alone provides the fluorescence response, *i.e.* when the first and second terms of Eq. (10.24) are null. Addressing such a case requires progression beyond the two-level approximation, so that higher energy levels are accommodated. Consider, for instance, a system where (following optical excitation) the electronic population is efficiently transferred to a state $|\alpha\rangle$ that would normally decay non-radiatively, if transitions from $|\alpha\rangle$ to $|0\rangle$ are weak or entirely precluded – as, for example, through inherent geometric or symmetry constraints. In such a situation, terms that feature the transition dipole $\boldsymbol{\mu}^{0\alpha}$ in Eq. (10.24) do not contribute: radiative emission only occurs in a response to the off-resonant throughput beam, in a three-photon allowed transition.

Such a configuration may provide the basis for an all-optical molecular switch, whose operation would be as follows: (i) a molecule is indirectly excited to a 'dark' state (*i.e.* one whose direct dipolar excitation from the ground state is forbidden); (ii) precluded by the one-photon dipole selection rules, fluorescence from the 'dark' state occurs on application of the probe beam only; (iii) this activation of the emission occurs for molecular transitions that are three – photon allowed, but single-photon forbidden – examples are afforded by excited states of A_2 symmetry, in molecules of C_{2v} or C_{3v} symmetry, or states of A_u symmetry in D_{2h} species. In summary, switching action occurs since the throughput and absence of the input laser results in activation and deactivation of the fluorescence, respectively. Practically, it will be necessary for the radiation to be delivered in a pulse whose duration and delay, both with respect to the initial excitation, are sufficiently short that it can engage with the system before there is significant non-radiative dissipation of the excited state. For fluorescence output of this type, an eighth-rank rotational average will lead to the following expression:

$$\begin{aligned}\langle I'(\Omega')\rangle = \left(\frac{I^2}{84c^2\varepsilon_0{}^2}\right)K^{(1)}\big[&3T_{i(ijj)}\overline{T}_{k(kll)}(3\cos^2\theta\cos^2\varphi - \cos^2\theta - \cos^2\varphi + \cos^2\phi)\\ &+6T_{i(ijk)}\overline{T}_{j(kll)}(6\cos\theta\cos\varphi\cos\phi - \cos^2\theta - 2\cos^2\varphi - 2\cos^2\phi + 1)\\ &+3T_{i(ijk)}\overline{T}_{l(jkl)}(\cos^2\theta\cos^2\varphi - 4\cos\theta\cos\varphi\cos\phi + 5\cos^2\theta + \cos^2\varphi + 4\cos^2\phi - 3)\\ &+3T_{i(jjk)}\overline{T}_{i(kll)}(-\cos^2\theta\cos^2\varphi - 4\cos\theta\cos\varphi\cos\phi + \cos^2\theta + 5\cos^2\varphi + \cos^2\phi - 1)\\ &+T_{i(jkl)}\overline{T}_{i(jkl)}(-\cos^2\theta\cos^2\varphi + 4\cos\theta\cos\varphi\cos\phi - 5\cos^2\theta + \cos^2\varphi - 4\cos^2\phi + 7)\big].\end{aligned} \tag{10.33}$$

Here, the **T** tensors accommodate summation over products of transition dipole moments that specifically exclude $\boldsymbol{\mu}^{0\alpha}$, since decay transitions are symmetry-forbidden under one-photon selection rules. However, for simplicity, we retain the assumption of Kleinman index symmetry in the embedded $\boldsymbol{\chi}$ tensor (corresponding to the last three indices in each **T**).

For completeness, one may determine an entirely general result for the laser-modified fluorescence anisotropy. Such an expression accommodates all the terms of Eq. (10.24), including the higher-order contributions that are usually negligible (outside of the mentioned model), and is given by:

$$r' = \frac{3\cos^2\beta - 1 + KI|\boldsymbol{\mu}^{0\alpha}|^2(\cos^2\beta - 1) + \left(K^2I^2|\boldsymbol{\mu}^{0\alpha}|^4/21\right)(15\cos^2\beta - 17)}{5 + \left(KI|\boldsymbol{\mu}^{0\alpha}|^2/7\right)(20 - 11\cos^2\beta) + \left(K^2I^2|\boldsymbol{\mu}^{0\alpha}|^4/21\right)(43 - 30\cos^2\beta)}. \tag{10.34}$$

In cases where the absorption and emission transition moments are parallel or anti-parallel, we may then secure the simpler result:

$$r' = \frac{42 - 2K^2I^2|\boldsymbol{\mu}^{0\alpha}|^4}{105 + 27KI|\boldsymbol{\mu}^{0\alpha}|^2 + 13K^2I^2|\boldsymbol{\mu}^{0\alpha}|^4}. \tag{10.35}$$

With increasing intensity of the probe beam, the first departures from the laser-free result, $r' = 0.4$, can be anticipated in the linear-response regime. In fact, it is evident from Taylor series expansions of both Eqs. (10.34) and (10.35) that a plot of the measured anisotropy against I will exhibit a monotonic reduction taking the form $r' \approx \frac{1}{5}(3\cos^2\beta - 1)(1 - \eta I)$, whose constant of proportionality η can be interpreted in terms of the transition moments. Beyond the proposed model to optically switch 'on' and 'off' fluorescent emission using an off-resonance probe beam, the capacity to engage with and to optically control fluorescence also offers significant new grounds for the interrogation of fluorescent materials.

10.5 Multi-emitter Fluorescence

Until this point, fluorescence from an isolated single-molecule source has been considered. Whilst there are some imaging applications of fluorescence that resolve the emission of individual emitters, most fluorescence studies detect signals created through the uncorrelated emission from numerous fluorophores. Of course, optical processes can be appreciably modified by the presence of neighboring material – for example secondary co-doped chromophores – if they are in close proximity [72–75]. It is also known that plasmonic interactions of molecules coupled with a metallic nanoantenna may alter their rate of fluorescent emission [76–83]. However, in most fluorescence studies the net response from a system of emitters can be assumed to be representative of the mean signal from each component.

There are, nonetheless, conditions where such implicit reliance on the ergodic theorem fails at the quantum level, namely when two or more active fluorophores are cooperatively involved in each individual photon emission. To account for the effects of neighboring *molecules*, the focus of this section is the correlated fluorescence from molecular sources that experience mutual interactions. In contrast to the fluorescence phenomena considered in the previous sections, it is most appropriate for the emitters to now be considered immobile and locally correlated in position, as for example would be the case for fluorophores that are surface-adsorbed – or indeed surface-functionalized. Clearly, results will be dependent on the displacement of the detector from the emitter pair, the relative dipolar orientation of the detector with respect to the pair of emitters, and the coupling parameters of the nanoemitters and detector.

It transpires that novel electrodynamic mechanisms can operate between such electromagnetically coupled sources of emission. To begin, theory is developed for fluorescent emission from a pair of nanoemitters, and the effect of a phase difference between the excited state wavefunctions of the pair is considered. This possibility, as will emerge, proves to be of substantial importance when the number of nanoemitters is greater than two. More specialized systems, involving three or more nanoemitters, then afford a basis for generating optical vortex radiation, *i.e.* a form of 'twisted' light that involves an azimuthal progression of phase around a singularity.

10.5.1 Emission from a Nanoemitter Pair

Consider a system in which two molecular nanoemitters A and B in close proximity are so placed that significant electromagnetic coupling occurs between them. There are three distinct types of fluorescence that can occur, in which this coupling can manifest features that differ from the fluorescence of either individual component: (a) single-photon emission from an individual excited molecule whose states are influenced by electromagnetic coupling with its neighbor; (b) single-photon excitonic emission from the pair, in which a single initial electronic excitation is delocalized across both molecules; (c) correlated two-photon emission from the pair, in which both emitters are initially electronically excited [84]. To fulfil the initial conditions for the latter, third case is experimentally more demanding, and the phenomenon also presents less novelty; accordingly, we here focus on the first two scenarios.

First, for case (a) we consider the effects of fluorescent emission of a photon of energy $\hbar\omega = \hbar ck$, from a molecule labelled A under the influence of coupling with a neighbor B. To correctly account for features in the near-field, where some of the most distinctive features can be expected to arise, it will be helpful to begin by considering the full system to comprise three elements, one of which is a light detector, D (later, explicit reference to such a device is removed for simplicity). The matrix element for this system is represented by:

$$M^A = M^{DA} + M^{DAB} + M^{DBA} + M^{BDA}. \tag{10.36}$$

Here, the excitation is localized on molecule A prior to emission, so that the leading term represents direct coupling (photon propagation) between A and the detector, independent of the second emitter. This case is more concisely expressed as:

$$M^{DA} = \mu_i^{\nu 0}(D) V_{ij}(k, \mathbf{R}_{DA}) \mu_j^{0\alpha}(A), \tag{10.37}$$

where $\mathbf{R}_{DA}$ is the displacement between D and A and V_{ij} represents a second-rank intermolecular coupling tensor, acting between the electronically excited emitter and ground state detector in this instance, which is written as:

$$V_{ij}(k,\mathbf{R}_{DA}) = \frac{\exp(ikR)}{4\pi\varepsilon_0 R^3}\left[(1-ikR)\left(\delta_{ij}-3\hat{R}_i\hat{R}_j\right)-k^2R^2\left(\delta_{ij}-\hat{R}_i\hat{R}_j\right)\right]. \quad (10.38)$$

Written in this form, the matrix element accommodates both near- and far-field limits as asymptotes $kR_{DA} \ll 1$ and $kR_{DA} \gg 1$, respectively. The remaining higher-order (B-dependent) terms in Eq. (10.36) correspond to different combinations of the interacting nanoemitters and the detector, and relate to the following matrix element contributions:

$$\begin{aligned} M^{DAB} &= \mu_i^{\nu 0}(D)V_{ij}(k,\mathbf{R}_{DA})\alpha_{jk}^{0\alpha}(A,-k;0)V_{kl}(0,\mathbf{R}_{AB})\mu_l^{00}(B),\\ M^{DBA} &= \mu_i^{\nu 0}(D)V_{ij}(k,\mathbf{R}_{DB})\alpha_{(jk)}^{00}(B,-k;k)V_{kl}(k,\mathbf{R}_{AB})\mu_l^{0\alpha}(A),\\ M^{BDA} &= \mu_i^{00}(B)V_{ij}(0,\mathbf{R}_{DB})\alpha_{jk}^{\nu 0}(D,k;0)V_{kl}(k,\mathbf{R}_{DA})\mu_l^{0\alpha}(A), \end{aligned} \quad (10.39)$$

which feature both the index-symmetric molecular polarizability $\alpha_{(jk)}^{00}$ and non-index symmetric forms of the two-photon tensor defined by Eq. (10.12). In principle, all four terms of Eqs. (10.37) and (10.39) contribute to M^A, as each connects the same initial and final system states. Since it is the modulus square of M^A that relates to the observable rate of fluorescence (using Fermi's rule), the result will clearly include cross-terms signifying quantum interference between the various contributions. The lead term for the emission rate is the modulus square of Eq. (10.37), and the interferences between M^{DA} and each of the three higher-order amplitudes (M^{DAB}, M^{DBA} or M^{BDA}) will represent significant corrections. The exact nature of the leading correction term is primarily determined by considering the relative positions of the three components, although selection rules and molecular properties also need to be considered since, for example, the appearance of the static electric dipole $\boldsymbol{\mu}^{00}$ in terms M^{DAB} and M^{BDA} means that B cannot be non-polar if they are to contribute.

In case (*b*), significant additional effects may arise as result of delocalization of a single initial excitation across the pair, when both nanoemitters are identical. Such circumstances involve the formation of an exciton, and the corresponding quantum state (which follows the initial excitation) is described by a superposition of two localized-excitation states – in either a symmetric $|i^+\rangle$ or an antisymmetric $|i^-\rangle$ combination, *i.e.*:

$$|i^{\pm}\rangle = 2^{-1/2}|D_0\rangle\left(|A_\alpha\rangle|B_0\rangle \pm |A_0\rangle|B_\alpha\rangle\right), \quad (10.40)$$

where the subscripts label the excited or unexcited state of each participant chromophore. The nanoemitter pair may undergo fluorescent emission from either of the two excitonic states, and the latter corresponds to the following matrix elements;

$$M_{\pm} = 2^{-1/2}\left(M^A \pm M^B\right), \quad (10.41)$$

in which the superscripts designate the effective position of the localized excitation, and the difference in signs is equivalent to introducing a π-phase difference between the two emitters. In situations where molecular coupling is small, the result may be reduced to $|M^A|^2 + |M^B|^2$ which corresponds to independent (non-interfering) emitters. Otherwise, the fluorescence signals will relate to symmetric, $|M_+|^2$, or anti-symmetric, $|M_-|^2$, excitonic emission.

Cast in terms of the electric field created from the strongly coupled emitter pair, achieved by excluding $\mu_i^{\nu 0}(D)$ from the relevant matrix elements (*i.e.* either the symmetric or anti-symmetric case), the distribution in optical phase is found from the argument of the electric field – more detail is given later. Figure 10.4 show plots of the fluorescence emitted from a pair of coupled molecules in symmetric and anti-symmetric configurations, with the colors representing the optical phase distribution. For the purposes of the graphs, the transition dipole moments of A and B are perpendicular to the exhibited plane. On comparing the two types of pairwise excitonic emission, it is apparent that a nodal plane (corresponding to no fluorescence signal) appears for antisymmetric situations when the detector is equidistant from both A and B, which is visibly distinct from symmetric emission. In any case, such contour maps reveal striking departures from the known character of single-center emission [85].

10.5.2 Multi-emitter Systems: Generators of Vortex Light

Given a larger number of identical nanoemitters it is possible, by satisfying certain phase and symmetry constraints, to produce fluorescence whose phase distribution twists around an axis of phase singularity; this is the phenomenon known as an 'optical vortex' or 'twisted' beam. Vortex beams, whose existence was first entertained in a series of works [86–89], are characterized by a helical wavefront, based on the azimuthal progression of phase around a singular axis. The production of such beams, which is now experimentally routine, has proven that they convey orbital angular momentum (OAM) – an attribute that is separate from the more familiar spin angular momentum associated with circular polarizations [90]. Ongoing advances have allowed the quantum nature of such beams to be fully elicited [91]. The structure is primarily dependent on the topological charge, l (signifying an OAM of $l\hbar$ per photon) an integer that can be either positive or negative – denoting left- or right-handed gyration, respectively, of the light. A beam with a topological charge l has a field distribution in the form of l intertwined helices, each completing a turn of 2π radians about the axis over a span of l wavelengths [92–94]. Whereas the production of vortex light usually involves imparting OAM onto a laser beam with a more common mode structure [95–102], it has only recently emerged that

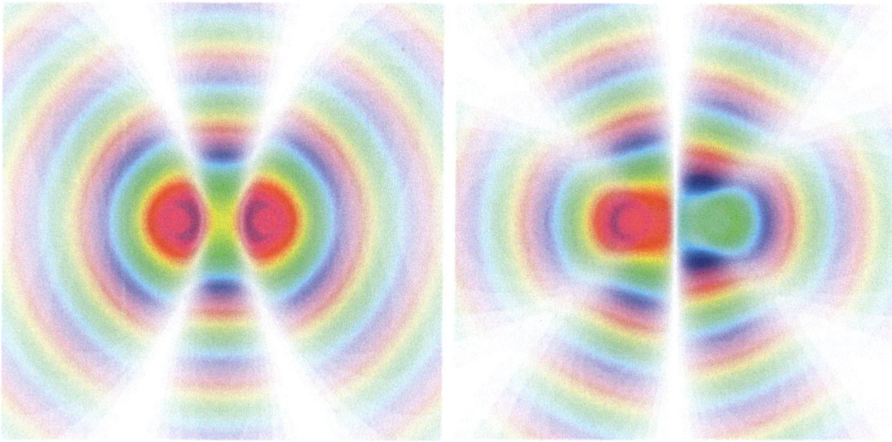

Fig. 10.4 Plots of the fluorescence measured at the detector, with colors representation of the optical phase distribution. Brightness and color hue relate to the modulus and complex argument of the fields emitted by symmetric (*left*) and antisymmetric (*right*) excitons

such light can indeed be directly produced in the fluorescent decay of a set of coupled nanoemitters, as illustrated in Fig. 10.5 [103–105].

For the sought effect to occur, the electromagnetically coupled nanoemitters once again have to sustain an exciton, delocalized across the array. The relative orientations of the molecular dipole moments are crucially important for producing vortex light. It transpires that the array must belong to one of the Schoenflies point groups C_n, C_{nh}, S_n, T, T_h, where n is the number of nanoemitters; the example shown in Fig. 10.5 has C_7 symmetry. Table 10.1 provides the possible integer values of OAM for vortex light emitted from a nanoarray belonging to one of the C_n and C_{nh} point groups [105].

In the decay transitions that accompany the relaxation of the excitonic states – assumed to terminate in a totally symmetric ground electronic state – the symmetry character of the initial exciton maps directly onto the vortex structure of the emitted light. It is therefore necessary to determine the form of the requisite excitons. A block diagonalized form of the array Hamiltonian is required, with the single delocalized excitonic state as a basis. In general, the matrix form of the array Hamiltonian is expressible as follows:

$$H_{rs} = E_u\delta_{rs} + (\boldsymbol{\mu}_r.\mathbf{V}(\mathbf{R}_{rs}).\boldsymbol{\mu}_s)\left(\delta_{r-1,s(\text{mod}n)} + \delta_{r(\text{mod}n),s-1}\right). \tag{10.42}$$

where each element of the n-square matrix relates to a pair of emitters $\{r, s\}$. Each diagonal element $E_u = \hbar ck_u$ signifies the energy of an isolated nanoemitter in its excited electronic level u. The off-diagonal elements, denoting pairwise interactions, involve the electrodynamic coupling $\mathbf{V}(\mathbf{R}_{rs})$ between neighboring transition dipole moments $\boldsymbol{\mu}_r$ and $\boldsymbol{\mu}_s$ – and all are identical, *i.e.*:

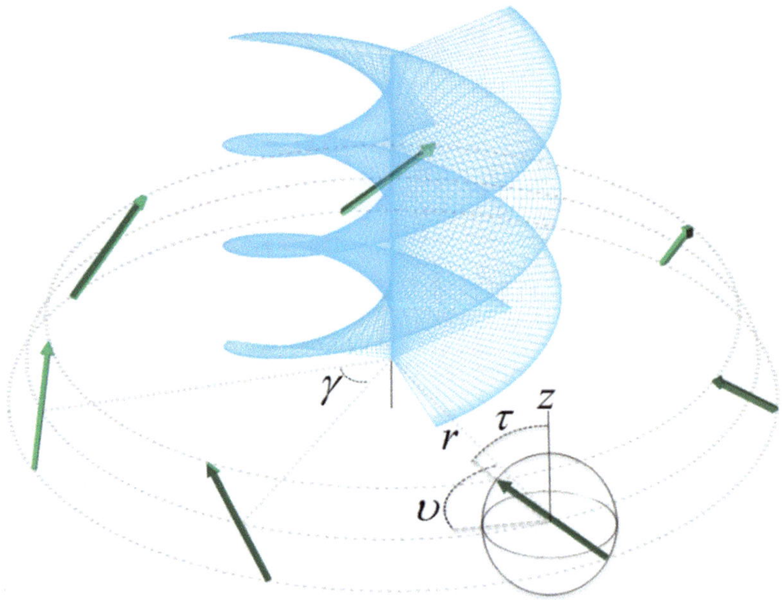

Fig. 10.5 Schematic depiction for a ring of seven molecular nanoemitters, positioned such that their transition moments form a ring whose normal lies in the z-direction. Emission from the lowest energy exciton state releases a photon with topological charge $l = 3$ along the normal axis, with a helicoid evolvement of each wavefront component signifying a surface of constant phase. The angle γ designates the azimuthal position in the plane; τ and υ are angles that specify the local orientation of each emitter. Dotted lines are guides for the eye

Table 10.1 Summary of permitted topological charges, l values, for OAM outputs from nanoarrays with the required symmetry

	Symmetry group	
Number of emitters	C_n, C_{nh}	S_n
3	1	–
4	1	1
5	1,2	–
6	1,2	1
7	1,2,3	–
8	1,2,3	1,2,3
9	1,2,3,4	–
10	1,2,3,4	1,2

$$\mathbf{V}(\mathbf{R}_{rs}) \equiv V_{r,(r+1)\mathrm{mod}n}\left(k_u, \mathbf{R}_{r,(r+1)\mathrm{mod}n}\right). \tag{10.43}$$

Here, $\mathbf{R}_{rs}$ is the vector displacement between the relevant nanoemitters: R is defined by $\mathbf{R}_r - \mathbf{R}_s \equiv \mathbf{R}_{rs} = R\hat{\mathbf{R}}_{rs}$. Under the described symmetry conditions, all the non-zero off-diagonal elements of Eq. (10.43) return the same scalar value, U, given by;

$$\begin{aligned} U &= \boldsymbol{\mu}_r.\mathbf{V}(\mathbf{R}_{rs}).\boldsymbol{\mu}_s \\ &\equiv \frac{e^{\mathrm{i}k_uR}}{4\pi\varepsilon_0R^3}\Big[\Big\{1 - \mathrm{i}k_uR - (k_uR)^2\Big\}\left(\boldsymbol{\mu}_1^{0u}\cdot\boldsymbol{\mu}_2^{0u}\right) \\ &-\Big\{3 - 3\mathrm{i}k_uR - (k_uR)^2\Big\}\Big\{\left(\boldsymbol{\mu}_1^{0u}\cdot\hat{\mathbf{R}}_{12}\right)\left(\boldsymbol{\mu}_2^{0u}\cdot\hat{\mathbf{R}}_{12}\right)\Big\}\Big]. \end{aligned} \tag{10.44}$$

for any specific pair arbitrarily labelled 1 and 2. The eigenfunctions now emerge as normalized superpositions of the basis states:

$$|\psi_p\rangle = \frac{1}{\sqrt{n}}\sum_{r=1}^{n}\varepsilon_n^{(r-1)p}|\xi^{r;u}\rangle\prod_{s\neq r}|\xi^{s;0}\rangle,\ p\in\{1,\ldots,n\}. \tag{10.45}$$

Here, $|\xi^{r;u}\rangle$ is a state function corresponding to an emitter r in electronic state u, and $\varepsilon_n = \exp(2\pi i/n)$. In every summand of Eq. (10.45), one molecule is in the electronically excited state u, while the others are in their ground states. The energy eigenvalues associated with the above exciton states are generally expressible in the form;

$$E_p = E_u + 2U\cos(2pq/n), \tag{10.46}$$

with the permissible range of values for the index q as indicated in Table 10.2 which also indicates, for the first few point groups C_n, the irreducible representation associated with each excitonic state.

Due to their differences in symmetry, the various exciton eigenstates with representations E_q, A (and also B, if present) will also exhibit differences in energy, manifest as line splittings centered upon the frequency of an isolated emitter. For example in a nanoarray with $n = 3$ emitters, the positioning of exciton levels leads to one non-degenerate state (belonging to the totally symmetric representation A), of energy $E_u - 2U$, and two doubly degenerate (E representation) states of energy $E_u + U$. Furthermore, the sign of the coupling U is readily shown to be positive for all $n \geq 3$, producing splitting patterns as exemplified in Fig. 10.6. Attention now focuses on exciton E states belonging to doubly degenerate irreducible representations. These excitons exist in the form of pairs, one with a left-handed and the other a right-handed progression of phase around the ring. These states exactly correspond with the sought distributions of phase about the symmetry axis, and are separated in energy from the A form – a principle that should enable the selective excitation of one symmetry type. The decay of each doubly degenerate exciton can therefore release a photon with a characteristic long wavelength; in most cases, the exciton pair with the lowest energy will be associated with the highest values of $|q|$ and, as it emerges, the largest topological charge for the emitted vortex.

Finally, to map the optical phase of the fluorescence, an expression for the electric field $\mathbf{E}_p$ $(\mathbf{R}_D)$ from each constituent nanoemitter is required, *i.e.*:

Table 10.2 List of the irreducible representations (irreps) of the C_n exciton states for $n = 3$–7

n	p	1	2	3	4	5	6	7
3	q	1	−1	0				
	Irrep	E_1	E_1	A				
4	q	1	2	−1	0			
	Irrep	E_1	B	E_1	A			
5	q	1	2	−2	−1	0		
	Irrep	E_1	E_2	E_2	E_1	A		
6	q	1	2	3	−2	−1	0	
	Irrep	E_1	E_2	B	E_2	E_1	A	
7	q	1	2	3	−3	−2	−1	0
	Irrep	E_1	E_2	E_3	E_3	E_2	E_1	A

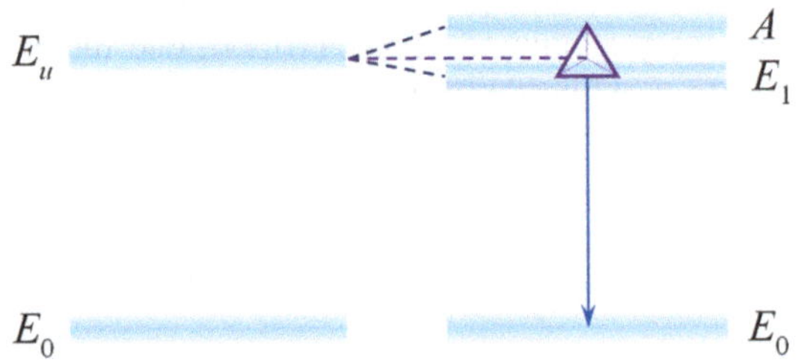

Fig. 10.6 Excitonic irreducible representations and Davydov energy level splitting for an array of C_3 point group symmetry. The splitting between A and E_1 equal to $3U$. The three-emitter case, as indicated by the relaxation arrow, will have a preferential $|q| = 1$ emission

$$\mathbf{E}_p(\mathbf{R}_D) = \sum_r^n \frac{e^{ikR_{Dr}}\varepsilon_n^{(r-1)p}}{4\pi\varepsilon_0 R_{Dr}^3}\left\{\left[\left(\hat{\mathbf{R}}_{Dr} \times \boldsymbol{\mu}_r^{0u}\right) \times \hat{\mathbf{R}}_{Dr}\right]k^2R_{Dr}^2 + \left[3\hat{\mathbf{R}}_{Dr}\left(\hat{\mathbf{R}}_{Dr} \cdot \boldsymbol{\mu}_r^{0u}\right) - \boldsymbol{\mu}_r^{0u}\right](1 - ikR_{Dr})\right\}, \tag{10.47}$$

where $\mathbf{R}_D$ signifies the displacement, relative to the ring center, of a point of measurement or detection. Notably, each term in Eq. (10.47) carries the phase factor $\varepsilon_n^{(r-1)p}$, from the corresponding emitter component in Eq. (10.45), thus delivering the sought progression in phase around the ring. At any point in space, the most appropriate measure of the phase for the emitted radiation is the function defined by:

$$\theta_i(\mathbf{R}) = \arg\left\{\mathbf{E}_{p;\,i}(\mathbf{R}_D)\right\}, \tag{10.48}$$

the principle argument of the complex electric field vector. Typical maps of the electromagnetic phase distributions, shown in Fig. 10.7, exhibit the variation of the phase in planes parallel to the source array, for several combinations of molecular number and exciton symmetry. The panes in this figure show rings with three and six chromophore components, the latter having two topological charges. The phase properties of the electromagnetic fields emitted by the arrays in each case map

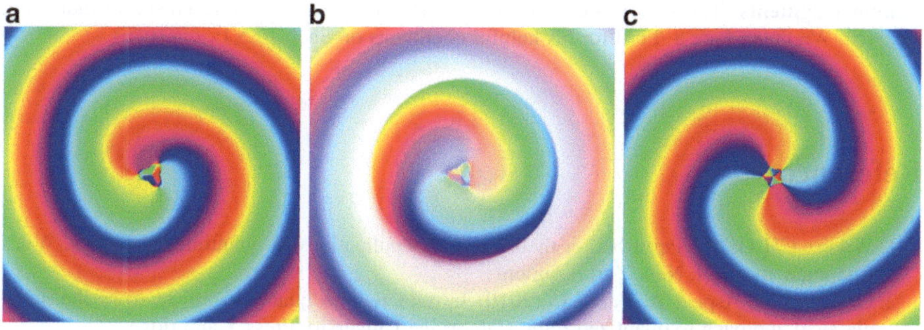

Fig. 10.7 Cross-sectional simulations of the scalar optical field in the array plane supporting optical vortex emission: (**a**) three chromophores, emission with topological charge $l = 1$; (**b**) also three chromophores $l = -1$, here also showing (by color intensity) the field magnitude; (**c**) five chromophores supporting $l = -2$ emission

exactly to the azimuthal phase dependence of a vortex mode. Such arrays thus represents systems whose fluorescence can naturally generate optical vorticity, without requiring any of the transformation optics that might otherwise need to be deployed [106–109].

10.6 Discussion

This chapter has endeavored to show that, by developing a fully photon-based representation of conventional molecular fluorescence, a variety of other closely related phenomena can readily be identified in the same quantum field framework of theory. The most familiar variant, multiphoton fluorescence, is sufficiently well established that it has already found advanced applications in optical diagnostics in biology and medicine. Laser-controlled and multi-emitter fluorescence represent more recent arrivals on the scene, and in these cases experimental work has not yet quite caught up with the developments in theory. Each topic holds promise for a range of new kinds of application. For example, the capacity to modify fluorescence with an auxiliary beam opens new ground for excited state lifetime manipulation, all-optical switching and the potential production of an all-optical transistor [54]. Equally, with advances in nanofabrication paving the way for the batch production of tailored nanoemitter arrays, the potential of materials that can directly generate optical vortex light is also becoming enticing, particularly in view of the associated capacity to convey significantly more information per photon [110, 111]. In other areas, too, the science is still advancing apace. For example, building on the well-known theory of circularly polarized luminescence [112], there is now fresh interest in establishing signatures of chirality in the fluorescence produced by chiral molecules [113]. It will be fascinating to see where the next advance will arise.

Acknowledgments The authors are grateful to EPSRC and the Leverhulme Trust for funding our research. We thank Jack S. Ford for his helpful comments on Sect. 10.5.1.

References

1. Wilcox CH (1966) Perturbation theory and its applications in quantum mechanics. Wiley Chapman and Hall, New York/London
2. Grynberg G, Aspect A, Fabre C (2010) Introduction to quantum optics: from the semi-classical approach to quantized light. Cambridge University Press, Cambridge
3. Andrews DL, Bradshaw DS, Coles MM (2011) Perturbation theory and the two-level approximation: a corollary and critique. Chem Phys Lett 503:153–156
4. Kmetic MA, Meath WJ (1985) Permanent dipole moments and multi-photon resonances. Phys Lett A 108:340–343
5. Kondo AE, Meath WJ, Nilar SH, Thakkar AJ (1994) Pump-probe studies of the effects of permanent dipoles in one- and two-colour molecular excitations. Chem Phys 186:375–394
6. Jagatap BN, Meath WJ (1996) On the competition between permanent dipole and virtual state two-photon excitation mechanisms, and two-photon optical excitation pathways, in molecular excitation. Chem Phys Lett 258:293–300
7. Spano FC, Mukamel S (1989) Nonlinear susceptibilities of molecular aggregates: enhancement of X(3) by size. Phys Rev A 40:5783–5801
8. Leegwater JA, Mukamel S (1992) Exciton-scattering mechanism for enhanced nonlinear response of molecular nanostructures. Phys Rev A 46:452–464
9. Mukamel S (1995) Principles of nonlinear optical spectroscopy. Oxford University Press, New York
10. Venkatramani R, Mukamel S (2005) Dephasing-induced vibronic resonances in difference frequency generation spectroscopy. J Phys Chem B 109:8132–8143
11. Zrenner A, Beham E, Stufler S, Findeis F, Bichler M, Abstreiter G (2002) Coherent properties of a two-level system based on a quantum-dot photodiode. Nature 418:612
12. Klimov VI (2003) Nanocrystal quantum dots. Los Alamos Sci 28:214
13. Kirkpatrick SM, Naik RR, Stone MO (2001) Nonlinear saturation and determination of the two-photon absorption cross section of green fluorescent protein. J Phys Chem B 105:2867–2873
14. Stufler S, Ester P, Zrenner A, Bichler M (2005) Quantum optical properties of a single $In_xGa_{1-X}As$ – GaAs quantum dot two-level system. Phys Rev B 72:121301
15. Drobizhev M, Marakov NS, Hughes T, Rebane AJ (2007) Resonance enhancement of two-photon absorption in fluorescent proteins. J Phys Chem B 111:14051–14054
16. Asselberghs I, Flors C, Ferrighi L, Botek E, Champagne B, Mizuno H, Ando R, Miyawaki A, Hofkens J, Van der Auweraer M, Clays K (2008) Second-harmonic generation in Gfp-like proteins. J Am Chem Soc 130:15713–15719
17. Drobizhev M, Tillo S, Makarov NS, Hughes TE, Rebane AJ (2009) Color hues in red fluorescent proteins are due to internal quadratic stark effect. J Phys Chem B 113:12860–12864
18. Beuerman E, Makarov NS, Drobizhev M, Rebane AJ (2010) Justification of two-level approximation for description of two-photon absorption in oxazine dyes. Proc SPIE 7599:75990X
19. Andrews DL, Dávila Romero LC, Meath WJ (1999) An algorithm for the nonlinear optical susceptibilities of dipolar molecules, and an application to third harmonic generation. J Phys B At Mol Opt Phys 32:1–17
20. Juzeliūnas G, Dávila Romero LC, Andrews DL (2003) Eliminating ground-state dipole moments in quantum optics via canonical transformation. Phys Rev A 68:043811

21. Lakowicz JR (1999) Principles of fluorescence spectroscopy, 2nd edn. Kluwer, New York
22. Valeur B, Berberan-Santos MN (2013) Molecular fluorescence: principles and applications, 2nd edn. Wiley-VCH, Weinheim
23. Inoué S, Shimomura O, Goda M, Shribak M, Tran PT (2002) Fluorescence polarization of green fluorescence protein. Proc Natl Acad Sci U S A 99:4272–4277
24. Brasselet S, Le Floc'h V, Treussart F, Roch J-F, Zyss J (2004) In situ diagnostics of the crystalline nature of single organic nanocrystals by nonlinear microscopy. Phys Rev Lett 92:207401
25. Gasecka A, Dieu L-Q, Brühwiler D, Brasselet S (2010) Probing molecular order in zeolite L inclusion compounds using two-photon fluorescence polarimetric microscopy. J Phys Chem B 114:4192–4198
26. Corry B, Jayatilaka D, Martinac B, Rigby P (2006) Determination of the orientational distribution and orientation factor for transfer between membrane-bound fluorophores using a confocal microscope. Biophys J 91:1032–1045
27. Gasecka A, Han T-J, Favard C, Cho BR, Brasselet S (2009) Quantitative imaging of molecular order in lipid membranes using two-photon fluorescence polarimetry. Biophys J 97:2854–2862
28. Empedocles SA, Neuhauser R, Bawendi MG (1999) Three-dimensional orientation measurements of symmetric single chromophores using polarization microscopy. Nature 399:126–130
29. Weiss S (2000) Measuring conformational dynamics of biomolecules by single molecule fluorescence spectroscopy. Nat Struct Biol 7:724–729
30. Weston KD, Goldner LS (2001) Orientation imaging and reorientation dynamics of single dye molecules. J Phys Chem B 105:3453–3462
31. Vacha M, Kotani M (2003) Three-dimensional orientation of single molecules observed by far- and near-field fluorescence microscopy. J Chem Phys 118:5279–5282
32. Festy F, Ameer-Beg SM, Ng T, Suhling K (2007) Imaging proteins in vivo using fluorescence lifetime microscopy. Mol Biosyst 3:381–391
33. Levitt JA, Matthews DR, Ameer-Beg SM, Suhling K (2009) Fluorescence lifetime and polarization-resolved imaging in cell biology. Curr Opin Biotechnol 20:28–36
34. Gradinaru CC, Marushchak DO, Samim M, Krull UJ (2010) Fluorescence anisotropy: from single molecules to live cells. Analyst 135:452
35. Loudon R (2000) The quantum theory of light, 3rd edn. Oxford University Press, Oxford
36. Meath WJ, Power EA (1984) On the importance of permanent moments in multiphoton absorption using perturbation-theory. J Phys B At Mol Opt Phys 17:763–781
37. Andrews DL, Thirunamachandran T (1977) Polarization effects in nonlinear scattering. Opt Commun 22:312–314
38. Andrews DL, Blake NP (1989) Three-dimensional rotational averages in radiation molecule interactions: an irreducible Cartesian tensor formulation. J Phys A Math Gen 22:49–60
39. Friese DH, Beerepoot MTP, Ruud K (2014) Rotational averaging of multiphoton absorption cross sections. J Chem Phys 141:204103
40. McClain WM (1971) Excited state symmetry assignment through polarized two-photon absorption studies of fluids. J Chem Phys 55:2789–2796
41. Monson PR, McClain WM (1972) Complete polarization study of the two-photon absorption of liquid 1-chloronaphthalene. J Chem Phys 56:4817–4825
42. Andrews DL, Ghoul WA (1981) Eighth rank isotropic tensors and rotational averages. J Phys A Math Gen 14:1281–1290
43. Leeder JM, Andrews DL (2011) A molecular theory for two-photon and three-photon fluorescence polarization. J Chem Phys 134:094503
44. Lakowicz JR, Gryczynski I, Malak H, Schrader M, Engelhardt P, Kano H, Hell SW (1997) Time-resolved fluorescence spectroscopy and imaging of DNA labeled with Dapi and Hoechst 33342 using three-photon excitation. Biophys J 72:567–578

45. Kastrup L, Blom H, Eggeling C, Hell SW (2005) Fluorescence fluctuation spectroscopy in subdiffraction focal volumes. Phys Rev Lett 94:178104
46. Hell SW (2007) Far-field optical nanoscopy. Science 316:1153–1158
47. Willig KI, Harke B, Medda R, Hell SW (2007) Sted microscopy with continuous wave beams. Nat Methods 4:915–918
48. Zhou LC, Liu JY, Zhao GJ, Shi Y, Peng XJ, Han KL (2007) The ultrafast dynamics of near-infrared heptamethine cyanine dye in alcoholic and aprotic solvents. Chem Phys 333:179–185
49. Armoogum DA, Marsh RJ, Nicolaou N, Mongin O, Blanchard-Desce M, Bain AJ (2008) Stimulated emission depletion and fluorescence correlation spectroscopy of a branched quadrupolar chromophore. Proc SPIE 7030:70300S
50. Harke B, Keller J, Ullal CK, Westphal V, Schoenle A, Hell SW (2008) Resolution scaling in STED microscopy. Opt Express 16:4154
51. Hell SW (2009) Microscopy and its focal switch. Nat Methods 6:24–32
52. Andrews DL, Bradshaw DS, Jenkins RD, Rodríguez J (2009) Dendrimer light-harvesting: intramolecular electrodynamics and mechanisms. Dalton Trans 10006–10014
53. Bradshaw DS, Andrews DL (2010) All-optical control of molecular fluorescence. Phys Rev A 81:013424
54. Andrews DL, Bradshaw DS (2010) Off-resonant activation of optical emission. Opt Commun 283:4365–4367
55. Leeder JM, Bradshaw DS, Andrews DL (2011) Laser-controlled fluorescence in two-level systems. J Phys Chem B 115:5227–5233
56. Lakowicz JR (2005) Radiative decay engineering 5: metal-enhanced fluorescence and plasmon emission. Anal Biochem 337:171–194
57. Anger P, Bharadwaj P, Novotny L (2006) Enhancement and quenching of single-molecule fluorescence. Phys Rev Lett 96:113002
58. Tam F, Goodrich GP, Johnson BR, Halas NJ (2007) Plasmonic enhancement of molecular fluorescence. Nano Lett 7:496–501
59. Giannini V, Sánchez-Gil JA (2008) Excitation and emission enhancement of single molecule fluorescence through multiple surface-plasmon resonances on metal trimer nanoantennas. Opt Lett 33:899–901
60. Fort E, Grésillon S (2008) Surface enhanced fluorescence. J Phys D Appl Phys 41:013001
61. Lin C-Y, Chiu K-C, Chang C-Y, Chang S-H, Guo T-F, Chen S-J (2010) Surface plasmon-enhanced and quenched two-photon excited fluorescence. Opt Express 18:12807–12817
62. Lu G, Zhang T, Li W, Hou L, Liu J, Gong Q (2011) Single-molecule spontaneous emission in the vicinity of an individual gold nanorod. J Phys Chem C 115:15822–15828
63. Derom S, Berthelot A, Pillonnet A, Benamara O, Jurdyc AM, Girard C, Colas des Francs G (2013) Metal enhanced fluorescence in rare earth doped plasmonic core–shell nanoparticles. Nanotechnology 24:495704
64. Khatua S, Paulo PMR, Yuan H, Gupta A, Zijlstra P, Orrit M (2014) Resonant plasmonic enhancement of single-molecule fluorescence by individual gold nanorods. ACS Nano 8:4440–4449
65. Lu D, Kan JJ, Fullerton EE, Liu Z (2014) Enhancing spontaneous emission rates of molecules using nanopatterned multilayer hyperbolic metamaterials. Nat Nanotechnol 9:48–53
66. Allcock P, Jenkins RD, Andrews DL (2000) Laser-assisted resonance-energy transfer. Phys Rev A 61:023812
67. Andrews DL (2007) Optically switched energy transfer: twin-beam off-resonance control. Phys Rev Lett 99:023812
68. Bradshaw DS, Andrews DL (2008) Optically controlled resonance energy transfer: mechanism and configuration for all-optical switching. J Chem Phys 128:144506
69. Craig DP, Thirunamachandran T (1998) Molecular quantum electrodynamics: an introduction to radiation-molecule interactions. Dover Publications, Mineola

70. Andrews DL, Allcock P (2002) Optical harmonics in molecular systems. Wiley-VCH, Weinheim
71. Kleinman DA, Dielectric N (1962) Polarization in optical media. Phys Rev 126:1977–1979
72. Salam A (2012) Mediation of resonance energy transfer by a third molecule. J Chem Phys 136:014509
73. Andrews DL, Ford JS (2013) Resonance energy transfer: influence of neighboring matter absorbing in the wavelength region of the acceptor. J Chem Phys 139:014107
74. Coles MM, Leeder JM, Andrews DL (2014) Static and dynamic modifications to photon absorption: the effects of surrounding chromophores. Chem Phys Lett 595–596:151–155
75. Leeder JM, Andrews DL (2014) Enhancing optical up-conversion through electrodynamic coupling with ancillary chromophores. J Phys Chem C 118:23535–23544
76. Kühn S, Håkanson U, Rogobete L, Sandoghdar V (2006) Enhancement of single-molecule fluorescence using a gold nanoparticle as an optical nanoantenna. Phys Rev Lett 97:017402
77. Ringler M, Schwemer A, Wunderlich M, Nichtl A, Kürzinger K, Klar TA, Feldmann J (2008) Shaping emission spectra of fluorescent molecules with single plasmonic nanoresonators. Phys Rev Lett 100:203002
78. Bakker RM, Drachev VP, Liu Z, Yuan H-K, Pedersen RH, Boltasseva A, Chen J, Irudayaraj J, Kildishev AV, Shalaev VM (2008) Nanoantenna array-induced fluorescence enhancement and reduced lifetimes. New J Phys 10:125022
79. Vecchi G, Giannini V, Gómez Rivas J (2009) Shaping the fluorescent emission by lattice resonances in plasmonic crystals of nanoantennas. Phys Rev Lett 102:146807
80. Ward DR, Hüser F, Pauly F, Cuevas JC, Natelson D (2010) Optical rectification and field enhancement in a plasmonic nanogap. Nat Nanotechnol 5:732–736
81. Novotny L, van Hulst N (2011) Antennas for light. Nat Photonics 5:83–90
82. Agio M (2012) Optical antennas as nanoscale resonators. Nanoscale 4:692–706
83. Lee KG, Eghlidi H, Chen XW, Renn A, Götzinger S, Sandoghdar V (2012) Spontaneous emission enhancement of a single molecule by a double-sphere nanoantenna across an interface. Opt Express 20:23331–23338
84. Rice EM, Andrews DL (2012) Optical emission of a molecular nanoantenna pair. J Chem Phys 136:244503
85. Rice EM, Bradshaw DS, Saadi K, Andrews DL (2012) Identifying the development in phase and amplitude of dipole and multipole radiation. Eur J Phys 33:345–358
86. Nye J, Berry M (1974) Dislocations in wave trains. Proc Royal Soc A Math Phys Eng Sci 336:165–190
87. Coullet P, Gil L, Rocca F (1989) Optical vortices. Opt Commun 73:403–408
88. Saleh BEA, Teich MC (1991) Fundamentals of photonics. Wiley, New York
89. Allen L, Beijersbergen MW, Spreeuw RJC, Woerdman JP (1992) Orbital angular momentum of light and the transformation of Laguerre-Gaussian laser modes. Phys Rev A 45:8185–8189
90. Andrews DL, Babiker M (2013) The angular momentum of light. Cambridge University Press, Cambridge
91. Dávila Romero LC, Andrews DL, Babiker M (2002) A quantum electrodynamics framework for the nonlinear optics of twisted beams. J Opt B Quantum Semiclassical Opt 4:S66
92. Padgett MJ, Courtial J (1999) Poincaré-sphere equivalent for light beams containing orbital angular momentum. Opt Lett 24:430–432
93. Allen L (2002) Introduction to the atoms and angular momentum of light special issue. J Opt B Quantum Semiclassical Opt 4:S1–S6
94. Milione G, Sztul HI, Nolan DA, Alfano RR (2011) Higher-order Poincaré sphere, Stokes parameters, and the angular momentum of light. Phys Rev Lett 107:053601
95. Heckenberg NR, McDuff R, Smith CP, White AG (1992) Generation of optical-phase singularities by computer-generated holograms. Opt Lett 17:221–223
96. Bazhenov VY, Soskin M, Vasnetsov M (1992) Screw dislocations in light wavefronts. J Mod Opt 39:985–990

97. Beijersbergen MW, Allen L, Vanderveen HELO, Woerdman JP (1993) Astigmatic laser mode converters and transfer of orbital angular-momentum. Opt Commun 96:123–132
98. Beijersbergen MW, Coerwinkel RPC, Kristensen M, Woerdman JP (1994) Helical-wavefront laser-beams produced with a spiral phaseplate. Opt Commun 112:321–327
99. Marrucci L, Manzo C, Paparo D (2006) Optical spin-to-orbital angular momentum conversion in inhomogeneous anisotropic media. Phys Rev Lett 96:163905
100. Mirhosseini M, Magana-Loaiza OS, Chen C, Rodenburg B, Malik M, Boyd RW (2013) Rapid generation of light beams carrying orbital angular momentum. Opt Express 21:30196–30203
101. Sun J, Zeng J, Litchinitser NM (2013) Twisting light with hyperbolic metamaterials. Opt Express 21:14975–14981
102. Ostrovsky AS, Rickenstorff-Parrao C, Arrizón V (2013) Generation of the "perfect" optical vortex using a liquid-crystal spatial light modulator. Opt Lett 38:534–536
103. Coles MM, Williams MD, Andrews DL (2013) Second harmonic generation in isotropic media: six-wave mixing of optical vortices. Opt Express 21:12783–12789
104. Williams MD, Coles MM, Saadi K, Bradshaw DS, Andrews DL (2013) Optical vortex generation from molecular chromophore arrays. Phys Rev Lett 111:153603
105. Williams MD, Coles MM, Bradshaw DS, Andrews DL (2014) Direct generation of optical vortices. Phys Rev A 89:033837
106. García-García J, Rickenstorff-Parrao C, Ramos-García R, Arrizón V, Ostrovsky AS (2014) Simple technique for generating the perfect optical vortex. Opt Lett 39:5305–5308
107. Karimi E, Schulz SA, De Leon I, Qassim H, Upham J, Boyd RW (2014) Generating optical orbital angular momentum at visible wavelengths using a plasmonic metasurface. Light Sci Appl 3:e167
108. Andrews DL, Williams MD, Bradshaw DS, Lui R, Phillips DB, Franke-Arnold S, Padgett MJ (2014) Nanoarrays for the generation of complex optical wave-forms. Proc SPIE 9160:91601L
109. Liu R, Phillips DB, Li F, Williams MD, Andrews DL, Padgett MJ (2015) Discrete emitters as a source of orbital angular momentum. J Opt 17:045608
110. Franke-Arnold S, Jeffers J (2008) Orbital angular momentum in quantum communication and information. In: Andrews DL (ed) Structured light and its applications: an introduction to phase-structured beams and nanoscale optical forces. Academic, Amsterdam/Boston, pp 271–291
111. Mirhosseini M, Magaña-Loaiza OS, O'Sullivan MN, Rodenburg B, Malik M, Lavery MPJ, Padgett MJ, Gauthier DJ, Boyd RW (2015) High-dimensional quantum cryptography with twisted light. New J Phys 17:033033
112. Riehl JP, Richardson FS (1993) Circularly polarized luminescence. Methods Enzymol 226:539–553
113. Bradshaw DS, Leeder JM, Coles MM, Andrews DL (2015) Signatures of material and optical chirality: origins and measures. Chem Phys Lett 626:106–110

Chapter 11
Luminescent Lanthanide Sensors and Lanthanide Doped Upconversion Nanoparticles: Current Status and Future Expectations

Garima Sharma, Preeti Sehgal, and Anudeep Kumar Narula

Abstract Lanthanide ions exhibit fascinating optical properties with their potential applications largely governed by their interaction with light. This chapter deals with some relevant aspects concerning the electronic and coordination properties of lanthanides and the basic principles related to the design of efficient luminescent lanthanide complexes. The cleverly designed environment consisting of ligands containing adequate chromophoric groups provide a rigid and protective coordination shell to minimize non-radiative deactivation. Lanthanide doped upconversion nanoparticles (UCNPs) have attracted extensive attention in the field of biomedical applications due to their long luminescence lifetime, narrow emission bandwidth, high quantum yields and low toxicity. In this chapter the upconversion phenomenon is explained with emphasis on the mechanism of upconversion, selection of host materials and impurities in host matrices. The various chemical approaches for the synthesis of lanthanide doped UCNPs have also been discussed. Subsequently, some selected results of our recent work concerning the photoluminescence studies of Eu(III) and Yb (III) complexes are reported which exhibit the characteristic emission bands of Eu(III) ion corresponding to $^5D_o \rightarrow {}^7F_J$ (J = 0–4) transitions with intense red emission at 615 nm due to $^5D_0 \rightarrow {}^7F_2$ transition of the central Eu(III) ion. These complexes show long radiative lifetime and quantum efficiency which suggest that these complexes can be well utilized as fluorescent probes.

Keywords Fluoroprobe • Energy transfer • Lanthanide complexes • Quanttum efficiency • Radiative lifetime

G. Sharma • P. Sehgal • A.K. Narula (✉)
University School of Basic and Applied Sciences, Guru Gobind Singh Indraprastha University, Delhi 110078, India
e-mail: researchchemlab58@gmail.com

© Springer International Publishing Switzerland 2016
C.D. Geddes (ed.), *Reviews in Fluorescence 2015*,
DOI 10.1007/978-3-319-24609-3_11

11.1 Introduction

The lanthanide complexes attract intense attention due to their superior optical properties and wide variety of potential applications i.e. biomedical assays and imaging, luminescence devices, ionic conductors and sensors. Majority of Ln^{3+} ions are luminescent, but some lanthanide ions are more emissive than others which depends on the feasibility of its excited state and non radiative de-activation paths. This can be achieved by sensitization through the surrounding ion and the overall quantum yield can be achieved by:

$$Q_{Ln}^{L} = \eta_{sens} Q_{Ln}^{Ln} \tag{11.1}$$

Where Q_{Ln}^{L} and Q_{Ln}^{Ln} are the quantum yields from indirect and direct transitions respectively while η_{sens} represents the efficiency with which electromagnetic energy is transferred from the surroundings to the metal ion [1]. The quantum yield for direct transition Q_{Ln}^{Ln} depends upon the energy gap between the emissive state of the metal ion and the highest sublevel of its ground multiplet. If this energy gap is small, the non radiative deactivation process will be faster. In lieu of above, some lanthanide ions such as Eu^{3+}, Gd^{3+} and Tb^{3+} ions are the suitable candidates with $\Delta E = 512{,}300$ ($^5D_0 \rightarrow {}^7F_6$), 32 200 ($^6P_{7/2} \rightarrow {}^8S_{7/2}$), 14,800 ($^5D_4 \rightarrow {}^7F_0$) cm^{-1} respectively. Out of above these, Gd^{3+} has potential applications for mercury free fluorescent lamps because under vaccum – UV excitation Gd^{3+} can efficiently transfer energy onto Eu^{3+} which results emission of two red photons also known as down conversion effect; but it is not very useful for bioanalysis because it emits in the UV region and its luminescence interferes with the emission or absorption process in organic part of complexes [2–6]. The advantages of Eu^{3+} and Tb^{3+} ions are their tunable energy gap with their sizes, therefore these two ions received much attention. Other lanthanides ions such as Sm^{3+}, Dy^{3+} have very low quantum yield hence they appear to be less useful in the applications of luminescence. Pr^{3+} emits both in visible and NIR regions and is considered a suitable component for optical materials because of its ability of generating up conversion process as well. Moreover, Nd^{3+}, Ho^{3+} and Yb^{3+} have found applications in the field of lasers and telecommunication devices because they emit in NIR spectral range. Nd^{3+} (lies at 1.06 nm) and Yb^{3+}(lies under 1 and 1.6 nm) are less useful than Pr^{3+} but act as an efficient sensitizer of Er^{3+} which emits at 1.55 nm [7, 8].

Recently, three ions Nd^{3+}, Er^{3+} and Yb^{3+} have gained popularity because technical developments occurred which facilitated the detection of weak NIR emissions for which efficient sensitizing groups have been explored.

11.2 Principle of Lanthanide Luminescence

The most interesting feature of these lanthanide series is their photoluminescence as these complexes show luminescence in visible or near infrared spectral region [9, 10]. The color of the emitted light depends upon the lanthanide ion viz. Eu^{3}

$^{+}$emits red light, Tb^{3+} emits green light, Sm^{3+} orange light, Tm^{3+} blue light. Pr^{3+}, Sm^{3+}, Dy^{3+}, Ho^{3+}, Tm^{3+} shows transitions in the near infrared region while Yb^{3+}, Nd^{3+}, Er^{3+}emits near infrared luminescence. Gd^{3+} luminescence can be observed in absence of organic ligands as it emits in ultraviolet region. In lanthanides, the emissions are due to f-f transitions [11, 12]. The partially filled 4f shell is shielded by $5s^2$ and $5p^6$ orbitals and the ligands in the first and second coordination sphere perturb the coordination sphere of lanthanide ions. The narrow band emission and long lifetimes of excited states of lanthanides is observed due to this shielding. Ce^{3+} is a unique case among all lanthanides due to f-d transitions. The emission maxima largely depends upon the ligand environment around Ce^{3+} ion [13–15].

All the lanthanide ions suffer from weak light absorption as the molar absorption coefficients of most of the transitions are smaller than 10 L $mol^{-1}cm^{-1}$ and a limited amount of radiations is absorbed by direct excitation in the 4f levels because luminescence intensity is proportional to quantum yield and the amount of light absorbed, weak light absorption results in weak luminescence. However, this can be overcome by the antenna effect. Weissman [16] found that upon excitation of lanthanide complexes with organic ligands, intense metal centered luminescence can be observed due to intense absorption of organic chromophores. The excitation energy is transferred from ligand to lanthanide ion via intramolecular energy transfer [17–20]. The same phenomena for the europium complexes with salicylaldehyde, benzoylacetone, and dibenzoylmethane and meta- nitro benzoylacetone has been described.

The mechanism of energy transfer from organic ligands to lanthanide ions is proposed by Crosby and Whan [21]. Upon ultraviolet irradiation, the organic ligands of the complexes are excited to a vibrational level of first excited singlet state (S_1 or S_0), then molecule undergoes to lower vibrational level of S_1 state via internal conversion. The excited singlet state can be deactivated to the ground state ($S_1 \rightarrow S_0$) or can undergo nonradiative intersystem crossing from singlet state S_1 to triplet state T_1. The triplet state T_1 can be radiatively to the ground state S_0 by spin forbidden transitions $T_1 \rightarrow S_0$. The complex may undergo indirect excitation by a non radiation transition from the triplet state to an excited state of the lanthanide ion which may undergo a radiative transition to a lower 4f state by photoluminescence or may be deactivated by non radiative process [22–28]. According to them, vibronic coupling of lanthanide with the ligand and solvent molecule is the main cause of non radiative deactivation of the lanthanide ion (Fig. 11.1).

However, a mechanism was proposed for energy transfer from excited singlet state S_1 to the energy levels of the lanthanide ions. But this theory is not of great importance because it is not efficient due to short lifetime of the singlet excited state [29]. This type of excitation is seen in Tb^{3+} and Eu^{3+}. Luminescence by the lanthanide ions is done by resonance levels. The main resonance levels are $^4G_{5/2}$ for Sm^{3+} ion (17,800 cm^{-1}), $^5D_0$ for Eu^{3+}ion (17,250 cm^{-1}), $^5D_4$ for Tb^{3+} (20,430 cm^{-1}) and $^4F_{9/2}$ for Dy^{3+} (20,960 cm^{-1}) [30]. If the lanthanide is excited to a non emitting level either by direct excitation or indirect excitation, the excitation energy is dissipated via non radiative processes. To achieve the resonance level of the lanthanide, it is mandatory that the lowest triplet state of the

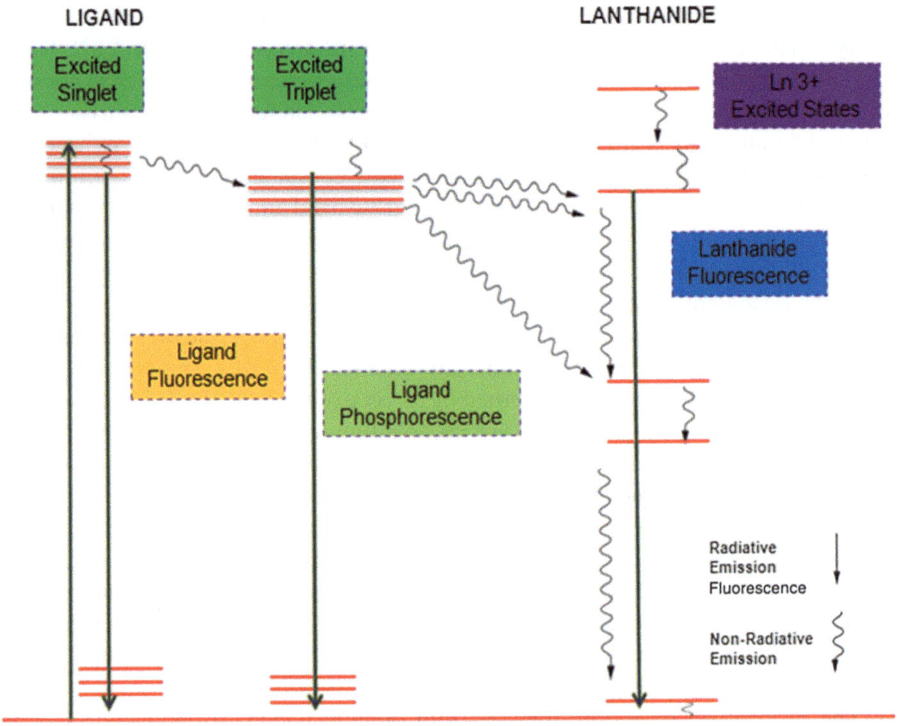

Fig. 11.1 Jablonski diagram representing the energy transfer in lanthanide complexes [28, 29]

complex is located at or equal to the resonance level of the lanthanide ion; if the energy levels of the organic ligands are present below the resonance level of lanthanide, molecular fluorescence or phosphorescence of the ligand is observed. Therefore it is a sensitive function of the lowest triplet level of the complex relative to a resonance level of the lanthanide ion [31–33].

Thus by the variation of ligands, it is possible to control the luminescence intensity for a given lanthanide ion, as the position of the triplet level depends on the nature of the ligands. Further, it is also temperature dependent, the luminescence caused by direct excitation of 4f level is less temperature sensitive than the luminescence caused by indirect excitation by organic ligands. The metal centered emission cannot be observed in case of Gd^{3+} ion as its 4f levels are located above the triplet levels. The presence of heavy paramagnetic ion results in mixing of triplet and singlet states and thereby enhancing the intersystem crossing from singlet state to triplet state [34]. The spin orbit coupling interaction helps the triplet state to acquire a partially singlet character which results in relaxation of selection rules.

The efficiency of energy transfer depends upon the overlap between the phosphorescence spectrum of the ligand and absorption spectrum of lanthanide ion. The back energy transfer can occur from the emitting 4f level of the lanthanide ion to the triplet state of the ligand, if the energy match between these is very close to each other [35]. The fluorescence and phosphorescence of the ligands are quenched if there is efficient energy transfer from ligand to lanthanide ion, but if the energy transfer is not effective it results in observation of ligand emission in addition to the lanthanide emission.

The energy transfer through charge transfer states is another possibility of lanthanide luminescence. In this case, the light is absorbed by the ligand to metal charge transfer states (LMCT) from where the energy can be transferred to the 4f levels of the lanthanide ion, if the LMCT state lies high enough to the lanthanide 4f level. Eu^{3+} ion shows an efficient sensitization through charge transfer state if the LMCT lies above 40,000 cm^{-1} [36]. If this energy gap is low it results in the quenching of luminescence and when this energy gap is lower than 25,000 cm^{-1} it leads to total quenching.

11.3 Design of Luminescent Lanthanide Complexes

11.3.1 General Criteria

The lanthanide ions need a cleverly designed environment which consists of ligands containing adequate chromophores, harvesting light and simultaneously providing a rigid and protective coordination shell to minimize non-radiative deactivation. The laporte-forbidden f-f transitions have weak oscillator strengths which results in quenching by the high energy oscillators such as O-H, N-H or C-H groups located in the inner and outer coordination spheres [37]. The design of luminescent lanthanide complexes requires particular care as the electronic, magnetic and photo physical properties of Ln(III) complexes strongly depends on the control of the coordination sphere of the metal. Among various parameters which govern the control over chemical, structural and thermodynamical properties of the complex are ligand topology, (dimensionality, connectivity, shape, size and chirality) and binding sites (electronic properties, nature, number, shape and arrangement), layer properties (rigidity/flexibility and lipophilicity/hydrophilicity ratio, thickness), environment properties and the nature of counter ions present [38]. The luminescent Ln(III) complex represents a multi-component system organized as a supramolecular structure comprising of metal cation and antenna. In general, the complex formation results from the attraction of ligand and metal ion, the resulting complex then undergoes partial to total desolvation [39]. The coordination of lanthanide ion with the ligand is substantially based on the Van der Walls electrostatic forces of attraction. These type of interactions are also observed in alkali-earth cations particularly Ca^{2+}. The ligand interacts with the surface of metal cation thus replacing the first solvation sphere either partially or totally. The desolvation step shows an increase in entropy however the enthalpy variation can be positive or negative which can be attributed to the difference in the energy of the bonds formed and the energy of the bonds broken between ligand and cation with solvent respectively [40]. In the complexation of Ln(III) in aqueous solution the dehydration step is endothermic ($\Delta H > 0$) which is an unfavourable energy contribution to the Gibbs free energy [41]. Therefore it is suggested to use polydentate ligands which show chelate effect and afford highly stable complexes in aqueous medium. In order to achieve the overall sensitization efficiency, the design of lanthanide complexes need certain factors to be considered which can be summarized as follows:

11.3.2 Choice of Lanthanide

The lanthanide ions show unique emission properties that cover the ultraviolet (Gd^{3+}) to the visible range (Sm^{3+}, Eu^{3+}, Dy^{3+}, Tb^{3+} and Tm^{3+}) and the near infrared range (Yb^{3+}, Nd^{3+} and Er^{3+}). The selection of the lanthanide ion depends on the intrinsic quantum yield which in turn depends on the energy gap between the lowest lying excited emissive state of metal ion and the highest sublevel of its ground multiplet. Accordingly, Eu^{3+}, Tb^{3+} and Gd^{3+} ions serve as good candidates for luminescent probes with energy gap values of 12,300 cm^{-1}, 14,800 cm^{-1} and 32,200 cm^{-1} however Gd^{3+} emits in ultraviolet region and used as contrast agents. The lanthanide ions possess relatively long lived excited states (microseconds to milliseconds) which can undergo energy transfer to high frequency vibrational oscillators such as O-H, N-H and to a lower extent C-H [42]. As a consequence, the presence of these groups in the proximity of the metal favours thermal dissipation of the energy (vibronic coupling) which gives rise to quenching of the luminescence (Fig. 11.2).

11.3.3 Choice of the Antenna

The term "antenna" refers to the chromophore which promotes the sensitization of lanthanide ion. It can be any aromatic or heteroaromatic highly π-conjugated system characterized by high efficiency of light absorption and high efficiencies of intersystem crossing and energy transfer process. The high yield of energy transfer from the antenna to the metal ion occurs until certain conditions are satisfied such as nature of ion, electronic structure of the donor, their relative positioning and the nature of interaction between metal ion and ligand [43]. The ligand geometric and electronic structure must be taken into account for various energy transfer pathways resulting from energy levels such as singlet $^1S_i(i = 1,2)$, triplet ^3T, intra ligand charge transfer (ILCT), ligand to metal charge transfer (LMCT). Generally, these energy pathways terminate at the emissive level of the lanthanide ion. The potential donor levels of the antenna should be chosen such that these are reasonably above the Ln(III) emissive level to avoid back transfer and is

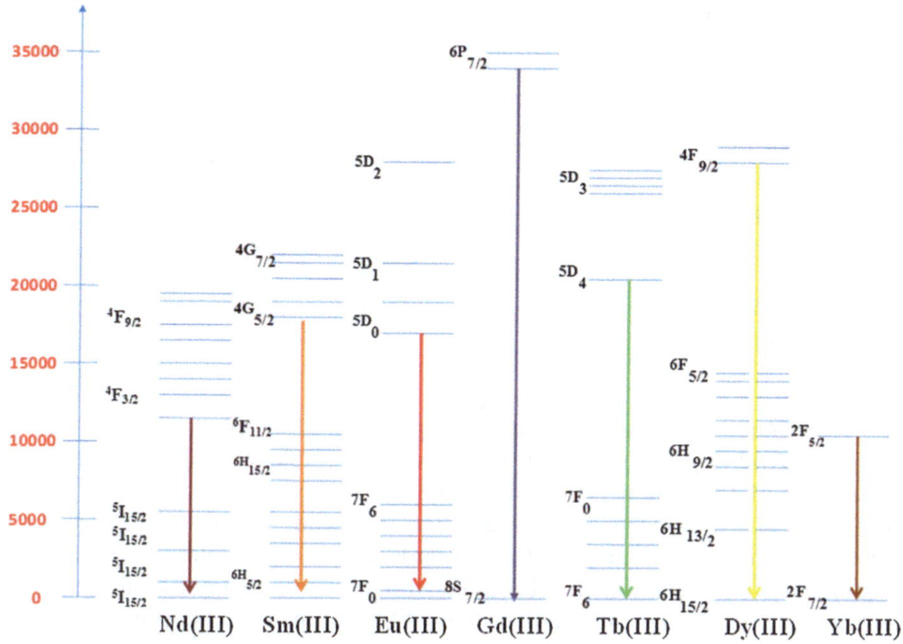

Fig. 11.2 Energy level diagram for the Ln(III) ions showing the main emitting levels and the transitions to the ground state levels [42]

close resonance with one of the higher 4f states. This energy gap should be atleast 1850 cm^{-1} higher than the lowest emitting level of Ln(III) ion. At higher value, the energy transfer from the triplet flows through the non-radiative excited state of the metal until it reaches the emissive levels and metal centered emission occurs. However, at lower energy gap strong thermal deactivation takes place due to back energy transfer and O_2 –quenching towards the chromophore triplet level [44]. The triplet state energies of some of the commonly used chromophores are reported in Table 11.1. The dependence of the luminescence of Ln(III) ion on the excitation wavelength also suggests that it should be above ca. 350 nm to facilitate the use of inexpensive excitation sources.

The lanthanides do not have restricted coordination number and geometry, these can therefore extend their coordination number ranging from 3 to 12, although 8 and 9 are considered the most common ones [46]. This property can be utilized by several auxillary N-or O- donor co-ligands for coordination and to make saturated complexes. In such complexes, the energy transfer is more effective from ligand to metal and complexes show intense photoluminescence which can be attributed to the increased anisotropy around the lanthanide ion. The introduced auxillary coligands make the complex more rigid and asymmetric around the central metal ion which overall improve the photophysical properties of the complexes [47]. The photophysical properties can also be tuned by substituting releasing or donating groups at various positions of the coordinated neutral auxillary ligands. The

Table 11.1 Energy levels of some commonly used chromophores [45]

Chromophore	Singlet (cm^{-1})	Triplet (cm^{-1})
1,10 Phenanthroline	29,200	22,100
Acetophenone	28,200	26,000
1,4-Napthoquinone		20,200
8-Hydroxyquinone	27,000	
7-Amino-4-methyl-2- hydroxyquinone		23,100
Tetrazatriphenylene	29,000	24,000
Napthalene	32,200	21,200
2-Hydroxyisophthalamide	24,200	23,350
1-Hydroxy pyrin-2-one		21,260

N-donor neutral auxillary ligands are 1,10-phenanthroline (phen), 2,2′-bipyridine (bipy), terpyridine (terpy) and their analogues form stable ternary complexes with wide range of 1:3 neutral lanthanide complexes [48–52]. The 1,3-diketonate tris complexes and adducts with some auxillary coligands explained the structural behavior of the coligand and its effect towards luminescence intensity stating that a rigid planar structural ligand behaves better in effective energy transfer which is further supported by the comparative study of $Eu(tta)_3$.phen and $Eu(tta)_3$.bipy. The important factor to choose a neutral ligand to enhance photoluminescence properties of lanthanides is the value of excited energy levels (singlet and triplet) of the corresponding auxillary ligand [53]. The corresponding singlet and triplet energy levels of bipy(29,900 cm^{-1}, 22,900 cm^{-1}) and phen(31,000 cm^{-1}, 22,100 cm^{-1}) were found suitable for effective energy transfer from ligand to metal and exhibit enhanced metal centered luminescence [54].

1,10-phenanthroline ligand behaves as a weak base in aqueous solution, the protonation constant being 4.95 log units. Its basicity is remarkably lower than aliphatic diamines, such as ethylenediamine(log $k = 10.65$ and 8.05 for successive addition of acidic protons) which shows its lower donor ability of its nitrogen atoms. [55] Compared to the parent 2,2′-bipyridine (bipy) and 2,2′,6′,6″ terpyridine systems, phen is characterized by two inward-pointing nitrogen donor atoms and therefore pre-organized for strong and entropically favoured metal binding [56]. The photophysical properties of phen have been a subject of a number of studies. Phen is characterized in aqueous solution by UV absorptions at 229 nm and 265 nm, the latter attributed to the $\Pi \rightarrow \Pi^*$ transition to the lowest energy excited singlet state. Excitation at 307 nm originates a fluorescence emission band at 380 nm, due to radiative decay of $\Pi \rightarrow \Pi^*$ state. These spectroscopic features are modulated by appropriate substituents on the phen framework and pH changes i.e. by protonation of the heteroaromatic nitrogen atoms. The UV-spectrum recorded at pH 4, where phen is monoprotonated form, $(phenH)^+$ shows the absorbance 8 nm red shifted to the absorption band at 265 nm attributed to the charge transfer of the $\Pi \rightarrow \Pi^*$ transition(i.e. the LUMO has a higher electron density on the nitrogen atoms than the HOMO. Monoprotonation of phen leads to the disappearance of the emission band at 380 nm and to the formation of a new emission broad red shifted band at ca. 410 nm accounting for stabilization of lowest energy state by protonation of the heteroaromatic nitrogen atoms because of its charge transfer character [57]. The electron deficiency of chelating agents phen and bipy is their electron deficiency that makes them excellent acceptors capable of stabilizing metal ions in lower oxidation states. Due to the presence of low energy Π* orbitals of the ligand, metal complexes can be characterized by strong metal-to-ligand charge transfer (MLCT) absorption bands in the visible spectrum and red shifted fluorescent emissions.

Another major class of co-ligands are phosphine oxides and their analogues which are widely studied and used as sensitizers now. The disappearance of broad absorption band in the region of 3000 cm^{-1} in the IR spectra of lanthanide ternary complexes of pyrazolonates confirms the substitution of water coordination by various phosphine oxide ligands. The increase in photoluminescence intensity of phosphine oxides coordinated Eu(III) complexes can be due to square antiprismatic structures of the complexes which promotes faster radiation rates and an increase in $^5D_0 \rightarrow {}^7F_2$ emissions related to odd parity. The substitution of auxillary sensitizing ligands in lanthanide complexes also increases the solubility of complexes in organic solvents depending on the substituted neutral ligand used.

The above discussion concludes that the ligand topology, binding sites, layer properties, and environment properties of the antenna plays an important role in the design of lanthanide complexes. In addition to this, the radiative lifetime and photoluminescence quantum yield of lanthanide complexes also influence the design of luminescent sensors.

Among the organic ligands, the pyridine derivatives are important as they form a diversity of coordinating structures including coordination polymers and also have numerous applications in catalysis, non linear optics, luminescence, ion exchange, material chemistry and magnetochemistry, the role of 1,10-phenanthroline as an auxillary ligand is well known and plays a significant role in absorption and transfer of energy to the emission center Eu(III) ion [58–60]. The auxillary ligand reduces the non radiative decay of the excited states of Eu(III) ion and increase the energy transfer efficiency from ligands to Eu(III) ion thereby increasing the stability of europium complexes. Similar observations have been made with the europium complexes $[Eu(2\text{-}ap)_3(phen)]Cl_3$**(1)**, $[Eu(2\text{-}ap)_2(phen)_2]Cl_3$**(2)** and $[Eu(2\text{-}ap)(phen)_3]Cl_3$**(3)** synthesized with 2-aminopyridine and 1,10-phenanthroline ligands in different molar ratios [61] (Fig. 11.3).

The complexes exhibited characteristic emissions of Eu(III) ion at around 537 nm, 592 nm, 615 nm, 652 nm and 700 nm. These five expected bands are attributed to the $^5D_o \rightarrow {}^7F_J$ ($J = 0\text{–}4$) transitions of Eu(III) ion. The emission bands at 537 nm and 652 nm are very weak since their corresponding transitions $^5D_o \rightarrow {}^7F_o$ and $^5D_o \rightarrow {}^7F_3$ are forbidden in both magnetic dipole and electric dipole fields. The relative emission of $^5D_o \rightarrow {}^7F_1$ transition at 592 nm are relatively strong as it is a magnetic dipole transition which is independent of the coordination environment of Eu(III) ion. The $^5D_o \rightarrow {}^7F_2$ transition at 615 nm is an induced electric dipole transition which is very sensitive to the coordination environment of the Eu(III) ion. This transition is responsible for the strong red emission of

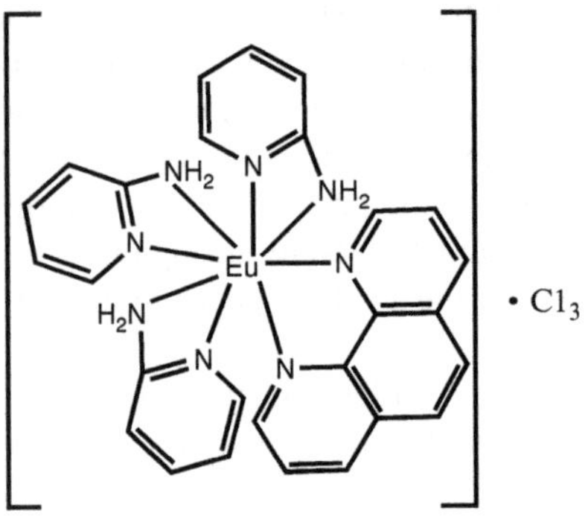

x:y = 3:1
[Eu(2–ap)$_3$(phen)].Cl$_3$

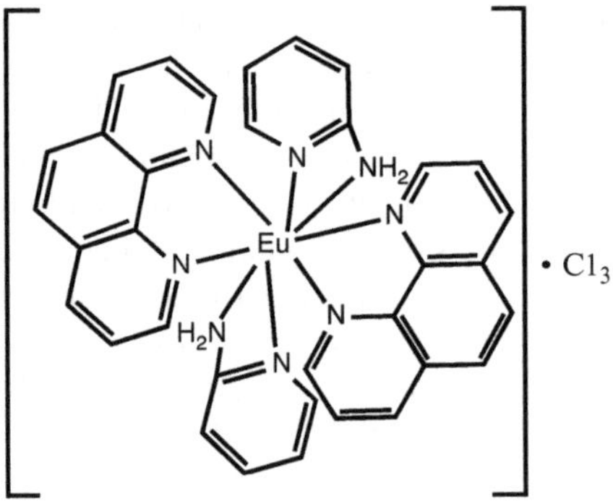

x:y = 2:2
[Eu(2–ap)$_2$(phen)$_2$].Cl$_3$

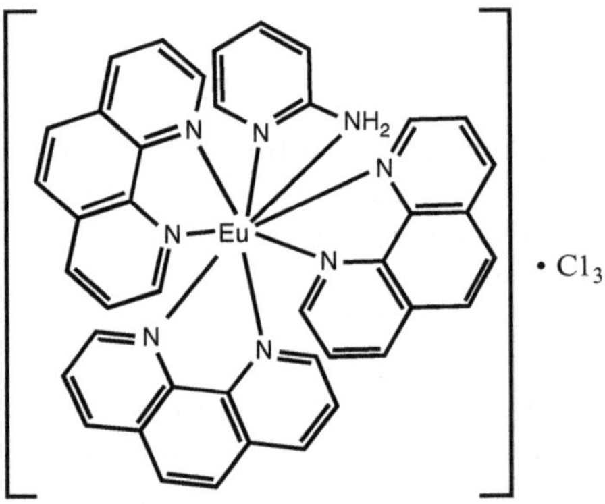

x:y = 1:3
[Eu(2–ap)(phen)$_3$].Cl$_3$

Fig. 11.3 Structure of europium complexes [Eu(2-ap)$_3$(phen)]Cl$_3$(1), [Eu(2-ap)$_2$(phen)$_2$] Cl$_3$(2) and [Eu(2-ap)(phen)$_3$]Cl$_3$ (3) (x = 2-aminopyridine, y = 1,10-phenanthroline) [61]

europium complexes and also indicates the presence of highly polarizable chemical environment around the Eu(III) ion. The band at 700 nm is relatively weak and corresponds to the $^5D_o \rightarrow {}^7F_4$ transition. It was also noted that in all the three complexes, the intensity ratio of $^5D_o \rightarrow {}^7F_2$ transition to $^5D_o \rightarrow {}^7F_1$ transition was around 4.0 which indicated the location of Eu(III) ion in the environment of low symmetry. The presence of only one line for $^5D_o \rightarrow {}^7F_o$ transition revealed the presence of similar chemical environment around Eu(III) ion. The photoluminescence results showed that the relative emission intensity of $^5D_o \rightarrow {}^7F_2$ transition was enhanced as the molar ratio of the ligand phen increased from complex **(1)** to complex **(3)** which revealed that the sensitization ability of ligand phen is better than that of ligand 2-ap (Fig. 11.4).

The decay profile of complex **(3)** shows monoexponential function indicating the presence of a single chemical environment around Eu(III) ion. The lifetime value

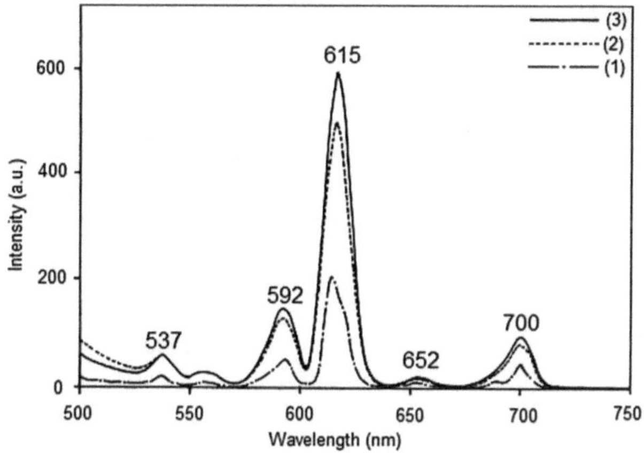

Fig. 11.4 Photoluminescence emission spectra of europium complexes in methanol solution (1×10^{-5} mol dm^{-3})[Eu(2-ap)$_3$(phen)]Cl$_3$ (1), [Eu(2-ap)$_2$(phen)$_2$]Cl$_3$ (2) and [Eu(2-ap)(phen)$_3$]Cl$_3$ (3) at the excitation wavelength of 247 nm [61]

was found to be 324 μs. According to the emission spectra and the lifetime of the Eu(III) excited state ($^5D_0$), the emission quantum efficiency (η) was found to be 10.33, which is good for a ligand-sensitized luminescent europium(III) complex (Fig. 11.5).

The above results indicate that the europium complex [Eu(2-ap)(phen)$_3$]Cl$_3$ with higher molar ratio of ligand 1,10-phenanthroline shows high luminescence as compared to other complexes [Eu(2-ap)$_3$(phen)]Cl$_3$ and [Eu(2-ap)$_2$(phen)$_2$]Cl$_3$. The quantum efficiency is found to be 10.33 which suggest the complex can act as a good fluorescent probe.

The pyridine carboxylates act as good candidates due to existence of N and O mixed donor atoms which can coordinate to the metal ion in a bidentate chelating, bidentate bridging and unidentate manner [62–64]. 2,6-pyridinedicarboxylic acid and α-picolinic acid have been found as promising ligands due to the stability of their Ln(III) complexes, strong fluorescence intensity with long excitation lifetimes [65–68]. The ligand 2,6-pyridinedicarboxylic acid is water soluble, commercially available having nitrogen and oxygen atoms to coordinate with the metal ion. Aminopyridine ligands also act as useful chelating agents for inorganic and organometallic applications and their derivatives can coordinate to the metal ions in a monodentate fashion through the N atom of the ring, however there are several works reported where the amino group also participates in coordination to the metal ion. The hydroxypyridine ligands can also coordinate effectively to the metal ion through the N and O donor atoms and are expected to show good luminescent properties [69–71]. The europium complexes [{Eu(dpa)(α-pc)(CH$_3$OH)].2CH$_3$OH}](**4**), [{Eu(dpa)(2-ap)(CH$_3$OH)].2CH$_3$OH}](**5**) and [{Eu(dpa)(2-hp)(CH$_3$OH)].2CH$_3$OH}](**6**) with dipicolinic acid (dpa) as a primary ligand and α-picolinic acid (α-pc), 2-aminopyridine (2-ap) and 2-hydroxypyridine (2-hp)

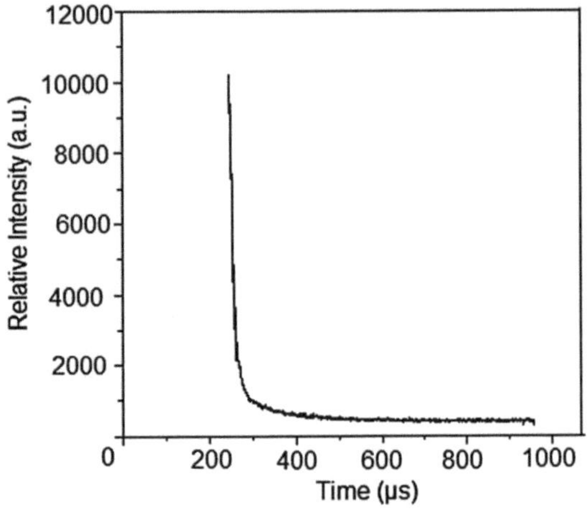

Fig. 11.5 Photoluminescence decay curve under laser pulse excitation of [Eu(2-ap)(phen)$_3$]Cl$_3$ complex [61]

as secondary ligands [72] shows strong characteristic emission bands of Eu(III) ions in the visible region, which attributes to the electronic transitions from the excited ^5D_o level to the ground ^7F_J (J = 0–4) levels of Eu(III) ion . The relative luminescent intensity of $^5D_o \rightarrow {}^7F_2$ transition is strongest in the complex based on 2-ap as secondary ligand, however the weakest luminescent intensity is observed in complex with 2-hp as secondary ligand. This can be explained by the non-radiative deactivation of energy of excited state as a result of their interaction with high frequency oscillators O-H group which act as efficient quencher of lanthanide ion luminescence in 2-hp ligand. These results showed that –NH$_2$ group in 2-ap ligand can strongly sensitize the luminescence of Eu(III) ion as compared to the 2-pc and 2-hp ligands (Fig. 11.6).

The luminescence decay curves of the complexes obtained at 298 K by monitoring the $^5D_0 \rightarrow {}^7F_2$ transition (615 nm) show monoexponential function indicating the presence of a single chemical environment around Eu(III) ion. The lifetime values are found to be 725 μs, 825 μs and 675 μs and the quantum efficiency values are found to be 21.60, 27.30 and 17.89 for complexes **(4)**, **(5)** and **(6)** respectively (Fig. 11.7).

The complexes show the characteristic emission bands of Eu(III) ion in visible region at 580 nm, 592 nm, 615 nm, 650 nm and 698 nm for $^5D_0 \rightarrow {}^7F_J$ (J = 0–4) transitions respectively. The photoluminescence properties of the complexes are influenced by the secondary ligands as studied by the photoluminescence spectra and decay measurements where the complex [{Eu(dpa)(2-ap)(CH$_3$OH)].2CH$_3$OH}] with 2-ap as secondary ligand exhibits strongest emission intensity and relatively longer luminescence lifetime, quantum efficiency as compared to the complex [{Eu(dpa)(α-pc)(CH$_3$OH)].2CH$_3$OH}] and complex [{Eu(dpa)(2-hp)(CH$_3$OH)].2CH$_3$OH}] (6).

Recently, the luminescent materials with the emissions in the near- infrared region such as Sm(III), Dy(III), Pr(III), Ho(III), Yb(III), Nd(III) and Er(III) have gained much interest due to their applications in telecommunication network as optical

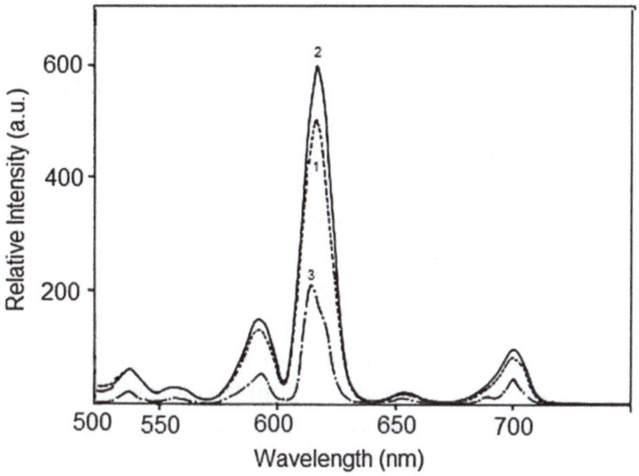

Fig. 11.6 Luminescent emission spectra of [{Eu (dpa)(α-pc) $(CH_3OH)].2CH_3OH$}] (4), [{Eu(dpa)(2-ap) $(CH_3OH)].2CH_3OH$}] (5) and [{Eu(dpa)(2-hp) $(CH_3OH)].2CH_3OH$}] (6) [72]

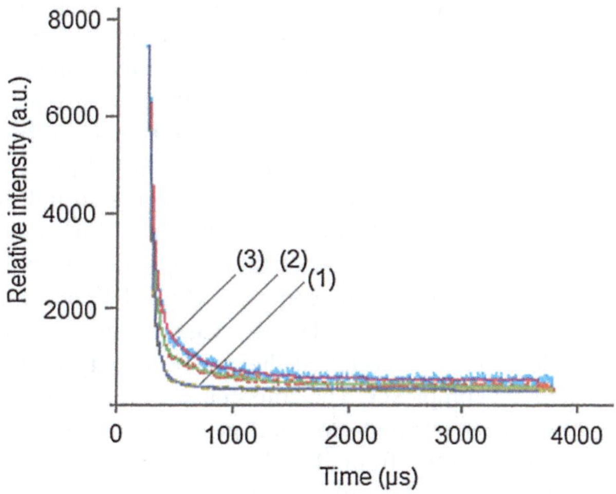

Fig. 11.7 Photoluminescent lifetime decay measurement of [{Eu (dpa)(α-pc) $(CH_3OH)].2CH_3OH$}] (4), [{Eu(dpa)(2-ap) $(CH_3OH)].2CH_3OH$}] (5) and [{Eu(dpa)(2-hp) $(CH_3OH)].2CH_3OH$}] (6) [72]

signal amplifier, probes for bioassays because human tissue is relatively transparent to near infrared light at 1000 nm. Yb^{3+} ion is usually a prime candidate to be chosen due to its luminescent efficiency close to 100 % and relatively simple electronic structure of two energy level manifolds: the $^2F_{7/2}$ ground state and $^2F_{5/2}$ excited state around at 10,000 cm^{-1} in the NIR region [73–77]. The rare earth ion couples such as RE^{3+}-Yb^{3+} can be efficiently utilized for near infrared (NIR) quantum cutting (QC) materials [78–80]. The energy gap of $^5D_2 - {}^7F_0$ transition in Eu^{3+} is approximately twice as large as that of $^2F_{5/2} - {}^2F_{7/2}$ transition in Yb^{3+} which facilitates the energy transfer between Eu^{3+} ion to Yb^{3+} ion [81]. The room temperature excitation spectra of complex $[Eu_{0.5}Yb_{0.5}$ $(sal)_3(phen)]$ [82] obtained by monitoring the longest emission wavelength of Eu^{3+} ion at 614 nm and Yb^{3+} ion at 980 nm exhibits a broad band at 290 nm attributable to the ligand to metal charge transfer (CT) transitions caused by the interaction between the organic ligands and the Eu^{3+} ions at 614 nm . The strong absorption band centered at 261 nm corresponded to the ligand to metal charge transfer transitions of Yb^{3+} ions. The presence of relatively strong CT absorption band in the excitation spectrum of Yb^{3+} ions reveals the energy transfer from Eu^{3+} ions to Yb^{3+} ions. The excitation spectrum also shows the sharp 4f-4f transitions of Eu^{3+} from the $^7F_0$ ground state to $^5D_{3,4,6,7,1,2}$ excited states. The difference in energy between the $^5D_2 - {}^7F_0$ transition of Eu^{3+} ions is twice the energy difference between $^2F_{5/2} - {}^2F_{7/2}$ transition of Yb^{3+} ions, which means that the $^5D_2$ excited state of Eu^{3+} ion can simultaneously transfer energy to two nearby Yb^{3+} ions and hence the Yb^{3+} ions can emit two infrared photons. This absorption is followed by rapid multi-phonon assisted relaxation from the populated $^5D_2$ energy levels to the metastable energy levels of Eu^{3+} ion (Fig. 11.8).

The emission spectra obtained by the excitation at 384 nm, it was observed that the complexes [Eu(sal)$_3$(phen)] and [Eu$_{0.5}$Yb$_{0.5}$(sal)$_3$(phen)] show the characteristic narrow emission peaks corresponding to the $^5D_0 - {}^7F_j$ (J = 0–4) transitions of Eu^{3+} ions. The emission peaks are well resolved and it was observed that the peak at 578 nm and 650 nm were weak since their corresponding transitions $^5D_0 - {}^7F_0$ and $^5D_0 - {}^7F_3$ were forbidden in magnetic and electric dipole fields. The peak at 591 nm is relatively strong and corresponds to $^5D_0 - {}^7F_1$ magnetic transition, the strongest emission observed at 614 nm ($^5D_0 - {}^7F_2$) is an induced electric dipole transition sensitive to the coordination environment of Eu^{3+} ion. This transition was responsible for the red emission of the europium complexes. The relative intensity of peaks are stronger in complex [Eu(sal)$_3$(phen)] as compared to complex [Eu$_{0.5}$Yb$_{0.5}$(sal)$_3$(phen)]. The decrease in intensity of all the emission peaks in complex can be considered due to the energy transfer from Eu^{3+} ions to Yb^{3+} ions in the complexes. The emission spectrum of complex [Yb(sal)$_3$(phen)] which exhibits near infrared emission peaks at 980 and 1030 nm associated with the transition of Yb^{3+} from $^2F_{5/2}$ level to $^2F_{7/2}$ energy level. The decay curve shows the lifetime quenching of Eu^{3+} emission due to the energy transfer from Eu^{3+} to Yb^{3+} with mean lifetime of 461 ps (Figs. 11.9, 11.10, and 11.11).

The above results show the study of excitation, emission and decay measurements of [Eu$_{0.5}$Yb$_{0.5}$ (sal)$_3$(phen)] complex which reveals the efficient energy transfer from Eu^{3+} to Yb^{3+} by cooperative energy transfer process, which leads to Eu^{3+} emission quenching and occurrence of near infrared emission at 980 and 1030 nm.

11.3.4 Photoluminescence Quantum Yield and Radiative Lifetime

The most important feature of luminescent lanthanide complexes is the photoluminescence quantum yield (PLQY). It is defined as the ratio of number of photons emitted with that of number of photons absorbed when excited at a particular

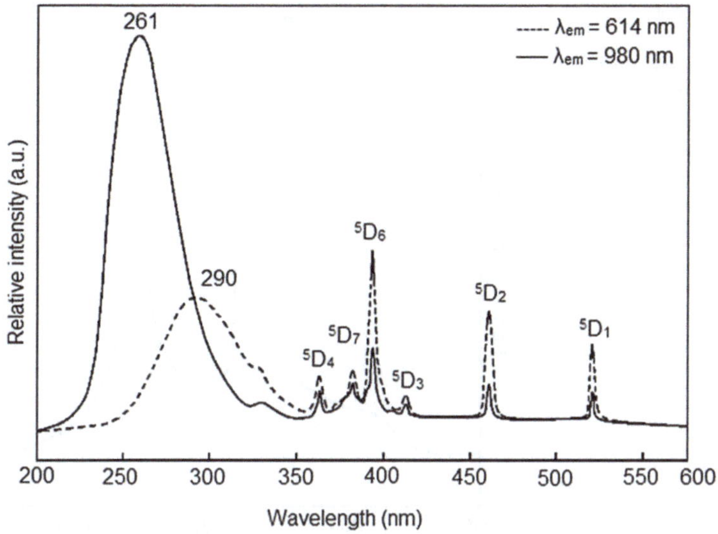

Fig. 11.8 Photoluminescent excitation spectra of Eu^{3+} emission monitored at 614 nm and Yb^{3+} emission monitored at 980 nm of [Eu $_{0.5}$Yb$_{0.5}$ (sal)$_3$(phen)] complex [82]

Fig. 11.9 Room temperature photoluminescent emission spectra of [Eu(sal)$_3$(phen)] and [Eu$_{0.5}$Yb$_{0.5}$(sal)$_3$(phen)] complexes excited at 384 nm [82]

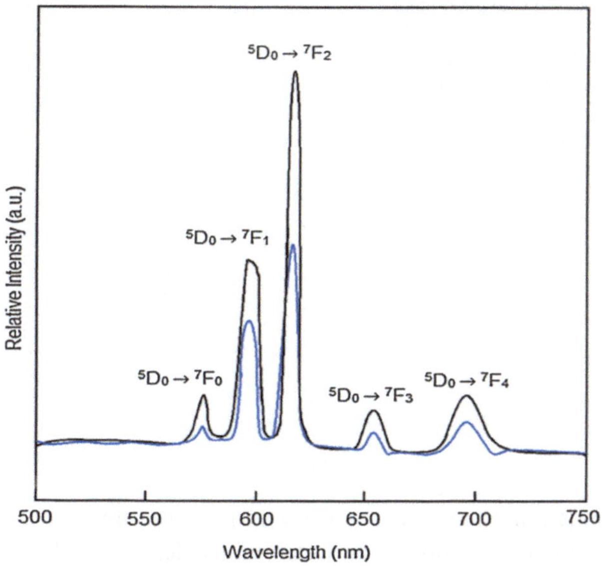

wavelength. The quantum yield of lanthanide complexes involves several factors such as ligand to Ln(III) energy transfer, multiphonon-relaxation, energy back transfer, crossover to charge transfer states etc. [83] Also, there are several competing processes such as fluorescence of antenna (competes with intersystem crossing), quenching of triplet state by dissolved molecular oxygen in case of

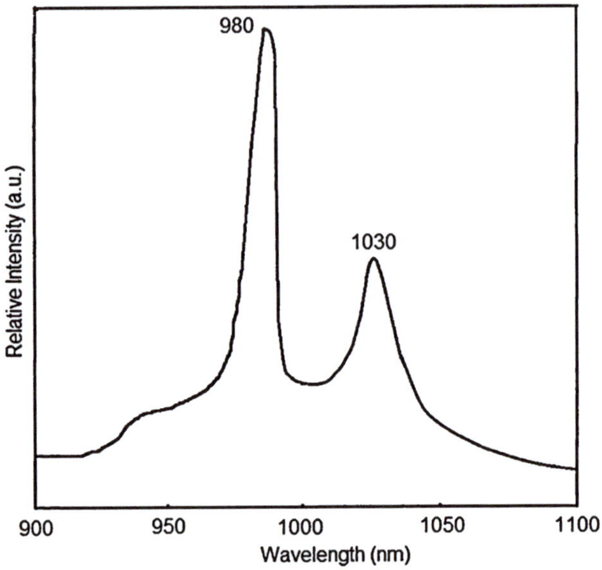

Fig. 11.10 Room temperature photoluminescent emission spectra of [Yb(sal)$_3$(phen)] excited at 384 nm [82]

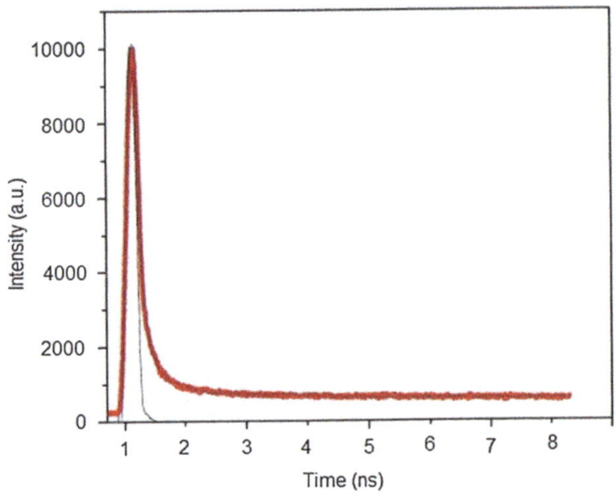

Fig. 11.11 Decay curve of Eu^{3+}, $^5D_0$ excited state (excitation wavelength: 384 nm; emission wavelength: 614 nm) of [$Eu_{0.5}Yb_{0.5}$ (sal)$_3$(phen)] complex [82]

NIR emitting ions (competes with energy transfer to the lanthanide ion) and the presence of solvent or water molecules which lowers the value of photoluminescence quantum yield. Hence, the overall quantum yield of sensitized emission (Φ_S) is the product of intersystem crossing quantum yield (Φ_{ISC}), the energy transfer quantum yield (Φ_{ET}) and the lanthanide luminescence quantum yield (Φ_{Lum})

$$\boldsymbol{\Phi}_{\mathbf{S}} = \boldsymbol{\Phi}_{\mathbf{ISC}} * \boldsymbol{\Phi}_{\mathbf{ET}} * \boldsymbol{\Phi}_{\mathbf{Lum}} \tag{11.2}$$

The overall quantum yield of the lanthanide complexes with organic ligands can be calculated as:

$$\Phi_{Ln}^{L} = \eta_{sens}.\Phi_{Ln}^{Ln} = \eta_{sens} \cdot \left(\frac{Krad}{Kobs}\right) = \eta_{sens} \cdot \left(\frac{\tau rad}{\tau obs}\right) \tag{11.3}$$

where Φ_{Ln}^{L} and Φ_{Ln}^{Ln} represent the quantum yield resulting from indirect and direct excitation respectively and η_{sens} is the quantum efficiency with which the electromagnetic energy if transferred from the organic ligands to the lanthanide ion. K_{rad} is the radiative rate constant, K_{obs} is the sum total of rates of various deactivation processes and τ_{rad} is the radiative lifetime. The intrinsic quantum yield Φ_{Ln}^{Ln} can be calculated as:

$$\Phi_{Ln}^{Ln} = K_{rad}/(K_{rad} + K_{nr}) \tag{11.4}$$

K_{nr} represents nonradiative rate constant which can be attributed to the back energy transfer to the sensitizer, quenching by matrix vibrations or by electron transfer mainly for the lanthanides having low reduction potential such as Eu(III) and Yb (III) ions [84]. The intrinsic quantum yield can be determined by two proposed mechanisms:

(a) The rapid diffusion enhanced resonance energy transfer after mixing the lanthanide complexes with a known quantum yield acceptor in solution and the efficiency of energy transfer between them can be calculated from both lifetime and intensity parameters.
(b) This mechanism is valid only for europium and relies on the fact that the intensity of purely magnetic dipolar transition $^5D_0 \rightarrow {}^7F_1$ transition is independent of the chemical environment of the lanthanide ion and the radiative lifetime can be calculated from the emission spectrum. The measurement of absolute quantum yield is a sophisticated task, it can be calculated by determination of relative quantum yield with that of standard sample of known absolute quantum yield. The absolute quantum yield of lanthanide complexes in solution can be calculated as:

$$\Phi_S = \Phi_{ST}(I_S/I_{ST})(RI_S^2/RI_{ST}^2)(F_{ST}/F_S) \tag{11.5}$$

where Φ_S represents the absolute quantum yield of sample, Φ_{ST} is the absolute quantum yield of reference compound, RI_S and RI_{ST} are the refractive index of sample solvent and reference compound respectively and represents the fraction of light absorbed by reference and sample. The accurate determination of quantum yields includes a number of factors e.g. internal filter effects, self-quenching and reliability of the standard value [85]. The other important parameter termed as luminescence lifetime can be defined as the time which an electron spent in excited state after excitation. It can be calculated as:

$$\boldsymbol{\tau}_{\mathbf{obs}} = 1/\mathbf{K}_{\mathbf{obs}} \tag{11.6}$$

The radiative rate (K_{rad}) and non-radiative rate (K_{nr}) can be calculated as:

$$\mathbf{K}_{\mathbf{rad}} = \boldsymbol{\Phi}/\boldsymbol{\tau} \tag{11.7}$$

$$1/\boldsymbol{\tau} = \mathbf{K}_{\mathbf{rad}} + \mathbf{K}_{\mathbf{nr}} \tag{11.8}$$

The overall absorbance of the lanthanide complexes is increased by the direct and indirect attachment of the chromophores to the lanthanide ion. The energy transfer process from the ligand to the metal ion is explained by the Forster and Dexter mechanism for direct chromophore attachment, while in case of indirect attachment of chromophore, the energy transfer proceeds through Forster dipole-dipole mechanism. The separation between two chromophores can be accessible through the determination of energy transfer efficiency (η_{et}) and can be calculated as:

$$\boldsymbol{\eta}_{\mathbf{et}} = 1 - (\boldsymbol{\tau}_{\mathbf{obs}}/\boldsymbol{\tau}_0) = \mathbf{K}_0/\mathbf{K}_{\mathbf{obs}} = 1/(1 + \mathbf{R}_{\mathbf{DA}}/\mathbf{R}_0)^6 \tag{11.9}$$

where τ_{obs} and τ_0 are the lifetime of donor chromophore in presence and absence of acceptor chromophore respectively, K_0 and K_{obs} are the decay rates of acceptor without and with donor respectively. R_{DA} and R_0 are the distance between donor-acceptor and critical distance respectively. The critical distance depends on the quantum yield of donor Q_D without acceptor, refractive index n of the medium, the overlap integral J_{ov} between the emission spectrum of donor and absorption spectrum of acceptor and orientation factor κ having isotropic limit of 2/3 and can be calculated as:

$$\mathbf{R}_{06} = 8.75 \times 10^{-25}\left(\boldsymbol{\kappa}^2.\mathbf{J}_{\mathbf{ov}}.\mathbf{Q}_{\mathbf{D}}.\boldsymbol{n}^{-4}\right) \tag{11.10}$$

The Judd-Ofelt analysis is a useful technique to estimate the population of odd-parity electron transitions for Eu(III) complexes. The interaction parameters of the ligand fields can be calculated from the Judd-Ofelt parameter, Ω_λ. The parameter Ω_2 is more sensitive to the symmetry and sequence fields and for anti-symmetrical Eu(III) complexes Ω_2 is large for faster radiative rates. The experimental intensity parameters Ω_λ where $\lambda = 2$ and 4, can be determined from the emission spectrum of Eu(III) complexes based on $^5D_0 \rightarrow {}^7F_2$ and $^5D_0 \rightarrow {}^7F_4$ transitions where the $^5D_0 \rightarrow {}^7F_1$ magnetic dipole-allowed transition is taken as reference and can be estimated as:

$$\mathbf{A}_{\mathbf{RAD}} = \left(4\mathbf{e}^2 2\boldsymbol{\omega}^3/3\hbar\mathbf{c}^3\right)\chi\sum\nolimits_\lambda \boldsymbol{\Omega}_\lambda < {}^7\mathbf{F}_\mathbf{J}\left|\mathbf{U}^{(\lambda)}\right|{}^5\mathbf{D}_0>^2.(1/2\mathbf{J}+1) \tag{11.11}$$

A_{RAD} is the correspondent coefficient of spontaneous emission, e is the charge on electron, ω is the angular frequency of the transition, ħ is the Planck's constant over 2π, c is velocity of light and χ is the Lorentz local field correction term which is given by $n(n^2+2)^2/9$ with a refraction index n = 1.43 and $< {}^7F_J \mid U^{(\lambda)} \mid {}^5D_0 >^2$ are the squared reduced matrix elements with values of 0.0032 and 0.0023 for J = 2 and 4 respectively. The Ω_6 parameter is difficult to determine as ${}^5D_0 \rightarrow {}^7F_6$ transition cannot be experimentally detected in most of Eu(III) complexes.

11.4 Lanthanide Doped Upconversion Nanoparticles

Lanthanide–doped upconversion nanoparticles (UCNPs) are mainly composed of three key components: a host matrix, sensitizer and activator. Generally, these UCNPs are fabricated by doping the trivalent lanthanides which have metastable excited states into an inorganic crystalline host lattice that can host these dopant ions. In case of lanthanide doped UCNPs, the sensitizer absorbs photons at 980 nm due to which it is excited to the higher energy state. The activator ions now obtain energy from the sensitizers to reach their corresponding excited states and show emission at short wavelengths (Fig. 11.12).

The selection of the host materials is important as it determines the lattice between the dopant ions, relative spatial position, coordination numbers and the type of anions surrounding the dopant. Several conditions need to be fulfilled while choosing the host lattice as reviewed by Wang and Liu (2009) [86, 87] and Ong *et. al.* (2010) [88] are (i) close lattice matches to dopant ions; (ii) low phonon vibration energies and (iii) good chemical stability. The host lattice must be able to accommodate the lanthanide dopant ions as it largely affects the luminescence output and emission intensity ratios of different transitions. The phonon energy of the host lattice must be low to ensure homogeneous doping and minimizing the lattice stress and non-radiative energy loss. The fluoride compounds serve as best host lattices due to their low phonon energy (350 cm^{-1}) and high chemical stability. These compounds are LaF_4, YF_4, $NaYF_4$ and $BaYF_4$ which are widely used as optimal host materials. The longer lifetimes are observed for the excited states of fluorides lattices due to the low phonon energies of the crystal lattice [89]. The lattice impurities increase the multi-phonon relaxation rates between the metastable states, thereby reducing the overall visible emission intensity. The halogenides also have small radiative losses but these materials have low chemical stability. The oxides have high chemical stability but their phonon energy is high (>500 cm^{-1}) due to stretching vibration of host lattice. Therefore fluorides prove to be ideal host candidates for the green and blue upconversion phosphors.

The upconverting luminescence efficiency depends upon nature of absorbers/emitters and the doping concentration of absorbers/emitters. The luminescence efficiency is highest when the doping concentration of absorbers/emitters shows well matching energy level. The improper absorber/emitter doping concentration ratio results concentration related cross-relaxation quenching or disables energy transfer. Among the lanthanide ions, Yb^{3+} ions are frequently used as absorbers due to its simple energy levels ${}^2F_{5/2}$ and ${}^2F_{7/2}$ and high absorption coefficient at 980 nm

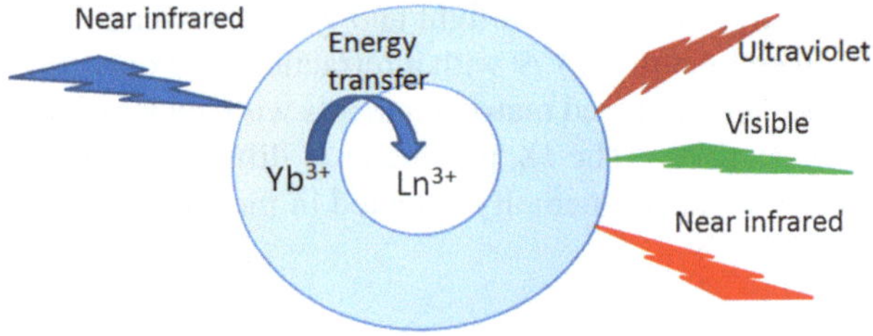

Fig. 11.12 The energy transfer process in up-conversion nanoparticles

and are used in relatively high doping ratios (20–30 %) [90]. Some other lanthanide ions like Er^{3+}, Tm^{3+} and Ho^{3+} have ladder-like energy levels and are often used as activators with low doping concentrations. The other lanthanide ions (Y^{3+}, La^{3+}, Gd^{3+}, Sc^{3+}), transition metal ions (Zr^{4+}, Ti^{4+}) and alkaline earth ions (Ca^{2+}, Sr^{2+}, Ba^{2+}) are also suitable ions that can be used. The most practical hosts are the halides ($NaYF_4$, YF_3, LaF_3), oxides (Y_2O_3, ZrO_2) and oxysulphides (Y_2O_2S, La_2O_2S) due to their low phonon-energy lattice. These hosts reduce multi-phonon relaxation and increase the lifetime of the intermediate states involved in upconversion process [91]. Thus to overcome the limitations of conventional fluoroprobes the lanthanide doped hexagonal phase $NaYF_4$ upconverting materials are appropriate substituents due to their superior upconverting properties.

The nano lanthanide doped oxide hosts are extensively reported with regard to Y_2O_3 as an important host material. It has broad transparency range (0.2–8 nm) with a band gap of 5.6 eV, high refractive index, better thermal conductivity and low phonon energy that makes it an attractive choice as the host material. The mostly used lanthanide host-dopant systems are listed in Table 11.2.

11.4.1 Mechanism of Upconversion

The conventional fluorophores exhibit the phenomenon of downconversion, i.e. higher energy photons are absorbed while lower energy ones are emitted due to energy loss (IEL). Compared with downconversion, upconversion is a process that causes the emission of higher energy photons through sequential absorption of lower energy photons. The mechanism of upconversion process has been extensively studied and is generally divided into three steps: excited state absorption (ESA), energy transfer upconversion (ETU) and photon avalanche (PA). In comparison with the other two processes, ETU has been widely employed to obtain high upconversion efficiency (emission density versus NIR excitation power), involving the absorption of a pump phonon of the same energy by each of the two neighboring ions. A subsequent non-radiative energy transfer promotes one of the ions to an upper energy level (EL2) while the other ion relaxes back to the ground state (GS). The relaxation from EL2 results in the emission of higher energy photons [98] (Fig. 11.13).

Table 11.2 Representative UCNPs with different host-dopant systems, excitation wavelengths and emission peaks [92–97]

Dopant ions			Major emissions (nm)		
Host lattice	Sensitizer	Activator	Blue	Green	Red
$NaYF_4$	Yb	Er		518, 537	652
	Yb	Er		540	660
	Yb	Er		510–530	635–675
	Yb	Er		521,539	651
LaF_3	Yb	Tm	475		
	Yb	Er		520, 545	659
CaF_2	Yb	Er		524	654
Y_2O_3	Yb	Er		550	660
Lu_2O_3	Yb	Er, Tm	490	540	662
$LuPO_4$	Yb	Tm	475		649

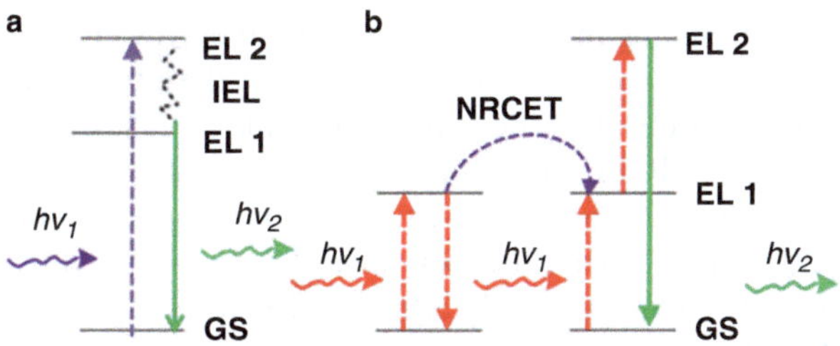

Fig. 11.13 Illustration of (**a**) downconversion and (**b**) energy transfer upconversion mechanism. *IEL* internal energy loss, *GS* ground state, *EL* energy level, *NRET* nonradiative energy transfer, *hυl* incident light, *hυ2* emission light [98]

11.4.2 Enhancing the Up Conversion Efficiency

The lanthanide–doped nanoparticles have great potential as lasers, phosphors and recently biolabels as well. The possibility to enhance the up-conversion luminescence has always been an interesting research topic. These nanoluminophores act as efficient biolabels when compared to organic dyes and quantum dots due to their excellent photostability, efficient near infrared to visible and long luminescence lifetimes suitable for time delayed detection. There is a need to enhance the up-conversion efficiency as these UCNPs suffer from low brightness when compared to conventional biolabels. There are certain ways to modify the upconversion efficiency which includes the dopant concentration, modifying the local chemical and structural environment, controlling the distribution of active ions in host material, varying the composition, structure and morphology of the up-converting nanoparticles. These factors reduce the quenching effect, enhance the absorption of NIR radiation or improve the sensitizer to activator energy transfer.

11.4.2.1 Concentration of Optically Active Ions

The up-conversion quantum yield can be enhanced by increasing the concentration of optically active ions within a single nanoparticle as the upconversion in singly doped NPs is very weak, therefore it is necessary to co-doping the activator ions with the sensitizer ions. The enhanced up-conversion in doubly doped material was first studied by F. Auzel in 1966 [99]. Yb^{3+} ion is most frequently used sensitizing lanthanide ion due to its simple energy levels, large absorption cross section at 940–990 nm in NIR region and large energy gap (10,000 cm^{-1}) and long luminescence lifetimes (1 ms). Yb^{3+} is usually chosen as it absorbes in NIR region and further capable of transferring its energy to the activator ions such as Tm^{3+}, Er^{3+} in the energy transfer up-conversion process which requires a single activator ions surrounded by atleast two or three Yb^{3+} ions.

The increase in the activator ions concentration results in the decrease in the effective distance between the ions and when the energy levels are quasi-resonant, the excited ions become non-radiatively depopulated. The concentration quenching manifests itself in shortening the luminescence lifetimes as well as in decreasing the effective luminescence quantum yield enhancement despite the increase in luminescence centres. Therefore the concentration of the activator ions must be carefully optimized. The NIR PL emission from the ultrasmall $NaYF_4$: 2 % Tm^{3+} nanocrystals to be 3.6 times more intense than the one from conventional 25–30 nm sized $NaYF_4$: 20 % Yb^{3+} / Tm^{3+} nanocrystals. These Yb^{3+} / Tm^{3+} doped UCNPs are particularly attractive due to highly efficient NIR-to-NIR conversion i.e. 980 nm excitation to 800 nm up conversion emission [100].

11.4.2.2 Selection of Host Material

The up conversion efficiency can be enhanced by the proper selection of the host material for lanthanide ion dopants. The low energy phonons cause low non-radiative and multiphonon losses and increase the luminescence lifetimes. Different hosts such as chlorides, bromides and iodides with phonon energies of ~144, 172 and 260 cm^{-1} and low multiphonon relaxation rates are suitable for up-conversion process as compared to fluorides and oxides [101]. Concluding, we can say that the selection of the host material is a balance between phonon properties and excitation spectra intensity which makes the host material appropriate for energy up-conversion in lanthanide doped materials.

11.4.2.3 Impurities in Host Matrix

The up-conversion efficiency can be enhanced by co-doping passive or active impurities in an efficient way. In case of passive impurities, the doping ions induce distortions to the local symmetry around activators or dissociate the lanthanide clusters in the nanocrystals and leads to increase in the up-conversion efficiency. These impurities do not participate in energy transfer within the matrix. On the other hand, the active impurities modify the energy transfer rates within the system and enhance the up-conversion efficiency. The enhancement of up-conversion efficiency in singly Er^{3+} -doped $YAlO_3$ phosphor by exchanging 40 % of Y^{3+} ions by larger Gd^{3+} ions (0.1159, 0.1193 nm) resulting in expanding the host lattice and distorting the local symmetry as studied by Li et.al [102]. The co-doping of β-$NaGdF_4$: Yb^{3+} /Er^{3+} nanoparticles with different amount of Li^+ ions also showed a higher up conversion enhancement as studied by Cheng et.al [103]. The green and red up-conversion emission yield intensities of the nanoparticles co-doped with 7 mol % Li^+ ions were enhanced 47 and 23 times respectively. The alkali ions also have an impact on the structure and spectral properties of up-converting fluoride nanocrystals. The phase, size, shape and stability of the Yb/Er co-doped $NaYF_4$ also get affected by the Li^+ and K^+ content, the green to blue ratio varied in the range ~2–6.5 and ~0.7–1.7 respectively and rose to 20–80 %. The passive impurities affect activator distribution within the UCNP, the active impurities also play a critical role in the energy transfer. This approach leads to the modification of energy distribution as well as the interaction between respective excited energy levels of the co-doping lanthanides. The improvement in the up-conversion quantum yield by the co-doping of Tm^{3+} and Er^{3+} activators with the Yb^{3+} sensitizers serves as an important example.

11.5 Synthesis of Lanthanide Doped UCNPs

The various chemical approaches for the synthesis of Ln-doped UCNPs have been studied by Wang and Liu in 2009 [104], these methods such as co-precipitation, thermal decomposition, sol-gel method and hydro thermal method are discussed as:

11.5.1 Co-precipitation Method

The co-precipitation method is simple as it does not require costly set up, complex procedures and severe reaction conditions. In this method, the nanoparticle growth can be controlled and stabilized by adding capping ligands such as polyvinylpyrrolidone (PVP), polyethylenimine (PEI) and ethylene diaminetetraacetic acid (EDTA) into the solvent. This method generally yields cubic phase $NaYF_4$: Yb, Er which is not an efficient up-convertor, subsequent calcinations at high temperatures results in sharpened crystal structure or partial phase transfer to hexagonal – phase $NaYF_4$: Yb, Er which has higher upconversion efficiency as studied by Yi et. al [105].

11.5.2 Thermal Decomposition Method

This method involves dissolving of organic precursors in high boiling point solvents (e.g. oleic acid (OA), oleylamine (OM), octadecene(ODE) for the synthesis of highly monodispersed UCNPs. The rare earth trifluoroacetates are thermolyzed in the presence of high boiling point solvents at a temperature usually exceeding 300 °C. Mai et.al [106]. synthesized Er^{3+}, Yb^{3+} and Tm^{3+}, Yb^{3+} doped monodispersed cubic-phase and hexagonal-phase $NaYF_4$ nanocrystals by the thermal decomposition of trifluoroacetate precursors in OA/OM/ODE solvents and OA/ODE solvents respectively and also synthesized $NaYF_4$ based UCNPs with different luminescent properties with similar synthetic methods. Yi and Chow [107] obtained hexagonal-phase $NaYF_4$:Yb, Er and $NaYF_4$:Yb, Tm nanoparticles with an average particle size of 10.5 nm by decomposing the precursors of Na (CF_3COO), $Y(CF_3COO)_3$, $Yb(CF_3COO)_3$ and $Er(CF_3COO)_3/Tm(CF_3COO)_3$ in OM solvent at 300 °C with much higher up-conversion fluorescence intensity. This method however, has some disadvantages such as use of expensive and air-sensitive metal precursors and the generation of toxic by-products.

11.5.3 Sol Gel Method

The sol-gel method is mainly based on the hydrolysis and polycondensation of metal alkoxides or metal acetate precursors to form extended networks with an oxide skeleton. Li et.al [108]. studied the preparation of metal oxide based Ln-doped UCNPs such as $YV0_4$:Yb, Er, $Lu_3Ga_5O_{12}$: Er, $BaTiO_3$: Er and ZrO_2:Er by the sol-gel method. Though this method is good for the synthesis of various Ln-doped UCNPs, however there are certain limitations such as particle size and particle aggregation may occur when dispersed in aqueous solution. Some post treatments are often needed to improve the crystalline phase purity for enhanced luminescence efficiency.

11.5.4 Hydrothermal Method

The synthesis of highly crystalline nanocrystals with tunable size, morphology, optical and magnetic properties via controlled reaction temperature, time, concentration, pH and precursors etc can be done by hydrothermal method. This method improves the solubility of solids under hydrothermal conditions, which accelerates the reactions between solids. The advantage of this method lies in the fact that it requires "one-pot process" with heat resistant polymer (PEI, PVP) added to the solvent. Wang et.al [109]. studied the preparation of uniform-sized nanoparticles with appropriate surface modification which were obtained by single reaction process. The only limitation with the hydrothermal method is the difficulty in observing the nanocrystals growth process.

11.6 Conclusion

Recently the luminescent lanthanide complexes are receiving the attention of scientific fraternity due to their high luminous lifetimes and extremely sharp emission bands arising from the electronic transitions between the 4f energy levels. The lanthanide complexes are widely used in many technologically interesting fields such as photoluminescent materials in display devices, fluorescence probes and labels in biological systems. In order to design the efficient luminescent fluorophores, it is necessary to understand the electronic and spectroscopic properties of rare earth elements, basic principles of photoluminescence which includes the basic concepts of antenna effect, coordination features of Ln^{3+} ions, radiative lifetime and photoluminescence quantum yield of lanthanide complexes. Apart from the luminescent lanthanide complexes, the lanthanide up-converting nanoparticles possess great potential as novel fluorophores for biological applications. These exhibit unique fluorescent property providing tremendous applications which have enormous advantages over conventional fluorophores. This review insights into the main aspects related to design of luminescent lanthanide complexes and lanthanide up-converting nanoparticles and also gives an overview of our recent work on photoluminescent properties of lanthanide complexes with nitrogen donor and oxygen donor ligands. The synthesis and photoluminescence properties of europium and ytterbium complexes with 2-aminopyridine, 1,10-phenanthroline, dipicolinic acid, α-picolinic acid, salicylic acid and 2-hydroxypyridine have been reported. The prepared complexes can act as good candidates for fluorescence probes and optoelectronic devices.

Acknowledgement The authors acknowledge Guru Gobind Singh Indraprastha University, New Delhi for providing financial support in the form of Indraprastha research fellowship (IPRF) for research work. Also, the authors are thankful to Ms. Shruti Peshoria for her contribution in preparation of the manuscript.

References

1. Di Bernardo P, Melchior A, Tolazzi M, Zanonato PL (2012) Thermodynamics of lanthanide (III) complexation in non-aqueous solvents. Coord Chem Rev 256(1–2):328–351
2. Bünzli J-CG, Chauvin A-S, Kim HK, Deiters E, Eliseeva SV (2010) Lanthanide luminescence efficiency in eight- and nine-coordinate complexes: role of the radiative lifetime. Coord Chem Rev 254(21–22):2623–2633

3. Li H-Y, Wu J, Huang W, Zhou Y-H, Li H-R, Zheng Y-X, Zuo J-L (2009) Synthesis and photoluminescent properties of five homodinuclear lanthanide (Ln^{3+}=Eu^{3+}, Sm^{3+}, Er^{3+}, Yb^{3+}, Pr^{3+}) complexes. J Photochem Photobiol A 208(2–3):110–116
4. Sengar RS, Nigam A, Geib SJ, Wiener EC (2009) Syntheses and crystal structures of gadolinium and europium complexes of AAZTA analogues. Polyhedron 28(8):1525–1531
5. Pietraszkiewicz O, Pietraszkiewicz M, Karpiuk J, Jesień M (2009) Eu(III) complexes involving 1,3,5-triazine diphosphine oxides. J Rare Earths 27(4):584–587
6. Vicentini G, Zinner LB, Zukerman-Schpector J, Zinner K (2000) Luminescence and structure of europium compounds. Coord Chem Rev 196(1):353–382
7. Eliseeva SV, Bunzli J-CG (2010) Lanthanide luminescence for functional materials and bio-sciences. Chem Soc Rev 39(1):189–227
8. Montgomery CP, Murray BS, New EJ, Pal R, Parker D (2009) Cell-penetrating metal complex optical probes: targeted and responsive systems based on lanthanide luminescence. Acc Chem Res 42(7):925–937
9. Tsukube H, Shinoda S (2002) Lanthanide complexes in molecular recognition and chirality sensing of biological substrates. Chem Rev 102(6):2389–2404
10. Pandya S, Yu J, Parker D (2006) Engineering emissive europium and terbium complexes for molecular imaging and sensing. Dalton Trans 23:2757–2766
11. Duke RM, Veale EB, Pfeffer FM, Kruger PE, Gunnlaugsson T (2010) Colorimetric and fluorescent anion sensors: an overview of recent developments in the use of 1,8-naphthalimide-based chemosensors. Chem Soc Rev 39(10):3936–3953
12. dos Santos CMG, Harte AJ, Quinn SJ, Gunnlaugsson T (2008) Recent developments in the field of supramolecular lanthanide luminescent sensors and self-assemblies. Coord Chem Rev 252(23–24):2512–2527
13. Shinoda S, Tsukube H (2011) Luminescent lanthanide complexes as analytical tools in anion sensing, pH indication and protein recognition. Analyst 136(3):431–435
14. Tsukube H, Yano K, Shinoda S (2009) Near-infrared luminescence sensing of glutamic acid, aspartic acid, and their dipeptides with tris(β-diketonato)lanthanide probes. Helv Chim Acta 92(11):2488–2496
15. Bünzli J-CG (2006) Benefiting from the unique properties of lanthanide ions. Acc Chem Res 39(1):53–61
16. Evans RC, Douglas P, Winscom CJ (2006) Coordination complexes exhibiting room-temperature phosphorescence: evaluation of their suitability as triplet emitters in organic light emitting diodes. Coord Chem Rev 250(15–16):2093–2126
17. Yongliang Z, Fengying Z, Qiang L, Deqing G (2006) Synthesis, characterization and fluorescence properties of europium, terbium complexes with biphenyl-4-carboxylic acid and o-phenanthroline. J. Rare Earths 24(1, Supplement 1):18–22
18. Bunzli J-CG, Piguet C (2005) Taking advantage of luminescent lanthanide ions. Chem Soc Rev 34(12):1048–1077
19. Kido J, Okamoto Y (2002) Organo lanthanide metal complexes for electroluminescent materials. Chem Rev 102(6):2357–2368
20. Faulkner S, Pope SJA, Burton-Pye BP (2005) Lanthanide complexes for luminescence imaging applications. Appl Spectrosc Rev 40(1):1–31
21. Whan RE, Crosby GA (1962) Luminescence studies of rare earth complexes: benzoylacetonate and dibenzoylmethide chelates. J Mol Spectrosc 8(1–6):315–327
22. Crosby GA, Whan RE, Alire RM (1961) Intramolecular energy transfer in rare earth chelates. Role of the triplet state. J Chem Phys 34(3):743–748
23. Martell AE, Hancock RD, Motekaitis RJ (1994) Factors affecting stabilities of chelate, macrocyclic and macrobicyclic complexes in solution. Coord Chem Rev 133:39–65
24. Bünzli J-CG, Piguet C (2002) Lanthanide-containing molecular and supramolecular polymetallic functional assemblies. Chem Rev 102(6):1897–1928
25. Piguet C, Bunzli J-CG (1999) Mono- and polymetallic lanthanide-containing functional assemblies: a field between tradition and novelty. Chem Soc Rev 28(6):347–358

26. Kuriki K, Koike Y, Okamoto Y (2002) Plastic optical fiber lasers and amplifiers containing lanthanide complexes. Chem Rev 102(6):2347–2356
27. Yanagida S, Hasegawa Y, Murakoshi K, Wada Y, Nakashima N, Yamanaka T (1998) Strategies for enhancing photoluminescence of Nd3+ in liquid media. Coord Chem Rev 171:461–480
28. Hasegawa Y, Wada Y, Yanagida S (2004) Strategies for the design of luminescent lanthanide (III) complexes and their photonic applications. J Photochem Photobiol C 5(3):183–202
29. Sivakumar S, van Veggel FCJM, Raudsepp M (2005) Bright white light through Up-conversion of a single NIR source from sol–gel-derived thin film made with Ln3+−doped LaF3 nanoparticles. J Am Chem Soc 127(36):12464–12465
30. Werts MHV, Jukes RTF, Verhoeven JW (2002) The emission spectrum and the radiative lifetime of Eu3+ in luminescent lanthanide complexes. Phys Chem Chem Phys 4 (9):1542–1548
31. Tamaki S, Hasegawa Y, Yajima H (2013) Factors influencing the luminescence intensity of europium(III) complexes prepared via synergistic extraction. Talanta 105:262–266
32. Pandya S, Yu J, Parker D (2006) Engineering emissive europium and terbium complexes for molecular imaging and sensing. Dalton Trans 23:2757–2766
33. Johnsson N, Johnsson K (2007) Chemical tools for biomolecular imaging. ACS Chem Biol 2 (1):31–38
34. Clapp AR, Medintz IL, Mattoussi H (2006) Förster resonance energy transfer investigations using quantum-dot fluorophores. ChemPhysChem 7(1):47–57
35. Bazin H, Trinquet E, Mathis G (2002) Time resolved amplification of cryptate emission: a versatile technology to trace biomolecular interactions. Rev Mol Biotechnol 82(3):233–250
36. Shiraishi Y, Furubayashi Y, Nishimura G, Hirai T (2007) Sensitized luminescence properties of dinuclear lanthanide macrocyclic complexes bearing a benzophenone antenna. J Lumin 127(2):623–632
37. Werts MHV, Woudenberg RH, Emmerink PG, van Gassel R, Hofstraat JW, Verhoeven JW (2000) A near-infrared luminescent label based on YbIII ions and its application in a fluoroimmunoassay. Angew Chem Int Ed 39(24):4542–4544
38. Zhang J, Badger PD, Geib SJ, Petoud S (2005) Sensitization of near-infrared-emitting lanthanide cations in solution by tropolonate ligands. Angew Chem Int Ed 44(17):2508–2512
39. Comby S, Gumy F, Bünzli J-CG, Saraidarov T, Reisfeld R (2006) Luminescent properties of an Yb podate in sol–gel silica films, solution, and solid state. Chem Phys Lett 432 (1–3):128–132
40. Comby S, Imbert D, Vandevyver C, Bünzli J-CG (2007) A novel strategy for the design of 8-hydroxyquinolinate-based lanthanide bioprobes that emit in the near infrared range. Chem Eur J 13(3):936–944
41. Bassett AP, Van Deun R, Nockemann P, Glover PB, Kariuki BM, Van Hecke K, Van Meervelt L, Pikramenou Z (2005) Long-lived near-infrared luminescent lanthanide complexes of imidodiphosphinate “shell” ligands. Inorg Chem 44(18):6140–6142
42. Imbert D, Cantuel M, Bünzli J-CG, Bernardinelli G, Piguet C (2003) Extending lifetimes of lanthanide-based near-infrared emitters (Nd, Yb) in the millisecond range through Cr(III) sensitization in discrete bimetallic edifices. J Am Chem Soc 125(51):15698–15699
43. Torelli S, Imbert D, Cantuel M, Bernardinelli G, Delahaye S, Hauser A, Bünzli J-CG, Piguet C (2005) Tuning the decay time of lanthanide-based near infrared luminescence from micro- to milliseconds through d→f energy transfer in discrete heterobimetallic complexes. Chem Eur J 11(11):3228–3242
44. Picot A, Malvolti F, Le Guennic B, Baldeck PL, Williams JAG, Andraud C, Maury O (2007) Two-photon antenna effect induced in octupolar europium complexes. Inorg Chem 46 (7):2659–2665
45. Fu L-M, Wen X-F, Ai X-C, Sun Y, Wu Y-S, Zhang J-P, Wang Y (2005) Efficient two-photon-sensitized luminescence of a europium(III) complex. Angew Chem Int Ed 44 (5):747–750

46. Manning HC, Goebel T, Thompson RC, Price RR, Lee H, Bornhop DJ (2004) Targeted molecular imaging agents for cellular-scale bimodal imaging. Bioconjug Chem 15 (6):1488–1495
47. Alzakhem N, Bischof C, Seitz M (2012) Dependence of the photophysical properties on the number of 2,2′-bipyridine units in a series of luminescent europium and terbium cryptates. Inorg Chem 51(17):9343–9349
48. Bornhop DJ, Hubbard DS, Houlne MP, Adair C, Kiefer GE, Pence BC, Morgan DL (1999) Fluorescent tissue site-selective lanthanide chelate, Tb-PCTMB for enhanced imaging of cancer. Anal Chem 71(14):2607–2615
49. Hanaoka K, Kikuchi K, Kojima H, Urano Y, Nagano T (2003) Selective detection of zinc ions with novel luminescent lanthanide probes. Angew Chem Int Ed 42(26):2996–2999
50. Hanaoka K, Kikuchi K, Kojima H, Urano Y, Nagano T (2004) Development of a zinc Ion-selective luminescent lanthanide chemosensor for biological applications. J Am Chem Soc 126(39):12470–12476
51. Parker D (2000) Luminescent lanthanide sensors for pH, pO2 and selected anions. Coord Chem Rev 205(1):109–130
52. Pal R, Parker D (2007) A single component ratiometric pH probe with long wavelength excitation of europium emission. Chem Commun 5:474–476
53. Song B, Wang G, Tan M, Yuan J (2006) A europium(III) complex as an efficient singlet oxygen luminescence probe. J Am Chem Soc 128(41):13442–13450
54. Bencini A, Lippolis V (2010) 1,10-phenanthroline: a versatile building block for the construction of ligands for various purposes. Coord Chem Rev 254(17–18):2096–2180
55. Stan CS, Rosca I, Sutiman D, Secula MS (2012) Highly luminescent europium and terbium complexes based on succinimide and N-hydroxysuccinimide. J Rare Earths 30(5):401–407
56. Pérez-Mayoral E, Soler-Padrós J, Negri V, Cerdán S, Ballesteros P (2007) Synthetic approaches to heterocyclic ligands for Gd-based MRI contrast agents. Molecules 12 (8):1771–1795
57. Strasser A, Vogler A (2004) Phosphorescence of gadolinium(III) chelates under ambient conditions. Inorg Chim Acta 357(8):2345–2348
58. Yuan W, Cui Y, Shi R, Tao D, Wang Y, Zhang W, Chen J, Sun L, Liu S, Xu Y (2011) Study on fluorescence properties of rare earth complexes influenced by steric effect. J Rare Earths 29(11):1013–1017
59. Wang Q-M, Yan B (2004) From molecules to materials: a new way to construct luminescent chemical bonded hybrid systems based with ternary lanthanide complexes of 1,10-phenanthroline. Inorg Chem Commun 7(10):1124–1127
60. Wang D, Pi Y, Zheng C, Fan L, Hu Y, Wei X (2013) Preparation and photoluminescence of some europium (III) ternary complexes with β-diketone and nitrogen heterocyclic ligands. J Alloys Compd 574:54–58
61. Sharma G, Narula AK (2015) Synthesis of Eu(III) complexes with 2-aminopyridine and 1,10-phenanthroline: structural, optical, thermal and morphological studies. Sens Actuators B Chem 215:584–591
62. Lin M, Wang X, Tang Q, Ling Q (2013) Luminescence properties of polymers containing europium complexes with 4-tert-butylbenzoic acid. J Rare Earths 31(10):950–956
63. Zhuravlev KP, Tsaryuk VI, Pekareva IS, Sokolnicki J, Klemenkova ZS (2011) Europium and terbium ortho-, meta-, and para-methoxybenzoates: structural peculiarities, luminescence, and energy transfer. J Photochem Photobiol A 219(1):139–147
64. Räsänen M, Takalo H, Rosenberg J, Mäkelä J, Haapakka K, Kankare J (2014) Study on photophysical properties of Eu(III) complexes with aromatic β-diketones – role of charge transfer states in the energy migration. J Lumin 146:211–217
65. Lahoud MG, Marques LF, da Silva PB, de Jesus CAS, da Silva CCP, Ellena J, Freitas RS, Davolos MR, Frem RCG (2013) Synthesis, crystal structure and photoluminescence of a binuclear complex of europium(III) containing 3,5-dicarboxypyrazolate and succinate. Polyhedron 54:1–7

66. Zucchi G, Maury O, Thuéry P, Ephritikhine M (2008) Structural diversity in neodymium bipyrimidine compounds with near infrared luminescence: from mono- and binuclear complexes to metal-organic frameworks. Inorg Chem 47(22):10398–10406
67. Armelao L, Quici S, Barigelletti F, Accorsi G, Bottaro G, Cavazzini M, Tondello E (2010) Design of luminescent lanthanide complexes: from molecules to highly efficient photo-emitting materials. Coord Chem Rev 254(5–6):487–505
68. de Sá GF, Malta OL, de Mello Donegá C, Simas AM, Longo RL, Santa-Cruz PA, da Silva Jr EF (2000) Spectroscopic properties and design of highly luminescent lanthanide coordination complexes. Coord Chem Rev 196(1):165–195
69. Azab HA, Duerkop A, Anwar ZM, Hussein BHM, Rizk MA, Amin T (2013) Luminescence recognition of different organophosphorus pesticides by the luminescent Eu(III)–pyridine-2,6-dicarboxylic acid probe. Anal Chim Acta 759:81–91
70. Zheng H, Gao D, Fu Z, Wang E, Lei Y, Tuan Y, Cui M (2011) Fluorescence enhancement of Ln3+ doped nanoparticles. J Lumin 131(3):423–428
71. Zhang L, An Y, Ahmad W, Zhou Y, Shi Z, Zheng X (2013) A new quaternary luminescence enhancement system of Eu–N-(3-methoxysalicylidene)-2-aminopyridine–1,10-phenanthroline–Zn and its application in determining trace amounts of Eu3+ and Zn2+. J Photochem Photobiol A 252:167–173
72. Sharma G, Narula A (2015) Synthesis and optoelectronic properties of three Eu(III)-dipicolinate complexes based on α-picolinic acid, 2-aminopyridine and 2-hydroxypyridine as secondary ligands. J Mater Sci Mater Electron 26(2):1009–1017
73. Yue Q, Yang J, Li G-H, Li G-D, Xu W, Chen J-S, Wang S-N (2005) Three-dimensional 3d–4f heterometallic coordination polymers: synthesis, structures, and magnetic properties. Inorg Chem 44(15):5241–5246
74. Bünzli J-CG (2010) Lanthanide luminescence for biomedical analyses and imaging. Chem Rev 110(5):2729–2755
75. Tsukube H, Shinoda S (2002) Lanthanide complexes in molecular recognition and chirality sensing of biological substrates. Chem Rev 102(6):2389–2404
76. de Lill DT, de Bettencourt-Dias A, Cahill CL (2007) Exploring lanthanide luminescence in metal-organic frameworks: synthesis, structure, and guest-sensitized luminescence of a mixed europium/terbium-adipate framework and a terbium-adipate framework. Inorg Chem 46(10):3960–3965
77. Ma D, Wang W, Li Y, Li J, Daiguebonne C, Calvez G, Guillou O (2010) In situ 2,5-pyrazinedicarboxylate and oxalate ligands synthesis leading to a microporous europium-organic framework capable of selective sensing of small molecules. CrystEngComm 12(12):4372–4377
78. Huang J, Xu Y, Chen X, Xu D, Xu Y, He Q (2012) Synthesis, characterization and properties of some rare earth complexes with 2,6-pyridine dicarboxylic acid and α-Picolinic acid. J Rare Earths 30(6):586–591
79. Mistri S, Zangrando E, Manna SC (2013) Cu(II) complexes of pyridine-2,6-dicarboxylate and N-donor neutral ligands: synthesis, crystal structure, thermal behavior, DFT calculation and effect of aromatic compounds on their fluorescence. Inorg Chim Acta 405:331–338
80. Jose SP, Mohan S (2006) Vibrational spectra and normal co-ordinate analysis of 2-aminopyridine and 2-amino picoline. Spectrochim Acta A Mol Biomol Spectrosc 64 (1):240–245
81. Dwivedi Y, Rai A, Rai SB (2009) Energy transfer in Er:Eu:Yb co-doped tellurite glasses: Yb as enhancer and quencher. J Lumin 129(6):629–633
82. Sharma G, Narula A (2015) Eu3+, Yb3+ and Eu3+—Yb3+ complexes with salicylic acid and 1,10-phenanthroline: synthesis, photoluminescent properties and energy transfer. J Fluoresc 25(2):355–360
83. Łyszczek R, Mazur L (2012) Polynuclear complexes constructed by lanthanides and pyridine-3,5-dicarboxylate ligand: Structures, thermal and luminescent properties. Polyhedron 41 (1):7–19

84. Beeby A, Clarkson IM, Dickins RS, Faulkner S, Parker D, Royle L, de Sousa AS, Gareth Williams JA, Woods M (1999) Non-radiative deactivation of the excited states of europium, terbium and ytterbium complexes by proximate energy-matched OH, NH and CH oscillators: an improved luminescence method for establishing solution hydration states. J Chem Soc Perkin Trans 2(3):493–504
85. Dwivedi Y, Thakur SN, Rai SB (2007) Study of frequency upconversion in Yb3+/Eu3+ by cooperative energy transfer in oxyfluoroborate glass matrix. Appl Phys B 89(1):45–51
86. Yang CH, Yang GF, Pan YX, Zhang QY (2009) Synthesis and spectroscopic properties of GdAl3(BO3)4 poly-crystals codoped with Yb3+ and Eu3+. J Fluoresc 19(1):105–109
87. Wang F, Liu X (2008) Upconversion multicolor fine-tuning: visible to near-infrared emission from lanthanide-doped NaYF4 nanoparticles. J Am Chem Soc 130(17):5642–5643
88. Wang F, Liu X (2009) Recent advances in the chemistry of lanthanide-doped upconversion nanocrystals. Chem Soc Rev 38(4):976–989
89. Ong LC, Gnanasammandhan MK, Nagarajan S, Zhang Y (2010) Upconversion: road to El Dorado of the fluorescence world. Luminescence 25(4):290–293
90. Qiu H, Chen G, Sun L, Hao S, Han G, Yang C (2011) Ethylenediaminetetraacetic acid (EDTA)-controlled synthesis of multicolor lanthanide doped BaYF5 upconversion nanocrystals. J Mater Chem 21(43):17202–17208
91. Yin A, Zhang Y, Sun L, Yan C (2010) Colloidal synthesis and blue based multicolor upconversion emissions of size and composition controlled monodisperse hexagonal NaYF4: Yb, Tm nanocrystals. Nanoscale 2(6):953–959
92. Yi G-S, Chow G-M (2007) Water-soluble NaYF4:Yb, Er(Tm)/NaYF4/polymer core/shell/shell nanoparticles with significant enhancement of upconversion fluorescence. Chem Mater 19(3):341–343
93. Liu C, Chen D (2007) Controlled synthesis of hexagon shaped lanthanide-doped LaF3 nanoplates with multicolor upconversion fluorescence. J Mater Chem 17(37):3875–3880
94. Wang L, Li Y (2006) Green upconversion nanocrystals for DNA detection. Chem Commun 24:2557–2559
95. Qin X, Yokomori T, Ju Y (2007) Flame synthesis and characterization of rare-earth (Er3+, Ho3+, and Tm3+) doped upconversion nanophosphors. Appl Phys Lett 90(7):073104
96. Heer S, Kömpe K, Güdel HU, Haase M (2004) Highly efficient multicolour upconversion emission in transparent colloids of lanthanide-doped NaYF4 nanocrystals. Adv Mater 16 (23–24):2102–2105
97. Yang J, Zhang C, Peng C, Li C, Wang L, Chai R, Lin J (2009) Controllable Red, green, blue (RGB) and bright white upconversion luminescence of Lu2O3:Yb3+/Er3+/Tm3+ nanocrystals through single laser excitation at 980 nm. Chem Eur J 15(18):4649–4655
98. Auzel F (2004) Upconversion and anti-stokes processes with f and d ions in solids. Chem Rev 104(1):139–174
99. Xu CT, Svensson N, Axelsson J, Svenmarker P, Somesfalean G, Chen G, Liang H, Liu H, Zhang Z, Andersson-Engels S (2008) Autofluorescence insensitive imaging using upconverting nanocrystals in scattering media. Appl Phys Lett 93(17):171103
100. Soukka T, Rantanen T, Kuningas K (2008) Photon upconversion in homogeneous fluorescence-based bioanalytical assays. Ann N Y Acad Sci 1130(1):188–200
101. Li C, Quan Z, Yang P, Huang S, Lian H, Lin J (2008) Shape-controllable synthesis and upconversion properties of lutetium fluoride (doped with Yb3+/Er3+) microcrystals by hydrothermal process. J Phys Chem C 112(35):13395–13404
102. Cheng L, Yang K, Li Y, Chen J, Wang C, Shao M, Lee S-T, Liu Z (2011) Facile preparation of multifunctional upconversion nanoprobes for multimodal imaging and dual-targeted photothermal therapy. Angew Chem 123(32):7523–7528
103. León-Luis SF, Rodríguez-Mendoza UR, Haro-González P, Martín IR, Lavín V (2012) Role of the host matrix on the thermal sensitivity of Er3+ luminescence in optical temperature sensors. Sens Actuators B Chem 174:176–186

104. Yi GS, Chow GM (2006) Synthesis of hexagonal-phase NaYF4:Yb, Er and NaYF4:Yb, Tm nanocrystals with efficient up-conversion fluorescence. Adv Funct Mater 16(18):2324–2329
105. Mai H-X, Zhang Y-W, Si R, Yan Z-G, Sun L-d, You L-P, Yan C-H (2006) High-quality sodium rare-earth fluoride nanocrystals: controlled synthesis and optical properties. J Am Chem Soc 128(19):6426–6436
106. Yi G-S, Chow G-M (2005) Colloidal LaF3:Yb, Er, LaF3:Yb, Ho and LaF3:Yb, Tm nanocrystals with multicolor upconversion fluorescence. J Mater Chem 15(41):4460–4464
107. Nabika H, Deki S (2003) Enhancing and quenching functions of silver nanoparticles on the luminescent properties of europium complex in the solution phase. J Phys Chem B 107 (35):9161–9164
108. Wang G, Peng Q, Li Y (2011) Lanthanide-doped nanocrystals: synthesis, optical-magnetic properties, and applications. Acc Chem Res 44(5):322–332
109. Demas JN, DeGraff BA (2001) Applications of luminescent transition platinum group metal complexes to sensor technology and molecular probes. Coord Chem Rev 211(1):317–351

Chapter 12
Interplay Between DNA-Binding/Catalytic Functions and Oligomerization of Retroviral Integrases Studied by a Combination of Time-Resolved Fluorescence Anisotropy, Fluorescence Correlation Spectroscopy and Resonance Energy Transfer

Olivier Delelis and Eric Deprez

Abstract Fluorescence based technologies are widely used to study enzyme properties and to understand their mode of action. In particular, fluorescence anisotropy (including both steady-state and time-resolved approaches), fluorescence correlation spectroscopy (FCS) and resonance energy transfer (FRET) were used for monitoring the DNA-binding and enzymatic activities of retroviral integrases. This protein is involved in the integration step of the viral genome that is crucial for retroviral replication. Thanks to these methodologies, important information has been obtained regarding the oligomeric status responsible for integrase catalytic activities as well as the positioning of integrase onto its DNA substrate. We present in this chapter a steady-state fluorescent anisotropy-based assay which monitors, in the same sample, both the DNA-binding step and the subsequent reaction catalysed by integrase allowing the separation of binding and catalytic parameters. The combination of this approach with the analysis of time-resolved fluorescence anisotropy decays is also particularly suitable for studying the relationship between the overall size of integrase-DNA complexes and the catalytic activity. The characterization of catalytically competent complexes by time-resolved fluorescence anisotropy led to the identification of the dimeric form of HIV-1 integrase as the optimal active form for catalytic activity. The protein:DNA ratio determines the aggregative properties of HIV-1 integrase, and high ratios led to aggregative and inactive forms. The better solubility of prototype foamy virus (PFV-1) integrase compared to HIV-1 integrase was compatible with protein labeling using an extrinsic fluorophore, allowing the study of the integrase/DNA interaction specificity by FRET between the fluorescently-labeled PFV-1 integrase and viral DNA

O. Delelis • E. Deprez (✉)
LBPA (Laboratoire de Biologie et Pharmacologie Appliquée), CNRS UMR8113, IDA FR3242, ENS Cachan, Université Paris-Saclay, 61 avenue President Wilson, F-94235 Cachan, France
e-mail: delelis@lbpa.ens-cachan.fr; deprez@lbpa.ens-cachan.fr

© Springer International Publishing Switzerland 2016
C.D. Geddes (ed.), *Reviews in Fluorescence 2015*,
DOI 10.1007/978-3-319-24609-3_12

ends. It has also been possible to investigate, by a combined approach of FCS and time-resolved fluorescence anisotropy, the oligomeric state of PFV-1 integrase. The results show a monomer-dimer equilibrium of PFV-1 integrase in solution and that DNA promotes protein dimerization although high protein:DNA ratios led to aggregated complexes, confirming results obtained with HIV-1 integrase. The use of fluorescence-based assays for the study of the mechanism of action of anti-integrase compounds is also discussed.

Keywords Fluorescence anisotropy • Fluorescence correlation spectroscopy • Multimerization • Protein-DNA interactions • Integrase

12.1 Introduction

Integration of a HIV-1 DNA copy genome into the host genome is crucial for efficient HIV-1 replication [1, 2]. The viral protein integrase (IN), coded by the *pol* gene of the virus, is involved in this step. HIV-1 IN (288 amino-acid residues) consists of three functional domains. The central or catalytic core domain (CC, from position 50 to 213) is strictly required for catalysis and contains the well-known Asp-Asp-Glu catalytic triad that coordinates one or two metallic cofactors (Mg^{2+} or Mn^{2+}), most likely a pair coordinated by three carboxylate groups of the triad in the catalytically competent form, based on the X-ray structure of the prototype foamy virus integrase (PFV-1 IN) [3]. The CC domain is flanked by the N-terminal domain (NTD, from pos. 1 to 49) and the C-terminal domain (CTD, from pos. 214 to 288). This domain establishes specific contacts with the viral DNA and, together with the NTD and CTD, contributes to complex stabilization [3–9]. The NTD contains a conserved non-conventional HHCC motif. The binding of zinc to this motif ensures a proper domain folding and promotes IN multimerization [10–12]. However, the integrity of the HHCC motif is crucial in the stringent Mg^{2+}-context but appears dispensable under the less stringent Mn^{2+} condition [13], suggesting a role of the NTD in the establishment of specific and physiologically relevant IN/DNA complexes.

Integration results from two consecutive reactions that constitute the overall integration process. The first reaction, named 3′-processing (3′-P), requires cleavage of the 3′-terminal GT dinucleotide at each viral DNA end. The hydroxyl groups of the newly recessed 3′-ends are then used in the second reaction, named strand transfer (ST) for the covalent joining of viral and target DNAs, resulting in full-site integration. For both reactions, IN functions as a multimer, most likely a dimer for 3′-P and a tetramer (dimer of a dimer) for full-site integration [3, 14–19]. Two other reactions occur *in vitro*, (i) the reversal of the half-site integration process, named disintegration [20], and (ii) a specific internal cleavage occurring on a symmetrical palindromic DNA site [21–23]. All reactions require a metallic cofactor, and, except for disintegration [24, 25], strictly require the full-length IN. The catalysis in the presence of Mg^{2+} exhibits weaker non-specific endonucleolytic cleavage and the tolerance of sequence variation at the viral DNA ends is greater in the presence of Mn^{2+} compared to Mg^{2+}, suggesting that Mg^{2+} is more physiologically relevant as a cofactor than Mn^{2+} [8, 26–28]. The ST reaction also leads to distinct profiles of products and the potency of inhibitors can be differently appreciated, depending on the nature of the cofactor used [29, 30].

IN is sufficient for 3′-P catalysis *in vitro*, using short-length oligodeoxynucleotides that mimic one viral long terminal repeat (LTR), in the presence of Mg^{2+}/Mn^{2+}. The 3′-P reaction generates two products: the viral DNA containing the recessed end and the released GT dinucleotide. The first product of the reaction, the processed viral DNA, serves as a substrate together with the target DNA for the subsequent joining reaction. The 3′-P reaction is highly specific and the terminal 13 base pairs of the viral DNA end are important for the reaction specificity in the presence of Mg^{2+} [8, 26]. In particular, the CA sequence preceding the GT dinucleotide is strictly required under both Mg^{2+} and Mn^{2+} conditions. Other positions are crucial for 3′-P activity under Mg^{2+} conditions [8, 26]. However, *in vitro*, there is no IN preference for the cognate U5 or U3 terminal viral DNA sequence at the DNA-binding level and no difference in affinity is generally observed in terms of DNA sequence [7–9, 26, 31]. Consequently, the reaction specificity is fully explained at the catalytic level. Due to poor solubility properties of HIV-1 IN, it is a difficult task to determine which enzymatic properties are related to intrinsic properties of retroviral INs, and which properties rely on aggregative properties. The use of detergent or site-directed mutagenesis may improve the solubility but may also cause changes to IN properties. For instance, detergent may be detrimental to 3′-P activity in the presence of Mg^{2+} [29] and one soluble mutant was found to be intrinsically resistant to an anti-IN compound [32]. PFV-1 IN, hortologue of the HIV-1 protein, shares common structural and functional features with other retroviral INs, such as three-domain organization as well as the catalytic activities, 3′-P, ST, disintegration and internal cleavage on symmetrical sequences [22, 33–35]. Importantly, PFV-1 IN displays a more soluble profile compared to HIV-1 IN [15]. To date, many X-ray structures of the full-length PFV-1 IN are available, including bound to viral DNA end with or without target DNA [3, 36] as well as in the presence of IN inhibitors [37, 38]. By contrast, no structure is available for the full-length HIV-1 protein, although several structures exist: DNA-free single- or double-domains, alone or complexed with the IN-binding domain of the cellular partner LEDGF [39–43]. Consequently, biochemical and structural studies of HIV-1 IN have been severely impeded for solubility reasons and several structural determinants of protein activity are unclear, in particular concerning its multimeric status with regard to each catalytic activity.

Based on topological considerations concerning the distance separating the active sites of the two protomers, which is more than double the distance separating the two joining sites on the target DNA (separated by 5 bp), a multimeric state of higher order than dimers, tetramers at least, is required for complete integration. However, the oligomeric state of IN (DNA-free in solution or bound to the viral DNA end) and the multimeric state responsible for the 3′-P reaction was uncertain, in particular regarding the modulation of the 3′-P activity by the self-association properties of IN. Thanks to a combination of fluorescence-based methodologies such as fluorescence anisotropy (real-time and steady-state), FCS (Fluorescence correlation spectroscopy) and FRET (Fluorescence resonance energy transfer), crucial information on DNA-binding and catalytic properties of IN were provided. The comparison between HIV-1 and PFV-1 INs was also an important point to highlight the proper self-association properties of retroviral integrases. More particularly, we present a fluorescence steady-state anisotropy assay for monitoring DNA-binding and subsequent 3′-P in the same sample, allowing the study of real-time 3′-P kinetics and the separation of binding and catalytic parameters [17]. The use of a DNA substrate containing a fluorescent probe covalently linked to the GT dinucleotide (released upon 3′-P) allowed the simultaneous monitoring of DNA-binding and GT release since both steps influence the size of the fluorescent moiety. The steady-state anisotropy-based technique in combination with the analysis of time-resolved fluorescence anisotropy decays [10, 44–47] was also particularly suitable for studies of the relationship between the overall size of IN-DNA complexes and 3′-P activity. A more precise characterization of catalytically competent complexes by time-resolved fluorescence anisotropy led to the identification of the dimeric form of IN as the optimal active form for 3′-P catalysis [15, 17]. Time-resolved fluorescence results also highlight that the IN:DNA ratio determines both the aggregation and catalytic properties of IN, and that DNA-binding stimulates the self-organization of IN to give a catalytically competent non-aggregative form, even though high IN:DNA ratios lead to aggregative and inactive forms for both HIV-1 and PFV-1 INs.

To get deeper insight into the multimerization properties of retroviral INs and IN/DNA recognition mechanism, we used transglutaminase (TGase)-mediated labeling of PFV-1 IN by an extrinsic fluorophore emitting in the visible part of the spectrum [15]. In contrast to conventional chemical labeling methods, such a strategy leads to a site-specific labeling of the protein of interest, avoiding multiply labeled proteins and heterogeneous samples, and then is particularly well suited for FRET or FCS approaches. This enabled us to successfully study self-assembly properties of IN by FCS and compared the results to those obtained by time-resolved fluorescence anisotropy using intrinsic fluorescence of tryptophan residues. Furthermore, FRET and DNA-binding anisotropy-based assay were conducted to assess the specificity of IN/DNA interactions. A specific mode of interaction in terms of sequence recognition was observed for PFV-1 IN, but not for HIV-1 IN. Altogether, the different approaches show that DNA promotes the catalytically competent dimeric state for accomplishing 3′-P. However, this dimerization process is in competition with other processes detrimental for activity such as IN polymerization and aggregation onto DNA, mediated by nonspecific IN-DNA and IN-IN interactions, respectively.

12.2 Monitoring Simultaneously DNA-Binding and 3′-Processing Activity of Integrase by Steady-State Fluorescence Anisotropy

Fluorescence anisotropy measurements are based on the principle of photoselective excitation of a fluorophore by a polarized light, providing information about rotational motions of the fluorophore or the fluorescently labeled molecule between photon absorption and emission [48]. The overall rotational diffusion or internal dynamics/flexibility of the fluorophore are major causes of light depolarization leading to low anisotropy level whereas high levels of fluorescence anisotropy correspond to large molecules or complexes characterized by slow rotational diffusion or low flexibility level of the fluorescent moiety. In order to monitor the binding of IN to viral DNA substrate and the subsequent 3′-P reaction, in the same assay, we used an extrinsic fluorophore (fluorescein) covalently linked to DNA. The fluorescein was attached to the thymidine group of the GT dinucleotide which is released during the 3′-P reaction [17].

Both DNA binding and 3′-P have a significant effect on the steady-state fluorescence anisotropy value since each step has a major effect on the molecular size of the fluorescent moiety: the binding of IN to DNA and the dinucleotide release are supposed to significantly and sequentially increase and decrease the fluorescence anisotropy, respectively. The principle underlying the anisotropy-based assay is summarized in Fig. 12.1. The binding of IN to a double-stranded 21-mer DNA mimicking one end of the viral DNA (U5) increases the steady-state anisotropy value (*r*), most likely by restricting flexibility, allowing to determine the fractional saturation function. Several 21-mer DNA substrates were tested, each labeled with a fluorescein group at one end, in all possible combinations (5′- or 3′-end of the *reactive strand (a)* or the *non-transferred* strand *(b)*), or directly labeled on the 2′-amino group of a 3′-terminal 2′-aminouridine on the *reactive strand* (Fig. 12.2a). The IN-free double-stranded DNAs were characterized by steady-state fluorescence anisotropy of 0.060–0.130 at 25 °C, depending on the location of fluorescein [17]; the value characterizing the fluorescein-labeled dinucleotide was 0.02. DNA-binding of IN led to a significant increase in steady-state anisotropy value, higher than 0.2. Equilibrium was reached after 15 min [49, 50]. The steady-state anisotropy value obtained after DNA-binding defines the initial value for catalysis ($= r_{t=0}$) (see below). In the presence of Mg^{2+}, all tested DNAs gave similar $r_{t=0}$ values (0.22), indicating that the position of fluorescein does not influence the binding of IN to DNA (Fig. 12.2b). To note, in the absence of Mg^{2+}, the *r* values were larger (0.25), showing that (i) the binding of IN to DNA is not strictly dependent on the presence of Mg^{2+} [6, 7, 51] and (ii) aggregation of IN is favored by an absence of divalent cation [10].

Following DNA-binding, the reaction was started by shifting the sample to a permissive temperature: 37 °C. Among the different DNA substrates, only HIV-*a*3F and HIV-aUF with the fluorescein directly attached to the dinucleotide reaction

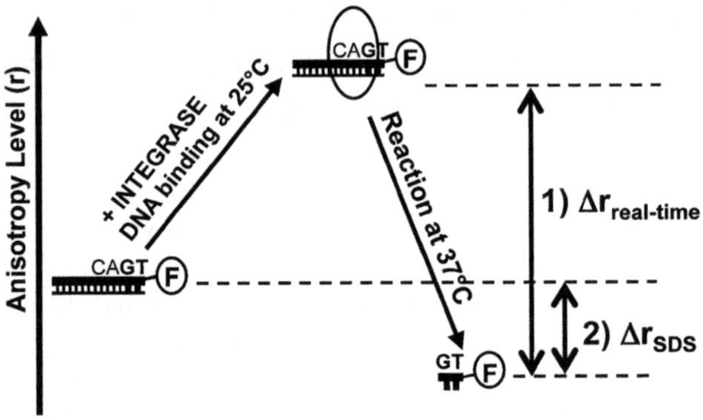

Fig. 12.1 Principle of the 3′-P activity assay based on the measurement of the steady-state fluorescence anisotropy. Upon IN binding to F-labeled DNA, the value of the steady-state anisotropy (*r*) increases. By shifting the temperature to 37 °C, the release of the GT dinucleotide (concomitant with the 3′-P activity) leads to *r* decrease. The fraction of released dinucleotide can be estimated under real-time or fixed-time conditions. In the latter case, SDS is used to quench the reaction (This research was originally published in The Journal of Biological Chemistry. Guiot et al. [17]. © the American Society for Biochemistry and Molecular Biology)

product (linked to the 3′-terminal GT or on the 2′-amino group of a 3′-terminal 2′-aminouridine, respectively), displays a significant decrease in the steady-state anisotropy value while in other cases, *i.e.* when the fluorophore is linked to other DNA ends, no significant decrease of r is observed, confirming that the cleavage of the terminal dinucleotide is specific of the CAGT extremity (Fig. 12.2b). Indeed, HIV-*a*5F, HIV-*b*5F and HIV-*b*3F displayed no significant decrease under either real-time or fixed-time conditions. Moreover, the decrease observed with HIV-*a*3F or HIV-aUF is strictly related to the presence of the metallic cofactor. Considering real-time conditions, this result also suggests that the processed DNA product remains tightly bound to the enzyme after 3′-P and that no dissociation of the IN-DNA complex occurs during the reaction time-course. This explains, at least in part, the single-turnover property of IN, even under conditions of excess DNA substrate [52]. By contrast, the terminal GT dinucleotide is released from the complex after 3′-P. In conclusion, the steady-state anisotropy-based assay is particularly suitable for monitoring the 3′-P reaction, but only if fluorescein is attached to the GT dinucleotide.

A precise quantification of the 3′-P time course can be done:

(i) The fraction of released GT can be determined from the real-time decrease in *r* observed during the reaction; $\Delta r_{real\text{-}time}$ corresponds to the difference in fluorescence anisotropy between IN-DNA complex and free GT (Fig. 12.1).
(ii) After the disruption of all IN-DNA complexes by the addition of SDS; Δr_{SDS} corresponds to the difference in fluorescence anisotropy between the unprocessed double-stranded 21-mer DNA and free GT (Fig. 12.1).

These two approaches, real-time and fixed-time, respectively, allow quantification of the released GT dinucleotide. Unlike most standard gel-electrophoresis-

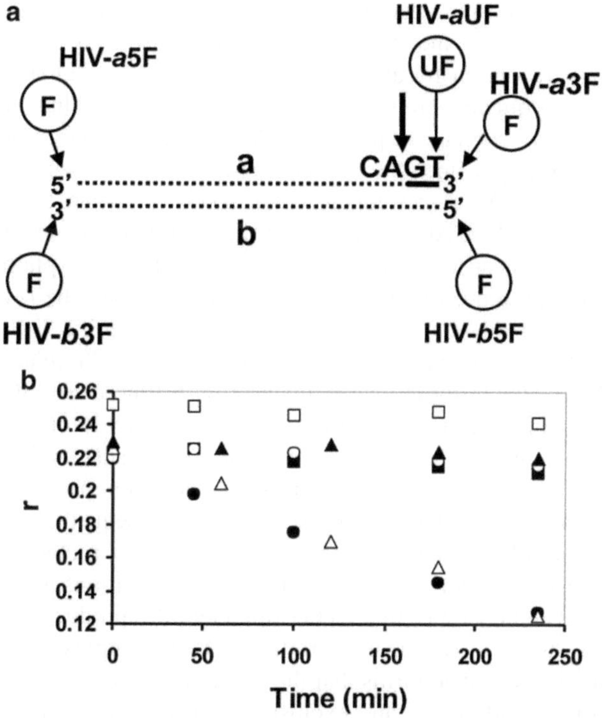

Fig. 12.2 Specific decrease in steady-state fluorescence anisotropy (r) during 3′-P. (**a**) Fluorescein was linked to the 3′- or 5′- extremity of each strand (*a* is the reactive strand containing the CAGT terminal sequence and *b* the complementary strand). HIV-*a*UF has a fluorescein-labeled uridine at the 3′-terminal position of strand *a*. (**b**) Decrease of r is observed only (i) when fluorescein is attached to the GT dinucleotide and (ii) in the presence of Mg^{2+}: *White squares*: HIV-*a*3F, no Mg^{2+}; black circles: HIV-*a*3F + Mg^{2+}; *white circles*: HIV-*b*3F + Mg^{2+}; *black squares*: HIV-*a*5F + Mg^{2+}; *black triangles*: HIV-*b*3F + Mg^{2+}; *white triangles*: HIV-*a*UF + Mg^{2+} (This research was originally published in The Journal of Biological Chemistry. Guiot et al. [17]. © the American Society for Biochemistry and Molecular Biology)

based 3′-P assays, which rely on the quantification of the radiolabeled processed strand (*e.g.* a 19-mer product starting from a 21-mer substrate), the fluorescence anisotropy-based assay allows direct measurement of the released dinucleotide. The steady-state anisotropy-based assay is sensitive (detection of low nanomolar concentrations of products), suitable for real-time kinetic studies of 3′-P activity, and does not require separation of the processed 19-mer product from the unprocessed 21-mer DNA substrate by gel electrophoresis.

The steady-state anisotropy-based assay was also used to study the sequence-dependent specificity of the 3′-P. Previous reports highlight the crucial requirement of the conserved CA sequence preceding the GT dinucleotide [8]. The 3′-P reaction was performed using various DNA sequences such as a variant sequence in which the 3′-terminal CAGT sequence is replaced by GTGT or several non-specific random sequences. We found that the DNA-binding step was not influenced by the DNA sequence. Indeed, the final steady-state anisotropy values (*i.e.* after DNA-binding) were similar for all DNAs, regardless of the sequence. By contrast, the nature of the sequence strongly influenced anisotropy values under catalytic conditions. Only the cognate sequence (HIV-*a*3F) gave a large decrease in the steady-state anisotropy value. Such a decrease was not observed when the 3-′-terminal sequence CAGT was replaced by GTGT or when using non-specific sequences, even with the CAGT sequence at the 3′ end. These results confirm that the CA dinucleotide is strictly required but not sufficient to explain 3′-P activity, particularly in the presence of Mg^{2+} [8, 26]. Taken together, the results show that the decrease in the steady-state anisotropy is related to specific 3′-P activity.

Fig. 12.3a displays one example of the time-course of GT release in real-time conditions. The 3′-P activity increases linearly with time over 180 min which is compatible with single-turnover conditions (excess of IN over DNA), when the kinetic rate constant of cleavage (or chemical constant) is intrinsically low ($k = 0.0035\ min^{-1}$). This time-course of product formation provides a measure of the actual cleavage step, which is not affected by subsequent turnovers [52]. A similar result was obtained using the fixed-time (SDS) method. The catalytic activities as measured under real-time or fixed-time conditions are well-correlated (Fig. 12.3b). This demonstrates that all GT dinucleotides produced by 3′-P reaction in the real-time assay were released from IN-DNA complexes and not trapped within the complexes, and confirms that the decrease in the steady-state anisotropy value in real-time condition is due exclusively to the GT release from IN-DNA complexes, rather than another phenomenon which could be the dissociation of IN from the DNA substrate/or product.

However, we found that the initial steady-state anisotropy value as obtained after the DNA-binding step ($r_{t=0}$) is not predictive of the 3′-P. This value is related to the fractional saturation function and the 3′-P activity is expected to increase with the fractional saturation function under single-turnover conditions, and the fraction of released GT should be proportional to the amount of pre-formed IN-DNA complexes. Unexpectedly, a plot of 3′-P activity against $r_{t=0}$ displays a bell-shaped response (Fig. 12.4). The increasing phase can be explained by two phenomena, not mutually exclusive: the amount of free DNA molecules which decreases with increasing IN:DNA ratio and self-assembly of IN upon DNA-binding which increases the size of IN-DNA complexes. The decreasing phase suggests that higher-order multimeric forms of IN on DNA, possibly aggregates, are less active than lower-order multimeric forms. These higher-order multimeric forms are favored under conditions of high IN:DNA ratio and lead to large steady-state anisotropy values. A more precise characterization of these complexes will be presented in the time-resolved fluorescence anisotropy section.

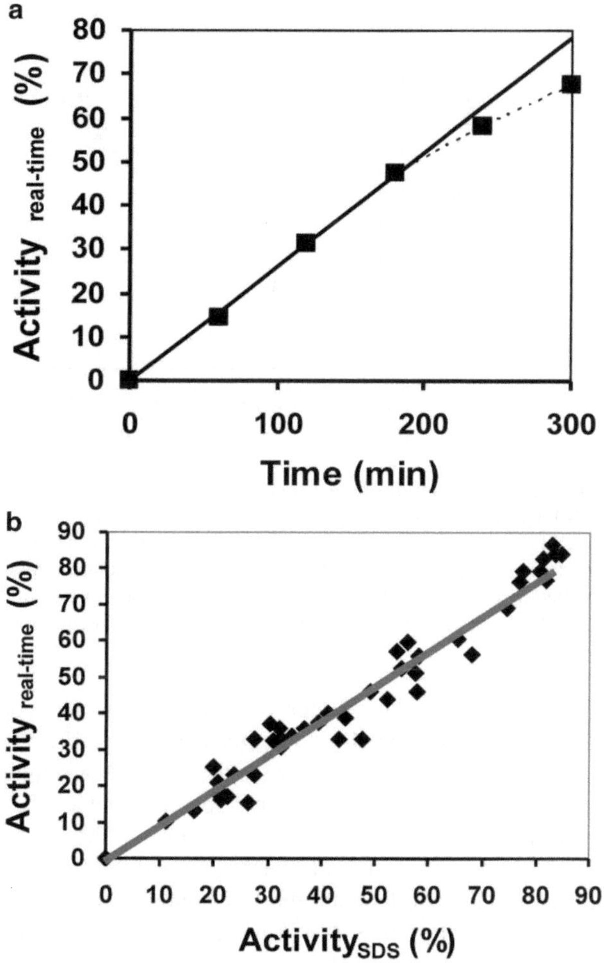

Fig. 12.3 Time-course of the 3′-P activity of IN. (**a**) Time-course of Mg^{2+}-dependent 3′-P activity obtained with the real-time assay using HIV-*a*3F DNA substrate (= $Activity_{real-time}$). (**b**) Correlation between fixed-time ($Activity_{SDS}$) and real-time activities. $Activity_{SDS}$ was measured after disruption of IN-DNA complexes by SDS (This research was originally published in The Journal of Biological Chemistry. Guiot. et al. [17]. © the American Society for Biochemistry and Molecular Biology)

12.3 Deeper Insight into the Specificity of IN-DNA Interaction Using Steady-State Fluorescence Anisotropy and FRET Approaches

A careful analysis of the DNA-binding step, as studied by steady-state fluorescence anisotropy, revealed that a cooperative binding mode of IN to DNA was associated with the formation of a specific and catalytically active complex. IN binding to a fluorescently labeled DNA mimicking the HIV-1 U5 DNA end as mentioned above, is cooperative in the presence of Mg^{2+} and characterized by a Hill coefficient ($\tilde{n}$) of 1.97 [53], consistent with both time-resolved fluorescence anisotropy ([15, 17, 44]; see below) and cross-linking [16] experiments showing that the dimeric IN corresponds to the catalytically active form for the 3′-P reaction. This result supports a

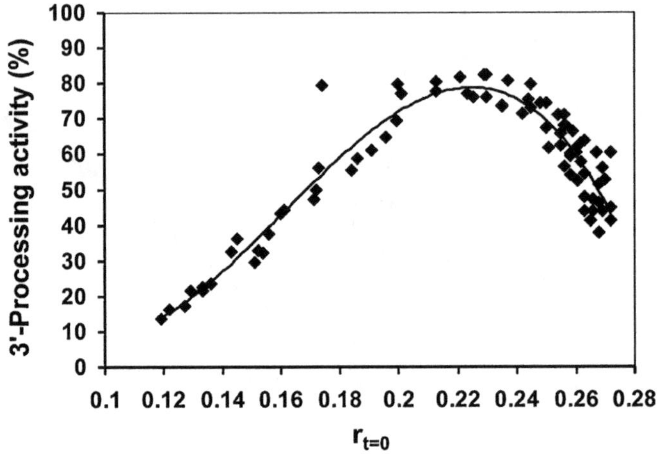

Fig. 12.4 3′-P activity as a function of the steady-state anisotropy characterizing the IN-DNA complex. Various IN/DNA ratios were incubated at 25 °C until *r* reached a plateau. The equilibrium $r_{t=0}$ value (= anisotropy before the start of the reaction) was recorded before shifting the temperature to 37 °C. 3′-P activity was then calculated from Δr_{SDS} values (see Fig. 12.1) and plotted against $r_{t=0}$ (This research was originally published in The Journal of Biological Chemistry. Guiot et al. [17]. © the American Society for Biochemistry and Molecular Biology)

model in which the dimer originates from the cooperative assembly of two protomeric units on the viral DNA end [15, 17, 49, 53]. It is known that 3′-P activity is much more sensitive to the viral DNA sequence in the Mg^{2+} context than the Mn^{2+} context, suggesting that specific contacts between IN and the viral DNA extremity are more favored in the presence of Mg^{2+} [8, 26]. Using the cognate viral DNA sequence, we found that Mg^{2+}- and Mn^{2+}-dependent DNA-binding display different Hill coefficients; the DNA-binding process was more cooperative in the presence of Mg^{2+} than Mn^{2+}: $\tilde{n} = 1.97$ and 1.29 for Mg^{2+} and Mn^{2+}, respectively. As mentioned above, IN does not require a metallic cofactor for DNA-binding [7, 17]. The corresponding $\tilde{n}$ value (= 1.34) in absence of divalent cation was found to be similar to the value obtained with Mn^{2+} or using non-cognate (random) sequences in the presence of Mg^{2+} ($\tilde{n} \approx 1.2$) showing that the cooperative DNA-binding mode requires both the cognate viral sequence and Mg^{2+} as a cofactor. To note, the modulation of the Hill coefficient is not correlated to the extent of IN oligomerization since the long rotational correlation times obtained for all the studied complexes (IN-Mg^{2+}-viral sequence, IN-Mn^{2+}-viral sequence and IN-Mg^{2+}-random sequence) are similar (≈40 ns [53]) and compatible with a dimeric IN bound to DNA (see the time-resolved fluorescence anisotropy section below).

Regarding the Hill coefficients characterizing truncated proteins in the presence of Mg^{2+} (Table 12.1), only NTD-CC displayed high cooperativity ($\tilde{n} = 1.9$) comparable to the full-length IN. By contrast, CC-CTD and CC proteins, lacking the NTD, were characterized by lower Hill numbers (1.38–1.45). This suggests that the

Table 12.1 Hill coefficients for the binding of the full-length and truncated IN to the HIV-1 U5 extremity in the presence or absence of DIBA-1 as determined by steady-state fluorescence anisotropy

	Hill coefficient					
	10 mM Mg^{2+}		10 mM Mn^{2+}		No divalent cation	
IN	DIBA –	DIBA +	DIBA –	DIBA +	DIBA –	DIBA +
Full-length	1.97 ± 0.10	1.47 ± 0.02	1.29 ± 0.09	1.20 ± 0.10	1.34 ± 0.08	1.37 ± 0.09
CC-CTD	1.45 ± 0.12	1.41 ± 0.10	1.20 ± 0.06	1.30 ± 0.13	nd	nd
NTD-CC	1.90 ± 0.10	1.43 ± 0.12	1.24 ± 0.07	1.21 ± 0.11	nd	nd
CC	1.38 ± 0.09	1.32 ± 0.06	1.26 ± 0.13	1.20 ± 0.05	nd	nd

Nucleic Acids Research. Copyright © 2010

nd not determined

NTD is responsible for the cooperative DNA-binding mode and thus plays a crucial role to form a specific and catalytically competent IN-Mg^{2+}/DNA complex. A zinc ejector compound, DIBA-1, which reacts with the HHCC motif in the NTD also significantly modulates the Hill coefficient (Table 12.1), suggesting that the specific cooperative DNA-binding mode strictly requires the cognate viral sequence, Mg^{2+} as a cofactor and zinc as a positive allosteric effector. The role of the NTD could be indirect via IN-IN interaction, consistent with the effect of zinc on the multimerization process as reported by others [11, 12, 54, 55]. In this context, a potential role of a NTD-NTD interaction in the cooperative response is unlikely [56, 57]. The X-ray structure of the full-length PFV-1 IN highlights an interaction between the NTD and CC domains, with the CC domain of one protomer in interaction (in trans) with the NTD of another protomer, suggesting that the zinc finger motif in the NTD mediates cooperativity but does not imply a NTD-NTD interaction [3]. Alternatively, the role of the NTD domain in the cooperative response could be related to a direct role in DNA-binding [3, 58]. Interestingly, the NTD is directly affected by conformational metal-induced changes [59, 60], suggesting that this region could be crucial for establishing specific IN-viral DNA contacts and mediating the differential effects of Mg^{2+}/Mn^{2+}.

To get deeper insight into the IN/DNA recognition mechanism, we performed FRET experiments to assess the specificity of IN/DNA interaction. We successfully applied transglutaminase (TGase)-mediated TAMRA (acceptor) labeling [15, 61] of PFV-1 IN that displays higher solubility properties compared to HIV-1 IN. This enzymatic labeling occurs thanks to a specific acyl transfer reaction mediated by TGase of a TAMRA-cadaverine compound on the γ-carboxyamide group of a Gln residue of the PKPQQFM peptide (fused in this example to the C-terminal IN end). This reaction leads to IN being labeled at a single position (confirmed by mass spectrometry analysis [15]). The labeling yield of purified TAMRA labeled-IN (IN^T) was estimated at 50 % by both absorbance spectroscopy and FCS [15]. DNAs mimicking the U5 viral DNA end of various lengths (from 21- to 300-mer) were fluorescein (donor)-labeled at the 5′-end, either at the processing end or at the opposite end (Fig. 12.5a). Figure 12.5b shows one example of the quenching of the donor (q_D) (measured at 520 nm) plotted against PFV-1 IN^T concentration. q_D increased as the IN^T concentration increased and reached a plateau (maximum quenching efficiency: 78 %). The plateau value is consistent with the theoretical value (at least 75 %), assuming a labeling yield of 50 % and a minimal IN:DNA stoichiometry of 2:1, as mentioned above. The decrease in the donor intensity actually corresponds to a resonance energy transfer since the $q_{D,max}$ value decreases continuously by decreasing the IN^T/unlabeled IN ratio (Fig. 12.5b).

The extent of q_D was found to be dependent on the DNA length (Fig. 12.5c): for a given IN concentration, q_D was systematically higher for short DNAs. This result demonstrates that IN positioning on DNA is the main factor influencing the dependence of q_D on the DNA length. Only positions of IN^T in close proximity to donor can significantly influence the energy transfer ($<$ 60–80 Å for the Fl-TAMRA pair). The decrease in quenching amplitude as obtained with longer

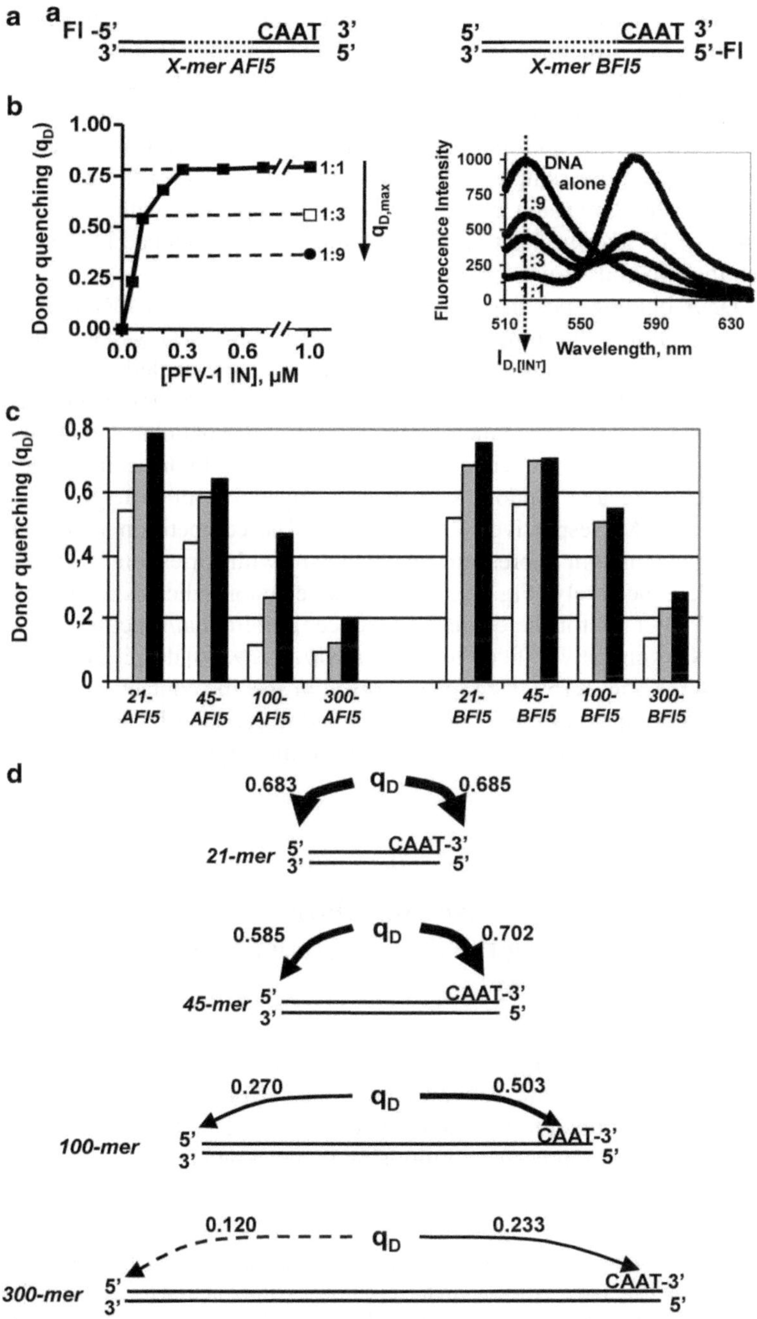

Fig. 12.5 FRET between viral DNA and PFV-1 IN. (**a**) Substrates used in the FRET study. The fluorescein (Fl) donor was attached at the 5′-end of either the reactive strand (a: *left*) or the non-transferred strand (b: *right*) leading to the X-mer AFl5 or BFl5 substrate, respectively (with X = 21, 45, 100 or 300 bp). (**b**) Quenching of the Fl donor (q_D) between the 21-mer AFl5 and IN^T (*left*). The plateau value is dependent on the labeled/unlabeled IN ratio (indicated on the plot).

DNA molecules shows that IN may occupy many internal positions in DNA leading to a pearl chain structure. In addition, except for the 21-mer DNA, q_D was systematically higher when fluorescein was attached on the terminal processing site compared to the opposite site, for a given DNA size (from 45-to 300-mer), up to two-fold higher for the 300-mer substrate (Fig. 12.5d), indicating a preference for IN binding at the processing end. The absence of any bias for the 21-mer DNA is explained by the dimension of DNA which is comparable with both the overall dimension of IN and the Förster distance, thus a net difference in the quenching between the two ends is not expected.

The DNA-binding specificity of PFV-1 IN as suggested by FRET experiments was next confirmed by competition experiments using steady-state fluorescence anisotropy. PFV-1 IN was incubated with fluorescein-labeled DNA, either the cognate 45-mer PFV-1 U5 sequence (Fig. 12.6a) or a 45-mer random sequence (Fig. 12.6b) in the presence of increasing concentrations of unlabeled DNA (45-mer PFV-1 U5 or random sequence). Using the cognate fluorescein-labeled DNA, unlabeled cognate and random sequences competed with different IC_{50} values (1.4 and >2.5 μM, respectively) (Fig. 12.6a). The competition was consistently easier using the random fluorescein-labeled DNA, with corresponding values of 0.5 and 1.6 μM, respectively (Fig. 12.6b). Such a behavior indicates a higher apparent affinity of PFV-1 IN for the cognate sequence. Preferential/specific DNA-binding of HIV-1 IN is generally difficult to assess *in vitro* [7, 9, 26] despite several residues (Q148, Y143, K156, K159, K14) are involved in specific contacts with DNA [8, 9, 58, 62, 63]. Similar experiments were performed with HIV-1 IN and confirm that no significant difference can be observed between the cognate and a random 45-mer DNA sequence (Fig. 12.6c, d).

12.4 Study of Competitive and Allosteric Integrase Inhibitors Using the Steady-State Fluorescence Anisotropy Assay

The emergence of viral strains resistant against available drugs and the dynamic nature of the HIV-1 genome support a continued effort towards the discovery and characterization of novel targets and anti-viral drugs. Due to its central role in the HIV-1 life cycle, IN represents a promising therapeutic target. In the past, *in vitro*

Fig. 12.5 (continued) Right, emission spectra for the various mixtures of labeled and unlabeled IN in the presence of the 21-mer AFl5 substrate. (**c**) q_D values depending on the size of different DNA substrates at three different IN concentrations: 100 nM (*white squares*), 200 nM (*gray squares*) and 300 nM (*black squares*). (**d**) Summarize of the quenching amplitudes between the two DNA ends as a function of DNA length (corresponding to IN = 200 nM) (This research was originally published in The Journal of Biological Chemistry. Delelis et al. [15]. © the American Society for Biochemistry and Molecular Biology)

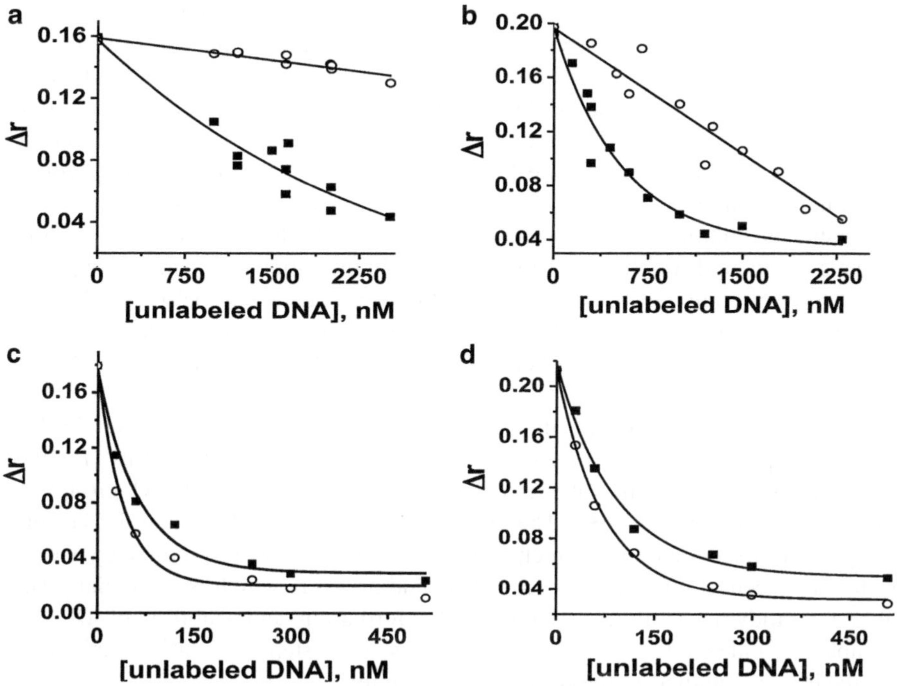

Fig. 12.6 Comparison of PFV-1 and HIV-1 DNA binding properties by steady-state fluorescence anisotropy. IN (*A* and *B* : PFV-1; *C* and *D* : HIV-1) was incubated with Fl-labeled 45-mer DNA (*A* and *C* : cognate sequence; *B* and *D* : random sequence) and varying concentrations of unlabeled 45-mer DNA competitor (*black square*: cognate sequence; *white circles*: random sequence). $\Delta r = r_{complex} - r_{free,DNA}$ (This research was originally published in The Journal of Biological Chemistry. Delelis et al. [15]. © the American Society for Biochemistry and Molecular Biology)

IN assays were extensively used to find IN inhibitors [64, 65]. Current inhibitors can be separated into three main classes, depending on their mechanisms of action: *(i)* Compounds that competitively prevent the binding of IN to the viral DNA and, consequently inhibit the 3′-P reaction. This group is referred to as IN DNA-binding inhibitors (INBI) and includes styrylquinoline compounds [49, 66–68]. *(ii)* Compounds that cannot bind to the DNA-free IN. They bind to the pre-formed IN-viral DNA complex only. These compounds inhibit ST over the 3′-processing reaction probably by displacing the viral DNA end from the active site or/and inhibiting the binding of IN to the target DNA [3, 38, 62, 69–73]; this group is referred to as IN strand transfer inhibitors (INSTI). *(iii)* Compounds that inhibit the interaction between IN and the cellular partner LEDGF which promotes and improve integration efficiency [74, 75]. This group is referred to as LEDGINs [72]. The well-known diketo acids (DKAs) and their more recent FDA-approved derivatives Elvitegravir (EVG), Raltegravir (RAL) and Dolutegravir (DTG) belong to the INSTI family and have good *ex vivo* activities against HIV replication.

For the two first groups, resistance mutations were identified [66, 70, 76–78]. Difficulties in deeply understanding and separating their mechanisms of action are closely related to the absence of structural data that clearly delineate the donor (viral DNA) and acceptor (target DNA) DNA-binding sites in the active site. INBI and INSTI are supposed to be coordinated to at least one or two divalent cations, respectively, in the IN active site [37, 38, 49]. The basis of drug action remains essentially unknown. In this context, the steady-state fluorescence anisotropy parameter is suitable to address the effect and the mechanism of action of small organic compounds on the DNA-binding step. In the following section, we present two family of compounds: (i) Styrylquinoline compounds that belong to the INBI family as mentioned above and (ii) DIBA-1, a zinc ejector compound which displays an intermediary profile between INBIs and INSTIs in terms of mechanism of action.

Styrylquinoline compounds inhibit *in vitro* 3′-P activity of HIV-1 IN, with IC_{50} values comprised between 0.5 and 5 μM. Some of them are efficient *ex vivo* with IC_{50} values in the micromolar range. The 7-COOH and the 8-OH groups are essential for inhibition [68]. The effects of FZ55, KHD161, and FZ41 three styrylquinoline compounds which inhibit 3′-P activity were studied by steady-state fluorescence anisotropy-based DNA-binding assay which is particularly suitable for studying inhibitor effects on the formation and stability of IN-DNA complexes, independently of catalysis. Below, both steps, *i.e.* DNA-binding and catalysis, were studied in the presence of inhibitor.

The fluorescein-labeled DNAs, free in solution, were characterized by a low anisotropy value, 0.048 and 0.060 at 37 °C and 25 °C, respectively. The binding of IN to DNA leads to an increase in the steady-state fluorescence anisotropy. The upper limits of anisotropy values were 0.232 and 0.240 at 37 and 25 °C, respectively, yielding a maximum amplitude of about 0.180. This value decreases in the presence of increasing concentrations of KHD161 accounting for a net decrease in the amount of IN-DNA complexes (Fig. 12.7) [49]. Data analysis of KHD161 curves yielded IC_{50} values of 1.50 and 0.98 μM at 37 °C and 25 °C, respectively. The IC_{50} characterizing the DNA-binding step at 37 °C is compatible with the value found in the 3′-P assay under similar experimental concentrations: $IC_{50} = 1.2$ μM. Corresponding DNA-binding IC_{50} values for FZ55 and FZ41 were 2.3 and 0.75 μM, respectively [49], consistent with 3′-P IC_{50} values determined either by standard gel-electrophoresis or steady-state fluorescence anisotropy as mentioned above [17, 68]. In contrast, concentrations of up to 10 μM of the structurally related compound, FZ117, but lacking the 8-OH group, did not lead to any inhibition of the DNA-binding step, in accordance with activity assays showing that FZ117 has no effect on 3′-P activity (Fig. 12.7). As the experimental conditions used in the DNA-binding assays were similar to those used in enzymatic assays, the IC_{50} values obtained in both cases can be directly compared. Consequently, DNA-binding inhibition accounts for the whole inhibition of enzyme activity, highlighting a competitive mechanism of action. The study of the DNA-binding step *per se* by steady-state fluorescence anisotropy suggest that styrylquinolines are pure competitive inhibitors of the 3′-P reaction and act by preventing the IN-DNA

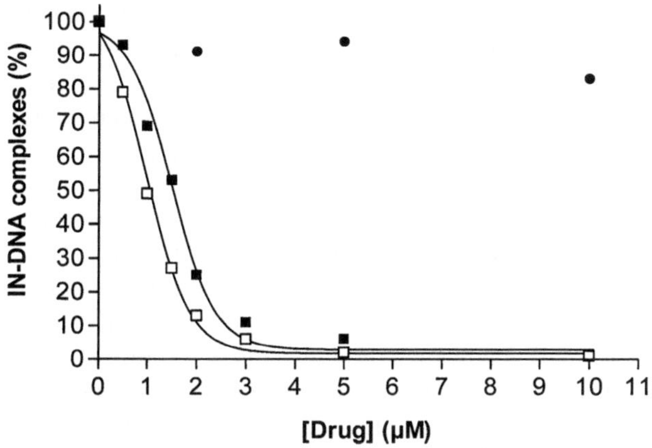

Fig. 12.7 Percentage of IN-DNA complexes (measured by steady-state fluorescence anisotropy) as a function of drug concentration. The styrylquinoline drug was KHD161 (at 25 °C: *white squares*; at 37 °C: *black squares*) or FZ117 (at 25 °C: *black circles*). Steady-state anisotropy values were recorded and normalized based on the anisotropy amplitude values (complex minus free DNA). The fit of KHD161 curves gave IC_{50} values of 0.98 ± 0.09 µM and 1.5 ± 0.13 µM at 25 and 37 °C, respectively (Reproduced from Ref. [49])

interaction, justifying that styrylquinoline compounds are classified in the IN binding inhibitors family or INBI.

We have mentioned above a cooperative binding of IN on the viral DNA which is related the zinc finger motif. Taking into account that zinc promotes the Mg^{2+}-dependent activity of IN and has no or a modest influence on the Mn^{2+}-dependent activity [11, 12], we addressed the question of whether zinc ejection, which influences the cooperative DNA-binding properties of IN, could prevent the formation of the specific and catalytically competent IN-DNA complex [53] using the zinc ejector DIBA-1 compound [79]. First, DIBA-1 is a very potent inhibitor of the 3′-P reaction. Moreover, it is more efficient when the reaction is performed under Mg^{2+} condition compared to Mn^{2+} ($IC_{50} = 14$ and 126 nM, respectively). The DNA-binding assay based on fluorescence anisotropy measurements was used to test whether DIBA-1 could inhibit HIV-1 IN activity via impairing the IN-DNA interaction (DIBA-1 behaves as a competitive inhibitor) as found for the INBI compounds (see above), or acts at the catalytic step (post-DNA-binding step event). Increasing concentrations of HIV-1 IN were added to a fluorescein-labeled viral DNA and fluorescence anisotropy was measured in the absence or presence of DIBA-1 (Fig. 12.8). Results clearly show that DIBA-1 efficiently inhibits the 3′-P activity under conditions that do not affect the formation of IN-DNA complexes. The apparent K_d values for the full-length IN were similar, regardless of the metallic cofactor used, and independent of the presence of DIBA-1, confirming that DIBA-1 allows IN to bind to its DNA substrate but inactivates the formed complexes. This behavior is in sharp contrast to the one observed with styrylquinoline compounds for which the inhibition of 3′-P was found to parallel the inhibition of the complex formation. Importantly, even though DIBA-1 had no measurable effect on the overall affinity, it significantly reduced the Hill coefficient from 2 to 1.47 (Table 12.1), suggesting that DIBA-1 interferes with the cooperative binding of IN to DNA. As mentioned above, such a cooperative binding was not observed in the presence of Mn^{2+} or in the absence of divalent cation ($\tilde{n} = 1.29$ and 1.34, respectively), and DIBA-1 did not further change these $\tilde{n}$ values ($\tilde{n} = 1.20$ and 1.37, respectively), suggesting that this differential effect accounts for the differential DIBA-1-mediated inhibition of Mn^{2+}- and Mg^{2+}-dependent 3′-P activities (Table 12.1). Altogether, our results show that the NTD and its HHCC motif do not play a direct role in terms of affinity/stability of the IN-DNA complex, but, most likely, account for the observed Mg^{2+}-dependent cooperative behavior of IN by mediating a conformational change through a cis or a trans NTD-CC interaction as mentioned above. The nature of the sequence and the simultaneous presence of zinc and Mg^{2+} are essential factors to mediate this effect. To note, in contrast to the situation in the presence of Mg^{2+}, the Mn^{2+}-dependent 3′-P activity which is less stringent and leads to non-specific products, is not sensitive to the presence of zinc, and this probably explains the differential inhibition effect of DIBA-1 under Mn^{2+}/Mg^{2+} conditions. In conclusion, DIBA-1 strongly inhibits 3′-P although it does not perturb the overall affinity of IN for the viral DNA end. This property is clearly different than INBI and INSTI compounds. INBIs prevent 3′-P by competitively inhibiting the binding of IN to viral DNA whereas INSTIs cannot fit in the active site of the free IN (they do not prevent 3′-P). Instead, they selectively target the pre-formed IN-viral DNA complex and are potent inhibitors of the ST reaction [69, 70]. In the case of DIBA-1, the compound blocks 3′-P, which is different from INSTIs, at a post DNA-binding step, which is different from INBIs. Since DIBA-1 does not prevent the formation of IN-DNA complexes, but significantly changes the Hill coefficient in the Mg^{2+} context, this suggests that the nature of the IN-IN interaction onto the DNA substrate is modulated by DIBA-1.

12.5 Study of the Modulation of the Multimeric State of IN by Time-Resolved Fluorescence Anisotropy and Fluorescence Correlation Spectroscopy (FCS)

FCS is used to study the hydrodynamic properties of proteins by the measure of the translational diffusion of molecules and is well adapted for studying protein-protein interactions, including multimerization process and aggregative properties [48, 80–85]. The temporal behavior of fluorescence fluctuations within a small excitation volume, described by the autocorrelation function (Eq. 12.1), allows the determination of translational diffusion times (τ_D):

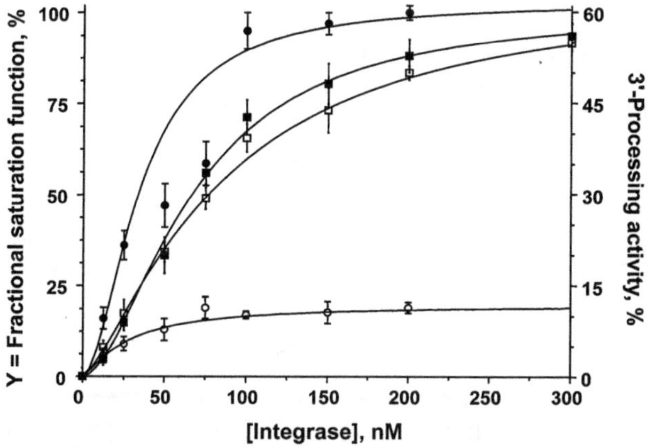

Fig. 12.8 DIBA-1 does not prevent formation of IN–DNA complexes. Increasing IN concentrations were incubated with DNA in the absence (*black*) or presence (*white*) of the zinc ejector DIBA-1 compound (DIBA-1:IN = 2:1). Left axis, the DNA-binding step (*squares*) was recorded by steady-state fluorescence anisotropy. Right axis: 3′-P activity (*circles*) as determined by Δr_{SDS} (see Fig. 12.1) (Nucleic Acids Research. Copyright © 2010)

$$g(\tau) = \frac{\langle I(t) \bullet I(t+\tau) \rangle}{\langle I(t) \rangle^2} = \frac{1}{N} \bullet \frac{1}{\left(1 + \frac{\tau}{\tau_D}\right) \bullet \sqrt{1 + \frac{\omega_0^2}{z_0^2} \bullet \frac{\tau}{\tau_D}}} \tag{12.1}$$

where I(t) is the number of photons per time unit, N is the mean number of fluorophores in the excitation volume, ω_0 and z_0 are the lateral and axial dimensions of excitation volume, respectively.

The TAMRA-labeled PFV-1 IN (or IN^T), as obtained using the TGase-mediated TAMRA labeling method (see description in the FRET section), was studied by FCS using a two-photon excitation set-up [15]. A typical autocorrelation curve corresponding to a concentration of 100 nM PFV-1 IN^T is shown in (Fig. 12.9). PFV-1 IN^T was characterized by a diffusion time (τ_D) of 565 μs at submicromolar concentrations. First, this study revealed a clearly distinct solubility properties between HIV-1 and PFV-1 INs. Indeed, when HIV-1 IN was labeled by TAMRA using the same procedure as described above for PFV-1 IN, it was not possible to measure accurately its diffusion time due to the presence of bright spikes originating from the presence of aggregates as independently found by others [86]. The frequency of bright spikes in HIV-1 samples was slightly lower but remains significant upon addition of detergent, indicating that irreversible aggregates are formed during the HIV-1 IN purification or labeling procedure. By contrast, PFV-1 IN gave satisfactory autocorrelation curves and no bright spike was detected during the acquisition time. Second, the oligomeric transition of PFV-1 IN in solution (DNA free) was studied by measuring the modulation of the diffusion time as a function of protein concentration. The FCS acquisition was then performed using a constant concentration of PFV-1 IN^T and varying the concentration of unlabeled IN

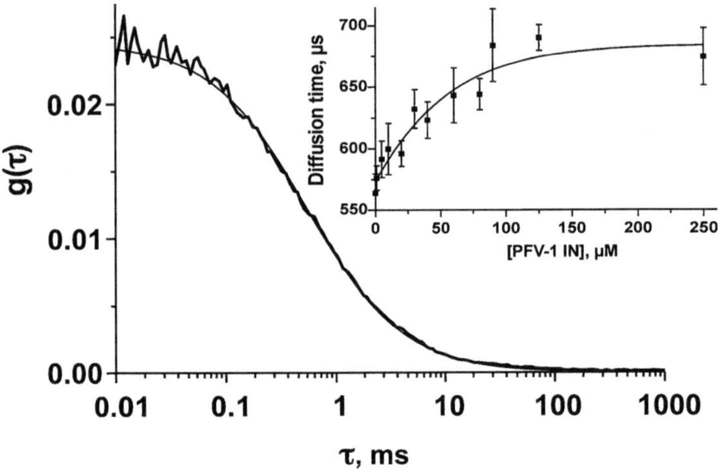

Fig. 12.9 FCS study of TAMRA-labeled PFV-1 IN. Autocorrelation analysis of DNA-free TAMRA-labeled PVF-1 IN (100 nM). Inset, average diffusion time as a function of IN concentration. IN concentration was varied using increasing concentrations of unlabeled IN and a constant concentration of TAMRA-labeled IN (100 nM) (This research was originally published in The Journal of Biological Chemistry. Delelis et al. [15]. © the American Society for Biochemistry and Molecular Biology)

(up to 250 µM) to measure the concentration dependence of τ_D (Fig. 12.9, inset). No bright spike was detected in the different mixtures of labeled/unlabeled PFV-1 IN and τ_D increased from 565 to 680 µs in this concentration range. The τ_D ratio was 1.2, close to the theoretical ratio (1.3) for a monomer-dimer or a dimer-tetramer transition. It is a difficult task in FCS to precisely identify the oligomeric status only based on the translational diffusion value. FCS experiments were then completed with time-resolved fluorescence anisotropy experiments.

Time-resolved fluorescence anisotropy allows the determination of the hydrodynamic properties of proteins by measuring the long rotational diffusion (θ_{long}) which is directly related to the hydrodynamic volume [48]. This approach is extensively used for characterizing molecular interactions, self-assembly properties of proteins or the size of protein-DNA complexes [87–89]. PFV-1 IN was studied by time-resolved fluorescence anisotropy using intrinsic Trp fluorescence and a time-correlated single photon counting set-up [10, 44]. Correlation times were calculated from the two polarized fluorescence decays, $I_{vv}(t)$ (Eq. 12.2) and $I_{vh}(t)$ (Eq. 12.3), analyzed by the maximum entropy method [90]:

$$\mathrm{Ivv(t)} = N \times \frac{1}{3} E\lambda(t) \otimes \left[\sum_{i=1}^{n} \alpha i_e^{-t/\tau i} \left(1 + 2 \sum_{j=1}^{m} \rho j_e^{-t/\theta j} \right) \right] \tag{12.2}$$

$$\mathrm{Ivh(t)} = \frac{N}{G} \times \frac{1}{3} E\lambda(t) \otimes \left[\sum_{i=1}^{n} \alpha i_e^{-t/\tau i} \left(1 - \sum_{j=1}^{m} \rho j_e^{-t/\theta j} \right) \right] \tag{12.3}$$

$$with \sum_{i=1}^{n} \alpha i = 1 \text{ and } \sum_{j=1}^{m} \rho j = \mathrm{r0}$$

where $E_\lambda(t)$ is the time profile of the excitation pulse, G is the G-factor, α_i is the proportion of molecules with lifetime τ_i, ρ_j is the initial anisotropy of molecules with rotational correlation time θ_j. The fluorescence anisotropy decay, r(t), is then defined by Eq. 12.4:

$$\mathrm{r(t)} = \frac{\mathrm{Ivv(t)} - \mathrm{G} \times \mathrm{Ivh(t)}}{\mathrm{Ivv(t)} + 2\mathrm{G} \times \mathrm{Ivh(t)}} \tag{12.4}$$

Distributions of θ, as recovered by maximum entropy method, are shown for two IN concentrations (Fig. 12.10a). $\theta_{\mathrm{long},20\ ^\circ\mathrm{C}}$ were approximately 25 and 53 ns for IN concentrations of 2 and 125 μM, respectively. Given a molecular weight of 46.4-kDa for monomeric PFV-1 IN, these values are consistent with a monomeric and a dimeric form, respectively [15]. Moreover, θ_{long} increased continuously from 25 to 53 ns as a function of protein concentration with the half-transition state occurring at 20 μM (Fig. 12.10b), consistent with the transition observed in FCS experiments. The intermediary θ_{long} values most likely represent average correlation times during the transition (see simulation section below) and the observed transition reasonably corresponds to a monomer-dimer equilibrium. The behavior of PFV-1 IN in solution significantly differs to those of HIV-1 IN. A submicromolar detergent-solubilized HIV-1 IN preparation is characterized by a θ of 16 ns at 25 °C [10]. This value is consistent with one expected for a monomeric form of IN (32-kDa). In contrast, when purified without any detergent [29], HIV-1 IN used up to 200 nM displayed a much longer correlation time (90 ns) indicating a higher-order multimeric organization, most likely a tetrameric form. The addition of unlabeled 21-mer DNA (from 0 to 3.0 μM) to HIV-1 IN results in a progressive shift from a large θ value (90 ns; as measured using the intrinsic Trp fluorescence signal) to a smaller one (θ = 24 ns) corresponding to a monomer or a monomer-dimer mixture, depending on the temperature [44]. Above 200 nM, HIV-1 IN was mainly characterized by correlation time distributions above 100 ns, suggesting the presence of aggregated forms. Taking into account that the detergent has a detrimental effect on the Mg^{2+}-dependent activity of IN, the experiments presented below were obtained with the detergent-free HIV-1 IN.

We next addressed the modulation of the oligomeric state of IN when bound to the viral DNA end using fluorescein-labeled DNA substrates. As previously described in the steady-state fluorescence anisotropy section, the *r* value is dependent on the number of IN/DNA complexes and the size of these complexes. It was shown that the 3′-P activity displays a bell-shaped response with regard to the steady-state value obtained after the DNA-binding step ($r_{t=0}$) (Fig. 12.4). The

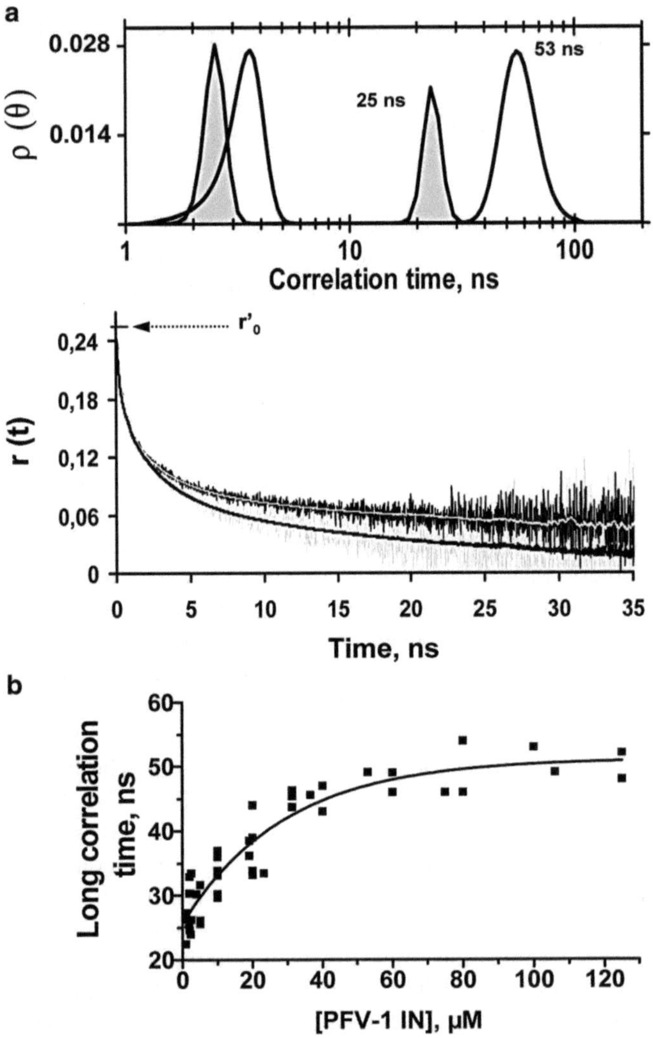

Fig. 12.10 Rotational diffusion of PFV-1 IN. (**a**) Top panel, rotational correlation time (θ) distribution of IN (*gray*, 2 μM; *white*, 125 μM) at 20 °C. Fluorescence anisotropy decays, as shown in the *bottom* panel, were obtained by monitoring tryptophan fluorescence (*gray*: 2 μM (line fit in black); *black*: 125 μM (line fit in white)). (**b**) Long correlation time value as a function of IN concentration (This research was originally published in The Journal of Biological Chemistry. Delelis et al. [15]. © the American Society for Biochemistry and Molecular Biology)

decreasing phase may account for the formation of aggregative DNA-bound forms of IN characterized by a lower specific activity. The molecular sizes of HIV-1 IN-DNA complexes were characterized in greater detail by time-resolved fluorescence anisotropy study of representative samples corresponding to different *r* values by (Fig. 12.11). A fluorescein-labeled double-stranded 21-mer DNA that

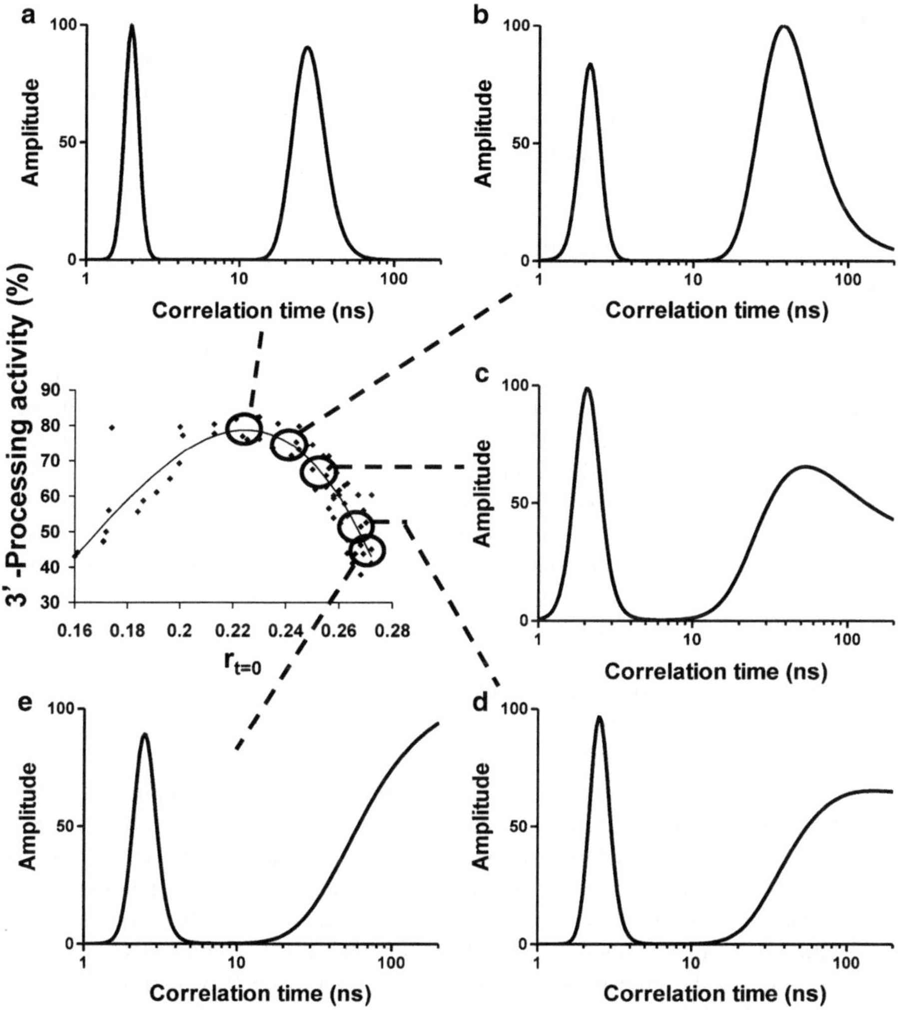

Fig. 12.11 Correlation time distributions of the IN/DNA complexes corresponding to various IN:DNA ratios. Samples corresponding to different *r* values in the decreasing phase of Fig. 12.4 were studied by time-resolved fluorescence anisotropy at 37 °C. IN:DNA ratios were 40:1 (**a**), 67:1 (**b**), 100:1 (**c**), 200:1 (**d**) and 400:1 (**e**). Complete data (θ and ρ_i/ρ_0) are reported in Table 12.3 (This research was originally published in The Journal of Biological Chemistry. Guiot et al. [17]. © the American Society for Biochemistry and Molecular Biology)

mimics the viral DNA U5 end was used to measure the long correlation time of the IN-DNA complex. In the absence of IN, double-stranded DNA of various lengths (10-, 13-, 21-, 36- and 45-mer) are characterized by short θ values (θ_1 and θ_2), which do not strongly vary with the size and can reasonably be assigned to rotation of the fluorescein moiety around the linker at the 5′ end of the DNA and the flexibility of the linker itself (Table 12.2). The longer values (θ_3 and θ_4) display a

Table 12.2 Correlation times (θ) obtained for fluorescein-labeled DNAs of various lengths at 25°C

DNA	Single-stranded	Double-stranded
10-mer	$\theta_1 = 0.13 \pm 0.02$ ns (0.385)	$\theta_1 = 0.23 \pm 0.05$ ns (0.48)
	$\theta_2 = 0.37 \pm 0.06$ ns (0.385)	$\theta_2 = 1.05 \pm 0.51$ ns (0.31)
	$\theta_3 = 1.48 \pm 0.08$ ns (0.23)	$\theta_3 = 3.30 \pm 1.17$ ns (0.21)
13-mer	$\theta_1 = 0.14 \pm 0.01$ ns (0.50)	$\theta_1 = 0.20 \pm 0.02$ ns (0.49)
	$\theta_2 = 0.38 \pm 0.02$ ns (0.285)	$\theta_2 = 0.84 \pm 0.02$ ns (0.26)
	$\theta_3 = 1.64 \pm 0.05$ ns (0.215)	$\theta_3 = 3.83 \pm 0.38$ ns (0.25)
21-mer	$\theta_1 = 0.15 \pm 0.04$ ns (0.28)	$\theta_1 = 0.16 \pm 0.03$ ns (0.425)
	$\theta_2 = 0.42 \pm 0.08$ ns (0.405)	$\theta_2 = 0.47 \pm 0.11$ ns (0.26)
	$\theta_3 = 1.48 \pm 0.25$ ns (0.16)	$\theta_3 = 1.75 \pm 0.56$ ns (0.19)
	$\theta_4 = 4.27 \pm 0.63$ ns (0.155)	$\theta_4 = 8.58 \pm 2.31$ ns (0.125)
36-mer	$\theta_1 = 0.12 \pm 0.03$ ns (0.35)	$\theta_1 = 0.17 \pm 0.02$ ns (0.46)
	$\theta_2 = 0.39 \pm 0.06$ ns (0.39)	$\theta_2 = 0.65 \pm 0.17$ ns (0.22)
	$\theta_3 = 1.62 \pm 0.48$ ns (0.17)	$\theta_3 = 2.70 \pm 0.56$ ns (0.17)
	$\theta_4 = 6.60 \pm 1.91$ ns (0.09)	$\theta_4 = 18.1 \pm 2.82$ ns (0.15)
45-mer	$\theta_1 = 0.16 \pm 0.02$ ns (0.35)	$\theta_1 = 0.26 \pm 0.03$ ns (0.44)
	$\theta_2 = 0.47 \pm 0.09$ ns (0.40)	$\theta_2 = 1.00 \pm 0.33$ ns (0.29)
	$\theta_3 = 1.81 \pm 0.31$ ns (0.15)	$\theta_3 = 4.07 \pm 0.75$ ns (0.17)
	$\theta_4 = 11.1 \pm 3.06$ ns (0.09)	$\theta_4 = 26.4 \pm 4.43$ ns (0.10)

This research was originally published in The Journal of Biological Chemistry. Guiot et al. [17]. © the American Society for Biochemistry and Molecular Biology
ρ_i/ρ_0 ratios are indicated in brackets

more significant dependence on the DNA size and thus account for the overall tumbling motions of DNAs. The 21-mer DNA was chosen because shorter DNAs (10- or 13-mer) exhibit a low binding affinity [17, 44] and longer DNAs (36- or 45-mer) are characterized by θ_4 values (18.1–26.4 ns) which could complicate the further analysis of complexes since they are compatible with the expected long correlation time of a monomeric IN. In other words, the 21-mer DNA substrate represents the best compromise a good affinity and a lower θ_4 value: 8.58 ns.

Time-resolved fluorescence anisotropy analysis was then performed on different samples corresponding to various IN:DNA ratios along the bell-shaped curve. The HIV-*a*5F (fluorescein is attached on the opposite site with regard to the processed end) was used to prevent analytical problems due to the release of the fluorescent moiety occurring during the acquisition if fluorescein is linked to the GT dinucleotide. The analysis was performed on samples corresponding to the peak of the bell-shaped curve and to the decreasing phase (the increasing phase is characterized by a mixture of free and IN-bound DNA) (Fig. 12.11). At 37 °C, two short θ (<2.5 ns) were resolved upon the addition of IN to DNA (Table 12.3). Their values were not significantly dependent on the IN:DNA ratio. In contrast, the characteristic long $\theta_{25\ °C}$ of 8.6 ns found for free 21-mer DNA was replaced by a $\theta_{25\ °C} \geq 37.8$ ns (26.8 ns at 37 °C), depending on the IN:DNA ratio (Fig. 12.11 and Table 12.3). The sample corresponding to the peak of the bell-shaped curve (IN:DNA 40:1;

Table 12.3 Correlation times (37 °C) obtained for IN-DNA complexes using 21-mer double-stranded fluorescein-labeled DNA

Correlation times	IN:DNA ratio				
	40:1	67:1	100:1	200:1	400:1
θ_1	0.29 ns (0.31)	0.19 ns (0.38)	0.17 ns (0.33)	0.15 ns (0.30)	0.19 ns (0.26)
θ_2	1.93 ns (0.23)	2.13 ns (0.14)	2.14 ns (0.17)	2.6 ns (0.15)	2.5 ns (0.15)
$\theta_{3,\text{ maximum}}$	26.8 ns (0.46)	38.5 ns (0.48)	53 ns (0.50)	nd (0.55)	nd (0.59)
$\theta_{3,\text{ center of gravity}}$	31.4 ns	53.3 ns	> 90 ns	> 120 ns	> 150 ns
$^{a}\theta_{25\,°C}$	37.8 ns	54.3 ns	74.7 ns	–	–

This research was originally published in The Journal of Biological Chemistry. Guiot et al. [17]. © the American Society for Biochemistry and Molecular Biology
[a]Normalization of the θ_3 value for a temperature of 25 °C. ρ_i/ρ_0 ratios are indicated in brackets. nd, not determined

$r = 0.226$) was rather mono-disperse, as suggested by the small difference at 37 °C between the maximum (26.8 ns) and the center of gravity (31.4 ns) of the peak in the θ distribution. These values are consistent with hydrodynamic properties of a dimeric form of IN on the DNA [10, 44, 91]. The shift of the center of gravity toward longer θ values accounts for the presence of a small amount of aggregated forms of IN on DNA. Further increases in the IN:DNA ratio displaced the distribution toward longer θ values (Fig. 12.11b–e). Our data confirmed that the sample corresponding to the highest *r* value (0.272) consisted essentially of aggregated forms of IN on DNA, as demonstrated by the broad distribution above 100 ns (Fig. 12.11e).

The optimal condition for 3′-P activity, corresponding to an IN:DNA ratio of about 40:1 ($r = 0.226$), gave a final activity of 92 % under single-turnover conditions, suggesting a situation which is very close to the saturation of DNA sites. This also indicates that complete saturation of DNA sites can be achieved at steady-state anisotropy values much lower than 0.272 (the highest steady-state anisotropy value obtained for IN:DNA ratios above 200:1). The further increase of r after saturation of the DNA sites can be explained by the formation of aggregative forms of IN (IN_{agg}) onto DNA. To note, although the samples contained only IN_{agg}-DNA complexes at ratios above 200:1, these complexes seem to retain a significant level of 3′-P activity (35 % of the optimal activity by the dimeric form). Taking into account (i) the fractional saturation function, (ii) the proper activity of IN_{agg}-DNA complexes as well as (iii) the additivity law of anisotropy, we can estimate that at the peak of the bell-shaped curve, the steady-state anisotropy value accounts for 11–15 % of total complexes. Further analysis indicates that the bell-shaped curve actually results from two opposing trends, the fractional saturation function and the amount of IN_{agg}-DNA complexes which both increase as a function of IN: DNA ratio, are beneficial and detrimental for 3′-P, respectively [17]. Optimal conditions for 3′-P were obtained for a majority of dimeric forms bound to the

DNA, consistent with complementation studies between individually inactive IN mutants, suggesting that at least the dimeric form is catalytically active for 3′-P [55, 92]. Steady-state fluorescence anisotropy is particularly helpful to establish direct relationships between DNA-binding parameters, the specific activity of the complexes and their hydrodynamic properties as measured by time-resolved fluorescence anisotropy. However, the ability of the tetrameric species to perform 3′-P is still unclear [16, 93] and, from this point of view, fluorescence anisotropy data do not indicate whether a tetrameric form of IN may exist at one viral DNA end since the intermediate θ distributions between the dimeric and aggregated forms most likely correspond to polydisperse solutions of different multimeric forms.

Higher-order HIV-1 IN multimers/aggregates are detrimental to 3′-P activity and account for characteristic bell-shaped curves when plotting activity versus IN concentration or the $r_{t=0}$ parameter. A similar behavior was observed with PFV-1 IN (Fig. 12.12a, b). Although DNA-free HIV-1 and PFV-1 INs are characterized by distinct self-association states, they display comparable enzymatic features for the 3′-P catalytic reaction. Interestingly, the 3′-P activity at the maximum of the bell-shaped curve (Act_{max}) increased by increasing ionic strength, suggesting that less catalytically competent and large complexes are more sensitive to ionic strength than catalytically competent complexes (Fig. 12.12a, b). In other words, ionic strength differentially affects the two types of complexes and delays the formation of large complexes. Moreover, Act_{max} was strongly DNA length-dependent and better 3′-P activity was obtained using shorter DNA substrates (*e.g.* 3-fold higher for 15-mer DNA than that of 65-mer DNA) (Fig. 12.12c); the decreasing phase was also more pronounced for longer DNA substrates. Altogether, these results indicate that short DNA substrates favor catalytically competent complexes over less active large complexes. Therefore, a selection of specific/competent complexes may be achieved using short DNA substrates under conditions of high ionic strength. The determination of long correlation times by time-resolved fluorescence anisotropy of PFV-1 IN-DNA complexes (using 15-mer and 21-mer fluorescein-labeled DNAs) confirms a dimeric form of DNA-bound PFV-1 IN under optimal conditions for 3′-P activity (Table 12.4). DNA promotes dimerization as dimers are present at low micromolar concentrations in which free PFV-1 IN is monomeric. As observed for HIV-1 protein, higher IN:DNA ratios (corresponding to the decreasing phases of bell-shaped curves) lead to the appearance of large complexes ($\theta_{long} > 100$ ns). Again, the dimeric form appears to be the most catalytically competent form for 3′-P while higher-order multimers or aggregates lead to suboptimal activity.

PFV-1 IN is characterized by a monomer-dimer equilibrium in the low micromolar range. This equilibrium is shifted toward the dimeric form upon DNA-binding under optimal conditions for activity. The same final state is observed for HIV-1 IN, a dimeric form bound to the viral DNA end, although samples of the DNA-free protein are characterized by the presence of a majority of higher-order multimers (tetramers and/or aggregates). This explains why addition of DNA in a sample containing HIV-1 IN decreases the long correlation time as measured using time-resolved fluorescence anisotropy of Trp as mentioned above.

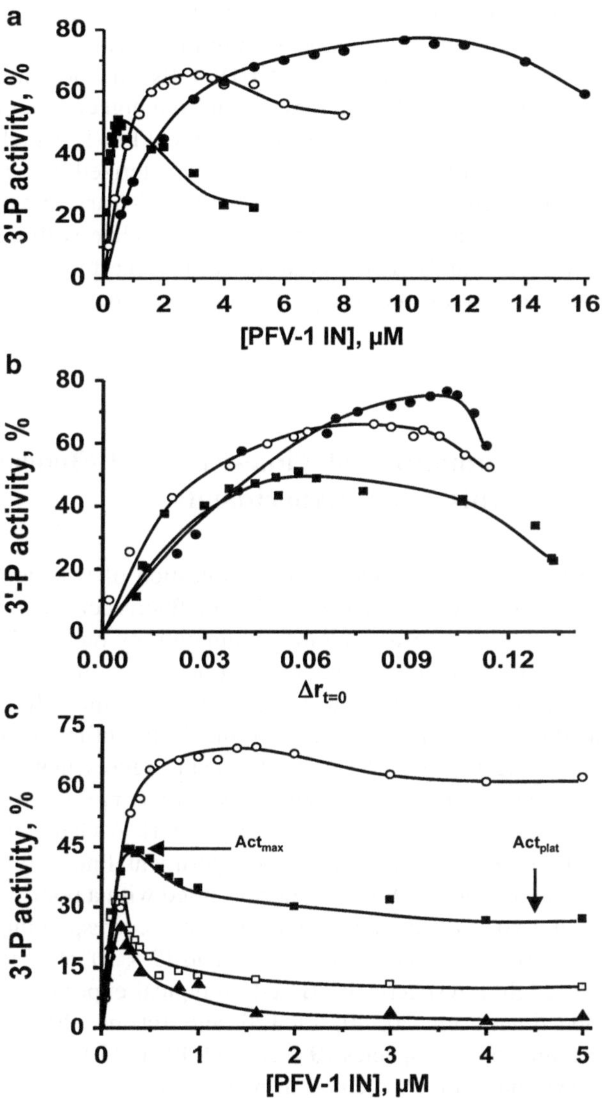

Fig. 12.12 Influences of DNA size and ionic strength on the PFV-1 IN 3′-P activity. (**a**, **b**) Influence of ionic strength ([NaCl] = 50 mM (*black squares*), 150 mM (*white circles*) or 200 mM (*black circles*)). 3′-P activity using a 21-mer DNA substrate was plotted against IN concentration (**a**) or the $\Delta r_{t=0}$ value (B). $\Delta r_{t=0} = r_{complex,t=0} - r_{free\ DNA}$. (**c**) Influence of DNA size using a 15-mer (*white circles*), a 21-mer (*black squares*), a 45-mer (*white squares*), or 65-mer DNA substrate (*black triangles*) (This research was originally published in The Journal of Biological Chemistry. Delelis et al. [15]. © the American Society for Biochemistry and Molecular Biology)

Table 12.4 Long correlation times of PFV-1 IN-DNA complexes

DNA size, bp	[PFV-1 IN], nM	θ_{long}, ns (20 °C)
15	1200	63 ± 6
21	400	53 ± 8
21	600	61 ± 5
21	1200	> 100

This research was originally published in The Journal of Biological Chemistry. Delelis et al. [15]. © the American Society for Biochemistry and Molecular Biology

It is important to note that the HIV-1 IN DNA-binding step is very slow ($k_{on} = 0.23\ min^{-1}$) while no such limiting-step is observed for PFV-1 IN [15, 52], suggesting that, in the case of HIV-1 IN, the slow kinetic of complex formation accounts for a pre-binding transition from higher-order multimeric states to smaller molecular species such as monomers. PFV-1, which is mainly monomeric at low-micromolar concentrations, is characterized by a rapid DNA-binding step. As a consequence of DNA-binding, both PFV-1 and HIV-1 proteins are characterized by DNA-bound dimeric forms which correlate with optimal 3′-P activity [17, 44, 52]. Accordingly, Faure et al. found that the monomer/DNA precedes the dimer/DNA forms by analyzing the time-dependent formation of LTR cross-linked species [16].

12.6 Simulation of Time-Resolved Fluorescence Anisotropy Experiments

In this section, we addressed the question of whether time-resolved fluorescence anisotropy experiments using Trp or fluorescein as fluorophores (having average lifetimes below 5 ns) and maximum entropy method have the ability to recover in an accurate manner long correlation times in the range of 20–200 ns, which correspond to typical values obtained experimentally. Synthetic data were generated by a convolution product of the flash profile and a sum of exponential decays according to Eqs. 12.2 and 12.3, using representative experimental fluorescence lifetimes of Trp characterizing HIV-1 IN (: $\tau_1 = 0.6$ ns (35 %); $\tau_2 = 2.1$ ns (40 %); $\tau_3 = 4.7$ ns (25 %)) and $r_o = 0.25$ [10, 94]. A Gaussian random noise (approximation of a Poisson statistic) or an experimental noise (extracted from the analysis of two experimental polarized decays fitted with a poor χ^2) were used for simulations, with two photon counting conditions, corresponding to 1.2×10^6 ($I_{vv}(t)$) and 1.6×10^6 ($I_{vh}(t)$) counts (= "low") and 12×10^6 ($I_{vv}(t)$) and 16×10^6 ($I_{vh}(t)$) counts (= "high"), respectively. Based on typical experimental values, the input correlation times were 20, 40, 80, 120, and 160 ns. We also simulated an "equitable" mixture of two species ($\theta = 20$ and 40 ns (50 %:50 %)) to assess the ability of maximum entropy method to recover two correlation times in the θ distribution. The maximum entropy method allowed the recovery of the θ_{20ns} input value, regardless of the counting statistic and the noise (Table 12.5). The recovery of the θ_{40ns} value is reasonable under low counting conditions and is significantly improved under high counting conditions while a satisfying recovery of the θ_{80ns} value strictly requires a high counting statistic (Table 12.5). No accurate recovery of $\theta_{>100ns}$ values was obtained even under high counting conditions (Table 12.5), showing that, as mentioned in the experimental section, oligomeric forms higher than tetramer cannot be distinguished from each other and from aggregated forms. Finally, the simulation results also show that only an average correlation time (29–31 ns) can be recovered when two input values are considered mimicking a

Table 12.5 Simulation data analyses

	Output value of θ (ns) [width (ns)]			
	Random noise		Experimental noise	
Input value of θ (ns)	Low counting	High counting	Low counting	High counting
20	20.6 ± 1.4 (3.5)	20.2 ± 0.4 (1.7)	21.2 ± 1.7 (3.9)	20.7 ± 0.9 (3.2)
40	45 ± 19 (20)	40.6 ± 2.2 (5.4)	48 ± 24 (24)	42 ± 5 (9.3)
80	118 ± 48 (71)	88 ± 21 (30)	113 ± 48 (65)	94 ± 29 (41)
120	151 ± 59 (70)	134 ± 41 (52)	165 ± 60 (70)	148 ± 46 (62)
160	188 ± 58 (66)	175 ± 43 (58)	205 ± 58 (62)	190 ± 44 (61)
20/40 (50 %/50 %)	29 ± 59 (70)	28.8 ± 3.2 (7.5)	31 ± 8.5 (11)	31 ± 8.6 (13)

Reproduced from Ref. [10]

mixture (50 %:50 %) of two equal populations characterized by rotational correlation times of 20 and 40 ns, respectively, regardless the nature of the noise or the counting statistic (Table 12.5). This is in accordance with the typical monomer-dimer transition of PFV-1 IN studied by time-resolved fluorescence anisotropy showing that a single average correlation time is obtained along the transition (Fig. 12.10b).

12.7 Conclusion

We used various fluorescence methodologies to address the problem of IN solubility and to study the interplay between the endonucleolytic catalytic of the protein and its oligomeric state. HIV-1 and PFV-1 INs are clearly characterized by different solubility properties. PFV-1 IN, free in solution, was found to be more soluble than HIV-1 IN. The correlation time distribution is characteristic of a monomer-dimer transition of PFV-1 IN in the low micromolar range while HIV-1 IN was mainly aggregated above 200 nM. In the presence of detergent, this critical concentration for HIV-1 IN was found in the low-micromolar range, whereas PFV-1 IN was found mainly monomeric in the low-micromolar range, in the absence of any detergent and DNA. In FCS, bright spikes were detected with TAMRA-labeled HIV-1 IN, while TAMRA-labeled PFV-1 IN resulted in satisfactory autocorrelation curves with no spike, confirming that HIV-1 IN has a higher propensity for aggregation than PFV-1 IN. FCS study confirms the time-resolved fluorescence anisotropy results with a clear monomer-dimer transition characterized by a similar half-transition concentration (20–30 μM). However, although more soluble, PFV-1 IN also leads to higher-order multimers or aggregates when bound to its DNA substrate under conditions of high IN:DNA ratio, leading to suboptimal activity, as found for HIV-1 IN. Furthermore, DNA size and ionic strength were found to be important parameters that modulate the number of higher-order complexes. Moreover, fluorescence anisotropy and FRET experiments highlight sequence-specific IN/DNA recognition that is measurable only with PFV-1 IN. The sequence influences both the positioning of IN onto the DNA (with a bias toward the processing DNA end) and the affinity (cognate viral sequence versus a non-specific random sequence). This suggests that, using PFV-1 IN, it is possible to distinguish between specific and nonspecific complexes, not only at the catalytic level, as typically found for HIV-1 [8, 17], but also at the DNA-binding level. The reason could be that the aggregative properties of HIV-1 IN mask an intrinsic ability of the protein to bind preferentially to its cognate sequence. Although HIV-1 IN does not display a differential affinity for specific (cognate) and non-specific (random) DNA sequences, we have shown that a cooperative assembly of HIV-1 IN/DNA complexes is important for the Mg^{2+}-dependent activity, although dispensable in the Mn^{2+} context. The NTD and zinc are important for this cooperative binding mode. However, FRET and steady-state fluorescence anisotropy experiments suggest that the specificity is modest (preference is a more suitable word), even for PFV-1 IN, most likely due to the inherent non-specific DNA-binding mode (responsible for the binding to the target DNA). Finally, we found that steady-state anisotropy values obtained after DNA binding and before catalysis were fully predictive of subsequent 3′-P activity for low IN: DNA ratios, according to the fractional saturation function. Nevertheless, for high ratios, anisotropy continued to increase, but the activity decreased, due to the presence of higher-order multimers or aggregated states of IN detrimental to 3′-P activity. 3′-P was highest for non-aggregative smaller species and the dimeric form, as assembled on DNA in a cooperative manner, corresponds to the most catalytically competent oligomeric form for 3′-P. The tetrameric form, a dimer of dimer, is responsible for concerted integration when two LTR ends are simultaneously present in the same synaptic complex [3, 18].

References

1. Delelis O, Carayon K, Saib A, Deprez E, Mouscadet JF (2008) Integrase and integration: biochemical activities of HIV-1 integrase. Retrovirology 5:114
2. Craigie R, Bushman FD (2012) HIV DNA integration. Cold Spring Harb Perspect Med 2: a006890
3. Hare S, Gupta SS, Valkov E, Engelman A, Cherepanov P (2010) Retroviral intasome assembly and inhibition of DNA strand transfer. Nature 464:232–236
4. Heuer TS, Brown PO (1998) Photo-cross-linking studies suggest a model for the architecture of an active human immunodeficiency virus type 1 integrase-DNA complex. Biochemistry-Us 37:6667–6678
5. Lutzke RAP, Plasterk RHA (1998) Structure-based mutational analysis of the C-terminal DNA-binding domain of human immunodeficiency virus type 1 integrase: critical residues for protein oligomerization and DNA binding. J Virol 72:4841–4848
6. Vink C, Lutzke RAP, Plasterk RHA (1994) Formation of a stable complex between the human-immunodeficiency-virus integrase protein and viral-DNA. Nucleic Acids Res 22:4103–4110
7. Engelman A, Hickman AB, Craigie R (1994) The core and carboxyl-terminal domains of the integrase protein of human-immunodeficiency-virus type-1 each contribute to nonspecific DNA-binding. J Virol 68:5911–5917
8. Esposito D, Craigie R (1998) Sequence specificity of viral end DNA binding by HIV-1 integrase reveals critical regions for protein-DNA interaction. EMBO J 17:5832–5843
9. Jenkins TM, Esposito D, Engelman A, Craigie R (1997) Critical contacts between HIV-1 integrase and viral DNA identified by structure-based analysis and photo-crosslinking. EMBO J 16:6849–6859
10. Deprez E, Tauc P, Leh H, Mouscadet JF, Auclair C, Brochon JC (2000) Oligomeric states of the HIV-1 integrase as measured by time-resolved fluorescence anisotropy. Biochemistry-Us 39:9275–9284
11. Lee SP, Xiao JM, Knutson JR, Lewis MS, Han MK (1997) Zn2+ promotes the self-association of human immunodeficiency virus type-1 integrase in vitro. Biochemistry 36:173–180
12. Zheng RL, Jenkins TM, Craigie R (1996) Zinc folds the N-terminal domain of HIV-1 integrase, promotes multimerization, and enhances catalytic activity. Proc Natl Acad Sci U S A 93:13659–13664
13. Khan E, Mack JP, Katz RA, Kulkosky J, Skalka AM (1991) Retroviral integrase domains: DNA binding and the recognition of LTR sequences. Nucleic Acids Res 19:851–860
14. Baranova S, Tuzikov FV, Zakharova OD, Tuzikova NA, Calmels C, Litvak S, Tarrago-Litvak L, Parissi V, Nevinsky GA (2007) Small-angle X-ray characterization of the

nucleoprotein complexes resulting from DNA-induced oligomerization of HIV-1 integrase. Nucleic Acids Res 35:975–987
15. Delelis O, Carayon K, Guiot E, Leh H, Tauc P, Brochon JC, Mouscadet JF, Deprez E (2008) Insight into the integrase-DNA recognition mechanism – a specific DNA-binding mode revealed by an enzymatically labeled integrase. J Biol Chem 283:27838–27849
16. Faure A, Calmels C, Desjobert C, Castroviejo M, Caumont-Sarcos A, Tarrago-Litvak L, Litvak S, Parissi V (2005) HIV-1 integrase crosslinked oligomers are active in vitro. Nucleic Acids Res 33:977–986
17. Guiot E, Carayon K, Delelis O, Simon F, Tauc P, Zubin E, Gottikh M, Mouscadet JF, Brochon JC, Deprez E (2006) Relationship between the oligomeric status of HIV-1 integrase on DNA and enzymatic activity. J Biol Chem 281:22707–22719
18. Li M, Mizuuchi M, Burke TR, Craigie R (2006) Retroviral DNA integration: reaction pathway and critical intermediates. EMBO J 25:1295–1304
19. Li M, Jurado KA, Lin S, Engelman A, Craigie R (2014) Engineered hyperactive integrase for concerted HIV-1 DNA integration. PLoS One 9:e105078
20. Chow SA, Vincent KA, Ellison V, Brown PO (1992) Reversal of integration and DNA splicing mediated by integrase of human-immunodeficiency-virus. Science 255:723–726
21. Delelis O, Parissi V, Leh H, Mbemba G, Petit C, Sonigo P, Deprez E, Mouscadet JF (2007) Efficient and specific internal cleavage of a retroviral palindromic DNA sequence by tetrameric HIV-1 integrase. PLoS One 2:e608
22. Delelis O, Petit C, Leh H, Mbemba G, Mouscadet JF, Sonigo P (2005) A novel function for spumaretrovirus integrase: an early requirement for integrase-mediated cleavage of 2 LTR circles. Retrovirology 2:31
23. Zhang DW, He HQ, Guo SX (2014) Hairpin DNA probe-based fluorescence assay for detecting palindrome cleavage activity of HIV-1 integrase. Anal Biochem 460C:36–38
24. Gerton JL, Brown PO (1997) The core domain of HIV-1 integrase recognizes key features of its DNA substrates. J Biol Chem 272:25809–25815
25. Laboulais C, Deprez E, Leh H, Mouscadet JF, Brochon JC, Le Bret M (2001) HIV-1 integrase catalytic core: molecular dynamics and simulated fluorescence decays. Biophys J 81:473–489
26. Agapkina J, Smolov M, Barbe S, Zubin E, Zatsepin T, Deprez E, Le Bret M, Mouscadet JF, Gottikh M (2006) Probing of HIV-1 integrase/DNA interactions using novel analogs of viral DNA. J Biol Chem 281:11530–11540
27. Engelman A, Craigie R (1995) Efficient magnesium-dependent human-immunodeficiency-virus type-1 integrase activity. J Virol 69:5908–5911
28. Skinner LM, Sudol M, Harper AL, Katzman M (2001) Nucleophile selection for the endonuclease activities of human, ovine, and avian retroviral integrases. J Biol Chem 276:114–124
29. Leh H, Brodin P, Bischerour J, Deprez E, Tauc P, Brochon JC, LeCam E, Coulaud D, Auclair C, Mouscadet JF (2000) Determinants of Mg2+-dependent activities of recombinant human immunodeficiency virus type 1 integrase. Biochemistry 39:9285–9294
30. Molteni V, Rhodes D, Rubins K, Hansen M, Bushman FD, Siegel JS (2000) A new class of HIV-1 integrase inhibitors: The 3,3,3′,3′-tetramethyl-1,1′-spirobi(indan)-5,5′,6,6′-tetrol family. J Med Chem 43:2031–2039
31. Mazumder A, Neamati N, Pilon AA, Sunder S, Pommier Y (1996) Chemical trapping of ternary complexes of human immunodeficiency virus type I integrase, divalent metal, and DNA substrates containing an abasic site – implications for the role of lysine 136 in DNA binding. J Biol Chem 271:27330–27338
32. Marchand C, Johnson AA, Karki RG, Pais GC, Zhang X, Cowansage K, Patel TA, Nicklaus MC, Burke TR Jr, Pommier Y (2003) Metal-dependent inhibition of HIV-1 integrase by beta-diketo acids and resistance of the soluble double-mutant (F185K/C280S). Mol Pharmacol 64:600–609
33. Lee HS, Kang SY, Shin CG (2005) Characterization of the functional domains of human foamy virus integrase using chimeric integrases. Mol Cell 19:246–255

34. Oh YT, Shin CG (1999) Comparison of enzymatic activities of the HIV-1 and HFV integrases to their U5 LTR substrates. Biochem Mol Biol Int 47:621–629
35. Pahl A, Flugel RM (1995) Characterization of the human spuma retrovirus integrase by site-directed mutagenesis, by complementation analysis, and by swapping the zinc-finger domain of Hiv-1. J Biol Chem 270:2957–2966
36. Maertens GN, Hare S, Cherepanov P (2010) The mechanism of retroviral integration from X-ray structures of its key intermediates. Nature 468:326–329
37. Hare S, Smith SJ, Metifiot M, Jaxa-Chamiec A, Pommier Y, Hughes SH, Cherepanov P (2011) Structural and functional analyses of the second-generation integrase strand transfer inhibitor dolutegravir (S/GSK1349572). Mol Pharmacol 80:565–572
38. Hare S, Vos AM, Clayton RF, Thuring JW, Cummings MD, Cherepanov P (2010) Molecular mechanisms of retroviral integrase inhibition and the evolution of viral resistance. Proc Natl Acad Sci U S A 107:20057–20062
39. Cherepanov P, Ambrosio ALB, Rahman S, Ellenberger T, Engelman A (2005) Structural basis for the recognition between HIV-1 integrase and transcriptional coactivator p75. Proc Natl Acad Sci U S A 102:17308–17313
40. Goldgur Y, Dyda F, Hickman AB, Jenkins TM, Craigie R, Davies DR (1998) Three new structures of the core domain of HIV-1 integrase: an active site that binds magnesium. Proc Natl Acad Sci U S A 95:9150–9154
41. Wang JY, Ling H, Yang W, Craigie R (2001) Structure of a two-domain fragment of HIV-1 integrase: implications for domain organization in the intact protein. EMBO J 20:7333–7343
42. Chen JCH, Krucinski J, Miercke LJW, Finer-Moore JS, Tang AH, Leavitt AD, Stroud RM (2000) Crystal structure of the HIV-1 integrase catalytic core and C-terminal domains: a model for viral DNA binding. Proc Natl Acad Sci U S A 97:8233–8238
43. Maignan S, Guilloteau JP, Zhou-Liu Q, Clement-Mella C, Mikol V (1998) Crystal structures of the catalytic domain of HIV-1 integrase free and complexed with its metal cofactor: high level of similarity of the active site with other viral integrases. J Mol Biol 282:359–368
44. Deprez E, Tauc P, Leh H, Mouscadet JF, Auclair C, Hawkins ME, Brochon JC (2001) DNA binding induces dissociation of the multimeric form of HIV-1 integrase: a time-resolved fluorescence anisotropy study. Proc Natl Acad Sci U S A 98:10090–10095
45. Bloom LB, Turner J, Kelman Z, Beechem JM, O'Donnell M, Goodman MF (1996) Dynamics of loading the beta sliding clamp of DNA polymerase III onto DNA. J Biol Chem 271:30699–30708
46. Ciubotaru M, Ptaszek LM, Baker GA, Baker SN, Bright FV, Schatz DG (2003) RAG1-DNA binding in V(D)J recombination – specificity and DNA-induced conformational changes revealed by fluorescence and CD spectroscopy. J Biol Chem 278:5584–5596
47. Perez-Howard GM, Weil PA, Beechem JM (1995) Yeast TATA binding protein interaction with DNA: fluorescence determination of oligomeric state, equilibrium binding, on-rate, and dissociation kinetics. Biochemistry 34:8005–8017
48. Lakowicz JR (2006) Principles of fluorescence spectroscopy, 3rd edn. Springer, New York
49. Deprez E, Barbe S, Kolaski M, Leh H, Zouhiri F, Auclair C, Brochon JC, Le Bret M, Mouscadet JF (2004) Mechanism of HIV-1 integrase inhibition by styrylquinoline derivatives in vitro. Mol Pharmacol 65:85–98
50. Pinskaya M, Romanova E, Volkov E, Deprez E, Leh H, Brochon JC, Mouscadet JF, Gottikh M (2004) HIV-1 integrase complexes with DNA dissociate in the presence of short oligonucleotides conjugated to acridine. Biochemistry 43:8735–8743
51. Hazuda DJ, Wolfe AL, Hastings JC, Robbins HL, Graham PL, Lafemina RL, Emini EA (1994) Viral long terminal repeat substrate-binding characteristics of the human-immunodeficiency-virus type-1 integrase. J Biol Chem 269:3999–4004
52. Smolov M, Gottikh M, Tashlitskii V, Korolev S, Demidyuk I, Brochon JC, Mouscadet JF, Deprez E (2006) Kinetic study of the HIV-1 DNA 3′-end processing – single-turnover property of integrase. FEBS J 273:1137–1151

53. Carayon K, Leh H, Henry E, Simon F, Mouscadet JF, Deprez E (2010) A cooperative and specific DNA-binding mode of HIV-1 integrase depends on the nature of the metallic cofactor and involves the zinc-containing N-terminal domain. Nucleic Acids Res 38:3692–3708
54. Lee SP, Han MK (1996) Zinc stimulates Mg2+-dependent 3′-processing activity of human immunodeficiency virus type 1 integrase in vitro. Biochemistry 35:3838–3844
55. van den Ent FMI, Vos A, Plasterk RHA (1999) Dissecting the role of the N-terminal domain of human immunodeficiency virus integrase by trans-complementation analysis. J Virol 73:3176–3183
56. Hare S, Shun MC, Gupta SS, Valkov E, Engelman A, Cherepanov P (2009) A novel co-crystal structure affords the design of gain-of-function lentiviral integrase mutants in the presence of modified PSIP1/LEDGF/p75. PLoS Pathog 5:e1000259
57. Michel F, Crucifix C, Granger F, Eiler S, Mouscadet JF, Korolev S, Agapkina J, Ziganshin R, Gottikh M, Nazabal A, Emiliani S, Benarous R, Moras D, Schultz P, Ruff M (2009) Structural basis for HIV-1 DNA integration in the human genome, role of the LEDGF/P75 cofactor. EMBO J 28:980–991
58. Zhao Z, Mckee CJ, Kessl JJ, Santos WL, Daigle JE, Engelman A, Verdine G, Kvaratskhelia M (2008) Subunit-specific protein footprinting reveals significant structural rearrangements and a role for N-terminal Lys-14 of HIV-1 integrase during viral DNA binding. J Biol Chem 283:5632–5641
59. Asante-Appiah E, Seeholzer SH, Skalka AM (1998) Structural determinants of metal-induced conformational changes in HIV-1 integrase. J Biol Chem 273:35078–35087
60. AsanteAppiah E, Skalka AM (1997) A metal-induced conformational change and activation of HIV-1 integrase. J Biol Chem 272:16196–16205
61. Taki M, Shiota M, Taira K (2004) Transglutaminase-mediated N- and C-terminal fluorescein labeling of a protein can support the native activity of the modified protein. Protein Eng Des Sel 17:119–126
62. Dicker IB, Samanta HK, Li ZF, Hong Y, Tian Y, Banville J, Remillard RR, Walker MA, Langley DR, Krystal M (2007) Changes to the HIV long terminal repeat and to HIV integrase differentially impact HIV integrase assembly, activity, and the binding of strand transfer inhibitors. J Biol Chem 282:31186–31196
63. Johnson AA, Santos W, Pais GC, Marchand C, Amin R, Burke TR Jr, Verdine G, Pommier Y (2006) Integration requires a specific interaction of the donor DNA terminal 5′-cytosine with glutamine 148 of the HIV-1 integrase flexible loop. J Biol Chem 281:461–467
64. Semenova EA, Marchand C, Pommier Y (2008) HIV-1 integrase inhibitors: update and perspectives. Adv Pharmacol 56:199–228
65. Mesplede T, Wainberg MA (2014) Is resistance to dolutegravir possible when this drug is used in first-line therapy? Viruses 6:3377–3385
66. Bonnenfant S, Thomas CM, Vita C, Subra F, Deprez E, Zouhiri F, Desmaele D, d'Angelo J, Mouscadet JF, Leh H (2004) Styrylquinolines, integrase inhibitors acting prior to integration: a new mechanism of action for anti-integrase agents. J Virol 78:5728–5736
67. Mekouar K, Mouscadet JF, Desmaele D, Subra F, Leh H, Savoure D, Auclair C, d'Angelo J (1998) Styrylquinoline derivatives: a new class of potent HIV-1 integrase inhibitors that block HIV-1 replication in CEM cells. J Med Chem 41:2846–2857
68. Zouhiri F, Mouscadet JF, Mekouar K, Desmaele D, Savoure D, Leh H, Subra F, Le Bret M, Auclair C, d'Angelo J (2000) Structure-activity relationships and binding mode of styrylquinolines as potent inhibitors of HIV-1 integrase and replication of HIV-1 in cell culture. J Med Chem 43:1533–1540
69. Espeseth AS, Felock P, Wolfe A, Witmer M, Grobler J, Anthony N, Egbertson M, Melamed JY, Young S, Hamill T, Cole JL, Hazuda DJ (2000) HIV-1 integrase inhibitors that compete with the target DNA substrate define a unique strand transfer conformation for integrase. Proc Natl Acad Sci U S A 97:11244–11249

70. Hazuda DJ, Felock P, Witmer M, Wolfe A, Stillmock K, Grobler JA, Espeseth A, Gabryelski L, Schleif W, Blau C, Miller MD (2000) Inhibitors of strand transfer that prevent integration and inhibit HIV-1 replication in cells. Science 287:646–650
71. Munir S, Thierry S, Subra F, Deprez E, Delelis O (2013) Quantitative analysis of the time-course of viral DNA forms during the HIV-1 life cycle. Retrovirology 10:87
72. Demeulemeester J, Chaltin P, Marchand A, De Maeyer M, Debyser Z, Christ F (2014) LEDGINs, non-catalytic site inhibitors of HIV-1 integrase: a patent review (2006–2014). Expert Opin Ther Pat 24:609–632
73. Li Y, Xuan S, Feng Y, Yan A (2015) Targeting HIV-1 integrase with strand transfer inhibitors. Drug Discov Today 20:435–449
74. Engelman A, Cherepanov P (2008) The lentiviral integrase binding protein LEDGF/p75 and HIV-1 replication. PLoS Pathog 4:e1000046
75. Maillot B, Levy N, Eiler S, Crucifix C, Granger F, Richert L, Didier P, Godet J, Pradeau-Aubreton K, Emiliani S, Nazabal A, Lesbats P, Parissi V, Mely Y, Moras D, Schultz P, Ruff M (2013) Structural and functional role of INI1 and LEDGF in the HIV-1 preintegration complex. PLoS One 8:e60734
76. Delelis O, Malet I, Na L, Tchertanov L, Calvez V, Marcelin AG, Subra F, Deprez E, Mouscadet JF (2009) The G140S mutation in HIV integrases from raltegravir-resistant patients rescues catalytic defect due to the resistance Q148H mutation. Nucleic Acids Res 37:1193–1201
77. Delelis O, Thierry S, Subra F, Simon F, Malet I, Alloui C, Sayon S, Calvez V, Deprez E, Marcelin AG, Tchertanov L, Mouscadet JF (2010) Impact of Y143 HIV-1 integrase mutations on resistance to raltegravir in vitro and in vivo. Antimicrob Agents Chemother 54:491–501
78. Munir S, Thierry E, Malet I, Subra F, Calvez V, Marcelin AG, Deprez E, Delelis O (2015) G118R and F121Y mutations identified in patients failing raltegravir treatment confer dolutegravir resistance. J Antimicrob Chemother 70:739–749
79. Rice WG, Supko JG, Malspeis L, Buckheit RW Jr, Clanton D, Bu M, Graham L, Schaeffer CA, Turpin JA, Domagala J, Gogliotti R, Bader JP, Halliday SM, Coren L, Sowder RC 2nd, Arthur LO, Henderson LE (1995) Inhibitors of HIV nucleocapsid protein zinc fingers as candidates for the treatment of AIDS. Science 270:1194–1197
80. Bacia K, Haustein E, Schwille P (2014) Fluorescence correlation spectroscopy: principles and applications. Cold Spring Harb Protoc 2014:709–725
81. Haustein E, Schwille P (2007) Fluorescence correlation spectroscopy: novel variations of an established technique. Annu Rev Biophys Biomol Struct 36:151–169
82. Kanno DM, Levitus M (2014) Protein oligomerization equilibria and kinetics investigated by fluorescence correlation spectroscopy: a mathematical treatment. J Phys Chem B 118:12404–12415
83. Paul SS, Sil P, Haldar S, Mitra S, Chattopadhyay K (2015) Subtle change in the charge distribution of surface residues may affect the secondary functions of cytochrome c. J Biol Chem 290:14476–14490
84. Ries J, Schwille P (2012) Fluorescence correlation spectroscopy. Bioessays 34:361–368
85. Tipping KW, Karamanos TK, Jakhria T, Iadanza MG, Goodchild SC, Tuma R, Ranson NA, Hewitt EW, Radford SE (2015) pH-induced molecular shedding drives the formation of amyloid fibril-derived oligomers. Proc Natl Acad Sci U S A 112:5691–5696
86. Vercammen J, Maertens G, Gerard M, De Clercq E, Debyser Z, Engelborghs Y (2002) DNA-induced polymerization of HIV-1 integrase analyzed with fluorescence fluctuation spectroscopy. J Biol Chem 277:38045–38052
87. Clark TB, Ziolkowski M, Schatz GC, Goodson T 3rd (2014) Two-photon and time-resolved fluorescence spectroscopy as probes for structural determination in amyloid-beta peptides and aggregates. J Phys Chem B 118:2351–2359
88. Lamprou P, Kempe D, Katranidis A, Buldt G, Fitter J (2014) Nanosecond dynamics of calmodulin and ribosome-bound nascent chains studied by time-resolved fluorescence anisotropy. Chembiochem 15:977–985

89. Samatanga B, Klostermeier D (2014) DEAD-box RNA helicase domains exhibit a continuum between complete functional independence and high thermodynamic coupling in nucleotide and RNA duplex recognition. Nucleic Acids Res 42:10644–10654
90. Brochon JC (1994) Maximum-entropy method of data-analysis in time-resolved spectroscopy. Numer Comput Methods Pt B 240:262–311
91. Podtelezhnikov AA, Gao K, Bushman FD, McCammon JA (2003) Modeling HIV-1 integrase complexes based on their hydrodynamic properties. Biopolymers 68:110–120
92. Engelman A, Bushman FD, Craigie R (1993) Identification of discrete functional domains of Hiv-1 integrase and their organization within an active multimeric complex. EMBO J 12:3269–3275
93. Bosserman MA, O'Quinn DF, Wong I (2007) Loop(202-208) in avian sarcoma virus integrase mediates tetramer assembly and processing activity. Biochemistry 46:11231–11239
94. Henry E, Deprez E, Brochon JC (2014) Maximum entropy analysis of data simulations and practical aspects of time-resolved fluorescence measurements in the study of molecular interactions. J Mol Struct 1077:77–86

Chapter 13
Dual Fluorescence Phenomenon in 'Push-Pull' Stilbenes

Dina Pines, Ehud Pines, Terry W.J. Steele, and Vladislav Papper

Abstract A short review of past and present experimental and theoretical work on the phenomenon of dual fluorescence in strong donor-acceptor (push-pull) stilbenes is presented. Various time resolved techniques such as femtosecond transient absorption, Kerr ellipsometry, fluorescence up conversion and time resolved single photon counting (TRSPC) has been utilized to confirm the presence of dual fluorescence. As the transition rate between the two emitting states is ultrafast one may be able to observe it only with the appropriate time resolution and experimental technique. The association of conformational change or twisting of the electronically excited push-pull stilbenes with dual fluorescence was supported by the dependence of the formation rate of the new emitting state on the solvent viscosity and on hydrodynamic factors of the molecules. The intramolecular charge transfer (ICT) transition may occur by changing the molecular configuration around a single bond and it competes with the *trans-cis* isomerisation around the double bond of the stilbene molecule. We discuss how to discern the two reactions as well as how to discern an ICT transition from a solvent dependent fluorescence Stokes shift as all the three processes may happen on a similar time scale. The dual fluorescence of the newly synthesized *trans*-4-dimethylamino-4′-carbomethoxystilbene serves for demonstrating an ICT transition and how it depends on the stilbene structure, the donor and acceptor properties and solvent polarity.

Keywords Stilbene • Dual fluorescence • Push-pull stilbene • Intramolecular charge transfer • TICT • Trans-cis photoisomerisation • Fluorescence probe

---

D. Pines • E. Pines
Department of Chemistry, Ben-Gurion University of the Negev, Beer-Sheva, Israel

T.W.J. Steele • V. Papper (✉)
School of Materials Science and Engineering, Nanyang Technological University, Singapore, Singapore
e-mail: vladp6@gmail.com

© Springer International Publishing Switzerland 2016

C.D. Geddes (ed.), *Reviews in Fluorescence 2015*,
DOI 10.1007/978-3-319-24609-3_13

13.1 Introduction

Optical irradiation of molecules with a stilbene backbone involves various excited-state processes occurring on multi-timescales. We have previously discussed these rate processes using a linear free-energy correlation approach [1, 2]. We found that different stilbenes exhibit different sensitivity to intramolecular donor-acceptor effects of substituents and medium polarity.

The non-radiative *trans-cis* photoisomerisation process $^1t^* \rightarrow {}^1p^*$ of the excited *trans*-stilbene molecule competes with its radiative fluorescence decay. In other words, the twisting $^1t^* \rightarrow {}^1p^*$ transition may be visualised as a quenching funnel for the *trans*-stilbene fluorescence. Relatively weak donor-acceptor pairs of *para*-substituted stilbenes ("push-pull" stilbenes) usually increase the rate of the $^1t^* \rightarrow {}^1p^*$ transition due to stabilisation of more polar $^1p^*$ state and the concomitant reduction of the intrinsic barrier to this reaction. Solvent polarity affects this transition in a similar way, and the rate of the *trans-cis* photoisomerisation for relatively weak push-pull stilbenes is increased by both polar solvents and more polar substituents [1, 2].

The strong donor *para*-dimethylamino substituent was found to participate efficiently in a charge delocalisation of the $^1t^*$ state and to stabilize it compared to the $^1p^*$ state [1, 2]. This electronic resonance interaction in the reactant $^1t^*$ state destabilises the activated transition $^1t^* \rightarrow {}^1p^*$ by increasing the activation barrier and retards the photoisomerisation rate making it slower compared to that of the weak donor-acceptor substituted stilbenes [1, 2]. Actually, the quantum-chemical calculations for the strong donor-acceptor substituted stilbenes predicted the low polarity of the $^1p^*$ state, in comparison to the highly polar $^1p^*$ state observed in the non-polar stilbenes [3, 4]. This switching from high to low polarity of the $^1p^*$ state was attributed to the phenomenon of "sudden polarisation" [5].

Moreover, the fluorescence quantum yield of the stilbenes bearing the *para*-dimethylamino group was found to be much more sensitive to the *para'*-substitution (on the opposite aromatic ring) effect than their fluorescence lifetime [1]. We concluded that the *para'*-substitution must affect both the non-radiative and the radiative decay channels of the *para*-dimethylamino-stilbenes.

Besides *trans-cis* photoisomerisation, both experimental work and quantum chemical calculations reported a common phenomenon of dual fluorescence observed in polar solvents but not in non-polar solvents in regards to donor - acceptor stilbenes. The new emitting state is characterized by large Stokes shift typical of a large dipole moment of the excited state. This indicates a large charge separation between the donor and acceptor groups. Based on time-resolved emission spectrum results, a precursor-successor relationship between the two emitting states was also observed. Among the donor-acceptor stilbene derivatives, push-pull *para*-dimethylamino stilbenes such as dimethylamino-cyano stilbene (DCS) and dimethylamino-nitro stilbene (DNS) (with their inherent strong donor ability) have been extensively studied to verify the formation of the new emitting state and formation pathway. The proposed formation is associated with a twisted [3, 4, 6–11] or partially twisted [12–14] relaxed conformation around the single bonds in the stilbene molecule. The exact single bond that is responsible for the rotation depends on the other substituent and is still experimentally unresolved. Meanwhile, the development of theoretical quantum calculations makes it possible to examine the electronic and geometrical structure during the lifetime of the excited electronic state of the stilbenes when present in polar and non-polar solvents. Hence, the excited-state twisting of both the single bond and double bond (*trans-cis* photoisomerisation) can be combined in a unified picture, which is of importance in identifying the appearance of the new emitting state.

In this manuscript, we review the phenomenon of the dual fluorescence in the *para*-dimethylamino-stilbenes, which includes unpublished results taken from our own study of this phenomenon.

13.2 Dual Fluorescence Phenomenon in *para*-dialkylamino-stilbenes

As the double bond twisting is the major non-radiative energy dissipation channel in stilbenes, less attention has been paid to the relaxation of the excited molecule via rotation of the adjacent single bonds. Nevertheless, the study of the ring-bridged "stiff stilbenes", where the twisting of phenyl group is prevented, clearly showed that the decay rate to the phantom-singlet P* (^{1}p*) state is faster [3, 12, 15–18]. This experimental observation could hardly be explained in terms of the rotational volumes of the stiff or flexible twisting moieties. Therefore, the photoisomerisation pathway involving only the double bond twisting is arguable, and the single bond rotation or twisting must have certain effect on the shape of the excited state hypersurface. Combined with the observed dual fluorescence phenomenon, already apparent in *para*-dimethylamino-benzonitrile [19–21], it was proposed that a second excited state in the molecules bearing the *para*-dialkylamino-phenyl moiety may formed after the initial excitation to the Franck-Condon state, and the new state originates from a more relaxed state involving the single bond twisting [10, 22, 23].

Time-resolved fluorescence studies conducted by Rettig and Majenz [6] for *trans*-4-dimethylamino-4′-cyanostilbene (DCS) and 4-(dicyanomethylene)-2-methyl-6-(4-dimethylaminostyryl)-4H-pyran (DCM) showed that the non-radiative decay rate constant (k_{nr}) in these compounds is strongly dependent on solvent polarity. And with decreasing solvent polarity, the value of k_{nr} was reported to increase – the trend was particularly obvious in homologous solvent series where any specific solute-solvent interaction is weakly altered.

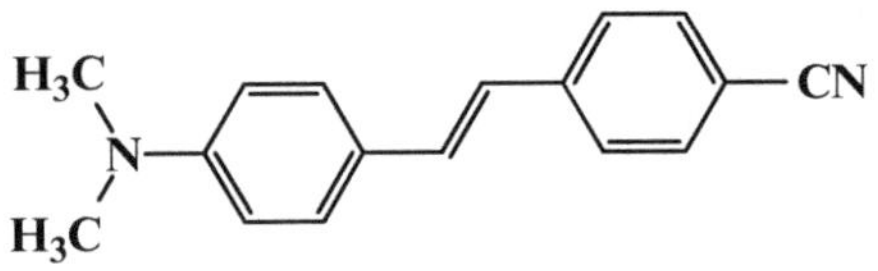

DCS

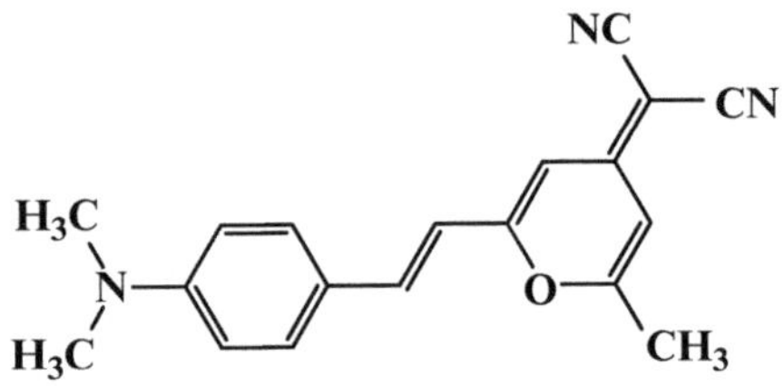

DCM

In addition, solvatochromic measurements implied a large dipole moment of the excited state. Based on the above observations, the three-state kinetic scheme shown in Fig. 13.1, involving a low-lying T* state resulting from a photochemical reaction, namely TICT (twisted-intramolecular charge-transfer) state (resulted from the single-bond twist) formation was proposed in addition to the low-lying stilbene-type "phantom-singlet" state P* (^{1}p*) (resulted from the double-bond twist) [6].

Gilabert et al. [7] further studied the photophysical properties of DCS using picosecond time-resolved spectroscopy. The presence of two time-resolved emission bands was observed in polar solvents while in a non-polar solvent, only the presence of a single emission band was observed. Furthermore, a precursor-successor relationship occurred between the two species responsible for the two emission bands. A kinetic scheme similar to the scheme shown in Fig. 13.1 was suggested, and a normal RICT (relaxed intramolecular charge transfer) and a TICT state was assigned to the observed two emission bands respectively. However, there has been no direct evidence for the twisting character of the more stabilized fluorescent state in both works above as dual fluorescence can also be observed when the locally excited (LE) state relaxes to an intramolecular charge transfer (ICT) state in stilbene compounds like the bridged (stiff) stilbenes mentioned above [3, 11, 13–16]. A second issue which has remained unresolved concerns the identification of the single bond involved in the ICT transition as several single bonds may be involved in the excited state conformational change in polar solvents.

Lapouyade et al. [3] and Gilabert et al. [24] reported a possible additional reaction route. They have studied three compounds (DCS, DCS-B134 and DCS-B24):

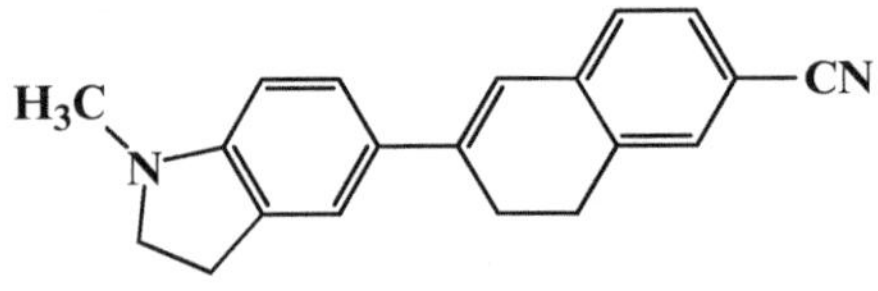

DCS-B134

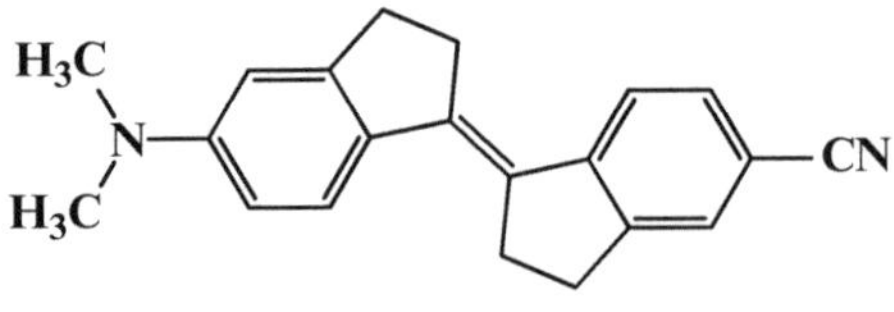

DCS-B24

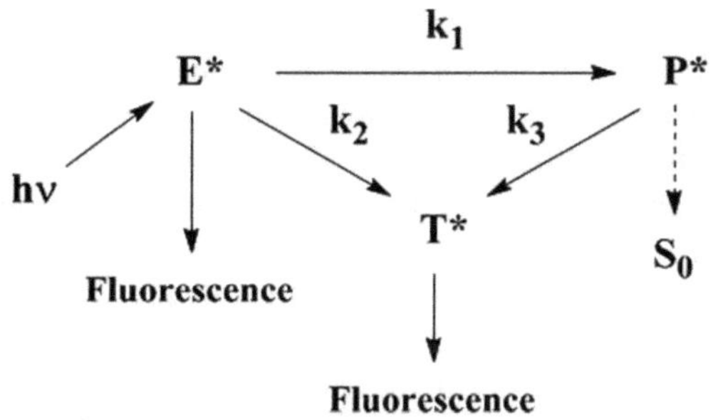

Fig. 13.1 Three-state photochemical reaction scheme involving the initial excitation to the localized excited Franck-Condon state E* (^{1}t*), a low-lying T* state resulted from the single-bond twist, and a low-lying stilbene-type "phantom-singlet" state P* (^{1}p*) resulted from the double-bond twist [6]

Under low concentration and/or low excitation intensity conditions, dual fluorescence emission was absent. At high concentration and excitation intensity the appearance of a new red-shifted emission band was recorded with a precursor-successor relationship between the blue and the red part of the spectra [3, 24]. They hypothesize a photocyclisation mechanism where association of two electronically excited molecules resulted in the creation of a 'bicimer' in which at least one of the excited units was assumed to be twisted. In this case of highly concentrated stilbene solutions, both the dependence on concentration and excitation intensity was observed. However, DCS and DCS-B134 have both a flexible methylamino-phenyl moiety, while it is not so for DCS-B24 where the phenyl group is rigid. Therefore, it was proposed that there should be another emitting species formed, which is not necessary connected to the single-bond twist of the methylamino-phenyl moiety [3, 22].

In response to this report, Eilers-Konig et al. [14] conducted their experiments under low concentration and low excitation intensity conditions to avoid bicimer formation. Dual fluorescence was observed for DCS in two polar solvents (methanol and acetonitrile) using femtosecond time resolution and suggested a successor-precursor relationships for the LE ('locally' excited) → CT (charge transfer) transition [14]. They reported the formation of the new electronic state and identified it as an intramolecular charge transfer state. Dual fluorescence of DCS in polar solvents was also reported by Abraham et al. [25] using pump-probe femtosecond Kerr ellipsometry under low concentration and low excitation intensity conditions. Combining all these observations, it is reasonable to conclude that dual fluorescence involving the formation of a new emitting state in polar solvent is a reality. However, the transition to a new emitting state may be very fast and escape detection without having sufficient time-resolution and sensitivity.

As dual fluorescence is observed only in polar solvents but not in non-polar solvents, it is proposed that the formation of the new emitting state is associated with conformational change through twisting or partial twisting which enhances the electron decoupling and makes it a more stabilized state. Based on the photophysical hypersurface kinetics scheme, the transition to the relaxed ICT state is barrierless, so the twisting should not be temperature dependent, but should be affected by hydrodynamic factors such as volume of the rotating moieties and solvent viscosity. Indeed, it has been shown that change from ACS to DCS to JCS significantly slows the reaction by a factor of two, and for DCS alone, the reaction is retarded in butanol compared to methanol [16]. Hence, rotation is involved in the formation of the relaxed state (ICT), and the twisting is not from the cyano-phenyl moiety for DCS and its derivatives. This hypothesis is supported with results obtained by Rafiq et al. [26], in which they rationalized the involvement of torsional motion of molecular fragments in the excited state based on the fluorescence intensity dependence of *trans*-4-dimethylamino-4′-nitrostilbene (DNS) on the solvent viscosity. Rotation around either the dimethylanilino and nitrophenyl group was proposed for the TICT state based on the ground and excited state potential surface energy from the solvent dependent quantum chemical calculations.

Combining the observations from [16] and the experimental results for various selectively bridged stilbenes [24], it is reasonable to conclude that the formation of ICT for DCS originates from the twisting of the dimethylanilino moiety. This conclusion is backed by independent theoretical calculations in Amatatsu's work [27, 28], who modelled potential energy curves and dipole moments with respect to various torsional angles, and found that DCS is stabilized in polar solvents by the torsional motion of the 4-diemthylanilino group. However, the single bond twisting responsible for the ICT state formation in DNS is not so straight forward to identify.

Lin et al. [29] claimed that the torsion of the nitro group is the main non-radiative pathway. However, Singh et al. [30] used theoretical calculations together with femtosecond transient absorption measurements and concluded that the twisted intramolecular charge transfer process involves the rotation of the nitrophenyl group, but not the dimethyl aniline or the nitro group in the excited state of DNS in polar solvents. Although the conclusions of the two groups are in disagreement, they both exclude the possibility of the dimethylanilino group movement around the single bond being responsible for the ICT state formation in DCS. This is unlike the situation in DMABN (4-N′N-dimethylamino-benzonitrile), where the dimethylamino moiety accounts for the twisting rotation. Hence, it makes sense to conclude that the single bond responsible for the ICT formation depends on the electron-donating and electron-withdrawing properties of the substituents as well as on the structure of the stilbene derivatives. This important conclusion was further supported by Yang et al. [31]. They studied the photophysical properties, such as fluorescence quantum yield and fluorescence lifetime, in a series of arylamino-cyano disubstituted *trans*-stilbenes and one arylamino-nitro analogue (DNS). They concluded that the investigated stilbenes can be divided into four mechanisms based on the nature of their ICT state:

1. a stilbenyl-anilino C-N bond twist,
2. a DMABN-like twist,
3. a styryl-anilino C-C bond twist, and
4. a planar intramolecular charge transfer (PICT) state.

For the forth type of ICT state, they proposed a planar intramolecular charge transfer (PICT) state, which has no dramatic charge decoupling and assumes a planar geometry depending strongly on the specific donors and acceptors in the stilbene molecule [32].

Dual fluorescence emission and formation of a new emitting state by twisting of the molecular fragment have also been observed in our own studies of the substituent and solvent effects on the photophysical behaviour of substituted stilbene compounds [1, 2]. When strong donor and acceptor groups are substituted onto the stilbene phenyl rings, a net increase of the fluorescence quantum yield and fluorescence lifetimes with increase of the solvent polarity were clearly observed. At that time, we concluded the formation of the red-shifted TICT state having a higher quantum yield than the LE state in relatively polar solvents, but postponed the detailed study of this phenomenon.

13.3 Review of Our Studies

We have previously divided all the investigated 4,4′-substituted stilbenes into three groups according to their intramolecular stabilization of the excited $^1t^*$ state (see Table 13.1).

The third group has been formed by three *trans*-4-dimethylamino-stilbenes having strong acceptor *para*-substituent on the second phenyl ring: DACS, DACMS and DANS (*trans*-4-dimethylamino-4′-nitrostilbene) [1, 2]. These three stilbenes exhibited very large Stokes shifts, compared to the first and second groups. In this case, the highly polarized excited state, which creates a large dipole moment, is stabilized extensively by polar interactions (Stokes shift value ΔE is about 22–23 kcal/mole).

Although these stilbenes were categorised into one group, they do not necessarily conform to the same *trans-cis* photoisomerisation pathway. They deviated from the linear dependence of log τ_{fl}^{-1} and log Φ_f on the Hammett σ-constants only in polar media. We have then assumed that the formation of the ICT state is responsible for this behaviour. In case of DANS, we assumed a non-emissive ICT state, which was attributed to a specific interaction of the nitro-group, which quenches the charge-transfer state emission [4, 8]. Actually, non-emissive ICT states were observed in polar solvents when the acceptor group is very powerful and has a low-lying anti-bonding orbital [4].

As noted above, the twisted excited intermediate $^1p^*$ acts as a photochemical funnel toward the ground state. The primary excited emissive state $^1t^*$ of the planar *trans*-isomer is separated from this funnel by an activation barrier of the order of 12 kJ in saturated hydrocarbon solvents [33]. This activation barrier decreases for polar solvents. As $^1p^*$ is non-emissive, its rapid formation acts as quenching process on the emission from $^1t^*$ state with correspondingly low fluorescence quantum yields in low-viscosity solvents of both low and high polarities [34].

Table 13.1 Comparison table of the substituent and solvent effects on the ^{1}t* and ^{1}p* stabilization and following *trans-cis* photoisomerisation for different groups of *trans*-4,4′-disubstituted stilbenes

	Group I	Group II	Group III
	Trans-4,4′-disubstituted stilbenes having relatively weak donor-acceptor substituents.	*Trans*-4,4′-disubstituted stilbenes with the $(CH_3)_2N$ group at one phenyl ring and relatively weak donor or acceptor substituent at another phenyl ring	'Push-pull' stilbenes with the $(CH_3)_2N$ group at one phenyl ring and strong acceptor substituent at another phenyl ring
Charge separation in the excited singlet ^1t state*	Localization in the small space around the zwitterion in the ^{1}t* state.	$(CH_3)_2N$ group facilitates the charge delocalization in the ^{1}t* state.	Large charge delocalization between the donor and the acceptor aromatic moieties, large dipole moment.
Stabilization of ^1t and ^1p* states*	Both polar solvents and substituents stabilize the ^{1}p*, which is more polar compared to the ^{1}t* state.	Stabilizing effect of the strong donor $(CH_3)_2N$ group on the ^{1}t* state is more pronounced than the polar stabilization of the ^{1}p* state.	The ^{1}p* state is of low polarity. The preferential stabilization of ^{1}t* compared to the less polar ^{1}p* state.
Activation energy and reaction rate of $^1t^* \rightarrow {}^1p^*$	Low activation energy, high rate constant τ_{fl}^{-1}.	High activation energy, slow transition rate τ_{fl}^{-1}.	Very high activation barrier.
Sensitivity to the substituent effect (ρ)	High sensitivity, increase in the *trans-cis* isomerization rate (^{1}p* stabilization) with increase of the substituent σ-value.	Because the stabilizing effect of $(CH_3)_2N$ group on the ^{1}t* is more pronounced than the polar stabilization of the ^{1}p* state, the effect of the second substituent has a less influence on the transition state and activation energy, i.e. the reaction rate is less sensitive to the substituent effects (low ρ-value).	Unclear, poor experimental database.
Sensitivity to the solvent effect	Hammett-like correlation is similar in all investigated solvents, and both life-time and quantum yield have the similar substituent dependence, indicating the same mechanism inside the group. The primary effect of solvent polarity will be on the transition-state energy.	The substituent effects are most pronounced in non-polar solvents (high ρ-value). Intramolecular effect of substituents (stabilisation of less polar ^{1}t* state) dominates over the solvent effect (stabilisation of more polar ^{1}p* state). Increasing the solvent polarity leads to higher stabilisation of the ^{1}p* state, therefore the reaction is less sensitive to the substituent effects in polar solvents.	Net increase of the fluorescence quantum yield and fluorescence life-time with increase of solvent polarity indicates another reaction mechanism.

Donor-acceptor substituted stilbenes from the third group, however, behave differently: they show a net increase of fluorescence quantum yields and lengthening of fluorescence lifetimes with increasing solvent polarity, which indicates a different photochemical reaction mechanism. The experimental finding that fluorescence quantum yields are high in strongly polar solvents could be interpreted as indicating a strongly preferred pathway toward the charge transfer state formation in the competing photochemical processes starting from $^1t^*$. Assuming that this point holds in less polar solvents, too, the reduced fluorescence quantum yield and increased non-radiative decay rate may be due to the insufficient stabilization of the ICT in these solvents. This can be explained by speculating that $^1p^*$ is less polar than the ICT state and $^1t^*$ states, and thus increasing solvent polarity preferentially lowers in energy a highly polar ICT state with respect to $^1p^*$.

13.4 New Dual Fluorescent *trans*-4-dimethylamino-4′-carbomethoxystilbene (DACMS)

To support the existence of an ICT state in *para*-dialkylamino-stilbenes, we have synthesised a new stilbene compound *trans*-4-dimethylamino-4′-carbomethoxystilbene (DACMS) and investigated its photophysical behaviour. Figure 13.2 shows the time-resolved fluorescence emission spectra of DACMS in *iso*-propanol and Table 13.2 summarises the photophysical properties of this new molecule.

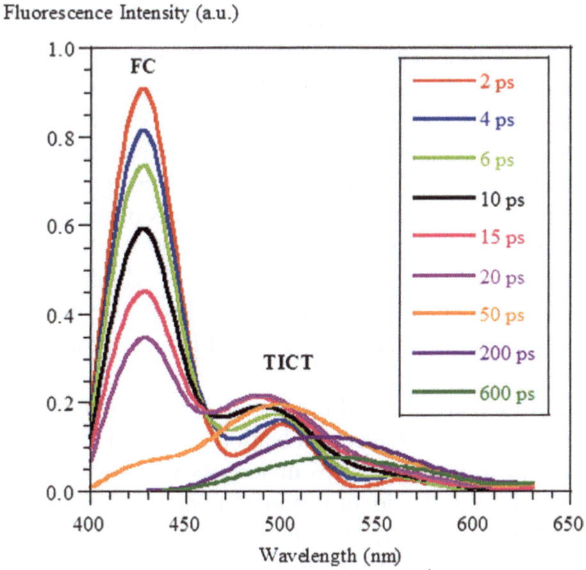

Fig. 13.2 Time-resolved emission spectra of *trans*-4-dimethylamino-4-carbomethoxystilbene in *iso*-propanol

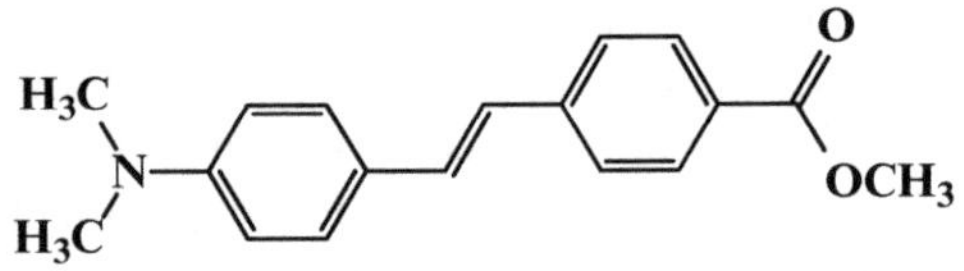

DACMS

There is a clear isosbestic point in the spectra of DACMS observed on a picosecond time scale by our experimental system that indicates the precursor-successor relationship between the two emissive states. Following the recent advances reviewed in this contribution we now conclude that the second emissive state is formed after the initial excitation due to further electron decoupling and charge separation.

The above considerations and time-resolved fluorescence measurements of DACMS, which extends the series of the dual fluorescence *para*-dialkylamino-stilbenes, support a three-state kinetic scheme. This scheme is shown in Fig. 13.3, which, in addition to states $^1t^*$ (planar geometry) and $^1p^*$ (double-bond twist), depicts a third state T* (ICT), which may be, for example, a TICT state.

In the non-polar media (---) the planar conformation $^1t^*$ is the lowest state, whereas in polar solvents (—) the perpendicular single-bond conformation (TICT) has a lower energy, because of its higher dipole moment, and competes with the $^1t^* \rightarrow {}^1p^*$ channel. The TICT-state corresponds to a perpendicular single-bond conformation where the donor and the acceptor groups are orbitally decoupled ($\varphi = 90°$), with strong orbital coefficients at the linking C-C atoms. In polar solvents, this state may be responsible for the main part of emission. If this state involves a strong charge transfer from the donor to the acceptor aromatic ring, its energy should be close to that of $^1t^*$ even in polar solvents, and hence, could overlap the $^1t^*$ fluorescence (see Table 13.2). A large charge separation between the strong donor and acceptor aromatic rings leads to a significant increase in the dipole moment of the excited molecule with respect to the ground state. Therefore, this ICT state is unusually sensitive to the solvent polarity, which stabilises it by dipolar interaction. In non-polar solvent, such as cyclohexane, and slightly polar chlorobenzene, only a single emission band is observed.

Figure 13.4 shows the logarithmic plot of the rate constant k_d (τ_{fl}^{-1}) versus the normalized solvent polarity value E_T^N yielding a linear dependence. Long-chain alcohols (octanol and decanol) visibly deviate from the linear correlation. Retarding of the rate in octanol and decanol is most probably due to high viscosity of these solvents.

An experimental difficulty associated with an affirmation of the reality of the ICT state is the dynamic solvation of both the $^1t^*$ and ICT state. As seen in Fig. 13.2, the spectrum after 20 ps undergoes emission shift with time, which considerably complicates the identification of the formed ICT state, particularly when these molecules are studied with the time-resolved fluorescence setups, which are not capable to cover the single-digit picosecond range. It happens that this phenomenon

Table 13.2 Normalized solvent polarity parameter (E_T^N), fluorescence emission wavelength, fluorescence lifetime of the excited state, weight of this state in a total emission and rise-time for *trans*-4-dimethylamino-4′-carbomethoxystilbene

Solvent	E_T^N	Wavelength, nm	Fluorescence life-time, ps		Weight of state in total fluorescence		Rise time, ps
			TICT	$^1t^*$	TICT	$^1t^*$	
CH	0.006	442	–	48	–	1.00	–
CB	0.188	478	–	298	–	1.00	–
MEK	0.327	498	–	549	–	1.00	–
DMSO	0.444	480	1343	6	0.25	0.75	–
		543	1365	–	1.00	–	–
Decanol	0.525	420	329	83	0.07	0.93	–
		440	570	130	0.08	0.92	–
		600	926	–	1.00	–	377
Octanol	0.537	440	464	90	0.04	0.96	–
		480	852	192	0.20	0.80	–
		500	883	224	0.41	0.59	–
		530	899	–	1.00	–	31
iso-Propanol	0.546	480	832	59	0.22	0.78	–
		500	830	82	0.51	0.49	–
		530	822	–	1.00	–	17
Pentanol	0.572	430	72	21	0.12	0.88	–
		480	868	102	0.18	0.82	–
		500	849	118	0.34	0.66	23
		530	849	93	0.65	0.35	41
		560	840	–	1.00	–	59
Butanol	0.586	430	–	11	–	1.00	–
		500	1040	121	0.35	0.65	–
		530	937	241	0.50	0.50	–

(continued)

Table 13.2 (continued)

Solvent	E_T^N	Wavelength, nm	Fluorescence life-time, ps		Weight of state in total fluorescence		Rise time, ps
			TICT	$^1t^*$	TICT	$^1t^*$	
Propanol	0.617	420	810	13	0.01	0.99	–
		440	844	23	0.03	0.97	–
		460	824	36	0.08	0.92	6
Ethanol	0.654	400	–	6	–	1.00	–
		420	955	11	0.01	0.99	–
		440	947	17	0.04	0.96	–
		460	891	27	0.15	0.85	–
		600	959	–	1.00	–	–
Methanol	0.762	440	1146	10	0.03	0.97	69
		460	1133	14	0.12	0.88	99

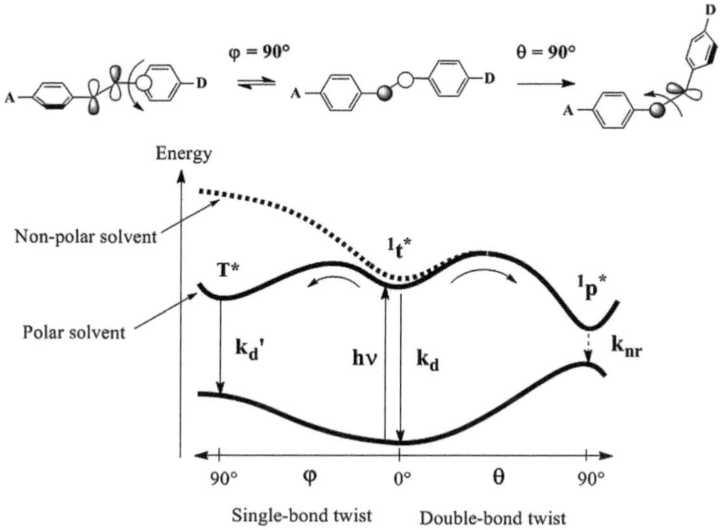

Fig. 13.3 Three-state kinetic scheme for the *trans-cis* photoisomerisation of strong donor-acceptor substituted stilbenes

is observed between 1 and 20 ps, so it is imperative that the time-resolved fluorescence setup would be tuned accordingly to conduct measurements in this range.

A series of recent investigations by Ernsting et al. [35–37] rejected the three-state kinetic model and the experimental evidence for the second emissive state formation in the push-pull stilbenes. Kovalenko et al. [35] showed that in a broad range of solvents, emission spectra on a 100 fs time scale appear to reflect only the dynamics of polar solvation. Quick et al. [37] in the most recent publication used femtosecond stimulated Raman spectroscopy to reconsider the stiff stilbenes. We speculate that the measurements of the push-pull stilbenes in polar solvents conducted by the group of Ernsting show only a continuous spectral shift over times characteristic of solvent relaxation. Their data did not investigate the aforementioned 1–20 ps range and does not observe the time-resolved dual fluorescence emission of these stilbenes.

13.5 Conclusion

Most of the previous and recent studies have abundantly substantiated the reality of the dual fluorescence phenomenon in *para*-dialkylamino-stilbenes measured in polar solvents and confirmed the presence of a more relaxed intramolecular charge transfer state, which is formed by twisting the single bond rather than a double

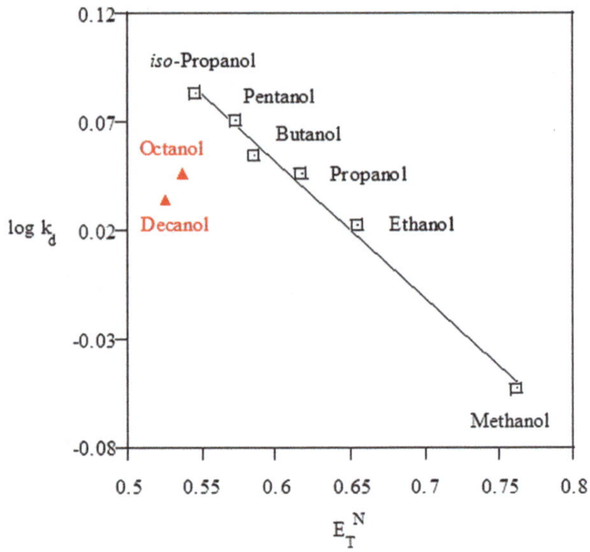

Fig. 13.4 Dependence of the fluorescence decay rate constant of *trans*-4-dimethylamino-4′-carbomethoxystilbene on E_T^N normalized solvent polarity parameter in aliphatic alcohols series

bond. The mechanism of the ICT state formation competes with *trans-cis* photoisomerisation in push-pull stilbenes and serves as an additional quenching funnel in the relaxation process of the excited molecule. It is a barrierless process, so the reaction rate of the entire process is temperature independent, but is strongly affected by the hydrodynamic factors, such as the size or volume of the substituent moieties and the viscosity of the surrounding medium.

The ICT state is likely formed by a single bond twist, however, which single bond is responsible for the twisting is ambiguous and depends on the donor-acceptor properties of the stilbene substituents. In addition, the relative reaction rate of the ICT state formation and *trans-cis* isomerisation depends on the relative polarity of the T* and P* states, which can further affect the fluorescence intensity and quantum yield of the specific molecule. Everything considered, this allows the *para*-dialkylamino-stilbenes to be employed as effective probes [38, 39] for further investigation of the dual fluorescence phenomenon in various chemical and biological applications.

References

1. Papper V, Pines D, Likhtenshtein GI, Pines E (1997) Photophysical characterization of trans-4,4'-disubstituted stilbenes. J Photochem Photobiol A Chem 111:87–96
2. Papper V, Likhtenshtein GI (2001) Substituted stilbenes: a new sight on well-known systems. New applications in chemistry and biophysics. J Photochem Photobiol A Chem 140:39–52
3. Lapouyade R, Czeschka K, Majenz W, Rettig W, Gilabert E, Rulliere C (1992) Photophysics of donor-acceptor substituted stilbenes. J Phys Chem 96:9643–9650

4. Lapouyade R, Kuhn A, Letard J-F, Rettig W (1993) Multiple relaxation pathways in photoexcited dimethylaminonitro- and dimethylaminocyano-stilbenes. Chem Phys Lett 208:48–64
5. Bonacic'-Koutecky' V, Bruckmann P, Hiberty P, Koutecky' J, Leforestier C, Salem L (1975) Sudden polarization in the zwitterionic Z excited states of organic intermediates. Photochemical implications. Angew Chem Int Ed Engl 14:575–587
6. Rettig W, Majenz W (1989) Competing adiabatic photoreaction channels in stilbene derivatives. Chem Phys Lett 154(4):335–341
7. Gilabert E, Lapouyade R, Rulliere C (1988) Dual fluorescence in 4-dimethylamino-4'-cyanostilbene revealed by picosecond time-resolved spectroscopy. Chem Phys Lett 145:262–267
8. Le Breton H, Bennetau B, Letard J-F, Lapouyade R, Rettig W (1996) Non-radiative twisted intramolecular charge transfer state in polar stilbenes. J Photochem Photobiol A Chem 95:7–20
9. Rechthaler K, Köhler G (1996) Photophysical properties of a highly fluorescent push-pull stilbenes. Chem Phys Lett 250:152–158
10. Abraham E, Oberle J, Jonusauskas G, Lapouyade R, Rulliere C (1997a) Photophysics of 4-dimethylamino-4'-cyanostilbene and model compounds: dual excited states revealed by sub-picosecond transient absorption and Kerr Ellipsometry. Chem Phys 214:409–423
11. Rettig W, Majenz W, Lapouyade R, Haucke G (1992) Multidimensional photochemistry in flexible dye systems. J Photochem Photobiol A Chem 62:415–427
12. Il'ichev YV, Kühnle W, Zachariasse KA (1996) Photophysics of 4-dimethylamino-4'-cyanostilbene and 4-azetidinyl-4'-cyanostilbene. Chem Phys 211:441–453
13. Il'ichev YV, Zachariasse KA (1997) Intramolecular charge transfer, isomerization and rotational reorientation of trans-4-dimethylamino-4'-cyanostilbene in liquid solution. Ber Bunsenges Phys Chem 101:625–635
14. Eilers-König N, Kühne T, Schwarzer D, Vöhringer P, Schroeder J (1996) Femtosecond dynamics of intramolecular charge transfer in 4-dimethylamino-4'-cyanostilbene in polar solvents. Chem Phys Lett 253:69–76
15. Rettig W, Majenz W, Herter R, Le'tard JF, Lapouyade R (1993) Photophysics of stilbenoid dye systems. A comparison of experiment and theory. Pure Appl Chem 65:1699–1704
16. Letard J-F, Lapouyade R, Rettig W (1994) Multidimensional photochemistry in 4-(N, N-dimethylamino) stilbene. Chem Phys 186:119–131
17. Le'tard JF, Lapouyade R, Rettig W (1993) Structure-photophysics correlations in a series of 4-(dialkylamino) stilbenes: intramolecular charge transfer in the excited state as related to the twist around the single bonds. J Am Chem Soc 115:2441–2447
18. Pines D, Pines E, Rettig W (2003) Dual fluorescence and excited state structural relaxations in donor-acceptor-stilbenes. J Phys Chem A 107:236–242
19. Lippert E, Lüder W, Boos H (1962) Fluoreszenzspektrum und franck-condon-prinzip in lösungen aromatischer verbindungen. In: Mangini A (ed) Advances in molecular spectroscopy. Pergamon, Oxford, pp 443–457
20. Grabowski ZR, Rotkiewicz K, Siemiarczuk A, Cowley DJ, Baumann W, Nou V (1979) Twisted intramolecular charge transfer states (TICT). A new class of excited states with a full charge separation. J Chim 3:443
21. Grabowski ZR, Dobkowski J (1983) Twisted intramolecular charge transfer (TICT) excited states: energy and molecular structure. Pure Appl Chem 55:245
22. Rettig W (1986) Charge separation in excited states of decoupled systems – TICT compounds and implications regarding the development of new laser dyes and the primary process of vision and photosynthesis. Angew Chem Int Ed Engl 25:971–988
23. Rettig W (1994) Photoinduced charge separation via twisted intramolecular charge transfer states. In: Mattay J (ed) Topics in current chemistry, electron-transfer I, vol 169. Springer, Berlin, pp 253–299

24. Gilabert E, Lapouyade R, Rulliere C (1991) Time-resolved dual fluorescence of push—pull stilbenes at high solute concentration and excitation intensity: evidence for an emitting bicimer. Chem Phys Lett 185:82–87
25. Abraham E, Oberle J, Jonusauskas G, Lapouyade R, Rulliere C (1997b) Dual excited states in 4-dimethylamino 4′-cyanostilbene (DCS) revealed by sub-picosecond transient absorption and Kerr ellipsometry. J Photochem Photobiol A 105:101–107
26. Rafiq S, Sen P (2013) Dielectric controlled excited state relaxation pathways of a representative push-pull stilbene: a mechanistic study using femtosecond fluorescence up-conversion technique. J Chem Phys 138:084308
27. Amatatsu Y (2001) Ab initio study on the photochemical behavior of 4-dimethylamino, 4′-cyanostilbene. Chem Phys 274:87–98
28. Amatatsu Y (2000) Ab initio configuration interaction study on electronically excited 4-dimethylamino-4′-cyanostilbene. Theor Chem Accounts 103:445–450
29. Lin CK, Wang YF, Cheng YC, Yang JSJ (2013) Multisite constrained model of trans-4-(N, N-Dimethylamino)-4′-nitrostilbene for structural elucidation of radiative and nonradiative excited states. Phys Chem A 117:3158–3164
30. Singh C, Ghosh R, Mondal JA, Palit DK (2013) Excited state dynamics of a push–pull stilbene: a femtosecond transient absorption spectroscopic study. J Photochem Photobiol A 263:50–60
31. Yang J-S, Liau K-L, Hwang C-Y, Wang C-M (2006) Photoinduced single- versus double-bond torsion in donor−acceptor−substituted trans-stilbenes. J Phys Chem A 110:8003–8010
32. Zachariasse KA (2000) Comment on "Pseudo-Jahn–Teller and TICT-models: a photophysical comparison of meta-and para-DMABN derivatives" [Chem. Phys. Lett. 305 (1999) 8]: The PICT model for dual fluorescence of aminobenzonitriles. Chem Phys Lett 305320:8–13
33. Schroeder J, Schwarzer D, Troe J, Voss F (1990) Cluster and barrier effects in the temperature and pressure dependence of the photoisomerization of trans-stilbene. J Phys Chem 93(4):2393–2404
34. Meier H (1992) The photochemistry of stilbenoid compounds and their role in materials technology. Angew Chem Int Ed Engl 31:1399–1420
35. Kovalenko SA, Schanz R, Senyushkina TA, Ernsting NP (2002) Femtosecond spectroscopy of p-dimethylaminocyanostilbene in solution – no evidence for dual fluorescence. Phys Chem Chem Phys 4:703–707
36. Ernsting NP, Breffke J, Yu D, Vorobyev DA, Duncan IP (2008) Sub-picosecond fluorescence evolution of amino-cyano-stilbenes in methanol: polar solvation obeys continuum theory without evidence of twisting. Phys Chem Chem Phys 10:2043–2049
37. Quick M, Berndt F, Dobryakov AL, Ioffe IN, Granovsky AA, Knie C, Mahrwald R, Lenoir D, Ernsting NP, Kovalenko SA (2014) Photoisomerization dynamics of stiff-stilbene in solution. J Phys Chem B 118:1389–1402
38. Papper V, Pokholenko O, Wu Y, Zhou Y, Jianfeng P, Steele TWJ, Marks RS (2014) Novel Photochrome Aptamer Switch Assay (PHASA) for adaptive binding to aptamers. J Fluoresc 24 (6):1581–1591. doi:10.1007/s10895-014-1441-9
39. Wu Y, Papper V, Pokholenko O, Kharlanov V, Zhou Y, Steele TWJ, Marks RS (2015) New photochrome probe allows simultaneous pH and microviscosity sensing. J Fluoresc 25 (34):961–972. doi:10.1007/s10895-015-1577-2

Chapter 14
Molecular Anatomy of an Ion Channel Explored Utilizing Fluorescence Spectroscopy

Arunima Chaudhuri and Amitabha Chattopadhyay

Abstract Ion channels are transmembrane proteins and represent important cellular components that connect the inside of the cell to its outside in a selective fashion. The linear ion channel peptide gramicidin serves as an excellent prototype for monitoring the organization, dynamics and function of membrane-spanning channels due to a variety of reasons. The fluorescent tryptophan residues in gramicidin channels are crucial for establishing and maintaining the structure and function of the channel in the membrane bilayer. In this review, we have highlighted a variety of representative fluorescence-based approaches to gain molecular insight into gramicidin conformations. Since gramicidin shares common structural features with more complex ion channels, the results from fluorescence-based studies with gramicidin could be relevant for more complex ion channels.

Keywords Ion channel • Gramicidin • REES • Tryptophan • Membrane interface

14.1 Introduction

Ion channels are important cellular nanomachines that regulate ionic permeability in cell membranes. They are integral membrane proteins with multiple transmembrane domains. Their ability to connect the inside of the cell to its outside in a selective fashion makes them crucial elements in cellular signaling and sensing. Defects in the function of ion channels result in diseases [1] such as cystic fibrosis [2]. Advances in DNA sequencing technology have linked many diseases to defects in ion channels and, the term 'channelopathy' has been coined [3]. Drugs acting on ion channels have long been used as therapeutics for treatment of a wide spectrum of disorders. This makes them a favorite target for the pharma industry and ~15 % of the world's 100 top-selling drugs are currently targeted to ion channels [4].

Although ion channels are important members in cellular physiology, detailed structure-function analysis of ion channels at high resolution has proved to be

A. Chaudhuri • A. Chattopadhyay (✉)
CSIR-Centre for Cellular and Molecular Biology, Uppal Road, Hyderabad 500007, India
e-mail: amit@ccmb.res.in

© Springer International Publishing Switzerland 2016
C.D. Geddes (ed.), *Reviews in Fluorescence 2015*,
DOI 10.1007/978-3-319-24609-3_14

a HCO-L-Val-Gly-L-Ala-D-Leu-L-Ala-D-Val-L-Val-D-Val-L-Trp-D-Leu-L-Trp-D-Leu-L-Trp-D-Leu-L-Trp-$NHCH_2CH_2OH$

9 11 13 15

b

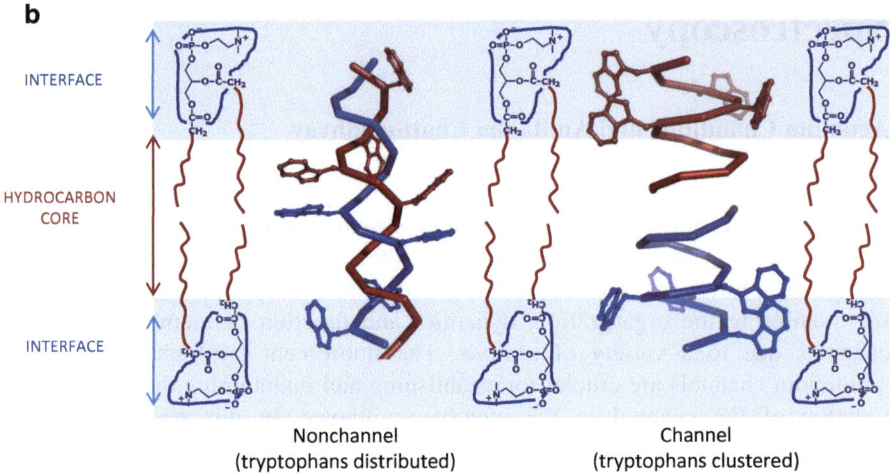

Fig. 14.1 (**a**) The amino acid sequence of gramicidin A with its unique alternating L- and D-chirality. Aromatic amino acids (tryptophans) are highlighted and their positions are indicated. (**b**) A schematic representation of the two predominant forms of gramicidin, the nonchannel and channel conformations, displaying the locations of tryptophan residues in the membrane bilayer. A hallmark of the channel conformation is the clustering of tryptophans toward the membrane interface. A distinctive difference is observed in the tryptophan distribution in the nonchannel conformation in membranes, where the tryptophans span the entire bilayer normal. See text for other details (Adapted and modified from refs. [5] and [26] with permission from Elsevier)

challenging till very recently. In the overall context of the complexity and difficulty involved in studying large ion channels at molecular resolution, the linear peptide gramicidin has proved to be a relevant model for ion channels. Gramicidin forms prototypical ion channels specific for monovalent cations and has been extensively used to monitor the organization, dynamics, and function of membrane-spanning channels and integral membrane proteins [5]. Gramicidin is a multi-tryptophan peptide (Trp-9, 11, 13, and 15) with alternating L- and D-chirality (see Fig. 14.1a). The advantages of gramicidin as ion channel include its small size, ready availability and the relative ease with which chemical modifications can be performed. These excellent features contribute to the usefulness of gramicidin and form the basis for its use to explore the principles that govern the folding and function of ion channels.

More importantly, gramicidin channels share important structural features involving ion selectivity with complex ion channels such as KcsA potassium channels [6]. The unique sequence of alternating L- and D-chirality allows gramicidin to assume a variety of environment-sensitive conformations. Among these, two major conformations are: (i) the single stranded $\beta^{6.3}$ helical dimer (the 'channel' form), and (ii) the double stranded intertwined helix (collectively known as the 'nonchannel' form) (see Fig. 14.1b) [5]. The amino terminal-to-amino terminal single-stranded $\beta^{6.3}$ helical dimer form is the thermodynamically preferred conformation in membranes and membrane-mimetic media. In this conformation, the tryptophan residues remain clustered at the membrane-water interface [7–10]. Interestingly, the membrane interfacial localization of tryptophan residues is absent in 'nonchannel' conformations and the tryptophan residues are distributed along the membrane axis [5, 7, 11]. Nonchannel conformations have been shown to exist in membranes with polyunsaturated lipids [12], and in membranes with increased acyl chain lengths under hydrophobic mismatch conditions [13, 14].

14.2 Tryptophan: A Uniquely Placed Amino Acid in Membrane Proteins

The presence of tryptophan residues as intrinsic fluorophores in membrane peptides and proteins makes them an attractive choice for fluorescence spectroscopic analyses [15–19]. Tryptophan residues play an important role in the structure and function of membrane proteins and peptides (recently reviewed in ref. 19). Tryptophan residues in membrane proteins and peptides are not uniformly distributed, but tend to be localized toward the membrane interface (see Fig. 14.2). The interfacial region in membranes has unique motional and dielectric characteristics distinct from both the bulk aqueous phase and the hydrocarbon-like interior of the membrane [18, 20, 21]. The interfacial localization of tryptophan in membrane proteins and peptides, along with the fact that the distribution and localization of tryptophans in the two major conformations of gramicidin are distinctly different (see Fig. 14.1b), allow us to explore conformations adopted by gramicidin and its tryptophan analogs using fluorescence spectroscopic readouts. In this review, we have highlighted representative fluorescence-based approaches to gain insight into gramicidin conformations. Since gramicidin is a prototypical ion channel [5] and shares common structural features with more complex ion channels, which are more challenging to work with [6], the results from these studies form the basis of addressing ion channel conformations under varying conditions.

14.3 Ion Channel Conformations Explored Using REES

Red edge excitation shift (REES) is a popular tool to explore organization, dynamics and conformation of membrane probes, proteins and peptides [17–20]. REES is defined as the shift in the wavelength of maximum fluorescence emission toward longer wavelengths, caused by a shift in the excitation wavelength toward the red edge of the absorption spectrum. REES becomes significant in case of fluorophores with a relatively large change in dipole moment upon excitation in restricted

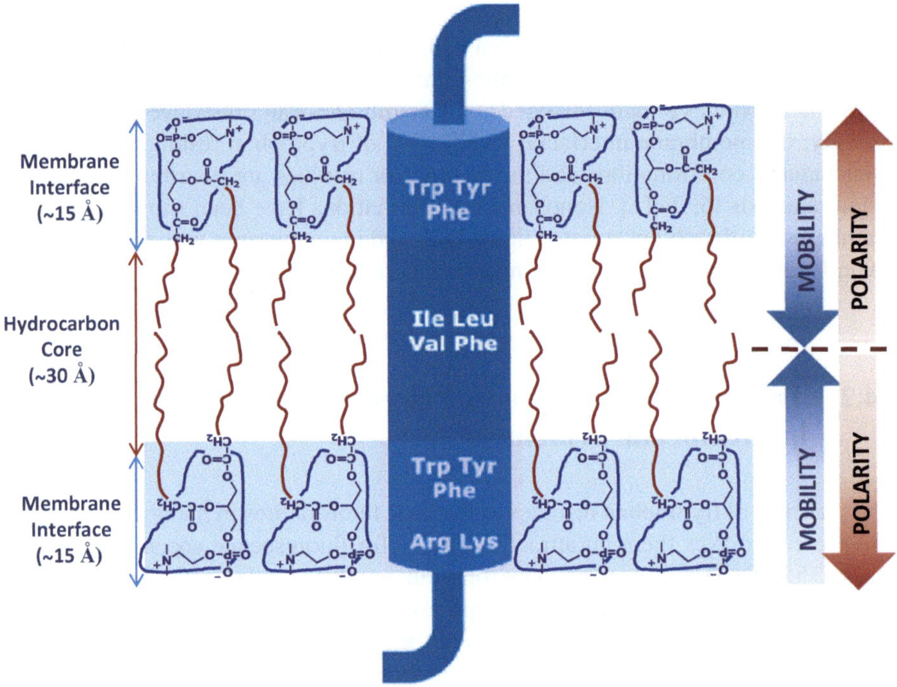

Fig. 14.2 A schematic representation of a typical transmembrane domain of a representative membrane protein (such as ion channels) in the bilayer showing distinct preferences of various amino acids for different parts of the membrane bilayer. The membrane lipids shown have two hydrophobic tails with a phosphatidylcholine headgroup. It should be noted that the aromatic amino acids, especially tryptophan and tyrosine residues, are localized in the membrane interface region, a feature shared by many ion channels [11]. The membrane interface, constituting ~ 50 % of the thickness of the bilayer, is represented by a heterogeneous environment characterized by relatively slow dynamics. This region also exhibits higher polarity relative to the hydrophobic core, predominantly due to the restricted water molecules (Adapted and modified with permission from ref. [19] (copyright (2014) American Chemical Society))

environment. Unlike other fluorescence approaches, REES data contain information on the rotational dynamics of excited state dipoles around the fluorophore, thereby providing a window to the dynamics of the environment. In our laboratory, we have utilized REES to monitor the conformation and dynamics of a variety of membrane peptides and proteins including gramicidin [7, 10, 22–26], melittin [27, 28], the pore-forming α-toxin from *S. aureus* [29], the N-terminal domain of CXC chemokine receptor (CXCR1) [30], and membrane-bound bovine α-lactalbumin. [31] A particularly attractive example is the application of REES to conformational analysis of gramicidin, a representative ion channel peptide.

As mentioned above, two major conformations adopted by gramicidin in various media are: (i) the single stranded $\beta^{6.3}$ helical dimer (the 'channel' form), and (ii) the double stranded intertwined helix (collectively known as the 'nonchannel' form) (see Fig. 14.1b). In the channel conformation in membranes, the tryptophan residues are clustered at the membrane-water interface [7, 9, 10]. We earlier showed that the tryptophan residues of gramicidin in the channel conformation exhibit REES, implying that the tryptophan residues are localized in the interfacial region and experience motional restriction [7, 10]. In an important application of REES to conformational analysis of membrane proteins, we demonstrated that various conformations of membrane-bound gramicidin could be distinguished using their REES signatures (see Fig. 14.3) [7]. The basic principle of such conformation-specific REES is the fact that the microenvironment of the tryptophans in these conformations are different, thereby giving rise to different REES readouts (see Fig. 14.3b). For example, while the tryptophan residues are clustered toward the membrane interface in the channel conformation of gramicidin, they are distributed along the membrane axis in the nonchannel conformation. We have previously demonstrated, using anthroyloxy probes, that the vertical location (depth) of a fluorophore in the membrane is an important parameter in their ability to exhibit REES [32]. Fluorophores localized at the shallow membrane interfacial region experience restricted dynamics due to the physicochemical nature of the interfacial region, and exhibit REES. On the other hand, fluorophores localized at deeper locations in the membrane experience much more dynamic environment and exhibit reduced REES. This principle has proved to be very useful in conformational analysis of gramicidin in membranes. In agreement with this, the tryptophans in the channel conformation of gramicidin (with tryptophans at the motionally restricted interfacial region) give rise to REES of 7 nm. In contrast, tryptophans in the nonchannel conformation of gramicidin (distributed along the membrane axis with varying degrees of motional restriction) give rise to REES of only 2 nm. More importantly, it was possible to monitor REES exhibited by conformational intermediates in the folding pathway of membrane-bound gramicidin from the initial nonchannel to the final channel conformation (denoted as intermediates I and II in Fig. 14.3b). The progressive increase in REES from the nonchannel conformation to the channel conformation through the intermediate folding conformations corresponds to the conversion of the nonchannel form to the channel form of gramicidin. This is due to gradual change in the location of tryptophan residues from a distribution along the bilayer normal to being clustered at the membrane interface.

14.4 Conformational Heterogeneity: Use of Fluorescence Lifetime Distribution Analysis

Conformational heterogeneity in membrane proteins can be assessed using fluorescence lifetime distribution analysis of tryptophan residues by the maximum entropy method (MEM). MEM represents a model-free robust approach for analyzing fluorescence lifetime distribution [21, 33–35]. The width of the lifetime distribution obtained by this method is correlated with the degree of heterogeneity of the environment sensed by the fluorophore.

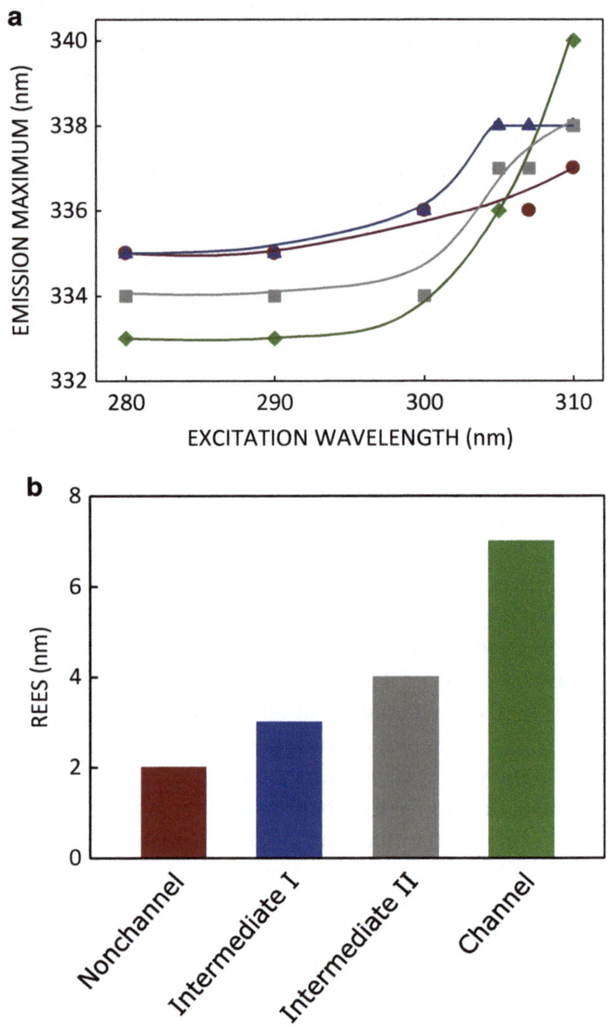

Fig. 14.3 Monitoring conformations of ion channels using REES as a reporter. (**a**) Effect of changing excitation wavelength on the wavelength of maximum emission for the nonchannel (●), intermediate I (▲), intermediate II (■), and channel (♦) forms of gramicidin. (**b**) The magnitude of REES is sensitive to the conformation of ion channel adopted by gramicidin. Intermediates I and II represent intermediates in the folding pathway from the nonchannel conformation to the channel conformation. The magnitude of REES corresponds to the total shift in emission maximum as the excitation wavelength was changed from 280 to 310 nm. The lines joining data points are provided merely as viewing guides (Adapted and modified with permission from ref. [7] with permission from Elsevier and ref. 19 (copyright (2014) American Chemical Society))

The differential conformational heterogeneity sampled by gramicidin conformers is shown in Fig. 14.4. The figure shows tryptophan lifetime distributions by MEM analysis in the channel and nonchannel conformations of gramicidin [36]. Interestingly, fluorescence lifetime distribution of tryptophan residues in the nonchannel form (characterized by a width of the distribution (w) of 3.40 ns, represented as full width at half maxima for the major lifetime peak) was found to be significantly broader relative to the corresponding width for the channel form (w = 0.96 ns). This indicates that the tryptophan residues in the nonchannel form experience relatively heterogeneous environment relative to the environment experienced in the channel form. This is in agreement with the fact that tryptophan residues are clustered at the membrane interfacial region in the channel form, while

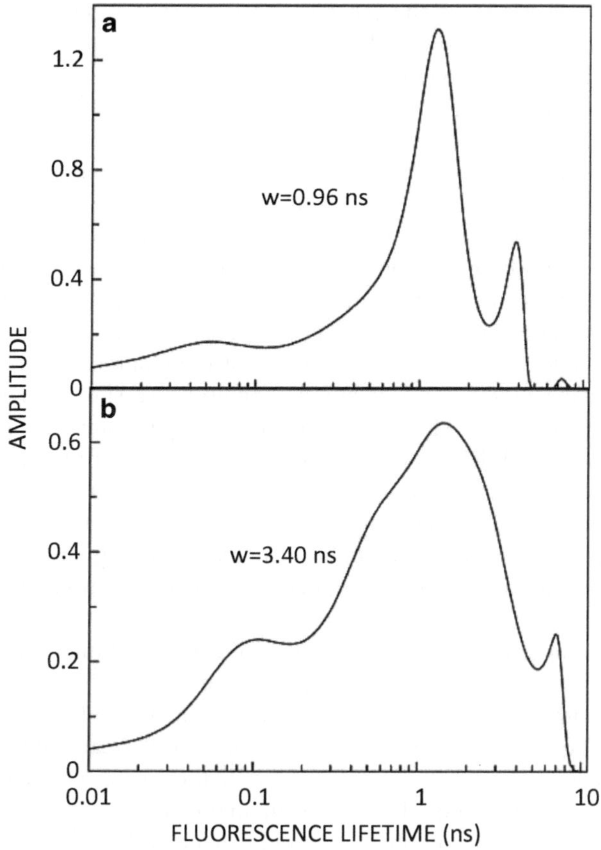

Fig. 14.4 Conformation-dependent heterogeneity profiles of the tryptophan residues in gramicidin derived from MEM lifetime distribution analysis for the (**a**) channel and (**b**) nonchannel conformations. Variation in the width of the tryptophan fluorescence lifetime distribution (w) for the major peaks indicates differential membrane heterogeneity experienced by the tryptophans localized at different regions of the membrane z-axis in the channel and nonchannel conformations (Reproduced from ref. [36] with kind permission from Springer Science and Business Media)

they are spread all across the bilayer normal in case of the nonchannel form (see Fig. 14.1b). Fluorescence lifetime distribution analysis by MEM therefore provides a novel window to monitor such conformational transitions in membrane proteins such as ion channels. Since ion channels require a variety of conformations for carrying out their function, this approach provides a unique way to sample the conformational plasticity associated with each conformation.

14.5 Understanding the Functional Role of Tryptophans in Gramicidin Channel: Insights from Tryptophan Analogs

Interestingly, tryptophans in gramicidin channels have been shown to be crucial for maintaining the structure and function of the channel [5]. The importance of gramicidin tryptophans is apparent from the observation that the cation conductivity of the ion channel decreases upon substitution of one or all of the tryptophan residues by phenylalanine, tyrosine or naphthylalanine [37–39], and also upon ultraviolet irradiation or chemical modification of the tryptophan side chains [40–42]. Additionally, it has been shown that gramicidins with Trp → Phe substitutions face greater difficulty in forming membrane-spanning dimeric channels [39, 43]. With an overall goal of understanding the structural basis of the role of tryptophans in maintaining the ion channel structure of gramicidin, we used single tryptophan gramicidin analogs with three Trp → Ser-*t*-butyl substitutions using a combination of fluorescence approaches which include REES and membrane penetration depth analysis [24]. The Ser-*t*-butyl side chain was chosen because it is approximately as hydrophobic as Phe, yet not aromatic. Figure 14.5a shows the amino acid sequence of these single tryptophan analogs. REES results using these single tryptophan analogs are shown in Fig. 14.5 (panels b and c). Figure 14.5c shows that single tryptophan analogs Trp-15, Trp-13, and Trp-11 exhibited REES of 4–7 nm. In contrast, Trp-9 displayed enhanced REES of 15 nm. Based on these results in combination with membrane penetration depth analysis [44], size-exclusion chromatography and backbone CD data, we demonstrated that the gramicidin analogs containing single tryptophan residues adopt a mixture of channel and nonchannel conformations, thereby reducing ion channel activity.

In order to address the basis of differential importance of tryptophan residues in gramicidin channel, we explored the effects of pairwise substitution of two of the four gramicidin tryptophans, the inner pair (Trp-9 and -11) and the outer pair (Trp-13 and -15), using a combination of steady state and time-resolved fluorescence approaches and circular dichroism spectroscopy (see Fig. 14.6a for the sequence of double tryptophan analogs) [25]. The normalized emission spectra of the double tryptophan gramicidin analogs ($Phe^{9,11}gA$ and $Phe^{13,15}gA$) are shown in Fig. 14.6b. The analog ($Phe^{13,15}gA$) containing the inner pair of tryptophans (Trp-9 and -11) displays an emission maximum of 331 nm, while the analog ($Phe^{9,11}gA$) containing the outer pair of tryptophans (Trp-13 and -15) displays a red shifted emission maximum of 338 nm (the emission maximum of gramicidin in the channel conformation is 333 nm, whereas the nonchannel conformation exhibits an emission maximum of 335 nm, when excited at 280 nm) [7]. The difference in emission maximum among the analogs is indicative of differential localization of the tryptophans along the bilayer normal. Figure 14.6c shows REES data for the double tryptophan analogs. In case of the analog $Phe^{9,11}gA$, containing the outer pair of tryptophans (Trp-13 and -15), we obtained a REES of 9 nm (shift in emission maximum from 338 to 347 nm when excitation wavelength was changed from 280 to 310 nm). In contrast, the emission maximum for the analog $Phe^{13,15}gA$ containing the inner pair of tryptophans (Trp-9 and -11), exhibited a shift from 331 to 349 nm as the excitation wavelength was changed from 280 to 310 nm, giving rise to an enhanced REES of 18 nm. Such large magnitudes of REES are indicative of ground state conformational heterogeneity. Further analysis employing circular dichroism and time-resolved anisotropy decay measurements established that these double tryptophan gramicidin analogs adopt different

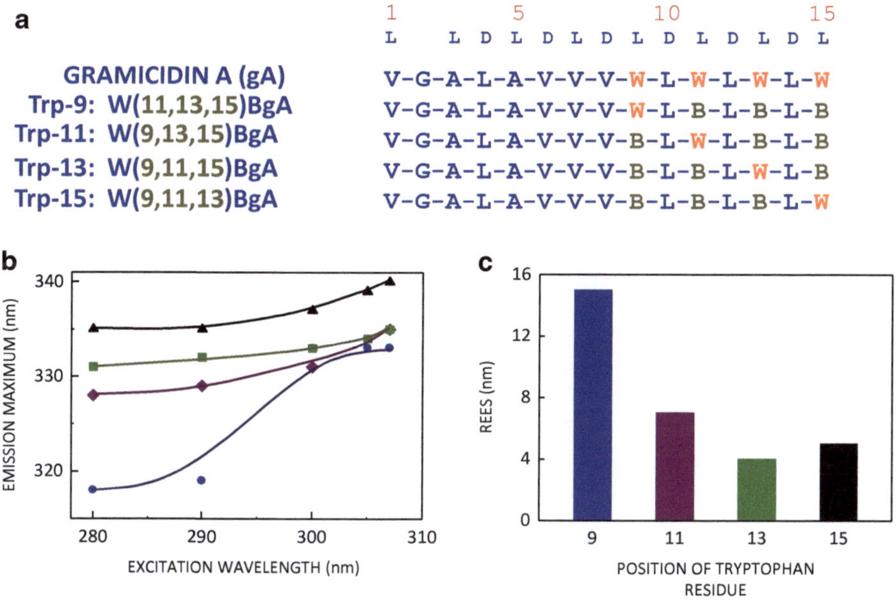

Fig. 14.5 Exploring the organization of single tryptophan analogs of gramicidin. (**a**) The positions of tryptophan and Ser-*t*-butyl (designated as B) in the amino acid sequence of gramicidin A and the single tryptophan analogs are highlighted, with alternating L- and D-residues indicated on the top. (**b**) Effect of changing excitation wavelength on the emission maximum for W(11,13,15) BgA (abbreviated as Trp-9) (●), W(9,13,15)BgA (Trp-11) (♦), W(9,11,15)BgA (Trp-13) (■), and W(9,11,13)BgA (Trp-15) (▲) in membranes. The lines joining data points are provided merely as viewing guides. (**c**) The magnitude of REES obtained for the single tryptophan analogs of gramicidin is sensitive to tryptophan depths from the centre of the bilayer. The magnitude of REES corresponds to the total shift in emission maximum when the excitation wavelength is changed from 280 to 307 nm (data shown in (**b**)) (Adapted and modified from ref. [24] with permission from Elsevier)

conformations in membranes, indicating that the conformational preference of double tryptophan gramicidin analogs is dictated by the positions of the tryptophans in the sequence. These results assume relevance in the context of the report that the inner pair of tryptophans (Trp-9 and -11) is more important for gramicidin channel formation and channel conductance [45].

In yet another study, we tested the importance of indole hydrogen bonding in gramicidin channels, by monitoring the effect of N-methylation of gramicidin tryptophans, using a combination of steady state and time-resolved fluorescence approaches [26]. As stated above, gramicidins with Trp → Phe or Tyr substitutions have greater difficulty in forming membrane-spanning dimeric channels [39, 43]. However, these results do not provide information on specific properties of tryptophan that contribute to the loss of channel structure and function. The loss in structure and function upon substitution of tryptophan with phenylalanine or tyrosine could be attributed to loss of dipole moment, hydrogen bonding ability, change in hydrophobicity, or a combination of these factors. In order to assess the

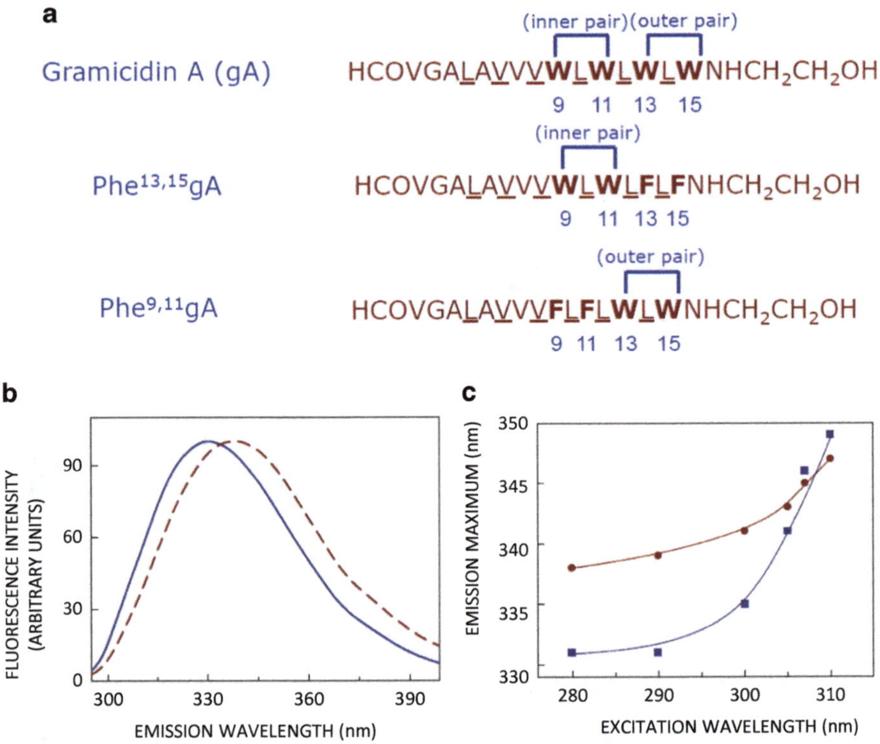

Fig. 14.6 Effect of pair-wise substitution of tryptophans in gramicidin. (**a**) To delineate the role of multiple tryptophans in the organization and dynamics of the gramicidin channel, the positions of tryptophans (four in number) in gramicidin A and the pair-wise substituted double tryptophan analogs (two in number) are shown. Tryptophan residues at positions 9 and 11 are denoted as the "inner pair", while tryptophan residues at positions 13 and 15 are denoted as the "outer pair". The analog in which the inner pair of tryptophans are substituted by phenylalanine is denoted as $Phe^{9,11}gA$, while the analog in which the outer pair of tryptophans are substituted is termed $Phe^{13,15}gA$. (**b**) Intensity-normalized fluorescence emission spectra of the two pair-wise substituted double tryptophan analogs, $Phe^{13,15}gA$ (——) and $Phe^{9,11}gA$ (---) in membranes. (**c**) Effect of changing excitation wavelength on the wavelength of maximum emission for $Phe^{13,15}gA$ (■) and $Phe^{9,11}gA$ (●). The magnitude of REES corresponds to 18 and 9 nm for the inner ($Phe^{13,15}gA$) and the outer ($Phe^{9,11}gA$) pairs, respectively. The lines joining data points are provided merely as viewing guides (Adapted with permission from ref. [25] (copyright (2014) American Chemical Society))

contribution of hydrogen bonding ability of tryptophans in maintaining the channel conformation of gramicidin, tryptophan residues were modified to 1-methyltryptophanm (see Fig. 14.7a for chemical structures of indole and 1-methylindole). This modification leads to loss of hydrogen bonding ability of tryptophans and yet, properties such as aromaticity and ring shape remain invariant. More importantly, the magnitude (~2.1 D for tryptophan and 2.2 D for 1-methyltryptophan) and direction of the dipole moment are not altered (see Fig. 14.7a) [46]. We therefore explored the membrane organization and dynamics

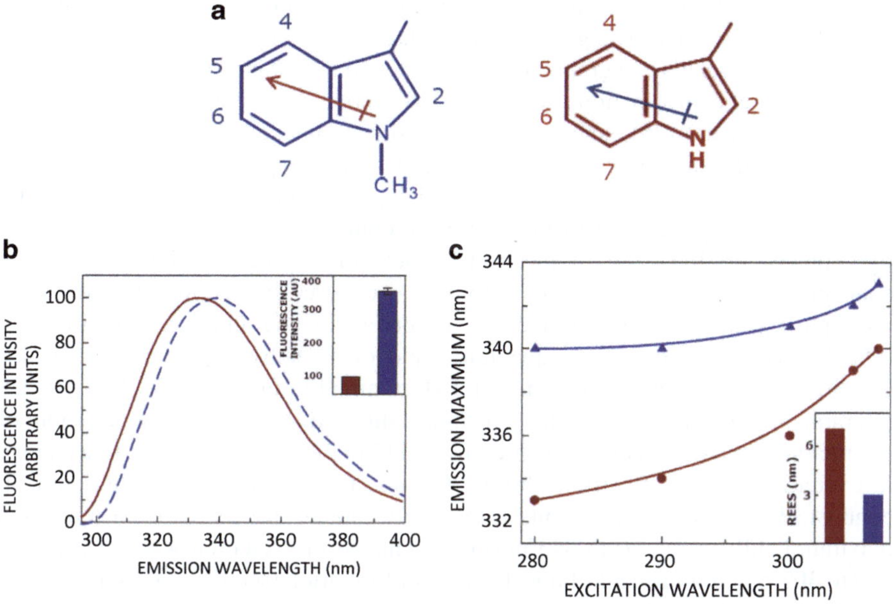

Fig. 14.7 Monitoring the effect of hydrogen bonding of the indole group of tryptophan in gramicidin. (**a**) Chemical structures of indole (*right*) and 1-methylindole (*left*) with the direction of the dipole moments are shown. The -NH group in indole can form hydrogen bond with the lipid carbonyl groups or the interfacial water molecules, while in 1-methylindole this ability is lost due the substitution with a methyl group. Importantly, aromaticity and ring shape are maintained in 1-methylindole and the dipole moment (shown as a vector) is similar in direction and magnitude (~2 D) to indole. The analog in which all four tryptophans are replaced by 1-methyltryptophan is designated as tetramethyltryptophan gramicidin (TM-gramicidin). (**b**) The intensity-normalized fluorescence emission spectra of gramicidin (——), and TM-gramicidin (- - -) in membranes are shown. The inset shows that the fluorescence intensity of TM-gramicidin (*right bar*) is higher relative to gramicidin at their respective emission maximum. (**c**) Effect of changing excitation wavelength on the wavelength of maximum emission for gramicidin (●), and TM-gramicidin (▲) in membranes. The lines joining data points are provided merely as viewing guides. The inset shows the magnitude of REES, which corresponds to the shift in emission maximum when the excitation wavelength was changed from 280 to 307 nm, is higher for gramicidin (*left bar*) (Adapted from ref. [26] with permission from Elsevier)

of the N-methylated tryptophan analog of gramicidin, *i.e.*, tetramethyltryptophan gramicidin (TM-gramicidin), a tryptophan analog of gramicidin in which all four tryptophans are replaced by 1-methyltryptophan residues. The normalized fluorescence emission spectra of gramicidin and TM-gramicidin are shown in Fig. 14.7b. The figure shows that while tryptophans in the channel form of gramicidin typically exhibit an emission maximum of 333 nm, the emission maximum of TM-gramicidin is significantly red shifted to 340 nm. The red-shifted emission maximum of TM-gramicidin is indicative of the average environment experienced by 1-methyltryptophans in TM-gramicidin due to conformational differences of gramicidin and TM-gramicidin. Interestingly, red-shifted emission maximum is characteristic of the nonchannel conformation of gramicidin [7]. The inset in Fig. 14.7b shows that TM-gramicidin displays considerable increase in fluorescence intensity relative to gramicidin when excited at 280 nm. This could be due to the nonchannel conformation adopted by TM-gramicidin (independently shown by CD measurements) [26], since the conformational change of gramicidin in membranes from the nonchannel to channel form is accompanied by a reduction in fluorescence intensity [7]. The nonchannel conformation is characterized by increased fluorescence due to the relatively nonpolar environment in which the tryptophans are localized in the nonchannel conformation and the release of quenching due to absence of aromatic-aromatic interaction between the fluorophores at positions 9 and 15 observed in the channel conformation [25, 26]. Higher fluorescence quantum yield of 1-methyltryptophan could also contribute to increased fluorescence of TM-gramicidin. Figure 14.7c shows REES of gramicidin and TM-gramicidin. The figure shows that the emission maximum of gramicidin is characteristically shifted from 333 to 340 nm in response to a change in excitation wavelength from 280 to 307 nm, amounting to REES of 7 nm. In contrast, TM-gramicidin, exhibits a relatively modest REES of 3 nm (emission maximum shift from 340 to 343 nm) upon change in excitation wavelength from 280 to 307 nm, reminiscent of nonchannel conformation (see Fig. 14.3b) [7]. These results clearly show the importance of tryptophan hydrogen bonding in maintaining the channel conformation of gramicidin in particular and ion channels in general. In addition, these results offer the possibility that fluorescence of 1-methyltryptophan could be effectively used as a tool to explore the hydrogen bonding ability of tryptophans in membrane proteins and peptides.

14.6 Conclusions and Future Perspectives

In this review, we have focused on the application of fluorescence-based approaches to gain insight into conformational plasticity of the ion channel peptide gramicidin. Since gramicidin shares common structural motifs with more complex ion channels, the results described in this review could be useful to study conformations of more complex ion channels. This review is not meant to be an exhaustive in nature. Rather, we have provided representative applications to illustrate a specific approach that would provide novel conformational insight. In case of multitryptophan proteins, analysis of fluorescence data could be complicated due to the complexity of fluorescence processes in such systems and lack of specific information. Site-specific incorporation of extrinsic fluorescent probes, accomplished by using unnatural amino acid mutagenesis [47, 48], could help avoid this complication.

It should be mentioned here that although we have focused mainly on ion channel peptide in this review, these fluorescence-based approaches are applicable to all membrane proteins. A particularly attractive application would be to monitor conformational plasticity of G protein-coupled receptors (GPCRs) using fluorescence-based approaches. GPCRs are involved in signal transduction from outside the cell to the cellular interior and constitute the largest family of current therapeutic targets [49]. GPCRs display remarkable structural plasticity, necessary for the functional diversity exhibited by them [50]. Unraveling conformational choices of GPCRs using fluorescence-based approaches therefore would be useful in deciphering GPCR function.

Acknowledgments Work in A.C.'s laboratory was supported by the Council of Scientific and Industrial Research and Department of Science and Technology (Govt. of India). Ar.C. thanks the Council of Scientific and Industrial Research for the award of a Research Associateship. A.C. is an Adjunct Professor of Tata Institute of Fundamental Research (Mumbai), RMIT university (Melbourne, Australia), Indian Institute of Technology (Kanpur), and Indian Institute of Science Education and Research (Mohali). A.C. gratefully acknowledges J.C. Bose Fellowship (Dept. of Science and Technology, Govt. of India). Some of the work described in this article was carried out by former members of A.C.'s research group whose contributions are gratefully acknowledged. We thank Sreetama Pal for help in making figures, and members of the Chattopadhyay laboratory for critically reading the manuscript.

References

1. Cooper EC, Jan LY (1999) Ion channel genes and human neurological disease: recent progress, prospects, and challenges. Proc Natl Acad Sci U S A 96:4759–4766
2. Stutts MJ, Canessa CM, Olsen JC, Hamrick M, Cohn JA, Rossier BC, Boucher RC (1995) CFTR as a cAMP-dependent regulator of sodium channels. Science 269:847–850
3. Niemeyer BA, Mery L, Zawar C, Suckow A, Monje F, Pardo LA, Stühmer W, Flockerzi V, Hoth M (2001) Ion channels in health and disease. 83rd Boehringer Ingelheim Fonds International Titisee Conference. EMBO Rep 2:568–573
4. England PJ (1999) Discovering ion-channel modulators – making the electrophysiologist's life more interesting. Drug Discov Today 4:391–392
5. Kelkar DA, Chattopadhyay A (2007) The gramicidin ion channel: a model membrane protein. Biochim Biophys Acta 1768:2011–2025
6. Chattopadhyay A, Kelkar DA (2005) Ion channels and D-amino acids. J Biosci 30:147–149
7. Rawat SS, Kelkar DA, Chattopadhyay A (2004) Monitoring gramicidin conformations in membranes: a fluorescence approach. Biophys J 87:831–843
8. O'Connell AM, Koeppe RE II, Andersen OS (1990) Kinetics of gramicidin channel formation in lipid bilayers: transmembrane monomer association. Science 250:1256–1259
9. Ketchem RR, Hu W, Cross TA (1993) High-resolution conformation of gramicidin A in a lipid bilayer by solid-state NMR. Science 261:1457–1460
10. Mukherjee S, Chattopadhyay A (1994) Motionally restricted tryptophan environments at the peptide-lipid interface of gramicidin channels. Biochemistry 33:5089–5097
11. Kelkar DA, Chattopadhyay A (2006) Membrane interfacial localization of aromatic amino acids and membrane protein function. J Biosci 31:297–302
12. Sychev SV, Barsukov LI, Ivanov VT (1993) The double $\pi\pi$5.6 helix of gramicidin A predominates in unsaturated lipid membranes. Eur Biophys J 22:279–288
13. Zein M, Winter R (2000) Effect of temperature, pressure and lipid acyl chain length on the structure and phase behaviour of phospholipid-gramicidin bilayers. Phys Chem Chem Phys 2:4545–4551

14. Kelkar DA, Chattopadhyay A (2007) Modulation of gramicidin channel conformation and organization by hydrophobic mismatch in saturated phosphatidylcholine bilayers. Biochim Biophys Acta 1768:1103–1113
15. Chattopadhyay A, Raghuraman H (2004) Application of fluorescence spectroscopy to membrane protein structure and dynamics. Curr Sci 87:175–180
16. Raghuraman H, Chattopadhyay A (2003) Organization and dynamics of melittin in environments of graded hydration: a fluorescence approach. Langmuir 19:10332–10341
17. Raghuraman H, Kelkar DA, Chattopadhyay A (2005) Novel insights into protein structure and dynamics utilizing the red edge excitation shift approach. In: Geddes CD, Lakowicz JR (eds) Reviews in fluorescence 2005. Springer, New York, pp 199–222
18. Haldar S, Chaudhuri A, Chattopadhyay A (2011) Organization and dynamics of membrane probes and proteins utilizing the red edge excitation shift. J Phys Chem B 115:5693–5706
19. Chattopadhyay A, Haldar S (2014) Dynamic insight into protein structure utilizing red edge excitation shift. Acc Chem Res 47:12–19
20. Chattopadhyay A (2003) Exploring membrane organization and dynamics by the wavelength-selective fluorescence approach. Chem Phys Lipids 122:3–17
21. Haldar S, Kombrabail M, Krishnamoorthy G, Chattopadhyay A (2012) Depth-dependent heterogeneity in membranes by fluorescence lifetime distribution analysis. J Phys Chem Lett 3:2676–2681
22. Kelkar DA, Chattopadhyay A (2005) Effect of graded hydration on the organization and dynamics of an ion channel: a fluorescence approach. Biophys J 88:1070–1080
23. Rawat SS, Kelkar DA, Chattopadhyay A (2005) Effect of structural transition of the host assembly on dynamics of an ion channel peptide: a fluorescence approach. Biophys J 89:3049–3058
24. Chattopadhyay A, Rawat SS, Greathouse DV, Kelkar DA, Koeppe RE II (2008) Role of tryptophan residues in gramicidin channel organization and function. Biophys J 95:166–175
25. Haldar S, Chaudhuri A, Gu H, Koeppe RE II, Kombrabail M, Krishnamoorthy G, Chattopadhyay A (2012) Membrane organization and dynamics of "inner pair" and "outer pair" tryptophan residues in gramicidin channels. J Phys Chem B 116:11056–11064
26. Chaudhuri A, Haldar S, Sun H, Koeppe RE II, Chattopadhyay A (2014) Importance of indole N-H hydrogen bonding in the organization and dynamics of gramicidin channels. Biochim Biophys Acta 1838:419–428
27. Ghosh AK, Rukmini R, Chattopadhyay A (1997) Modulation of tryptophan environment in membrane-bound melittin by negatively charged phospholipids: implications in membrane organization and function. Biochemistry 36:14291–14305
28. Raghuraman H, Chattopadhyay A (2004) Interaction of melittin with membrane cholesterol: a fluorescence approach. Biophys J 87:2419–2432
29. Raja SM, Rawat SS, Chattopadhyay A, Lala AK (1999) Localization and environment of tryptophans in soluble and membrane-bound states of a pore-forming toxin from *Staphylococcus aureus*. Biophys J 76:1469–1479
30. Haldar S, Raghuraman H, Namani T, Rajarathnam K, Chattopadhyay A (1798) Membrane interaction of the N-terminal domain of chemokine receptor CXCR1. Biochim Biophys Acta 2010:1056–1061
31. Chaudhuri A, Chattopadhyay A (1838) Lipid binding specificity of bovine α-lactalbumin: a multidimensional approach. Biochim Biophys Acta 2014:2078–2086
32. Chattopadhyay A, Mukherjee S (1999) Depth-dependent solvent relaxation in membranes: wavelength-selective fluorescence as a membrane dipstick. Langmuir 15:2142–2148
33. Brochon J-C (1994) Maximum entropy method of data analysis in time-resolved spectroscopy. Methods Enzymol 240:262–311
34. Swaminathan R, Periasamy N (1996) Analysis of fluorescence decay by the maximum entropy method: influence of noise and analysis parameters on the width of the distribution of lifetimes. Proc Indian Acad Sci Chem Sci 108:39–49

35. Mukherjee S, Kombrabail M, Krishnamoorthy G, Chattopadhyay A (1768) Dynamics and heterogeneity of bovine hippocampal membranes: role of cholesterol and proteins. Biochim Biophys Acta 2007:2130–2144
36. Haldar S, Kombrabail M, Krishnamoorthy G, Chattopadhyay A (2010) Monitoring membrane protein conformational heterogeneity by fluorescence lifetime distribution analysis using the maximum entropy method. J Fluoresc 20:407–413
37. Fonseca V, Daumas P, Ranjalahy-Rasoloarijao L, Heitz F, Lazaro R, Trudelle Y, Andersen OS (1992) Gramicidin channels that have no tryptophan residues. Biochemistry 31:5340–5350
38. Daumas P, Heitz F, Ranjalahy-Rasoloarijao L, Lazaro R (1989) Gramicidin A analogs: influence of the substitution of the tryptophans by naphthylalanines. Biochimie 71:77–81
39. Becker MD, Greathouse DV, Koeppe RE II, Andersen OS (1991) Amino acid sequence modulation of gramicidin channel function: effects of tryptophan-to-phenylalanine substitutions on the single-channel conductance and duration. Biochemistry 30:8830–8839
40. Andersen OS, Greathouse DV, Providence LL, Becker MD, Koeppe RE II (1998) Importance of tryptophan dipoles for protein function: 5-fluorination of tryptophans in gramicidin A channels. J Am Chem Soc 120:5142–5146
41. Barth C, Stark G (1991) Radiation inactivation of Ion channels formed by gramicidin a. Protection by lipid double bonds and by α-tocopherol. Biochim Biophys Acta 1066:54–58
42. Sobko AA, Vigasina MA, Rokitskaya TI, Kotova EA, Zakharov SD, Cramer WA, Antonenko YN (2004) Chemical and photochemical modification of colicin E1 and gramicidin A in bilayer lipid membranes. J Membr Biol 199:51–62
43. Salom D, Pérez-Payá E, Pascal J, Abad C (1998) Environment- and sequence-dependent modulation of the double-stranded to single-stranded conformational transition of gramicidin A in membranes. Biochemistry 37:14279–14291
44. Chattopadhyay A, London E (1987) Parallax method for direct measurement of membrane penetration depth utilizing fluorescence quenching by spin-labeled phospholipids. Biochemistry 26:39–45
45. Gu H, Lum K, Kim JH, Greathouse DV, Andersen OS, Koeppe RE II (2011) The membrane interface dictates different anchor roles for “inner pair” and “outer pair” tryptophan indole rings in gramicidin A channels. Biochemistry 50:4855–4866
46. Sun H, Greathouse DV, Andersen OS, Koeppe RE II (2008) The preference of tryptophan for membrane interfaces: insights from N-methylation of tryptophans in gramicidin channels. J Biol Chem 283:22233–22243
47. Cohen BE, McAnaney TB, Park ES, Jan YN, Boxer SG, Jan LY (2002) Probing protein electrostatics with a synthetic fluorescent amino acid. Science 296:1700–1703
48. Pless SA, Kim RY, Ahern CA, Kurata HT (2015) Atom-by-atom engineering of voltage-gated ion channels: magnified insights into function and pharmacology. J Physiol. 593:2627–2634
49. Chattopadhyay A (2014) GPCRs: lipid-dependent membrane receptors that act as drug targets. Adv Biol 2014:143023
50. Nygaard R, Zou Y, Dror RO, Mildorf TJ, Arlow DH, Manglik A, Pan AC, Liu CW, Fung JJ, Bokoch MP, Thian FS, Kobilka TS, Shaw DE, Mueller L, Prosser RS, Kobilka BK (2013) The dynamic process of β_2-adrenergic receptor activation. Cell 152:532–542

Index

A
AFM. *See* Atomic force microscopy (AFM)
Alzheimer's Disease, 150
Amyloid fibrils
  nanoscale resolution, 22
  oligomers, 3
  supramolecular amyloid fibrils, 3
  α-synuclein aggregation studies, 14
  tryptophans, 18
1-anilino-8-naphthalene sulfonate (ANS), 113
ANS. *See* 1-anilino-8-naphthalene sulfonate (ANS)
Atomic force microscopy (AFM), 3

B
Biomedical, 98, 99, 101, 102, 119
Biosensor
  definition, 151
  fluorescence lifetime, 160–164
Bovine serum albumin (BSA), 17
BSA. *See* Bovine serum albumin (BSA)

C
CAC. *See* Critical aggregation concentration (CAC)
Carbonic anhydrase II (Q92A), 162
CD. *See* Circular dichroism (CD)
Channelopathy, 353
Chelation therapy, 151
Circular dichroism (CD), 3
CMC. *See* Critical micelle concentration (CMC)
Conformational heterogeneity, 357
Copper
  biological roles, 147
  blue copper proteins, 149
  in cancer, 149
  CH_3CN complex, 148
  chaperones, 149
  classification, 149, 150
  Ctr1, 149
  Cu(I) (*see* Cu(I))
  Cu(II) (*see* Cu(II))
  enzymes, 149, 150
  fluorescent indicators and sensors (*see* Fluorescent sensors)
  in human diseases, 150, 151
  hydroxyl radical, 148
  inorganic and bioinorganic chemistry, 147
  lysyl oxidase, 149
  metallothioneins, 149
  mutagenic catalyst, 149
  reducing agents and anoxic conditions, 148
  transition element, Group 11 (IB), 147
Coppersensor-1 (CS1), 157
Co-precipitation method, 291
Creutzfeld-Jakob Disease, 151
Critical aggregation concentration (CAC), 16
Critical micelle concentration (CMC), 105
  ample Rh6G molecules, 112
  ANS, 114
  behavior, 112
  DLS, 112
  fluorophore photophysics, 112
  FRAP, 113
  macroscopic behavior, 111
  PNIPAm brushes, 113
  polymer and probe concentrations, 112

© Springer International Publishing Switzerland 2016
C.D. Geddes (ed.), *Reviews in Fluorescence 2015*,
DOI 10.1007/978-3-319-24609-3

Critical micelle concentration (CMC) (*cont.*)
polymer concentrations, 112
SMD, 112
Cu(I)
biological systems, 148
facile oxidation in water, 148
indicators, 156–159
analytical advantages of transducing, 158
BCP-Cu(I) complexes, 158
catalytic approach, 158
cells' morphology, 157
CS-1, 157
Fahrni's group, 158
fluorescence lifetime-based group, 158
FRET-based quenching acts, 158
oxygen tension, 157
PET, 158
tetrahedral, 148
$3d^{10}$ electronic structure, 147
Cu(II)
charge transfer absorbance bands, 148
electronic structure of $3d^9$, 148
fluorescence lifetime-based biosensing (*see* Fluorescence lifetime)
fluorescence quencher, 148
fluorescent indicators/sensors, 154
indicators, 155–157
CXC chemokine receptor (CXCR1), 356
CXCR1. *See* CXC chemokine receptor (CXCR1)

D

DAB. *See* Diaminobenzidine (DAB)
Dansyl, 117
DCS. *See* Dimethylamino cyano stilbene (DCS)
DCVJ. *See* 4-dicyanovinyl-julolidine (DCVJ)
DDEM. *See* Donor-donor energy migration (DDEM)
Debye model, 51
Detection of water impurities
diazepam and digoxin mixture, water (EEM), 198, 199
excitation emission wavelengths, maxima, 198, 200
indole-3-acetic acid, EEM, 200, 201
3-D and 2-D EEM, insecticide, 198, 199
3-D image, 200
total luminescence spectrum measurement, 198
Diaminobenzidine (DAB), 130
4-dicyanovinyl-julolidine (DCVJ), 7
Dimethylamino cyano stilbene (DCS), 338
Dimethylamino nitro stilbene (DNS), 338
Dimethylamino-4′-nitrostilbene (DNS), 342
Dithiane (propane-1,3-dithiol), 156
DLS. *See* Dynamic light scattering (DLS)
DMABN. *See* 4-N′N-dimethylamino-benzonitrile (DMABN)
DNA sensing, 85–87
Donor-donor energy migration (DDEM), 173
Drugs, body fluids, 201, 202, 204
Dual fluorescence
donor *para*-dimethylamino, 338
electronic and geometrical structure, 339
electronic resonance interaction, 338
fluorescence quantum yield, 338
intramolecular donor-acceptor effects, 338
linear free-energy correlation approach, 338
optical irradiation, 338
para-dialkylamino-stilbenes
compounds, 340
concentration and excitation intensity, 341
dimethylanilino moiety, 342
DMABN, 342
donor and acceptor groups, 343
hydrodynamic factors, 342
hypothesis, 342
mechanisms, 342
nitro group, 342
phenyl group, 341
photocyclisation mechanism, 341
photoisomerisation pathway, 339
photophysical hypersurface kinetics scheme, 341
photophysical properties, 342
picosecond time-resolved spectroscopy, 340
PICT, 343
polar solvents, 341
precursor-successor relationship, 340
pump-probe femtosecond Kerr ellipsometry, 341
RICT, 340
rotation, 342
solvatochromic measurements, 340
stiff stilbenes, 339
three-state photochemical reaction scheme, 340, 341
time-resolution and sensitivity, 341
time-resolved fluorescence studies, 339
para-dimethylamino-stilbenes, 339

photochemical reaction mechanism, 345
quantum-chemical calculations, 338
quenching funnel, 338
saturated hydrocarbon solvents, 343
stilbenes, 343
sudden polarisation, 338
time-resolved emission spectrum, 338
trans-4,4′-disubstituted stilbenes, 343, 344
trans-cis photoisomerisation process, 338
Dynamic light scattering (DLS), 3, 112

E
Electric field assisted surface plasmon coupled directional emission (E-SPCDE), 86
Electronic energy transfer
bicelles, 173
cholera toxin/NBD-DPPE, 173
chromophores, 173
DDEM, 173
homogenous bilayer, 178
lipid domains
DAET/FRET, 176
energy transfer, 177
lipid phase geometries, 175
parallel planes, 177
phase separation boundary, 175
simulation procedure, 176
translational diffusion, 177
lipid phase separation, 172
MC simulations, 177
microscopy techniques, 172
optical microscopy, 173
parameters, 177
phase diagram, 178
plasma/model membranes, 172
super-resolution spectroscopy, 172
Electrospray ionization mass spectrometry (ESI-MS), 3
ESI-MS. *See* Electrospray ionization mass spectrometry (ESI-MS)
E-SPCDE. *See* Electric field assisted surface plasmon coupled directional emission (E-SPCDE)
Euler angle, 46

F
Fahrni's group, 158
FBM. *See* Fluorescence blob model (FBM)
FCS. *See* Fluorescence correlation spectroscopy (FCS)
Fehling's test, 151
Fingerprints
blood, 127
bloody print development, 134–136
chemiluminescence, 127
classical print detection reagents
ABTS, 130
acid black 1 and acid violet 17, 129
acid fuchsin and acid yellow 7, 129
amido black and ninhydrin, 130
amino acid reagents, 129
benzidine dyes, 128, 129
chemiluminescence/fluorescence, 129
chemiluminogens and fluorogens, 128
common print reagents, 131
DAB, 130
1,8-diazafluoren-9-one (DFO), 130
DNA damage, 130
fluorescence, 128
fluorescin, 129
fluorophores, 128
hemoglobin catalyzes, 128
leuco dyes, 128
luminol, 129, 130
red blood cells, 128
redox or ionization state, 128
dark and multi-colored substrates, 143
facile one-pot, tri-component naphthoxanthene syntheses 136–138
HP/H_2O_2/pig blood/H_2O_2, 143
luminol and fluorescin, 127
naphthoxanthene, 138–140
NBD- and benzidine-basic dyes-horseradish peroxidase (HP) activity, 130–134
Fluoimager
CCD detector, 193
cell holder, 193
detector unit, 193
excitation monochromator, 192
geometry, 194
layout, 193, 194
M50 & M50C model, 192
microcontroller, 194
Xe-lampFluoimager
Xe-lamp, 192
Fluorescence
chromophores, 236
coupled nanoemitters, 237
initial and final system states, 237, 239
intermediate molecular states, 239
intermediate system states, 237
luminescence, 235

Fluorescence (*cont.*)
multiphoton
crystalline structures, 239
dimensions, 239
input laser beam, 240
optical signal, 240
technique's, 240
multiphoton induced
annihilation operator, 242
dipole, 243
electric polarisation, 242
intermediate system state, 242
virtual intermediate states, 243
optical devices, 235
optical mode, 238
photoexcitation, 239
photons, 237
single-photon absorption, 246
Fluorescence anisotropy
Cu(II)
description, 163
instruments, 163
and intensities, 164, 165
lifetime, 163
measurements, 163
metal ion, 164
Perrin's equation, 163
ratiometric measurement, 163
Fluorescence blob model (FBM), 115
Fluorescence correlation spectroscopy (FCS), 112
Aβ protein aggregates, 16
autocorrelation curve, 319
autocorrelation function, 15
binding and catalytic parameters, 304
catalysis, 302
central/catalytic core domain, 302
conventional chemical labeling methods, 304
correlation time distributions, 322–325
diffusion coefficient measurements, 16
diffusion-induced intensity fluctuations, 15
dimerization process, 304
and DNA-binding anisotropy-based assay, 304
fast conformational dynamics and triplet-state dynamics, 15
fluorescein-labeled DNA substrates, 321
fluorescence anisotropy decay, 321
GT dinucleotide, 324
HIV-1 DNA, 302
HIV-1 IN DNA-binding, 328
hydrodynamic properties, 318
hydroxyl groups, 302
image correlation spectroscopy, 16
intrinsic fluorescence, 304
joint fluctuations, 16
long correlation times, 326, 327
maximum entropy method, 320, 321
molecule, 15
monomer-dimer equilibrium, 326
multimerization properties, 304
multimers/aggregates, 326
nanomolar concentration, 15
normalized autocorrelation function, 15
oligomeric state, 303
optimal condition, 325, 326
3′-P reaction, 303
protomers, 303
rotational diffusion, PFV-1 IN, 321, 322
statistical analysis, 15
TAMRA-labeled PFV-1 IN, 320
temperature, 321
TGase-mediated TAMRA labeling method, 319
time-dependent formation, 328
time-resolved fluorescence anisotropy, 304, 320, 324, 328–330
translational diffusion time, 15, 318
tryptophan, 16
Fluorescence imaging
Fluoimager (*see* Fluoimager)
identification software package
Fluoimager M50C package, 194
identification process, 195
iteration module, 194, 195
modules, 195
SPECTRALYZER, pattern recognition and analysis unit, 195
units, 195
impurities, water (*see Detection of water impurities*)
instrument calibration, 197
lifetime measurements, 201, 205–208
materials, 196
measurements, 197
street narcotics and drugs determination, body fluids
fluoimaging reliability, 201, 202
phenol in water, quantitative determination, 201, 203
quantitative determination, individual fluorophores, 201, 204
Fluorescence lifetime
Cu (II)
CAII, 160

carbonic anhydrase II (Q92A), 162
carbonic anhydrase II Cu(II) sensor, 161
dynamic quenching, 160
FLIM, 162
free Cu(II)concentration-dependent intensities, 161, 162
frequency domain fluorometry, 161
FRET, 160
in situ calibration, 162
microscopy and fiber optic sensing applications, 162
phase angle, 161
polarization (anisotropy), 163, 164
static quenching, 160
Fluorescence lifetime imaging microscopy (FLIM), 153
Fluorescence lifetime-based (FLIM) microscopes
FRET-based measurements, 162
Fluorescence recovery after photobleaching (FRAP), 113
Fluorescence resonance energy transfer (FRET), 13–14
Fluorescence sensor
copper
attributes, 152
Cu(II), 152
Fehling's test, 151
human apocarbonic anhydrase II (CA II), 152
intensity ratios, 152
macromolecule, 159–160
merit, 151
permits continuous/quasi-continuous determination, 151
pH, 151
properties, 151
proportion of fluorescence emitters, 152
reagentless, 151
selectivity, 152
zinc interference in Cu (II) measurement, 152, 153
Fluorescence spectroscopy
amorphous aggregates and amyloid fibrils, 2
chemistry influences, 104
fibril stability, 2
fluorescence polarization, 10–12
FRET, 13–14
intermolecular interactions, 2
κ-Casein, 19–20, 22
lysozyme aggregation, 17, 18
lysozyme, serum albumins and casein, 2
neurodegenerative disorders, 2
oligomers, 2
physicochemical parameters
application, 109
dioxane/water mixtures, 110, 111
dipole-induced dipole interactions, 110
ethanol and 1-propanol, 110
hydrogen-bonding effects, 110
polarity sensitive fluorescent probes, 110
Py scale values, 110
pyrene, 109
quinoxaline, 111
rhodamine X, 110
vibronic band intensities, 109
physiochemical/physical characterization, 104
polymer characterisation, 103–104
polypeptide chains, 2
protein aggregation and amyloid assembly, 2
protein folding funnel, 1
quenching, 12–13
serum albumin aggregation, 17–19, 21
single-molecule fluorescence studies, 14
Fluorescent 3-hydroxyflavone (3-HF), 105
Fluorescent indicator
copper
definition, 151
macromolecule, 159–160
metal ions, 153
photostable, 152
quenching, 154
Fluorophores, 5
Franck-Condon factor, 34
Free space emission (FSE), 74
FRET
bicelles, 183–184
bilayers containing pores, 181–182
concentration, cholera toxin, 180
Fluorescence resonance energy transfer (FRET))
fluorophore-labelled cholera toxin, 178
steady-state (SS)/time-resolved (TR), 179
and z-scan fluorescence correlation spectroscopy, 180
FSE. *See* Free space emission (FSE)

G
Galactosamine, 155
Gramicidin, 354
circular dichroism and time-resolved anisotropy decay measurements, 360
emission maximum, 360
fluorescence approaches, 360

Gramicidin (*cont.*)
- hydrogen bonding, 362, 363
- indole hydrogen bonding, 361
- membrane penetration depth analysis, 360
- membrane proteins and peptides, 364
- membrane-spanning dimeric channels, 361
- 1-methyltryptophan, 364
- pair-wise substitution, 360, 362
- steady state and time-resolved fluorescence approaches, 360
- tetramethyltryptophan gramicidin, 363
- and TM-gramicidin, 363
- tryptophan analogs, 360, 361

H

HBD. *See* Hydrogen-bond donor (HBD)
HCST. *See* Higher critical solution temperature (HCST)
3-HF. *See* Fluorescent 3-hydroxyflavone (3-HF)
Higher critical solution temperature (HCST), 100
Hydrogen-bond donor (HBD), 108
Hydrothermal method, 292–293

I

ICT. *See* Intramolecular charge transfer (ICT)
Intramolecular charge transfer (ICT), 340
Intramolecular mechanisms
- anomalous emission, 214
- aromatic molecules
  - buffer gases, 215
  - collision-free conditions, 215
  - internal conversion, 217
  - photophysical property, 215
  - S_1 fluorescence, 215
  - S_2 fluorescence, 214, 215, 218
  - spectrum, 217
  - static vapor phase, 215, 217
  - *vide infra*, 214
- fluorescence, 213
- linear polyenes
  - absorption bands, 218
  - buffer gas, 221
  - carotenoids, 223, 225
  - chemical structure, 218
  - decatetraene vapor, 220
  - fast radiative process, 226
  - intensity, 218
  - octatetraene, 220
  - photophysics & dynamical behavior, 219
  - $S_1 - S_2$ energy separations, 220
  - S_1 fluorescence, 219, 223, 225, 226
  - $S_1(2^1A_g)$ fluorescence, 219
  - S_2 fluorescence, 219, 220, 224, 226
  - singlet state (S_1), 218
  - static vapor phase, 223
- nonradiative transitions, 214
- photoreactions, 214
- singlet state (S1), 213
- upper excited states, 226, 228

Ion channel
- cellular nanomachines, 353
- cellular signaling and sensing, 353
- channelopathy, 353
- conformational heterogeneity, 357
- cystic fibrosis, 353
- drugs, 353
- gramicidin, 354
- monitoring conformations, REES, 357, 358
- REES, 355–357

L

Langmuir-Blodgett (LB) films, 80
Lanthanide–doped upconversion nanoparticles (UCNPs)
- absorbers/emitters, 287
- component, 287
- co-precipitation method, 291
- crystal lattice, 287
- energy transfer process, 287, 288
- host lattice, 287
- host material, 290
- host matrix, 290–291
- host-dopant systems, excitation wavelengths and emission peaks, 288, 289
- hydrothermal method, 292–293
- multi-phonon relaxation, 288
- optically active ions, 290
- sol gel method, 292
- thermal decomposition method, 292

Laser-controlled fluorescence, 236 248–253, 259
- effects, 253, 255
- mechanism
  - optical frequency, 250
  - optical phase, 259
  - probe laser beam, 249
  - radiation states, 249
  - single-photon emission, 248
  - virtual states, 251
  - Wavy lines, 249
- plasmonic surface, 248
- probe beam
  - directions, 253

electric dipole moments, 252
emitted light, 252
photophysical processes, 251
wavelength, 252
quantum dots, 248
stimulated emission, 247
LCST. *See* Lower critical solution temperature (LCST)
LCV. *See* Leucocrystal violet (LCV)
Leakage radiation microscope (LRM), 82
LED. *See* Light-emitting diode (LED)
Leucocrystal violet (LCV), 130
Light-emitting diode (LED), 72
Lipid bilayer
Baumann-Fayer (B-F) approach, 175
energy migration, 175
energy transfer/migration, 174
fluorescence relaxation, 175
fluorescent label, 173
intra- and interlayer processes, 174
intralayer energy transport, 175
joint probability, 174
monolayers, 174
rotational correlation times, 175
Localized surface plasmon resonance (LSPR), 79
Lower critical solution temperature (LCST), 100
acenaphthylene (ACE), 116
factors, 117
mesoglobular phases, 116
monomers and excimers, 115
pyrene, 115
quantum dot (QD), 117
sensitive fluorescence methods, 115
steady-state emission spectra, 115
LRM. *See* Leakage radiation microscope (LRM)
LSPR. *See* Localized surface plasmon resonance (LSPR)
Luminescent lanthanide sensors
alkali-earth cations, 273
antenna
aminopyridine ligands, 279
auxillary coligands, 275
characteristic emissions, Eu(III) ion, 277
chromophore, 274
chromophores, 275
decay curve, 282, 284
electron deficiency, 277
emission spectra, 280, 281
europium complexes, 277, 278
2-hp ligand, 280
hydroxypyridine ligands, 279
intersystem crossing and energy transfer process, 274
ligand geometric and electronic structure, 274
ligand topology, 277
magnetic and electric dipole fields, 282
magnetic dipole and electric dipole fields, 277
metal ion and ligand, 274
N-donor neutral auxillary ligands, 275
organic ligands, 277
1,10-phenanthroline ligand, 276
phosphine oxide ligands, 277
photoluminescence decay curve 279, 280
photoluminescence emission spectra, 278, 279
photoluminescence properties, 280
photoluminescent excitation spectra, Eu^{3+} emission, 282, 283
photoluminescent lifetime decay measurement, 280, 281
photophysical properties, 275
protonation, 276
pyridine carboxylates, 279
2,6-pyridinedicarboxylic acid, 279
room temperature excitation spectra, 281
room temperature photoluminescent emission spectra, 282–284
telecommunication network, 280
UV absorptions, 276
contrast agents, 274
electromagnetic energy, 270
emitting levels and transitions, 274, 275
Eu^{3+} and Tb^{3+} ions, 270
Gd^{3+}, 270
Gibbs free energy, 274
laporte-forbidden f-f transitions, 273
metal cation and antenna, 273
Nd^{3+}, Er^{3+} and Yb^{3+}, 270
non radiative deactivation process 270, 273
potential applications, 270
principle, 270–273
quantum yield, 270
sensitization, 270
unique emission properties, 274

M

MA-SPCE. *See* Microwave-accelerated surface plasmon-coupled directional emission (MA-SPCE)

Mass spectrometry
stoichiometry, metal-ligand complex speciation, 155
Membrane interface, 355, 357
Menke's Disease, 150
Metal-to-ligand charge transfer (MLCT), 277
Microscope technique, 81–82
Microwave-accelerated surface plasmon-coupled directional emission (MA-SPCE), 86
MLCT. *See* Metal-to-ligand charge transfer (MLCT)
Monte Carlo simulations, 176, 184
Multi-emitter
generators of Vortex light
constant phase, 260
electrodynamic coupling, 259
electromagnetic phase, 262
energy level splitting array, 262
line splittings, 261
optical phase, 261
optical vortex/twisted beam, 258
orbital angular momentum (OAM), 258–260
scalar optical field, 263
metallic nanoantennas, 255
nanoemitter pair
direct coupling, 256
electric field, 258
electromagnetic coupling, 256
fluorescence, types, 256
initial & final system states, 257
light detector, 256
static electric dipole, 257
transition dipole moments, 258
nanoemitters & detector, 255
plasmonic interactions, 255
quantum level, 255
wavefunctions, 256
Multimerization, 302, 304, 312, 318
Multiphoton
freely tumbling molecules
angular disposition, 246
dipole, 244
input laser light, 245
single-value indicator, 247
virtual intermediate states, 245
one-photon induced
input beam, 241
interactions, 241
vacuum radiation field, 241

N
Nafion matrix, 158
Naphthol, 155
Naphthoxanthene
pH conditions, 138
pH profile, oxidation reaction, 138, 139
pig blood fingerprint, 140, 142
UV–vis spectra, 140
Neuronal network
SPECTRALYZER recognition software, 195
NIPAm/NtBA. *See* N-isopropylacrylamide/N-tert-butylacrylamide (NIPAm/NtBA)
N-isopropylacrylamide/N-tert-butylacrylamide (NIPAm/NtBA), 109
4-N′N-dimethylamino-benzonitrile (DMABN), 342
Nonlinear Stark Effect, 36

O
Octadecyltriethoxysilane (OTE), 113
Oligomers
amyloid fibrils, 3
and amyloids, 4–16
and amyloid-sensitive fluorescent probes, 4
α-synuclein, 9
Optical diagnostics, 263
Optical switching, 263
OTE. *See* Octadecyltriethoxysilane (OTE)

P
Pattern recognition software
neuronal network-pattern recognition software, 195
PCS. *See* Photon correlation spectroscopy (PCS)
Peroxidase, 128, 134, 138
Perrin's equation, 163
Perturbation theory, 36
Photoinduced electron transfer (PET), 158
Photoluminescence quantum yield (PLQY)
chromophores, 286
factors, 283
fluorescence, 283
Forster dipole-dipole mechanism, 286
intrinsic quantum yield, 285
Judd-Ofelt analysis, 286
organic ligands, 285
parameter, 285
spontaneous emission, 286
Photon correlation spectroscopy (PCS), 118
PICT. *See* Planar intramolecular charge transfer (PICT)
Planar intramolecular charge transfer (PICT), 343
Point spread function (PSF), 82, 83

Polarization, 72, 74, 75, 82, 84
Prion-mediated diseases, 151
Protein aggregation
amyloid aggregation, 5
and amyloid fibril formation, 4, 5
amyloid fibrils, 3, 4
biophysical techniques, 3
cysteine, 4
extrinsic fluorophores
amyloid fibrils, 7
chemical structures, 7, 8
DCVJ, 7
excited-state intramolecular proton transfer (ESIPT), 9
fluorophore labeling procedure, 8
maleimide/iodoacetamide derivatives, 9
oligomerization and fibril formation, 8
p-FTAA π-aggregates, 7
prefibrillar oligomers and mature amyloid fibrils, 9
probing prefibrillar and fibrillar species, 8
pyrene microenvironment, 9
site-specific fluorophore labeling approach, 8
thioflavin-T fluorescence assay, 7
α-synuclein, 9
fluorophores, 5
in vitro spectroscopic techniques, 3
intrinsic fluorophores, 6, 7
multiparametric technique, 4
oligomer- and amyloid-sensitive fluorescent probes, 4
parameters, 4
steady-state fluorescence spectroscopy, 5
X-ray diffraction pattern, 3
β-sheet-rich amyloids, 3
Protein misfolding, 1–2
PSF. *See* Point spread function (PSF)
Pyrazoline fluorescent indicator, 158
Pyrene molecules, 155

R

Radiative lifetime. *See* Photoluminescence quantum yield (PLQY)
Raster image correlation spectroscopy (RICS), 16
Ratiometric indicator, 155, 156
Red edge excitation shift (REES), 355–357
Redox reactions, 149
REES. *See* Red edge excitation shift (REES)
Relaxed intramolecular charge transfer (RICT), 340
RICS. *See* Raster image correlation spectroscopy (RICS)
RICT. *See* Relaxed intramolecular charge transfer (RICT)

S

SAF. *See* Supercritical angle fluorescence (SAF)
SCM. *See* Solvatochromic comparison method (SCM)
S_2 fluorescence, 225
Single molecule detection (SMD), 112
SMD. *See* Single molecule detection (SMD)
Sol-gel method, 292
Solvatochromic comparison method (SCM), 108
Solvatochromism
dimethyl sulfoxide (DMSO) and cyclohexane (c-C_6H_{12}), 108
$E_T(30)$ scale, 108
H-bond donor, 106
hypsochromic, 107
optical spectroscopy, 108
PNIPAm, 106
polarity and hydrogen bonding changes, 107
polarity scales, 107
UV-visible spectra, 109
water vapor pressures, 109
α and β scales, 108
Solvent relaxation, 55, 57, 60
SPCDE. *See* Surface plasmon-coupled directional emission (SPCDE)
SPPs. *See* Surface plasmon polaritons (SPPs)
SPR. *See* Surface plasmon resonance (SPR)
Steady-state fluorescence anisotropy
competitive and allosteric integrase inhibitors
compounds, 316
DIBA-1, 317, 319
diketo acids (DKAs), 315
DNA-binding assay, 317
drug concentration, 316, 317
fluorescein-labeled DNAs, 316
Hill coefficient, 318
HIV-1 life cycle, 314
IN DNA-binding inhibitors (INBI) and styrylquinoline compounds, 315
mechanisms, 315
Mn^{2+}-dependent activity, 317
styrylquinoline, 316
DNA-binding and 3′-processing activity
fluorescent moiety, 305

Steady-state fluorescence anisotropy (*cont.*)
gel electrophoresis, 307
GT dinucleotide, 306, 307
IN-DNA complex, 308, 310
metallic cofactor, 306
photon absorption and emission, 305
principle, 305, 306
quantification, 306
real-time and fixed-time, 306
single-turnover property, 306
temperature, 305
time-course, 305, 309
IN-DNA interaction
fluorescein (Fl) donor, 312, 313
Hill coefficients, 309, 311, 312
Mg^{2+}- and Mn^{2+}, 310
NTD-NTD interaction, 312
PFV-1 and HIV-1 DNA binding properties, 314, 315
quenching amplitudes, 314
quenching, Fl donor (q_D), 312, 313
transglutaminase (TGase), 312
Stimuli-responsive polymers
biomedical applications, 99
classification, 99, 100
LCST, 100, 101
poly(N-Isopropylacrylamide), PNIPAm, 101–103
smart polymers, 99
UCST, 100
Stokes shift, 35
Sugar aza-crowns, 155
Supercritical angle fluorescence (SAF), 82
Surface plasmon polaritons (SPPs), 82
Surface plasmon resonance (SPR), 72
Surface plasmon-coupled directional emission (SPCDE), 72
Surface plasmon-coupled emission (SPCE)
applications
aptamers-based protein sensing, 87, 89
DNA sensing, 85–87
fluorescence analysis, 90
immunological detection, 88–89
chemiluminescence and electrochemiluminescence species, 72
directional emission, 73, 74
distance-determined coupling, 76
fluorophores, 72
FSE, 74
hemisphere prism, 72
in vivo detection, 71
Kretschmann (KR) configuration, 72, 73
LED, 72
metal film, 74
optical imaging
bio-images, 84
cardiac myofibrils, 84
incident and emission angle, 84
microscope technique, 81–82
microscopic RK-SPCE-FCS, 84
PSF, 82, 83
shallow detection volume, 84
skeletal myofibrils, 84, 85
two-photon excitation, 84
P-polarized emission, 74–75
Rhodamine B, 74
SPCDE, 72, 73
spectrofluorometer, 72
strategies
applicability and sensitivity, 78–79
optical window, 76–77
spatial resolution, 79–81
surface plasmons, 72
Synchronous fluorescence spectroscopy (SFS)
chemical selectivity, 191
luminescent organic-aromatic compounds detection, 192
3D fluoimaging method, 190

T

TEM. *See* Transmission electron microscopy (TEM)
Thermal decomposition method, 292
Thermoresponsive polymers
4-acrylamidofluorescein-modified poly, 119
applications, 98
dansyl, 117
glutaraldehyde, 117
K^+ ion concentration, 118
microgel, 119
microscale thin polymer films, 98
optical spectroscopy, 98
peak emission wavelength, 117
physiochemical properties, 98
PNIPAm/water/methanol ternary system, 117
polyelectrolyte multilayers, 119
sensing applications, 118
stimuli-responsive polymers, 98, 99
swelling behavior, 118
water sorption
behavior, 106
emission measurements, 106
3-HF probes, 106
medical devices, 105
physicochemical properties, 105
PNIPAm films, 105, 106

solvent polarity and hydrogen, 106
wavelengths, 106
3D fluoimaging method, 190
TICT. *See* Twisted-intramolecular charge-transfer (TICT)
Time- or frequency-domain fluorometry, 160
Time-dependent spectral shift (TDSS)
amplification, 31
angular momentum, 30
atomic motions, 49
biological macromolecules, 32
biological water layer, 32
charge difference, 50
Coulomb's equation, 33
dielectric continuum model, 33
dielectric relaxation
conformal transformation, 53
curve, 53, 54
Debye model, 51, 53
electric field and the electric displacement, 53
fluorophores, 51
hydrophobic core, 54
irregular-shaped nonpolar dielectric, 55
organic hydrogen-bonding, 55
parallel-plate capacitor, 53
polar liquid, 54
polarization, 52
solvatochromic dye, 55
solvatochromic fluorophore jumps, 52
spherical cavity, 52
transverse and longitudinal, 51
transverse relaxation time, 51, 52
dielectric relaxation modes, 33
direct-response vs. linear-response method
autocorrelation, 45
charge jump, 48
Coulombic interaction, 42
electric field, 47
equilibrium trajectory, 42
Euler angle, 46
finite-length trajectory, 43, 44
fluctuation-dissipation theorem, 42
low-frequency noise, 44
nanosecond-scale relaxation, 46
non-equilibrium MD, 42
panels 3A and 3B, 46
random and systematic errors, 44
random noise patterns and autocorrelation functions, 44, 45
random switch, 44
red and blue line, 47, 48
systematic errors, 46
water molecule, 48
drawback, 48
electric field, 50
electronic energy levels
constant energies, 38
electric dipole operator, 37
electric field, 37
first-order estimate, energy, 38
Hamiltonian, 37
inner-shell electrons, 38
matrix elements, 38
nonlinear Stark effect, 36
non-uniform electric field, 38
partial atomic charge, 39
peak frequencies and center-of-gravity frequencies, 39
perturbation theory, 36, 37
quantum-mechanical perturbation theory, 36
radius-vector, 38
vectors, 37
empirical forcefields, 29
explicit and implicit assumptions, 33
fluorophore, 49
hybrid QM-MD *vs.* cassical MD, 40–43
hypothetical situation, 50
microwave spectroscopy, 31
molecular dynamics (MD), 30
motion of water molecules, 32
non-equilibrium MD and linear-response method, 33
nonlinear stark effect
convergence boundary, 40
off-diagonal matrix, 39
second-order correction term, 39
tryptophan (Trp), 40
vacuum, 40
nuclear motion, 31
partial sums, 48
picosecond time scale, 30
protein and solvent atoms, 50
protein atoms, 50
protein molecule, 49
proteins and biomolecules, 30
quantum mechanics - molecular dynamics (QM-MD) simulations, 33
radius-vectors, 49
solvatochromic dye, 32
spectral shifts and electronic energy levels
center of gravity frequency, 34
and dynamic stokes shift (DSS), 36
emission center of gravity, 35
energy gap, 34

Time-dependent spectral shift (TDSS) (*cont.*)
Franck-Condon factor, 34
frequency intervals, 35
mean frequencies, 34, 35
pump-dump-probe experiments, 35
Stokes shift, 35
UV-visible absorption and emission spectra, 34
spectrum, electromagnetic signal, 31
TDSS, 32
time scales
amplitudes, 58, 61, 63
bond and dihedral angles, 59
dielectric continuum model, 63
ensemble-averaged component, 58
global fitting, 58
hybrid QM-MD simulations, 61
hydrogen bonds, 60
hydrogen nuclei, 63
hypothesis, 64
motion of protein atoms, 60
negative correlation, 62
nonequilibrium MD trajectories, 58
relaxation time, 58
representative relaxed excited-state structure, 62
transient Glu-42 side chain conformations, 62, 64
time variation, 50
vibrational and conformational dynamics, 31
water and protein matrix, 32
water atoms, 51
water molecules
bulk solvent, 57
dipole moment, 56
dipole-dipole interaction energy, 57
electric dipole moment, 57
electric field, 56
hydrogen-bonded, 56
intrinsic relaxation time, 57
linear chain, 57
nonpolar dielectric, 57
rotational motion, 56
south and north pole, 56
translational and rotational motion, 55
wavefunction, 31
Time-resolved fluorescence anisotropy experiments, 328–330
TIRF system. *See* Total internal reflection fluorescence (TIRF) system
Total internal reflection fluorescence (TIRF) system, 81
Trans-4-dimethylamino-4′-carbomethoxystilbene (DACMS)
C-C atoms, 346
femtosecond stimulated Raman spectroscopy, 349
logarithmic plot, 346, 350
long-chain alcohols, 346
three-state kinetic scheme, 346, 349
time-resolved emission spectra, 345
Transmission electron microscopy (TEM), 3
Tryptophan
fluorescence spectroscopic analyses, 355
interfacial localization, 355
membrane bilayer, 355, 356
membrane-water interface, 355
nonchannel conformations, 355
Turn-on indicators, 152
Twisted-intramolecular charge-transfer (TICT), 340

U
UCST. *See* Upper critical solution temperature (UCST)
Upper critical solution temperature (UCST), 100

V
VCD. *See* Vibrational circular dichroism (VCD)
Vibrational circular dichroism (VCD), 3

W
Wilson's Disease, 150

X
Xanthenes, 136, 159

MIX
Papier aus verantwortungsvollen Quellen
Paper from responsible sources
FSC® C105338
FSC
www.fsc.org

If you have any concerns about our products,
you can contact us on
ProductSafety@springernature.com

In case Publisher is established outside the EU,
the EU authorized representative is:
Springer Nature Customer Service Center GmbH
Europaplatz 3, 69115 Heidelberg, Germany

Printed by Libri Plureos GmbH
in Hamburg, Germany